Variants of Evolutionary Algorithms for Real-World Applications

Raymond Chiong, Thomas Weise,
and Zbigniew Michalewicz (Eds.)

Variants of Evolutionary Algorithms for Real-World Applications

Springer

Editors

Raymond Chiong
Faculty of ICT
Swinburne University of Technology
Melbourne, VIC 3122, Australia
E-mail: rchiong@swin.edu.au

Zbigniew Michalewicz
School of Computer Science
University of Adelaide
Adelaide, SA 5005, Australia
E-mail: zbyszek@cs.adelaide.edu.au

Thomas Weise
Nature Inspired Computation and
Applications Laboratory
School of Computer Science and
Technology
University of Science and
Technology of China (USTC)
Hefei 230027, Anhui, China
E-mail: tweise@ustc.edu.cn

ISBN 978-3-642-23423-1 e-ISBN 978-3-642-23424-8

DOI 10.1007/978-3-642-23424-8

Library of Congress Control Number: 2011935740

Typeset & Cover Design: Scientific Publishing Services Pvt. Ltd., Chennai, India.

Printed on acid-free paper

9 8 7 6 5 4 3 2 1

springer.com

Preface

Started as a mere academic curiosity, Evolutionary Algorithms (EAs) first came into sight back in the 1960s. However, it was not until the 1980s that the research on EAs became less theoretical and more practical. As a manifestation of population-based, stochastic search algorithms that mimic natural evolution, EAs use genetic operators such as crossover and mutation for the search process to generate new solutions through a repeated application of variation and selection.

Due to their ability to find excellent solutions for conventionally hard and dynamic problems within acceptable time, EAs have attracted interest from many researchers and practitioners in recent years. The general-purpose, black-box character of EAs makes them suitable for a wide range of real-world applications. Standard EAs such as Genetic Algorithms (GAs) and Genetic Programming (GP) are becoming more and more accepted in the industry and commercial sectors. With the dramatic increase in computational power today, an incredible diversification of new application areas of these techniques can be observed. At the same time, variants and other classes of evolutionary optimisation methods such as Differential Evolution, Estimation of Distribution Algorithms, Co-evolutionary Algorithms and Multi-Objective Evolutionary Algorithms (MOEAs) have been developed.

When applications or systems utilising EAs reach the production stage, off-the-shelf versions of these methods are typically replaced by dedicated algorithm variants. These specialised EAs often use tailored reproduction operators, search spaces differing significantly from the well-known binary or tree-based encodings, non-trivial genotype-phenotype mappings, or are hybridised with other optimisation algorithms. This book aims to promote the practitioner's view on EAs by giving a comprehensive discussion of how EAs can be adapted to the requirements of various applications in the real-world domains. It comprises 14 chapters, which can be categorised into the following four sections:

- Section I: Introduction
- Section II: Planning & Scheduling
- Section III: Engineering
- Section IV: Data Collection, Retrieval & Mining

The first section contains only one single chapter – the introductory chapter. In this chapter, *Blum et al.* re-visit the fundamental question of "what is an EA?" in an attempt to clearly define the scope of this book. In this regard, they systematically explore and discuss both the traditional and the modern views on this question by relating it to other areas in the field. That is, apart from discussing the main characteristics of conventional EAs they also extend their discussion to Memetic Algorithms (MAs) and the Swarm Intelligence algorithms. It appears that establishing semantic borders between the different algorithm families is never easy, nor necessarily useful. In this book, however, the focus will be on the traditional set of EAs like GAs, GP, and their variants.

The second section of the book deals with planning and scheduling problems. Planning and scheduling activities are among the most important tasks in Business and Industry. Once orders are placed by a customer, it is necessary to schedule the purchase of raw materials and to decide which machines are going to be used in order to create the ordered product in the desired quality. Often, multiple different client requests need to be facilitated at the same time and the goal is to satisfy all of them in a timely and cost-effective manner. However, it is not only the production steps that need to be scheduled. In fact, the whole behaviour of a supply chain as well as the work assignments for employees can be subject to planning. This section contains six chapters, with different groups of researchers presenting efficient EA approaches to a variety of real-world planning and scheduling problems.

The first chapter in this section by *Mohais et al.* introduces a tailor-made EA for the process of bottling wine in a mass-production environment. Time-varying (dynamic) constraints are the focus of this chapter. That is, scheduling for job shop problems rarely starts with a blank sheet of paper. Instead, some production processes will already be in progress. Hence, there is typically a set of scheduled operations that are fixed and cannot be modified by optimisation, yet will influence the efficiency and feasibility of new plans. Mohais et al. successfully approach the wine bottling problem with their tailor-made evolutionary method.

Following which, *Toledo et al.* present a similar real-world problem for soft-drink manufacturing plants known as the synchronised and integrated two-level lot sizing and scheduling problem. Here, the first production level has tanks storing the soft drink flavours and the second level corresponds to the bottling lines. The problem involves capacity limits, different costs and production times depending on the raw materials involved as well as the inventory costs. In order to derive production schedules with low associated costs in this scenario, Toledo et al. propose the use of an MA. This algorithm has a population structured as tree of clusters. It uses either Threshold Accepting or Tabu Search as local search, and utilises different operators. These variants have shown to outperform both the GA and a Relax approach based on some real-world data sets. In particular, the Tabu Search variant has turned out to be very efficient and robust.

The third chapter of the section by *Lässig et al.* considers simulation-based optimisation of hub-and-spoke inventory systems and multi-location inventory systems with lateral transshipments. Such systems are very common in the industry, but it is extremely challenging to find the optimal order and transshipment policies for them in an analytical way. Lässig et al. therefore suggest a simulation-based evolutionary approach, where the utility of rules is estimated by simulating the behaviour of the system applying them. This simulation process is used to compute the fitness of the policies. Lässig et al. show that Threshold Accepting, Particle Swarm Optimisation, and especially GAs can effectively tackle the resulting optimisation problems.

Subsequently, *Schellenberg et al.* present a fuzzy-evolutionary approach for optimising the behaviour of a multi-echelon supply chain network of an Australian ASX Top 50 company. They use an EA for synthesising fuzzy rules for each link of the supply chain in order to satisfy all demands while adhering to system constraints (such as silo capacity limits which must not be exceeded due to overproduction further down the chain). Their experimental studies show that the evolution of behaviour rules that can issue commands based on the current situation is much more efficient than trying to generate complete plans scheduling each single supply and production event.

The following chapter by *Dasgupta et al.* provides a new solution to the task-based sailor assignment problem faced by the US Navy. That is, a sailor in active duty is usually reassigned to a different job around every three years. Here, the goal is to pick new jobs for the sailors currently scheduled for reassignment in a way that is most satisfying for them as well as the commanders. In the work presented by Dasgupta et al., these assignments have been broken further down to different tasks for different timeslots per sailor. For this purpose, Dasgupta et al. use a parallel implementation of a hybrid MOEA which combines the NSGA-II and some intelligent search operations. The experimental results show that the proposed solution is promising.

In the final chapter of the section, *Ma and Zhang* discuss how a production planning process can be optimised with a GA using the example of CNC-based work piece construction. A customisable job shop environment is presented, which can easily be adapted by the users. The optimisation approach then simultaneously selects the right machines, tools, commands for the tools, and operation sequences to manufacture a desired product. The applied GA minimises a compound of the machine costs, the tool costs and the machine, setup, and tool change cost. It is embedded into a commercial computer-aided design system and its utility is demonstrated through a case study.

The work of Ma and Zhang leads us to the third section of this book, addressing another crucial division of any industrial company: R & D (Research and Development) and Engineering. In this area, EA-based approaches again have shown huge potential for supporting the human operators in creating novel and more efficient products. However, there are two challenges. On one hand, the evaluation of an engineering design usually involves complex simulations and hence, takes quite a long time to complete. This decreases the utility of common EAs that often require numerous fitness evaluations. On the other hand, many engineering problems have a high-dimensional search space, i.e., they involve many decision variables. In this section, three chapters showcase how these challenges can be overcome and how EAs are able to deliver excellent solutions for hard, real-world engineering problems.

In mechanical design problems, the goal is to find structures with specific physical properties. The Finite Element Method (FEM) can for example be used to assess the robustness of composite beams, trusses, airplane wings, and piezoelectric actuators. If such structures are to be optimised, as is the case in the chapter presented by *Davarynejad et al.*, the FEM represents an indispensable tool for assessing the utility of the possible designs. However, each of its invocations requires a great amount of runtime and thus slows down the optimisation process considerably. To this end, Davarynejad et al. propose an adaptive fuzzy fitness granulation approach – a method which allows approximating the fitness of new designs based on previously tested ones. The proposed approach is shown to be able to reduce the amount of FEM invocations and speed up the optimisation process for these engineering problems significantly.

In the next chapter, *Turan and Cui* introduce a hybrid evolutionary approach for ship stability design, with a particular focus on roll on/roll off passenger ships. Since the evaluation of each ship design costs much runtime, the MOEA (i.e., NSGA-II) utilised by Turan and Cui is hybridised with Q-learning to guide the search directions. The proposed approach provides reasonably good results, where Turan and Cui are able to discover ship designs that represent significant improvements from the original design.

The chapter by *Rempis and Pasemann* presents a new evolutionary method, which they called the Interactively Constrained Neuro-Evolution (ICONE) approach. ICONE uses an EA for synthesising the walking behaviour of humanoid robots. While bio-inspired neural control techniques have been highly promising for robot control, in the case when many sensor inputs have to be processed and many actuators need to be controlled the search space size may increase rapidly. Rempis and Pasemann therefore propose the use of both domain knowledge and restrictions of the possible network structures in their approach. As the name suggests, ICONE is interactive, thus allows the experimenter to bias the search towards the desired structures. This leads to excellent results in the walking-behaviour synthesis experiments.

The final section of the book concerns data collection, retrieval, and mining. The gathering, storage, retrieval and analysis of data is yet another essential area not just in the industry but also the public sectors, or even military. Database systems are the backbone of virtually every enterprise computing environment. The extraction of information from data such as images has many important applications, e.g., in medicine. The ideal coverage of an area with mobile sensors in order to gather data can be indispensible for, e.g., disaster recovery operations. This section covers four chapters dealing with this line of real-world applications from diverse fields.

A common means to reduce cost in the civil construction industry is to stabilise soil by mixing lime, cement, asphalt or any combination of these chemicals into it. The resulting changes in soil features such as strength, porosity, and permeability can then ease road constructions and foundation. In the chapter presented by *Alavi et al.*, a Linear GP (LGP) approach is used to estimate the properties of stabilised soil. GP evolves program-like structures, and its linear version represents programs as a sequential list of instructions. Alavi et al. apply LGP in its original (purely evolutionary) version as well as a version hybridised with Simulated Annealing. Their experimental studies confirm that the accuracy of the proposed approach is satisfactory.

The next chapter by *Bilotta et al.* discusses the segmentation of MRI images for (multiple sclerosis) lesion detection and lesion tissue volume estimation. In their work, Bilotta et al. present an innovative approach based on Cellular Neural Networks (CNNs), which they synthesise with a GA. This way, CNNs can be generated for both 2D and 3D lesion detection, which provides new perspectives for diagnostics and is a stark improvement compared to the currently used manual lesion delineation approach.

Databases are among the most important elements of all enterprise software architectures. Most of them can be queried by using Structured Query Language (SQL). Skyline extends SQL by allowing queries for trade-off curves concerning two or more attributes over datasets, similar to Pareto frontiers. Before executing such a query, it is typically optimised via equivalence transformations for the purpose of minimising its runtime. In the penultimate chapter of this section (also of this book), *Goncalves et al.* introduce an alternative approach for Skyline Query Optimisation based on an EA. They show that the variants of their proposed approach are able to outperform the commonly-used dynamic programming, especially as the number of tables increases.

Distributing the nodes of Mobile Ad-hoc Networks (MANETs) as uniformly as possible over a given terrain is an important problem across a variety of real-world applications, ranging from those for civil to military purposes. The final chapter by *Şahin et al.* shows how a Force-based GA (FGA) can provide the node executing it with movement instructions which accomplish this objective. Here, one instance of the FGA is executed on each node of the MANET, and only local knowledge obtained from within the limited sensor and communication range of a node is utilised. The simulation experiments confirm that the FGA can be an effective mechanism for deploying mobile nodes with restrained communication capabilities in MANETs operating in unknown areas.

To sum up, we would like to extend our gratitude to all the authors for their excellent contributions to this book. We also wish to thank all the reviewers involved in the review process for their constructive and useful review comments. Without their help, this book project could not have been satisfactorily completed. A special note of thanks goes to Dr. Thomas Ditzinger (Engineering Senior Editor, Springer-Verlag) and Ms Heather King (Engineering Editorial, Springer-Verlag) for their editorial assistance and professional support. Finally, we hope that readers would enjoy reading this book as much as we have enjoyed putting it together!

June 2011

Raymond Chiong
Thomas Weise
Zbigniew Michalewicz

Editorial Review Board

Contents

Section I: Introduction

Evolutionary Optimization 1
Christian Blum, Raymond Chiong, Maurice Clerc, Kenneth De Jong, Zbigniew Michalewicz, Ferrante Neri, Thomas Weise

Section II: Planning and Scheduling

An Evolutionary Approach to Practical Constraints in Scheduling: A Case-Study of the Wine Bottling Problem 31
Arvind Mohais, Sven Schellenberg, Maksud Ibrahimov, Neal Wagner, Zbigniew Michalewicz

A Memetic Framework for Solving the Lot Sizing and Scheduling Problem in Soft Drink Plants 59
Claudio F.M. Toledo, Marcio S. Arantes, Paulo M. França, Reinaldo Morabito

Simulation-Based Evolutionary Optimization of Complex Multi-Location Inventory Models 95
Jörg Lässig, Christian A. Hochmuth, Stefanie Thiem

A Fuzzy-Evolutionary Approach to the Problem of Optimisation and Decision-Support in Supply Chain Networks 143
Sven Schellenberg, Arvind Mohais, Maksud Ibrahimov, Neal Wagner, Zbigniew Michalewicz

A Genetic-Based Solution to the Task-Based Sailor Assignment Problem 167
Dipankar Dasgupta, Deon Garrett, Fernando Nino, Alex Banceanu, David Becerra

Genetic Algorithms for Manufacturing Process Planning 205
Guohua Ma, Fu Zhang

Section III: Engineering

A Fitness Granulation Approach for Large-Scale Structural Design Optimization 245
Mohsen Davarynejad, Jos Vrancken, Jan van den Berg, Carlos A. Coello Coello

A Reinforcement Learning Based Hybrid Evolutionary Algorithm for Ship Stability Design 281
Osman Turan, Hao Cui

An Interactively Constrained Neuro-Evolution Approach for Behavior Control of Complex Robots 305
Christian Rempis, Frank Pasemann

Section IV: Data Collection, Retrieval and Mining

A Genetic Programming-Based Approach for the Performance Characteristics Assessment of Stabilized Soil ... 343
Amir Hossein Alavi, Amir Hossein Gandomi, Ali Mollahasani

Evolving Cellular Neural Networks for the Automated Segmentation of Multiple Sclerosis Lesions 377
Eleonora Bilotta, Antonio Cerasa, Pietro Pantano, Aldo Quattrone, Andrea Staino, Francesca Stramandinoli

An Evolutionary Algorithm for Skyline Query Optimization 413
Marlene Goncalves, Ivette Martínez, Gabi Escuela, Fabiola Di Bartolo, Francelice Sardá

A Bio-Inspired Approach to Self-Organization of Mobile Nodes in Real-Time Mobile Ad Hoc Network Applications 437
Cem Şafak Şahin, Elkin Urrea, M. Ümit Uyar, Stephen Gundry

Author Index 463

Evolutionary Optimization

Christian Blum, Raymond Chiong, Maurice Clerc, Kenneth De Jong, Zbigniew Michalewicz, Ferrante Neri, and Thomas Weise*

Abstract. The emergence of different metaheuristics and their new variants in recent years has made the definition of the term *Evolutionary Algorithms* unclear. Originally, it was coined to put a group of stochastic search

Christian Blum
ALBCOM Research Group, Universitat Politècnica de Catalunya, Barcelona, Spain
e-mail: cblum@lsi.upc.edu

Raymond Chiong
Faculty of Information & Communication Technologies, Swinburne University of Technology, Victoria 3122, Australia
e-mail: rchiong@swin.edu.au

Maurice Clerc
Independent Consultant, Groisy, France
e-mail: Maurice.Clerc@WriteMe.com

Kenneth De Jong
Department of Computer Science, George Mason University, Fairfax, VA 22030, USA
e-mail: kdejong@gmu.edu

Zbigniew Michalewicz
School of Computer Science, University of Adelaide, South Australia 5005, Australia; also at the Institute of Computer Science, Polish Academy of Sciences, ul. Ordona 21, 01-237 Warsaw, Poland, and the Polish-Japanese Institute of Information Technology, ul. Koszykowa 86, 02-008 Warsaw, Poland
e-mail: zbyszek@cs.adelaide.edu.au.

Ferrante Neri
Department of Mathematical Information Technology, P. O. Box 35 (Agora), 40014 University of Jyväskylä, Finland
e-mail: ferrante.neri@jyu.fi

Thomas Weise
Nature Inspired Computation and Applications Laboratory (NICAL), School of Computer Science and Technology, University of Science and Technology of China (USTC), Héféi 230027, Ānhuī, China
e-mail: tweise@ustc.edu.cn

* Corresponding author.

algorithms that mimic natural evolution together. While some people would still see it as a specific term devoted to this group of algorithms, including Genetic Algorithms, Genetic Programming, Evolution Strategies, Evolutionary Programming, and to a lesser extent Differential Evolution and Estimation of Distribution Algorithms, many others would regard "Evolutionary Algorithms" as a general term describing population-based search methods that involve some form of randomness and selection. In this chapter, we re-visit the fundamental question of "*what is an Evolutionary Algorithm?*" not only from the traditional viewpoint but also the wider, more modern perspectives relating it to other areas of Evolutionary Computation. To do so, apart from discussing the main characteristics of this family of algorithms we also look at Memetic Algorithms and the Swarm Intelligence algorithms. From our discussion, we see that establishing semantic borders between these algorithm families is not always easy, nor necessarily useful. It is anticipated that they will further converge as the research from these areas cross-fertilizes each other.

1 Introduction

Almost any design or decision task encountered in business, industry, or public services is, by its nature, an optimization problem. How can a ship be designed for *highest* safety and *maximum* cargo capacity at the same time? How should the production in a factory be scheduled in order to satisfy all customer requests *as soon* and timely *as possible*? How can multiple sclerosis lesions on an MRI be identified with the *best* precision? ... Three completely different questions and scenarios, three optimization problems as encountered by practitioners every day.

From the management perspective, an optimization problem is a situation that requires one to decide for a choice from a set of possible alternatives in order to reach a predefined/required benefit at minimal costs. From a mathematical point of view, solving an optimization problem requires finding an input value $x^\star$ for which a so-called objective function f takes on the smallest (or largest) possible value (while obeying to some restrictions on the possible values of $x^\star$). In other words, every task that has the goal of approaching certain configurations considered as optimal in the context of pre-defined criteria can be viewed as an optimization problem.

Many optimization algorithms for solving complex real-world problems nowadays are based on metaheuristic methods as opposed to traditional operations research techniques. The reason is simple – this is due to the *complexity* of the problems. Real-world problems are usually difficult to solve for several reasons, some of which include:

1. The number of possible solutions may be too large so that an exhaustive search for the best answer becomes infeasible.
2. The objective function f may be noisy or varies with time, thereby requiring not just a single solution but an entire series of solutions.
3. The possible solutions are so heavily constrained that constructing even one feasible answer is difficult, let alone searching for an optimum solution.

Naturally, this list could be extended to include many other possible obstacles. For example, noise associated with the observations and measurements, uncertainties about the given information, problems that have multiple conflicting objectives, just to mention a few. Moreover, computing the objective values may take much time and thus, the feasible number of objective function invocations could be low. All these reasons are just some of the aspects that can make an optimization problem difficult (see [76]; and also [106] for an in-depth discussion on this topic).

It is worth noting that every time a problem is "solved", in reality what has been discovered is only the solution to a *model* of the problem – and all models are simplification of the real world. When trying to solve the Travelling Salesman Problem (TSP), for example, the problem itself is usually modeled as a graph where the nodes correspond to cities and the edges are annotated with costs representing, e.g., the distances between the cities. Parameters such as traffic, the weather, petrol prices and times of the day are typically omitted.

In view of this, the general process of solving an optimization problem hence consists of two separate steps: (1) creating a model of the problem, and (2) using that model to generate a solution.

Problem $\Rightarrow$ Model $\Rightarrow$ Solution.

Again, the "solution" here is only a solution in terms of the model. If the model has a high degree of fidelity, this "solution" is more likely to be meaningful. In contrast, if the model has too many unfulfilled assumptions and rough approximations, the solution may be meaningless, or worse. From this perspective, there are at least two ways to proceed in solving real-world problems:

1. Trying to simplify the model so that conventional methods might return better answers.
2. Keeping the model with all its complexities and using non-conventional approaches to find a near-optimum solution.

So, the more difficult the problem is, the more appropriate it is to use a metaheuristic method. Here, we see that it will anyway be difficult to obtain precise solutions to a problem, since we have to approximate either the model or the solution. A large volume of experimental evidence has shown that the latter approach can often be used to practical advantages.

In recent years, we have seen the emergence of different types of metaheuristics. This gives rise to many new variants and concepts, making some of the fundamental views in the field no longer clear-cut. In this chapter, our focus is to discuss what Evolutionary Algorithms (EAs) – one of the most popular metaheuristic methods – are and how they differ from other metaheuristics. The aim is not to give a definitive answer to the question "What is an EA?" – it is almost impossible for anyone to do so. Instead, we will systematically explore and discuss the traditional and modern views of this topic. We start by describing what metaheuristics are, followed by the core question of what EAs are. We then present some of the most well-known EAs, such as Genetic Algorithms (GAs), Genetic Programming (GP), Evolution Strategies (ES) and Evolutionary Programming (EP). After that, we extend our discussion to Memetic Computing, taking a look at the relevance/connection between EAs and Memetic Algorithms (MAs). Finally, we also discuss the similarities and differences between EAs and the Swarm Intelligence algorithms such as Particle Swarm Optimization (PSO) and Ant Colony Optimization (ACO).

2 Metaheuristics

The field of metaheuristics has a rich history. During the second half of the 20th century, with the development of computational devices and demands of industrial processes, the necessity to solve some optimization problems arose despite the fact that there was not sufficient prior knowledge (hypotheses) on the optimization problem for the application of an exact method. In fact, in the majority of industrial cases, the problems are highly nonlinear, or characterized by a noisy fitness, or without an explicit analytical expression as the objective function might be the result of an experimental or simulation process. In this context, the earliest metaheuristics have been designed. The term metaheuristic, from the greek *meta-euriskein* which means beyond the search, refers to a computational method which progressively attempts to improve one or more candidate solutions while searching for the optimum.

Whenever an optimization problem is to be solved, we expect that there is some kind of utility measure which defines how good a solution is or how high the costs are. Usually this measure is given in the form of a mathematical function *f*. Then, as stated before, the inputs for which the function takes on the minimal (or maximal) value is sought. Sometimes, multiple such functions have to be optimized simultaneously.

A metaheuristic is a method for solving a *general* class of optimization problems. It combines utility measures in an abstract and hopefully efficient manner, typically without utilizing deeper insights into their inner structure. Metaheuristics do not require hypotheses on the optimization problem nor any kind of prior knowledge on the objective function. The treatment of objective functions as "black boxes" [11, 42, 45, 102] is the most prominent and attractive feature of metaheuristics. Metaheuristics obtain knowledge about the structure of an optimization problem by utilizing statistics obtained from the possible solutions (i.e., candidate solutions) evaluated in the past.

This knowledge is used to construct new candidate solutions which are likely to have a high utility.

Many different types of metaheuristics emerged during the last 30 years, and the majority of them have been inspired by some aspects of the nature (see [19] for a recent collection of nature-inspired algorithms). These include a variety of Hill Climbing techniques (deterministic and stochastic), the Swarm Intelligence algorithms (PSO and ACO), Artificial Immune Systems, Differential Evolution, Simulated Annealing, Tabu Search, Cultural Algorithms, Iterated Local Search, Variable Neighborhood Search, and – of course – Evolutionary and co-Evolutionary Algorithms.

Metaheuristics can be classified based on different criteria. For example, some of them process a single solution (e.g., Simulated Annealing), whereas some others process a set of solutions and are called population-based methods (e.g., EAs). Some metaheuristics are deterministic (e.g., Tabu Search), others are stochastic (e.g., Simulated Annealing). Some generate complete solutions by *modifying* complete solutions (e.g., EAs), while some others *construct* new solutions at every iteration (e.g., ACO). Many of these metaheuristics offer unique features, but even within a single approach, there are many variants which incorporate different representation of solutions and different modification or construction techniques for new solutions.

3 What Are "Evolutionary Algorithms"?

So, what are EAs? Perhaps the best place to start in answering the question is to note that there are at least two possible interpretations of the term *evolution*. It is frequently used in a very general sense to describe something that changes incrementally over time, such as the software requirements for a payroll accounting system. The second meaning is its narrower use in biology, where it describes an evolutionary system that changes from generation to generation via reproductive variation and selection. It is this Darwinian notion of evolutionary change that has been the core idea in EAs.

3.1 Principles Inspired by Nature

From a conventional point of view, an EA is an algorithm that simulates – at some level of abstraction – a Darwinian evolutionary system. To be more specific, a standard EA includes:

1. One or more populations of individuals competing for limited resources.
2. These populations change dynamically due to the birth and death of individuals.
3. A notion of fitness which reflects the ability of an individual to survive and reproduce.
4. A notion of variational reproduction: offspring closely resemble their parents, but are not identical.

In a nutshell, the Darwinian principles of evolution suggest that, on average, species improve their fitness over generations (i. e., their capability of adapting to the environment). A simulation of the evolution based on a set of candidate solutions whose fitness is properly correlated to the objective function to optimize will, on average, lead to an improvement of their fitness and thus steer the simulated population towards the solution.

3.2 The Basic Cycle of EAs

In the following, we try to introduce a very simple EA consisting of a single population of individuals exist in an environment that presents a time-invariant notion of fitness. We will do this from a general perspective, comprising most of the conventional EAs.

Like in nature, an individual may have two different representations: the data structure which is processed by the genetic search procedures and the format in which it is assessed by the environment (and finally handed to the human operator). Like in biology, in the context of EAs, the former representation is referred to as *genotype* and the latter as *phenotype*. EAs usually proceed in principle according to the scheme illustrated in Fig. 1. Its steps can be described as follows:

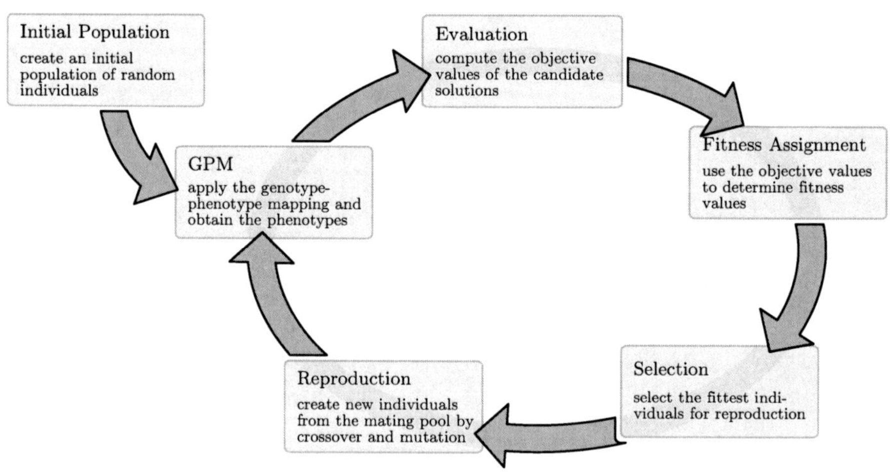

Fig. 1. The basic cycle of EAs

1. In the first generation, a population of $n > 0$ individuals is created. Usually, these individuals have random genotypes but sometimes, the initial population is *seeded* with good candidate solutions either previously known or created according to some other methods.

2. The genotypes, i. e., the points in the search space, are then translated to phenotypes. In the case that search operations directly work on the solution data structures, this genotype-phenotype mapping is the identity mapping.
3. The values of the objective functions are then evaluated for each candidate solution in the population. This evaluation may incorporate complicated simulations and calculations.
4. With the objective functions, the utility of different features of the candidate solutions has been determined. If there is more than one objective function, constraint, or other utility measure, then a scalar fitness value is assigned to each of them.
5. A subsequent selection process filters out the candidate solutions with poor fitness and allows those with good fitness to enter the mating pool with a higher probability.
6. In the reproduction phase, offspring are derived from the genotypes of the selected individuals by applying the search operations (which are called *reproduction operations* in the context of EAs). There are usually two different reproduction operations: mutation, which modifies one genotype, and crossover, which combines two genotypes to a new one. Whether the whole population is replaced by the offspring or whether they are integrated into the population as well as which individuals to recombine with each other depends on the applied population handling strategy.
7. If the termination criterion is met, the evolution stops here. Otherwise, the evolutionary cycle continues with the next generation at point 2.

Of course, such an algorithm description is too abstract to be executed directly.

3.3 Do All EAs Fit to the Basic Cycle?

According to our discussion so far, a simple answer to the question "What are EAs?" would be that EAs are those based on the concepts gleaned from natural evolution and which roughly adhere to the principles and the basic cycle introduced in the previous sections. From a high-level perspective, however, the definition is not so clear.

When solving a new challenging problem, often a new optimization method is designed. It is necessary to specify how the individuals in the population represent the problem solutions, how the fitness is calculated, how parents are selected, how offspring are produced, and how individuals are selected for removal from the population (i. e., to die[1]). Each of these decisions results in an EA variant with different computational properties.

[1] The "death" of an individual, candidate solution, or agent in terms of metaheuristic optimization means that it is removed from the set of elements under investigation and deleted from memory, possibly due to being replaced by a better element.

Will these design decisions result in an EA? Before you answer, let us recall that the (1 + 1) EA does not require a population of solutions but processes just one individual (and is comparing it with its only offspring). Many "Evolutionary Algorithms" assume deterministic selection methods, many "Evolutionary Algorithms" take advantage of smart initialization and problem-specific operators. Some "Evolutionary Algorithms" have been extended with memory structures (e.g., when they operate in dynamic environments) or by a parameter called *temperature* (to control mutation rates). The list could go on.

While there is a well-researched set of "default" EAs which we will introduce in the next section, for many real-world applications it is necessary to derive new, specialized approaches. Examples for this can be found in [103–105] as well as the collection in this book [20].

3.4 Conventional EAs

Historically, computer scientists and engineers had started to consider drawing inspiration from evolutionary principles for solving optimization problems as early as the 1950s (see [4, 5], and [39]). In the 1960s and 1970s, three research lines were developed in parallel [73]: EP [38], ES [91], and GAs [55]. De Jong's PhD thesis [25] further increased the interest in this field, and his PhD student Grefenstette, in turn, started the *International Conference on Genetic Algorithms and their Applications* (ICGA) [52] in 1985. At the 1991 ICGA [6], the three original research streams came together, with Hans-Paul Schwefel presenting the ES. At the same venue, Koza [64] introduced the new concept of Standard GP and Zbigniew Michalewicz outlined the concepts of different data structures which can undergo the evolutionary process in the so-called *Evolutionary Programs* [75]. This was considerably the first time all major areas of Evolutionary Computation were represented at once. As a result, the *Evolutionary Computation* Journal by MIT Press was established, later followed by the *IEEE Transactions on Evolutionary Computation.* The idea of unifying concepts, such as "Evolutionary Algorithms" (or the more general idea of Evolutionary Computation [73]), was then born. Thereafter, the first *IEEE Congress on Evolutionary Computation* [74] was initiated in 1994.

From the 1990s onwards, many new ideas have been introduced. One of the most important developments is the discovery that EAs are especially suitable for solving problems with multiple, possibly conflicting optimization criteria – the *Multi-Objective Evolutionary Algorithms* (MOEAs) [22, 29]. Today, the second, improved versions of NSGA [30, 98] and SPEA [113, 114] may be the most popular members of this MOEA family.

There is also growing interest in co-evolutionary EAs, originally introduced by Hillis [53] back in 1989. Potter and De Jong [88, 89] developed *cooperative co-evolution*, which is now regarded as one of the key approaches for tackling large-scale optimization problems because it provides a viable way to decompose the problems and co-evolve solutions for the problem parts which together make up a solution for the original task [18]. Other parallel developments include the works of Grefenstette [46], Deb and Goldberg [28] as well as De Jong [26] who considered the issues of *deception*.

Books such as [22, 43, 75] and [27] have always played a major role in opening the field of Evolutionary Computation to a wide audience, with the *Handbook of Evolutionary Computation* [2] one of the most prominent examples.

3.4.1 Genetic Algorithms

GAs are the original prototype of EAs. Here, the genotypes of the search space are strings of primitive elements (usually all of the same type) such as bits, integers or real numbers. Because of the simple structure of the search space of GAs, a genotype-phenotype mapping is often used to translate the genotypes to candidate solutions [43, 54, 55, 108].

The single elements of the string genotypes in GAs are called *genes*. GAs usually apply both the mutation and crossover operators. The mutation operator modifies one or multiple genes whereas the crossover operator takes two genotypes and combines them to form a new one, either by merging or by exchanging the values of the genes. The most common reproduction operations used in GAs, single-point mutation and single-point crossover, are sketched in Fig. 2 [43, 56].

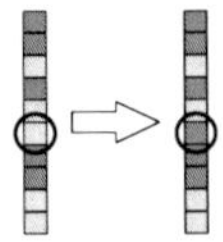

Fig. 2.a: Single-gene mutation.

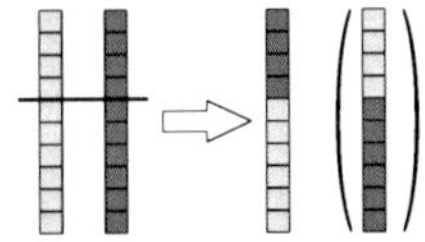

Fig. 2.b: Single-point Crossover (SPX).

Fig. 2. Mutation and Crossover in GAs

3.4.2 Genetic Programming

The term *Genetic Programming* [54, 64, 87] has two possible meanings. First, it can be viewed as a set of EAs that breed programs, algorithms, and similar constructs. Second, it is also often used to subsume EAs that have tree data structures as genotypes. Tree-based GP, usually referred to as the Standard GP, is the most widespread GP variant both for historical reasons and because of its efficiency in many problem domains. Here, the genotypes are tree data structures. Generally, a tree can represent a rule set [71, 101, 103], a mathematical expression [64], a decision tree [63, 101], or even the blueprint of an electrical circuit [65].

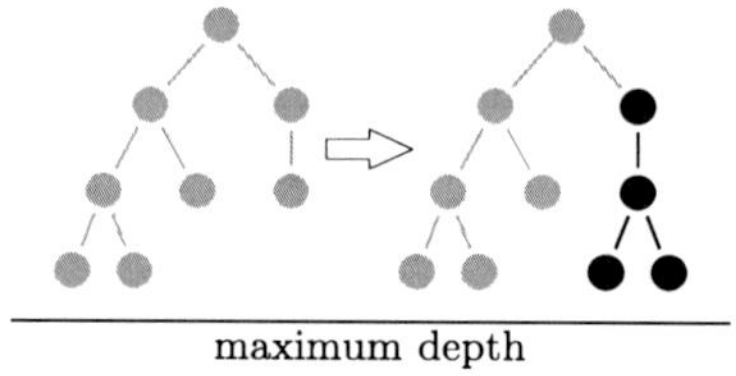

Fig. 3.a: Sub-tree replacement mutation.

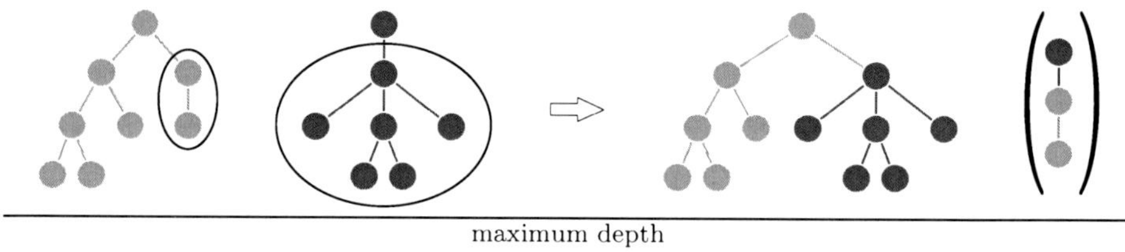

Fig. 3.b: Subtree exchange crossover.

Fig. 3. Mutation and Recombination in GP

Of course, mutation and crossover operators as used in GAs cannot be applied to tree data structures. Instead, new operations have been developed, such as the sub-tree replacement mutation which replaces a sub-tree of a genotype with a randomly created one and sub-tree exchange crossover which exchanges two sub-trees between two parental genotypes, as sketched in Fig. 3.

3.4.3 Evolution Strategies

ES, introduced by Rechenberg [90, 91, 92] and Schwefel [93, 94, 95], is a heuristic optimization technique based on the ideas of adaptation and evolution – a special form of EAs [1, 7, 8, 54, 90–92, 96]. The search space of today's ES usually consists of vectors from the $\mathbb{R}^n$, but bit strings or integer strings are common as well [8]. Mutation and selection are the primary reproduction operators and recombination is used less often in ES. Typically, normally distributed random numbers are used for mutation. The parameter of the mutation is the standard deviation of these random numbers. ES may either:

1. maintain a single standard deviation parameter and use identical normal distributions for generating the random numbers added to each element of the solution vectors,
2. use a separate standard deviation (from a standard deviation vector) for each element of the genotypes, i. e., create random numbers from different

normal distributions for mutations in order to facilitate different strengths and resolutions of the decision variables, or
3. use a complete covariance matrix for creating random vectors distributed in a hyperellipse and thus also taking into account binary interactions between the elements of the solution vectors.

The standard deviations are governed by self-adaptation [50, 66, 72] and may result from a stochastic analysis of the elements in the population [47–49, 58]. They are often treated as *endogenous* strategy parameters which can directly be stored in the individual records and evolve along with them [8].

3.4.4 Evolutionary Programming

EP is less precisely defined as other conventional EAs. There is a semantic difference though: while single individuals of a species are the biological metaphor for candidate solutions in other EAs, in EP a candidate solution is thought of as a species itself. Hence, mutation and selection are the only operators used in EP and recombination is usually not applied. The selection scheme utilized in EP is normally quite similar to the $(\mu + \lambda)$ method in ES.

EP was pioneered by Fogel [37] in his PhD thesis back in 1964. Fogel et al. [38] experimented with the evolution of finite state machines as predictors for data streams [35]. One of the most advanced EP algorithms for numerical optimization today has been developed by Yao et al. [110].

4 Memetic Computing

Memetic Computing is a growing area in computational intelligence closely related to EAs. During the creation of the initial population in an EA, a set of candidate solutions is generated, usually randomly within the decision space. Other sampling systems that include a certain degree of determinism for selecting the initial set of solutions are also widely used. The latter, usually known as intelligent initial sampling, is often considered as a memetic component within an EA framework [51].

4.1 MAs as an Extension of EAs

The main idea in 1980s and 1990s was to propose EAs with superior performance with respect to all the other algorithms present in the literature. This approach is visible in many famous texts and papers published in those years (see Section 3.4). After the publication of the No Free Lunch Theorem (NFLT) [109], however, researchers in the field have to dramatically change their view about the subject. The NFLT mathematically proves that the average performance of any pair of algorithms A and B across all possible problems with finite search spaces is identical. Thus, if an algorithm performs well on a certain class of problems, then it necessarily pays for that with degraded performance on other sets of problems. The concept that there is no universal optimizer has a significant impact on the scientific community.[2] In light of increasing interest in general-purpose optimization algorithms, it has become important to understand the relationship between the performance of an algorithm A and a given optimization problem f. The problem hence becomes the starting point for building up a suitable algorithm.

In this context, the term *Memetic Algorithms* was coined, representing an efficient alternative (or maybe a modification) of EAs. This term was first introduced in [77] with reference to an algorithm proposed in [82, 83] to indicate an approach that integrates a local search operator within an evolutionary structure. The metaphor of the term "memetic" was inspired by modern philosophy, more specifically by the meme's theory of Richard Dawkins [24]. The meme is an idea, a "unit of cultural transmission", the basic unit of knowledge. Although in Dawkins' studies the focus was to prove that evolution was based on the individual choices rather than collective choices (the selfish gene), in Computer Science another concept has been taken and adapted to computational problem-solving. By interpreting Dawkins' philosophy, it can be deduced that the collective culture is the result of an evolutionary process where ideas (memes) interact and diffuse over individuals modifying and getting enriched. Transferring this to the computing environment, different search operators (e.g., evolutionary framework and local search) compete and cooperate as different memes and process the solutions, by means of their harmonic interaction, towards the detection of the global optimum.

A definition which characterizes the structural features of MAs has been given in [51]. In general, an MA is a population-based hybrid optimization algorithm composed of an evolutionary framework and a list of local search algorithms activated within the generation cycle of the framework. In other words, MAs can be considered as specific hybrid algorithms which combine an EA framework and local search for enhancing the quality of some solutions of the population during the EA generation cycle. The sense of MAs is to compensate, for some specific problems, the limitations of EAs. As with all other metaheuristics, the functioning of an EA is due to the proper balance between exploration and exploitation. The generally optimal balance, in accordance with the NFLT, does not exist but it should be found for each fitness landscape. In addition, MAs contain multiple search components which can explore the fitness landscape from complementary perspectives and mitigate the typical undesired effects of stagnation and premature convergence.

Obviously, in MAs the coordination between the EA framework and local search operators can be hard to design. For this reason, a lot of research studies on MAs have been performed by paying great attention to the coordination logic of the various search operators. By updating the classification given in [85], MAs can be subdivided as: 1) Adaptive Hyper-Heuristic,

[2] Note, however, that it is possible to find algorithms which are best over large sets of (practically-relevant) problems; see [57].

see e.g., [14, 23, 59] and [61], where the coordination of the memes is performed by means of heuristic rules; 2) Self-Adaptive and Co-Evolutionary, see e.g., [97, 111] and [67], where the memes, either directly encoded within the candidate solutions or evolving in parallel to them, take part in the evolution and undergo recombination and selection in order to select the most promising operators; 3) Meta-Lamarckian Learning, see e.g., [62, 81, 84] and [69], where the success of the memes biases their activation probability, thus performing an on-line algorithmic design which can flexibly adapt to various optimization problems; 4) Diversity-Adaptive, see e.g., [16, 17, 78–80, 100] and [99], where a measure of the diversity is used to select and activate the most appropriate memes. In addition, it is worth to mention about the Baldwinian systems, i. e., those MAs that do not modify the solutions after the employment of local search, see [112] and [44]. The latter are basically EAs where the selection process is influenced by the search potential of each solution.

4.2 Can All MAs Be Considered EAs?

Generally speaking, MAs are population-based algorithms that evolve solutions under the same rules/logic as conventional EAs while additionally applying local search. In this sense, if the local search algorithms are to be considered as special operators, e.g., a hill-climb is seen as a mutation, then MAs can be considered as a subset of EAs. On the other hand, MAs can be considered as EAs that allow plenty of unconventional operators. To this end, MAs can be seen as an extension of EAs.

Regardless of the labeling, it is important to note that all these optimization algorithms are de facto the combination of two kinds of operators, i. e., search and selection, respectively. In conventional EAs, the search is performed by crossover and mutation operators, which are also known as variation operators, while the selection is included into the parent and survivor selection phases. Similarly, the combination of these two kinds of operators can be spotted within an MA by analyzing its evolutionary framework and each of its local search components. In this context, the more modern (and at the same time primitive) concept of Memetic Computing has been recently defined in a structured and systematic way. Specifically, Memetic Computing is a broad subject which studies complex and dynamic computing structures composed of interacting operators (memes) whose evolution dynamics is inspired by the diffusion of ideas.

Research in evolutionary optimization has always been closely tied to self-adaptation, i. e., the development of approaches which can adapt their parameters to the optimization problem at hand. An important research goal in this area would thus be to develop an intelligent unit which can choose, during the optimization run, the most suitable combination of operators for a given problem. Since a high degree of flexibility is necessary for solving a wide range of problems, Memetic Computing is strictly connected to the concept of modularity and an evolutionary structure that can be seen as a collection of interactive modules whose interaction, in an evolutionary sense, leads to the generation of the solution of the problem.

Concerning the structure of Memetic Computing approaches, there are two philosophies. On one hand, Memetic Computing can be seen as a broad subject which includes various kinds of algorithms. In order to solve optimization problems, a structure consisting of multiple operators, each of which performing a simple action, must be composed. Depending on the underlying algorithms used, a Memetic Computing approach may or may not be an EA (or its extension).

On the other hand, Memetic Computing can be considered from a bottom-up perspective. Here, the optimization algorithm would start as a blank slate to which components are added one by one. One interesting stream of research is the automatic design of algorithmic structures. Here, three aspects should be considered: 1) the memes should be simple operators, 2) the role and effect of each meme should be clearly understood so that this knowledge can be encoded and used by the automated designer in a flexible way, and 3) the algorithmic structure should be built up from scratch by means of the aforementioned bottom-up strategy which aims at including only the necessary memes and the simplest possible coordination rules.

5 Swarm Intelligence

Swarm Intelligence, another area closely related to EAs, is concerned with the design of algorithms or distributed problem-solving devices inspired by the collective behavior of social insects or animals. Two of the most popular Swarm Intelligence algorithms are PSO and ACO. Other representative examples include those for routing in communication networks based on the foraging behavior of bees [36], and those for dynamic task allocation inspired by the behavior of wasp colonies [15].

The natural role model of Particle Swarm Optimization, originally proposed by Eberhart and Kennedy [33, 34, 60], is the behavior of biological social systems like flocks of birds or schools of fish. PSO simulates a swarm of particles (individuals) in an n-dimensional search space, where each particle has its own position and velocity [40, 41, 86]. The velocity vector of an individual determines in which direction the search will continue and if it has an explorative (high velocity) or an exploitative (low velocity) character. This velocity vector represents an endogenous parameter – while the endogenous information in ES is used for an undirected mutation, the velocity in PSO is used to perform a directed modification of the genotypes (particles' positions).

ACO, developed by Dorigo et al. [31], is an optimization technique inspired by the capability of ant colonies to find short paths between encountered food sources and their ant hill [12, 13]. This capability is a result of the collective behavior of locally interacting ants. Here, the ants communicate indirectly via chemical pheromone trails which they deposit on the ground. This behavior can be simulated as a multi-agent system using a pheromone model in order to construct new solutions in each iteration.

In the following sections, we will take a closer look at PSO and ACO, and discuss their similarities and differences with EAs.

5.1 Particle Swarm Optimization

In what we refer to as the Standard PSO (SPSO), whose code is freely available on the Particle Swarm Central `http://www.particleswarm.info`, there is typically a unique swarm of agents, called particles, in which each particle P_i is defined as

$$P_i = (p_i, v_i, b_i) \tag{1}$$

where p_i is the position, v_i the velocity (more precisely the displacement), and b_i the best position ever found so far by the particle. Each particle is informed by a set $\mathcal{N} = \{P_j\}$ of other particles called "neighborhood". The metaphor is that each particle "moves", and the process at each time step can be described as follows:

1. each particle asks its neighbors, and chooses the best one (the one that has the best b_j)
2. it computes a new velocity v_i' by taking into account v_i, p_i, b_j; the precise formula is not important here, as it differs from version to version of SPSO (e.g., compare SPSO 2007 and SPSO 2011) – the most important key feature is that it contains some randomness and that its general form is

$$v_i' = a(v_i) + b(p_i) + c(b_i) + d(b_j) \tag{2}$$

3. each particle "moves", by computing the new position as $p_i' = p_i + v_i'$
4. if the new position is better, then b_i is updated, by $b_i' = p_i'$

There also exists another possible, formally equivalent but more flexible, point of view. That is, one may consider three kinds of agents:

1. position agents p_i
2. velocity agents v_i
3. memory agents m_i

Here, m_i is in fact the b_i of the previous process description. Now, there are three populations, $\mathcal{P} = \{p_i\}$, $\mathcal{V} = \{v_i\}$, and $\mathcal{M} = \{m_i\}$. Each v_i has a "neighborhood" of informants, which is a subset of $\mathcal{M}$, and the process at each time step can be described as follows:

1. the velocity agent v_i updates its components, thanks to the function a of Equation 2
2. then it combines them with some information coming from p_i, m_i, and from its best informant m_j, thanks to the functions b, c, d, in order to

define a new velocity v'_i (note that the order of the operations may be equivalently 2 then 1, as Equation 2 is commutative)
3. a new position agent p'_i is generated, by $p'_i = p_i + v'_i$
4. if the new agent is better than m_i, the agent m_i "dies", and is replaced by a better one, by using the formula $m'_i = p'_i$
5. p_i "dies"

Mathematically speaking, the behavior here is exactly the same as the previously described one, but as the metaphor is different, it is now easier to answer some of the relevant questions we want to address in this chapter.

5.2 Is PSO an EA?

A classical definition of an EA, given in Section 3, states that it uses mechanisms such as reproduction, mutation, recombination, and selection. Quite often, it is also added that some of these mechanisms have to make use of randomness. It is clear that randomness is used in all stochastic algorithms, including PSO, so we will not proceed on this point any further. In the following, let us consider, one by one, the four mechanisms of EAs – mutation, recombination, reproduction and selection – from a PSO point of view.

In molecular biology and genetics, mutations are changes in a genomic sequence. In a D-dimensional search space, a velocity agent can be written $v_i = (v_{i,1}, \cdots, v_{i,D})$. It is worth noting that, on a digital computer the search space is necessarily finite and discrete (even if the number of possible $v_{i,k}$ values is huge). Therefore, v_i can be seen as a "genomic sequence". According to point 1 in the algorithm description above, the velocity agent can be said to be "mutated". Here, however, the mutation rate is almost always equal to 100% (all components are modified). Also, mutation occurs before the reproduction.

Genetic recombination is a process by which a molecule of nucleic acid is broken and then joined to a different one. Point 2 in the PSO algorithm description can be seen as a recombination of the genomic sequences of three agents.

Reproduction (or procreation) is the biological process by which new "offspring" individual organisms are produced from their "parents". According to point 3 of the PSO description, a part of the process can be symbolically described by

$$(p_i, v'_i) \Rightarrow p'_i \tag{3}$$

which can be interpreted as procreation with two "parents".

Natural selection is the process by which traits become more or less common in a population due to consistent effects upon the survival or reproduction of their bearers. We can see that point 4 of the PSO algorithm is a selection mechanism: the agent m_i may die or survive, according to its "quality". Also, it can be proved (see more comments about this in [21]) that there is always convergence. It means that the m_i agents (and also the p_i agents)

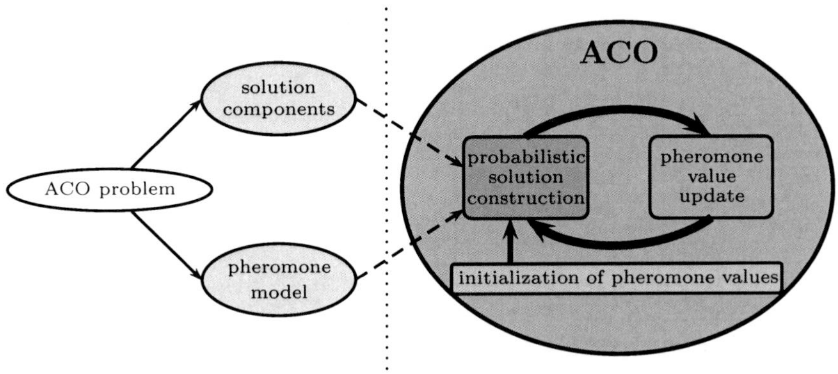

Fig. 4. A schematic view of ACO algorithms

become more and more similar. In the optimization context, this phenomenon is called stagnation, and is not very desirable. In other words, there is a kind of selection, but it has to be carefully controlled for good performance.

So, is PSO an EA or not? The answer to the question itself is not really interesting. It is just a matter of classification. By studying the question, however, a new point of view on PSO could be defined, which may suggest some fruitful variants (not studied in detail here). For instance:

1. The "mutation" rate may be smaller than 100%. In that case, not all velocity components are modified. In particular, if it is zero, there is no "generation", and, as the position agent "dies", the swarm size decreases.
2. Instead of being informed by always the same memory agent m_i, the velocity agent v_i may be informed by some others. The "combination" may make use of more than two memory agents, or even all (for this case, see [70]). Actually, we may also define a population $\mathcal{L}$ of "link agents". The existence of a (i, j) agent means there is an information link between the velocity agent v_i and the memory agent m_j. It is even possible to design an algorithm that works by co-evolution of the four populations $\mathcal{P}$, $\mathcal{V}$, $\mathcal{M}$, and $\mathcal{L}$.
3. The position agent may not die. In that case, and if the velocity agent is not null, the swarm size increases.

... and so on.

5.3 Ant Colony Optimization

Like EAs, ACO algorithms [9, 32] are bio-inspired techniques for optimization. A schematic view of ACO algorithms is shown in Fig. 4. They are based on a so-called pheromone model, which is a set of numerical values that are associated to opportunely defined solution components. In the case of the well-known TSP, for example, the edges of the underlying graph are the solution components. The pheromone model is used to generate – at each iteration – a fixed number of solutions to the considered problem. Again considering the case of the TSP, edges with a high pheromone value have a greater chance to be chosen during the solution construction process. In this way, the pheromone model – together with the mechanism for constructing solutions – implies a parameterized probability distribution over the search space. In general, the ACO approach attempts to solve an optimization problem by iterating the following two steps:

1. candidate solutions are constructed in a probabilistic way by using the pheromone model;
2. the candidate solutions are used to modify the pheromone values in a way that is deemed to bias future sampling toward high quality solutions.

In other words, the pheromone update aims to concentrate the search in regions of the search space containing high quality solutions. In particular, the reinforcement of solution components depending on the solution quality is an important ingredient of ACO algorithms. It implicitly assumes that good solutions consist of good solution components. To learn which components contribute to good solutions can help assembling them into better solutions.

5.4 Is ACO an EA?

While there are some similarities between EAs and ACO algorithms, there also exist some fundamental differences. Concerning the similarities, ACO algorithms are – just like EAs – population-based techniques. At each iteration a number of new solutions is generated. In both cases new solutions are generated based on the search experience. However, while most EAs store their search experience in the explicit form of a population of solutions, ACO algorithms store their search experience in the values of the pheromone model. Accordingly, there are also differences in updating the stored information. While standard EAs perform an explicit update of the population – that is, at each iteration some solutions are replaced by new ones – ACO algorithms use some of the generated solutions for making an update of the pheromone values.

Despite the differences, ACO algorithms and certain types of EAs can be studied under a common framework known as *model-based search* [115]. Apart from ACO algorithms, this framework also covers stochastic gradient ascent, the cross-entropy method, and EAs that can be labeled as Estimation of Distribution Algorithms (EDAs) [68]. According to [115], *"in model-based search algorithms, candidate solutions are generated using a parametrized probabilistic model that is updated using the previously seen solutions in such a way that the search will concentrate in the regions containing high quality solutions."*

The development of EDAs was initiated by mainly two observations. The first one concerns the fact that standard crossover operators were often observed to destroy good *building blocks*, i.e., partial solutions that are present in most, if not all, high quality solutions. The second observation is the one of *genetic drift*, i.e., a loss of diversity in the population due to its finite size. As a result of genetic drift, EAs may prematurely converge to sub-optimal solutions. One of the earliest EDAs is *Population-Based Incremental Learning* (PBIL) [3], developed with the idea of removing the genetics from the standard GA. In fact, for problems with independent decision variables, PBIL using only the best solution of each iteration for the update is equivalent to a specific version of ACO known as the *hyper-cube framework* with iteration-best update [10].

Summarizing, while ACO algorithms may be seen as model-based search algorithms, just like some EAs, ACO algorithms should not be labeled as "Evolutionary Algorithms".

6 Concluding Remarks

In this chapter, we have discussed the term "Evolutionary Algorithms" from various different perspectives. As seen in Section 3, there are at least two ways to define EAs. Traditionally, "Evolutionary Algorithms" is considered as a term identifying a set of algorithms (e.g., GAs, GP, ES and EP) which work according to the same basic cycle. Today, even these terms became mere names for large algorithm families which consist of many different sub-algorithms. The justification for such a variety of algorithms has been pointed out in Section 4.1: the NFLT which signifies that there may be an algorithm which is best for a certain family of optimization problems, but not for all possible ones. The variety of "Evolutionary Algorithms" has led to the controversy about what is an EA and what it is not.

One of the factors contributing to this situation is that there exist many new metaheuristics that share the characteristic traits of EAs but differ significantly in their semantics. Hybrid EAs incorporating local search algorithms and other Memetic Computing approaches, for instance, possess a different algorithmic structure. EDAs are population-based randomized algorithms and involve selection and possibly mutation – but are not related to any process in nature.

Another possible factor is that researchers nowadays tend to pay more efforts into defining common frameworks which can unite different algorithms, such as the already mentioned work in [115] or the recent framework proposed in [107] that unites both the traditional EAs and EDAs. Generally speaking, metaheuristics can be viewed as the combination of components for search and selection, i. e., a set of operations for generating one or more trial solutions and/or a set of operations to perform the selection of the solution and thus of the search directions.

Furthermore, the research communities working on particular algorithms pursue a process of generalization and formalization during which more similarities between formerly distinct approaches are discovered. These processes make it easier to construct versatile algorithms and also provide the chance of obtaining more generally applicable theoretical results.

Besides these reasons, there is the basic fact that researchers themselves are the ones who decide the name of their algorithms. It may indeed be argued whether a $(1 + 1)$ ES is actually an EA or just a Hill Climbing method, or whether those very first MAs were special EAs or not. Approaching this issue from the opposite direction, it is indeed possible to develop algorithms which improve a set of solutions with a process of choosing the best ones and slightly modifying them in an iterative way, e.g., by using unary and binary search operations, without utilizing any inspiration from nature. Would the term "Evolutionary Algorithm" appropriate for such an algorithm?

The meaning of the term is thus subject to interpretation, and we put three other metaheuristics, the MA, PSO and ACO, into the context of this controversy. The sections on PSO and ACO in particular symbolize very well how different researchers may either tend to generalize an algorithm's definition to make it more compatible to the evolutionary framework or may emphasize more on its individual features in favor of more distinct semantics.

A simple strategy to avoid ambiguity would be to use terms like *Nature-Inspired Algorithms* or *Evolutionary Computation Techniques* for general methods inspired by nature or evolution and to preserve the term "Evolutionary Algorithm" for GAs, GP, ES, EP and, to a lesser extent, Differential Evolution and EDAs.

Another idea would be to more strictly divide the theoretical algorithm structure from its inspirational roots and history, i. e., to totally abandon terms such as "genetics", "evolutionary", "mutation" or "crossover" from the naming conventions. Of course, this would probably not happen since these terms have already entered the folklore. However, more frequently using words such as "unary search operation" instead of "mutation" or "candidate solution" instead of "phenotype" in favor of a clearer ontology would lead to more precise definitions, inspire more rigorous analyses, and may reduce the quack aura sometimes wrongly attributed by industrial engineers to the so-called Evolutionary Computation techniques.

Yet, it is likely that "Evolutionary Algorithms" would suffer the same fate as the term "agent" and blur into a state of, on one hand, almost universally applicable and, on the other hand, lesser semantic values. Then again, this does not necessarily be bad – since it may open the door for even more cross-discipline interaction and cross-fertilization of ideas, as can be observed in the agent community during the past 20 years.

Acknowledgement. Ferrante Neri's work is supported by Academy of Finland, Akatemiatutkija 130600, Algorithmic Design Issues in Memetic Computing. Christian Blum acknowledges support from grant TIN2007-66523 (FORMLISM) and from the Ramón y Cajal program of the Spanish Government. Thomas Weise is supported by the Chinese Academy of Sciences (CAS) Fellowship for Young International Scientists 2011 and the China Postdoctoral Science Foundation Grant Number 20100470843.

References

1. Bäck, T., Hoffmeister, F., Schwefel, H.: A Survey of Evolution Strategies. In: Belew, R.K., Booker, L.B. (eds.) Proceedings of the Fourth International Conference on Genetic Algorithms (ICGA 1991), pp. 2–9. Morgan Kaufmann Publishers Inc., San Francisco (1991)
2. Bäck, T., Fogel, D.B., Michalewicz, Z. (eds.): Handbook of Evolutionary Computation. Oxford University Press, Inc., USA (1997)
3. Baluja, S., Caruana, R.A.: Removing the Genetics from the Standard Genetic Algorithm. In: Prieditis, A., Russell, S.J. (eds.) Proceedings of the Twelfth International Conference on Machine Learning (ICML 1995), pp. 38–46. Morgan Kaufmann Publishers Inc., San Francisco (1995)
4. Barricelli, N.A.: Esempi Numerici di Processi di Evoluzione. Methodos 6(21-22), 45–68 (1954)
5. Barricelli, N.A.: Symbiogenetic Evolution Processes Realized by Artificial Methods. Methodos 9(35-36), 143–182 (1957)
6. Belew, R.K., Booker, L.B. (eds.): Proceedings of the Fourth International Conference on Genetic Algorithms (ICGA 1991), July13–16, pp. 13–16. Morgan Kaufmann Publishers Inc., USA (1991)
7. Beyer, H.: The Theory of Evolution Strategies, Natural Computing Series, Springer, New York (May 27, 2001); ISBN: 3-540-67297-4
8. Beyer, H., Schwefel, H.: Evolution Strategies – A Comprehensive Introduction. Natural Computing: An International Journal 1(1), 3–52 (2002); doi10.1023/A:1015059928466
9. Blum, C.: Ant Colony Optimization: Introduction and Recent Trends. Physics of Life Reviews 2(4), 353–373 (2005); doi:10.1016/j.plrev.2005.10.001
10. Blum, C., Dorigo, M.: The Hyper-Cube Framework for Ant Colony Optimization. IEEE Transactions on Systems, Man, and Cybernetics – Part B: Cybernetics 34(2), 1161–1172 (2004); doi:10.1109/TSMCB.2003.821450
11. Blum, C., Roli, A.: Metaheuristics in Combinatorial Optimization: Overview and Conceptual Comparison. ACM Computing Surveys (CSUR) 35(3), 268–308 (2003); doi:10.1145/937503.937505
12. Bonabeau, E.W., Dorigo, M., Théraulaz, G.: Swarm Intelligence: From Natural to Artificial Systems. Oxford University Press, Inc., USA (1999); ISBN: 0195131592
13. Bonabeau, E.W., Dorigo, M., Théraulaz, G.: Inspiration for Optimization from Social Insect Behavior. Nature 406, 39–42 (2000); doi:10.1038/35017500
14. Burke, E.K., Kendall, G., Soubeiga, E.: A Tabu Search Hyperheuristic for Timetabling and Rostering. Journal of Heuristics 9(6), 451–470 (2003); doi:10.1023/B:HEUR.0000012446.94732.b6
15. Campos, M., Bonabeau, E.W., Théraulaz, G., Deneubourg, J.: Dynamic Scheduling and Division of Labor in Social Insects. Adaptive Behavior 8(2), 83–95 (2000); doi:10.1177/105971230000800201
16. Caponio, A., Cascella, G.L., Neri, F., Salvatore, N., Sumner, M.: A Fast Adaptive Memetic Algorithm for Online and Offline Control Design of PMSM Drives. IEEE Transactions on Systems, Man, and Cybernetics – Part B: Cybernetics 37(1), 28–41 (2007); doi:10.1109/TSMCB.2006.883271

17. Caponio, A., Neri, F., Tirronen, V.: Super-fit Control Adaptation in Memetic Differential Evolution Frameworks. Soft Computing – A Fusion of Foundations, Methodologies and Applications 13(8-9), 811–831 (2009); doi:10.1007/s00500-008-0357-1
18. Chen, W.X., Weise, T., Yang, Z.Y., Tang, K.: Large-Scale Global Optimization Using Cooperative Coevolution with Variable Interaction Learning. In: Schaefer, R., Cotta, C., Kołodziej, J., Rudolph, G., et al. (eds.) PPSN XI. LNCS, vol. 6239, pp. 300–309. Springer, Heidelberg (2010), doi:10.1007/978-3-642-15871-1-31
19. Chiong, R. (ed.): Nature-Inspired Algorithms for Optimisation, April 30. SCI, vol. 193. Springer, Heidelberg (2009); ISBN: 3-642-00266-8, 3-642-00267-6, doi:10.1007/978-3-642-00267-0
20. Chiong, R., Weise, T., Michalewicz, Z. (eds.): Variants of Evolutionary Algorithms for Real-World Applications. Springer, Heidelberg (2011)
21. Clerc, M., Kennedy, J.: The Particle Swarm – Explosion, Stability, and Convergence in a Multidimensional Complex Space. IEEE Transactions on Evolutionary Computation 6(1), 58–73 (2002); doi:10.1109/4235.985692
22. Coello Coello, C.A., Lamont, G.B., van Veldhuizen, D.A.: Evolutionary Algorithms for Solving Multi-Objective Problems. Genetic and Evolutionary Computation, vol. 5. Springer, Heidelberg (2002); doi:10.1007/978-0-387-36797-2
23. Cowling, P., Kendall, G., Soubeiga, E.: A Hyperheuristic Approach to Scheduling a Sales Summit. In: Burke, E., Erben, W. (eds.) PATAT 2000. LNCS, vol. 2079, pp. 176–190. Springer, Heidelberg (2001); doi:10.1007/3-540-44629-X-11
24. Dawkins, R.: The Selfish Gene, 1st, 2nd edn. Oxford University Press, Inc., USA (1976); ISBN:0-192-86092-5
25. De Jong, K.A.: An Analysis of the Behavior of a Class of Genetic Adaptive Systems, PhD thesis, University of Michigan: Ann Arbor, MI, USA (1975)
26. De Jong, K.A.: Genetic Algorithms are NOT Function Optimizers. In: Whitley, L.D. (ed.) Proceedings of the Second Workshop on Foundations of Genetic Algorithms (FOGA 1992), pp. 5–17. Morgan Kaufmann Publishers Inc., San Francisco (1992)
27. De Jong, K.A.: Evolutionary Computation: A Unified Approach. Complex Adaptive Systems, vol. 4. MIT Press, Cambridge (2006)
28. Deb, K., Goldberg, D.E.: Analyzing Deception in Trap Functions. In: Whitley, L.D. (ed.) Proceedings of the Second Workshop on Foundations of Genetic Algorithms (FOGA 1992), pp. 93–108. Morgan Kaufmann Publishers Inc., San Francisco (1992)
29. Deb, K.: Multi-Objective Optimization Using Evolutionary Algorithms. Wiley Interscience Series in Systems and Optimization, John Wiley & Sons Ltd., New York (2001)
30. Deb, K., Pratab, A., Agrawal, S., Meyarivan, T.: A Fast and Elitist Multi-objective Genetic Algorithm: NSGA-II. IEEE Transactions on Evolutionary Computation 6(2), 182–197 (2002); doi:10.1109/4235.996017
31. Dorigo, M., Maniezzo, V., Colorni, A.: The Ant System: Optimization by a Colony of Cooperating Agents. IEEE Transactions on Systems, Man, and Cybernetics Part B: Cybernetics 26(1), 29–41 (1996); doi:10.1109/3477.484436, ftp://iridia.ulb.ac.be/pub/mdorigo/journals/IJ.10-SMC96.pdf

32. Dorigo, M., Stützle, T.: Ant Colony Optimization. Bradford Books. MIT Press (July 2004); ISBN: 0-262-04219-3
33. Eberhart, R.C., Kennedy, J.: A New Optimizer Using Particle Swarm Theory. In: Proceedings of the Sixth International Symposium on Micro Machine and Human Science (MHS 1995), pp. 39–43. IEEE Computer Society, USA (1995); doi:10.1109/MHS.1995.494215
34. Eberhart, R.C., Shi, Y.: A Modified Particle Swarm Optimizer. In: Simpson, P.K. (ed.) The 1998 IEEE International Conference on Evolutionary Computation (CEC 1998), pp. 69–73. IEEE Computer Society, Los Alamitos (1998); doi:10.1109/ICEC.1998.699146
35. Eiben, Á.E., Smith, J.E.: Introduction to Evolutionary Computing, 1st edn. Natural Computing Series, ch. 10, pp. 173–188. Springer, New York (2003)
36. Farooq, M.: Bee-Inspired Protocol Engineering – From Nature to Networks. Natural Computing Series, vol. 15. Springer, New York (2009); ISBN: 3-540-85953-5, doi:10.1007/978-3-540-85954-3
37. Fogel, L.J.: On the Organization of Intellect. PhD thesis, University of California (UCLA): Los Angeles, CA, USA (1964)
38. Fogel, L.J., Owens, A.J., Walsh, M.J.: Artificial Intelligence through Simulated Evolution. John Wiley & Sons Ltd., USA (1966); ISBN: 0471265160
39. Fraser, A.S.: Simulation of Genetic Systems by Automatic Digital Computers. I. Introduction. Australian Journal of Biological Science (AJBS) 10, 484–491 (1957)
40. Gao, Y., Duan, Y.: An Adaptive Particle Swarm Optimization Algorithm with New Random Inertia Weight. In: Huang, D.-S., Heutte, L., Loog, M. (eds.) Advanced Intelligent Computing Theories and Applications. With Aspects of Contemporary Intelligent Computing Techniques - ICIC 2007. LNCS, vol. 2, pp. 342–350. Springer, Heidelberg (2007), doi:10.1007/978-3-540-74282-1-39
41. Gao, Y., Ren, Z.: Adaptive Particle Swarm Optimization Algorithm With Genetic Mutation Operation. In: Lei, J., Yao, J., Zhang, Q. (eds.) Proceedings of the Third International Conference on Advances in Natural Computation (ICNC'07), vol. 2, pp. 211–215. IEEE Computer Society Press, Los Alamitos (2007), doi:10.1109/ICNC.2007.161
42. Glover, F., Kochenberger, G.A. (eds.): Handbook of Metaheuristics. International Series in Operations Research & Management Science, vol. 57. Kluwer, Springer Netherlands, Dordrecht, Netherlands (2003), doi:10.1007/b101874
43. Goldberg, D.E.: Genetic Algorithms in Search, Optimization, and Machine Learning. Addison-Wesley Longman Publishing Co, USA (1989); ISBN: 0-201-15767-5
44. Gong, M., Jiao, L., Zhang, L.: Baldwinian Learning in Clonal Selection Algorithm for Optimization. Information Sciences – Informatics and Computer Science Intelligent Systems Applications: An International Journal 180(8), 1218–1236 (2010); doi:10.1016/j.ins.2009.12.007
45. Gonzalez, T.F. (ed.): Handbook of Approximation Algorithms and Metaheuristics, Chapmann & Hall/CRC Computer and Information Science Series. Chapman & Hall/CRC, Boca Raton, FL (2007)

46. Grefenstette, J.J.: Deception Considered Harmful. In: Whitley, L.D. (ed.) Proceedings of the Second Workshop on Foundations of Genetic Algorithms (FOGA 1992), pp. 75–91. Morgan Kaufmann Publishers Inc., USA (1992)
47. Hansen, N., Ostermeier, A.: Adapting Arbitrary Normal Mutation Distributions in Evolution Strategies: The Covariance Matrix Adaptation. In: Jidō, K., Gakkai, S. (eds.) Proceedings of IEEE International Conference on Evolutionary Computation (CEC 1996), pp. 312–317. IEEE Computer Society Press, Los Alamitos (1996); doi:10.1109/ICEC.1996.542381
48. Hansen, N., Ostermeier, A.: Convergence Properties of Evolution Strategies with the Derandomized Covariance Matrix Adaption: The $(\mu/\mu_I, \lambda)$-CMA-ES. In: Zimmermann, H. (ed.) Proceedings of the 5th European Congress on Intelligent Techniques and Soft Computing (EUFIT 1997), vol. 1, pp. 650–654. ELITE Foundation, Germany (1997)
49. Hansen, N., Ostermeier, A.: Completely Derandomized Self-Adaptation in Evolution Strategies. Evolutionary Computation 9(2), 159–195 (2001)
50. Hansen, N., Ostermeier, A., Gawelczyk, A.: On the Adaptation of Arbitrary Normal Mutation Distributions in Evolution Strategies: The Generating Set Adaptation. In: Eshelman, L.J. (ed.) Proceedings of the Sixth International Conference on Genetic Algorithms (ICGA 1995), pp. 57–64. Morgan Kaufmann Publishers Inc., San Francisco (1995)
51. Hart, W.E., Krasnogor, N., Smith, J.E.: Memetic Evolutionary Algorithms. In: Hart, W.E., Krasnogor, N., Smith, J.E. (eds.) Recent Advances in Memetic Algorithms. Studies in Fuzziness and Soft Computing, ch.1, vol. 166, pp. 3–27. Springer, Heidelberg (2005)
52. Grefenstette, J.J.: Proceedings of the 1st International Conference on Genetic Algorithms and their Applications (ICGA 1985), June 24-26, pp. 24–26. Carnegy Mellon University (CMU), Lawrence Erlbaum Associates, Hillsdale, USA (1985)
53. Hillis, W.D.: Co-Evolving Parasites Improve Simulated Evolution as an Optimization Procedure. Physica D: Nonlinear Phenomena 42(1-2), 228–234 (1990); doi:10.1016/0167-2789(90)90076-2
54. Hitch-Hiker's Guide to Evolutionary Computation: A List of Frequently Asked Questions (FAQ) (HHGT) (March 29, 2000)
55. Holland, J.H.: Adaptation in Natural and Artificial Systems: An Introductory Analysis with Applications to Biology, Control, and Artificial Intelligence. University of Michigan Press, USA (1975); ISBN: 0-472-08460-7
56. Holland, J.H.: Genetic Algorithms. Scientific American 267(1), 44–50 (1992)
57. Igel, C., Toussaint, M.: On Classes of Functions for which No Free Lunch Results Hold. Information Processing Letters 86(6), 317–321 (2003); doi:10.1016/S0020-0190(03)00222-9
58. Jastrebski, G.A., Arnold, D.V.: Improving Evolution Strategies through Active Covariance Matrix Adaptation. In: Yen, G.G., et al. (eds.) Proceedings of the IEEE Congress on Evolutionary Computation CEC 2006, pp. 9719–9726. IEEE Computer Society, Los Alamitos (2006); doi:10.1109/CEC.2006.1688662
59. Kendall, G., Cowling, P., Soubeiga, E.: Choice Function and Random Hyper-Heuristics. In: Tan, K.C., et al. (eds.) Recend Advances in Simulated Evolution and Learning – Proceedings of the Fourth Asia-Pacific Conference on Simulated Evolution And Learning (SEAL 2002). Advances in Natural Computation, vol. 2, pp. 667–671. World Scientific Publishing Co, Singapore (2002)

60. Kennedy, J., Eberhart, R.C.: Particle Swarm Optimization. In: Proceedings of the IEEE International Conference on Neural Networks (ICNN 1995), vol. 4, pp. 1942–1948. IEEE Computer Society Press, Los Alamitos (1995); doi:10.1109/ICNN.1995.488968
61. Kononova, A.V., Ingham, D.B., Pourkashanian, M.: Simple Scheduled Memetic Algorithm for Inverse Problems in Higher Dimensions: Application to Chemical Kinetics. In: Michalewicz, Z., Reynolds, R.G. (eds.) Proceedings of the IEEE Congress on Evolutionary Computation (CEC 2008), pp. 3905–3912. IEEE Computer Society Press, Los Alamitos (2008); doi:10.1109/CEC.2008.4631328
62. Korošec, P., Šilc, J., Filipič, B.: The Differential Ant-Stigmergy Algorithm. Information Sciences – Informatics and Computer Science Intelligent Systems Applications: An International Journal (2011)
63. Koza, J.R.: Concept Formation and Decision Tree Induction using the Genetic Programming Paradigm. In: Schwefel, H.-P., Männer, R. (eds.) PPSN 1990. LNCS, vol. 496, pp. 124–128. Springer, Heidelberg (1991); doi:10.1007/BFb0029742
64. Koza, J.R.: Genetic Programming: On the Programming of Computers by Means of Natural Selection, 1st edn. Bradford Books, MIT Press (1992); 2nd edn. (1993)
65. Koza, J.R., Andre, D., Bennett III, F.H., Keane, M.A.: Use of Automatically Defined Functions and Architecture-Altering Operations in Automated Circuit Synthesis Using Genetic Programming. In: Koza, J.R., et al. (eds.) Proceedings of the First Annual Conference of Genetic Programming (GP 1996), Complex Adaptive Systems, Bradford Books, pp. 132–149. MIT Press, Cambridge (1996)
66. Kramer, O.: Self-Adaptive Heuristics for Evolutionary Computation. SCI, vol. 147. Springer, Heidelberg (2008); doi:10.1007/978-3-540-69281-2
67. Krasnogor, N., Smith, J.E.: A Tutorial for Competent Memetic Algorithms: Model, Taxonomy, and Design Issues. IEEE Transactions on Evolutionary Computation 9(5), 474–488 (2005); doi:10.1109/TEVC.2005.850260
68. Larrañaga, P., Lozano, J.A. (eds.): Estimation of Distribution Algorithms – A New Tool for Evolutionary Computation. Genetic and Evolutionary Computation, vol. 2. Springer US, USA (2001)
69. Le, M.N., Ong, Y., Jin, Y., Sendhoff, B.: Lamarckian Memetic Algorithms: Local Optimum and Connectivity Structure Analysis. Memetic Computing 1(3), 175–190 (2009); doi:10.1007/s12293-009-0016-9
70. Mendes, R., Kennedy, J., Neves, J.: Fully Informed Particle Swarm: Simpler, Maybe Better. IEEE Transactions on Evolutionary Computation 8(3), 204–210 (2004); doi:10.1109/TEVC.2004.826074
71. Mendes, R.R.F., de Voznika, F.B., Freitas, A.A., Nievola, J.C.: Discovering Fuzzy Classification Rules with Genetic Programming and Co-evolution. In: Siebes, A., De Raedt, L. (eds.) PKDD 2001. LNCS (LNAI), vol. 2168, pp. 314–325. Springer, Heidelberg (2001), doi:10.1007/3-540-44794-6-26
72. Meyer-Nieberg, S., Beyer, H.: Self-Adaptation in Evolutionary Algorithms. In: Lobo, F.G., Lima, C.F., Michalewicz, Z. (eds.) Parameter Setting in Evolutionary Algorithms. SCI, ch. 3, vol. 54, pp. 47–75. Springer, Heidelberg (2007), doi:10.1007/978-3-540-69432-8-3

73. Michalewicz, Z.: A Perspective on Evolutionary Computation. In: Yao, X. (ed.) AI-WS 1993 and 1994. LNCS, vol. 956, pp. 73–89. Springer, Heidelberg (1995), doi:10.1007/3-540-60154-6-49
74. Michalewicz, Z., Schaffer, J.D., Schwefel, H.-P., Fogel, D.B., Kitano, H.: Proceedings of the First IEEE Conference on Evolutionary Computation (CEC 1994), June 27-29, pp. 27–29. IEEE Computer Society Press, Los Alamitos (1997)
75. Michalewicz, Z.: Genetic Algorithms + Data Structures = Evolution Programs. Springer, Heidelberg (1996)
76. Michalewicz, Z., Fogel, D.B.: How to Solve It: Modern Heuristics, 2nd extended edn. Springer, Heidelberg (2004)
77. Moscato, P.: On Evolution, Search, Optimization, Genetic Algorithms and Martial Arts: Towards Memetic Algorithms. Caltech Concurrent Computation Program C3P 826, California Institute of Technology (Caltech), Caltech Concurrent Computation Program (C3P), Pasadena (1989)
78. Neri, F., Tirronen, V., Kärkkäinen, T., Rossi, T.: Fitness Diversity based Adaptation in Multimeme Algorithms: A Comparative Study. In: Proceedings of the IEEE Congress on Evolutionary Computation (CEC 2007), pp. 2374–2381. IEEE Computer Society, Los Alamitos (2007); doi:10.1109/CEC.2007.4424768
79. Neri, F., Toivanen, J., Cascella, G.L., Ong, Y.: An Adaptive Multimeme Algorithm for Designing HIV Multidrug Therapies. IEEE/ACM Transactions on Computational Biology and Bioinformatics (TCBB) 4(2) (April 2007); doi:10.1109/TCBB.2007.070202
80. Neri, F., Toivanen, J., Mäkinen, R.A.E.: An Adaptive Evolutionary Algorithm with Intelligent Mutation Local Searchers for Designing Multidrug Therapies for HIV. Applied Intelligence – The International Journal of Artificial Intelligence, Neural Networks, and Complex Problem-Solving Technologies 27(3), 219–235 (2007); doi:10.1007/s10489-007-0069-8
81. Nguyen, Q.H., Ong, Y., Lim, M.H., Krasnogor, N.: Adaptive Cellular Memetic Algorithms. Evolutionary Computation 17(2), 231–256 (2009); doi:10.1162/evco.2009.17.2.231
82. Norman, M.G., Moscato, P.: A Competitive and Cooperative Approach to Complex Combinatorial Search. Caltech Concurrent Computation Program 790, California Institute of Technology (Caltech), Caltech Concurrent Computation Program (C3P), Pasadena (1989)
83. Norman, M.G., Moscato, P.: A Competitive and Cooperative Approach to Complex Combinatorial Search. In: Proceedings of the 20th Informatics and Operations Research Meeting (20th Jornadas Argentinas e Informática e Investigación Operativa) (JAIIO 1991), pp. 3.15–3.29 (1991); Also published as Technical Report Caltech Concurrent Computation Program, Report. 790, California Institute of Technology, Pasadena, California, USA (1989)
84. Ong, Y., Keane, A.J.: Meta-Lamarckian Learning in Memetic Algorithms. IEEE Transactions on Evolutionary Computation 8(2), 99–110 (2004); doi:10.1109/TEVC.2003.819944

85. Ong, Y., Lim, M.H., Zhu, N., Wong, K.: Classification of Adaptive Memetic Algorithms: A Comparative Study. IEEE Transactions on Systems, Man, and Cybernetics – Part B: Cybernetics 36(1), 141–152 (2006); doi:10.1109/TSMCB.2005.856143
86. Parrish, J.K., Hamner, W.M. (eds.): Animal Groups in Three Dimensions: How Species Aggregate. Cambridge University Press, Cambridge (1997); doi:10.2277/0521460247, ISBN: 0521460247
87. Poli, R., Langdon, W.B., McPhee, N.F.: A Field Guide to Genetic Programming. Lulu Enterprises UK Ltd., UK (2008)
88. Potter, M.A., De Jong, K.A.: A Cooperative Coevolutionary Approach to Function Optimization. In: Davidor, Y., Männer, R., Schwefel, H.-P. (eds.) PPSN 1994. LNCS, vol. 866, pp. 249–257. Springer, Heidelberg (1994); doi:10.1007/3-540-58484-6-269
89. Potter, M.A., De Jong, K.A.: Cooperative Coevolution: An Architecture for Evolving Coadapted Subcomponents. Evolutionary Computation 8(1), 1–29 (2000)
90. Rechenberg, I.: Cybernetic Solution Path of an Experimental Problem. Royal Aircraft Establishment, Farnborough (1965)
91. Rechenberg, I.: Evolutionsstrategie: Optimierung technischer Systeme nach Prinzipien der biologischen Evolution. PhD thesis, Technische Universität Berlin: Berlin, Germany (1971)
92. Rechenberg, I.: Evolutionsstrategie 1994. Werkstatt Bionik und Evolutionstechnik, vol. 1. Frommann-Holzboog Verlag, Germany (1994)
93. Schwefel, H.: Kybernetische Evolution als Strategie der exprimentellen Forschung in der Strömungstechnik. Master's thesis, Technische Universität Berlin: Berlin, Germany (1965)
94. Schwefel, H.: Experimentelle Optimierung einer Zweiphasendüse Teil I. Technical Report 35, AEG Research Institute: Berlin, Germany, Project MHD–Staustrahlrohr 11.034/68 (1968)
95. Schwefel, H.: Evolutionsstrategie und numerische Optimierung. PhD thesis, Technische Universität Berlin, Institut für Meß- und Regelungstechnik, Institut für Biologie und Anthropologie: Berlin, Germany (1975)
96. Schwefel, H.: Evolution and Optimum Seeking. Sixth-Generation Computer Technology Series. John Wiley & Sons Ltd., USA (1995); ISBN: 0-471-57148-2
97. Smith, J.E.: Coevolving Memetic Algorithms: A Review and Progress Report. IEEE Transactions on Systems, Man, and Cybernetics – Part B: Cybernetics 37(1), 6–17 (2007); doi:10.1109/TSMCB.2006.883273
98. Srinivas, N., Deb, K.: Muiltiobjective Optimization Using Nondominated Sorting in Genetic Algorithms. Evolutionary Computation 2(3), 221–248 (1994); doi:10.1162/evco.1994.2.3.221
99. Tang, J., Lim, M.H., Ong, Y.: Diversity-Adaptive Parallel Memetic Algorithm for Solving Large Scale Combinatorial Optimization Problems. Soft Computing – A Fusion of Foundations, Methodologies and Applications 11(9), 873–888 (2007); doi:10.1007/s00500-006-0139-6
100. Tirronen, V., Neri, F., Kärkkäinen, T., Majava, K., Rossi, T.: An Enhanced Memetic Differential Evolution in Filter Design for Defect Detection in Paper Production. Evolutionary Computation 16(4), 529–555 (2008); doi:10.1162/evco.2008.16.4.529

101. Wang, P., Weise, T., Chiong, R.: Novel Evolutionary Algorithms for Supervised Classification Problems: An Experimental Study. Evolutionary Intelligence 4(1), 3–16 (2011); doi:10.1007/s12065-010-0047-7
102. Weise, T.: Global Optimization Algorithms – Theory and Application. it-weise.de (self-published), Germany (2009), `http://www.it-weise.de/projects/book.pdf`
103. Weise, T., Tang, K.: Evolving Distributed Algorithms with Genetic Programming. IEEE Transactions on Evolutionary Computation (to appear, 2011)
104. Weise, T., Podlich, A., Gorldt, C.: Solving Real-World Vehicle Routing Problems with Evolutionary Algorithms. In: Chiong, R., Dhakal, S. (eds.) Natural Intelligence for Scheduling, Planning and Packing Problems. SCI, ch.2, vol. 250, pp. 29–53. Springer, Heidelberg (2009), doi:10.1007/978-3-642-04039-9-2
105. Weise, T., Podlich, A., Reinhard, K., Gorldt, C., Geihs, K.: Evolutionary Freight Transportation Planning. In: Giacobini, M., Brabazon, A., Cagnoni, S., Di Caro, G.A., Ekárt, A., Esparcia-Alcázar, A.I., Farooq, M., Fink, A., Machado, P. (eds.) EvoWorkshops 2009. LNCS, vol. 5484, pp. 768–777. Springer, Heidelberg (2009), doi:10.1007/978-3-642-01129-0-87
106. Weise, T., Zapf, M., Chiong, R., Nebro Urbaneja, A.J.: Why Is Optimization Difficult? In: Chiong, R. (ed.) Nature-Inspired Algorithms for Optimisation. SCI, ch. 1, vol. 193, pp. 1–50. Springer, Heidelberg (2009); doi:10.1007/978-3-642-00267-0-1
107. Weise, T., Niemczyk, S., Chiong, R., Wan, M.: A Framework for Multi-Model EDAs with Model Recombination. In: Proceedings of the 4th European Event on Bio-Inspired Algorithms for Continuous Parameter Optimisation (EvoNUM 2011), Proceedings of the European Conference on the Applications of Evolutionary Computation (EvoAPPLICATIONS 2011). LNCS, Springer, Heidelberg (2011)
108. Whitley, L.D.: A Genetic Algorithm Tutorial. Statistics and Computing 4(2), 65–85 (1994); doi:10.1007/BF00175354
109. Wolpert, D.H., Macready, W.G.: No Free Lunch Theorems for Optimization. IEEE Transactions on Evolutionary Computation 1(1), 67–82 (1997); doi:10.1109/4235.585893
110. Yao, X., Liu, Y., Lin, G.: Evolutionary Programming Made Faster. IEEE Transactions on Evolutionary Computation 3(2), 82–102 (1999); doi:10.1109/4235.771163
111. Yu, E.L., Suganthan, P.N.: Ensemble of Niching Algorithms. Information Sciences – Informatics and Computer Science Intelligent Systems Applications: An International Journal 180(15), 2815–2833 (2010); doi:10.1016/j.ins.2010.04.008
112. Yuan, Q., Qian, F., Du, W.: A Hybrid Genetic Algorithm with the Baldwin Effect. Information Sciences – Informatics and Computer Science Intelligent Systems Applications: An International Journal 180(5), 640–652 (2010), doi:10.1016/j.ins.2009.11.015
113. Zitzler, E., Thiele, L.: An Evolutionary Algorithm for Multiobjective Optimization: The Strength Pareto Approach. 43, Eidgenssische Technische Hochschule (ETH) Zürich, Department of Electrical Engineering, Computer Engineering and Networks Laboratory (TIK), Zürich, Switzerland (May 1995)

114. Zitzler, E., Laumanns, M., Thiele, L.: SPEA2: Improving the Strength Pareto Evolutionary Algorithm. 101, Eidgenssische Technische Hochschule (ETH) Zürich, Department of Electrical Engineering, Computer Engineering and Networks Laboratory (TIK), Zürich, Switzerland (May 2001)
115. Zlochin, M., Birattari, M., Meuleau, N., Dorigo, M.: Model-Based Search for Combinatorial Optimization: A Critical Survey. Annals of Operations Research 132(1-4), 373–395 (2004); doi:10.1023/B:ANOR.0000039526.52305.af

An Evolutionary Approach to Practical Constraints in Scheduling: A Case-Study of the Wine Bottling Problem

Arvind Mohais, Sven Schellenberg, Maksud Ibrahimov,
Neal Wagner, and Zbigniew Michalewicz

Abstract. Practical constraints associated with real-world problems are a key differentiator with respect to more artificially formulated problems. They create challenging variations on what might otherwise be considered as straightforward optimization problems from an evolutionary computation perspective. Through solving various commercial and industrial problems using evolutionary algorithms, we have gathered experience in dealing with practical dynamic constraints. Here, we present proven methods for dealing with these issues for scheduling problems. For use in real-world situations, an evolutionary algorithm must be designed to drive a software application that needs to be robust enough to deal with practical constraints in order to meet the demands and expectations of everyday use by domain specialists who are not necessarily optimization experts. In such situations, addressing these issues becomes critical to success. We show how these challenges can be dealt

---

Arvind Mohais · Neal Wagner
SolveIT Software Pty. Ltd., Level 1, 99 Frome Street,
South Australia 5000, Australia
e-mail: {am,nw}@solveitsoftware.com

Sven Schellenberg
SolveIT Software Pty. Ltd., Suite 201. 198 Harbour Esplanade,
Docklands, Victoria 3008, Australia
e-mail: ss@solveitsoftware.com

Maksud Ibrahimov · Zbigniew Michalewicz
School of Computer Science, University of Adelaide,
South Australia 5005, Australia
e-mail: {maksud.ibrahimov,zbigniew.michalewicz}@adelaide.edu.au

Zbigniew Michalewicz
Also at Institute of Computer Science, Polish Academy of Sciences, ul. Ordona 21, 01-237 Warsaw, Poland and Polish-Japanese Institute of Information Technology, ul. Koszykowa 86, 02-008 Warsaw, Poland

with by making adjustments to genotypic representation, phenotypic decoding, or the evaluation function itself. The ideas presented in this chapter are exemplified by the means of a case study of a real-world commercial problem, namely that of bottling wine in a mass-production environment. The methods described have the benefit of having been proven by a full-fledged implementation into a software application that undergoes continual and vigorous use in a live environment in which time-varying constraints, arising in multiple different combinations, are a routine occurrence.

1 Introduction

The application of evolutionary algorithms (EAs) to real-world problems brings us face-to-face with some challenging issues related to the time-varying nature of such problems. These issues are over and above the fundamental problems being solved, and in fact, they can be so critical from the standpoint of creating a robust and truly usable software application that their consideration significantly alters the approach taken to solve the problem.

From a pure-problem perspective, EAs have been used for quite a long time to solve scheduling problems [1, 2, 3]. A general scheduling problem requires that we make decisions on how to assign a number of jobs to a fixed set of machines able to perform those tasks. The assignment must be done in such a way as to minimize the amount of time required to complete all of the tasks (referred to as the makespan). There are many variations to this problem. For example, the processing of a job may require that it passes through multiple pre-defined phases, each possibly taking place on a different machine. All of the variants however are equivalent to the NP-Complete problem of *job shop scheduling* [4]. This of course makes it virtually unsolvable in a reasonable amount of time using currently known techniques. Thus, a usual approach is to look for an approximate solution using EAs.

The essence of a straightforward evolutionary approach is to represent a candidate solution as an ordered list of (machine, job) pairs, with the natural order of the list being representative of the sequence of job execution. Using a well-constructed fitness function that embodies the problem constraints and parameters, and perturbing the representation using simple operators will usually guide the population to a good solution.

This kind of approach is quickly seen to be insufficient when it comes to creating a fully-fledged solution that is to be used in a real-world business environment to manage the day-to-day scheduling needs of a large enterprise. There are many issues that intrude on the purity of a simple approach such as the one described above. These issues are usually critical and require careful consideration so that the business needs can be satisfied in a manner compatible with the architecture of the EA.

The issues in question arise because of the dynamic nature of real-world problems. One hardly ever encounters a situation that is static for any significant period of time. In everyday business life, things change almost constantly. The number of orders to be scheduled in a factory changes, the run rate of a machine varies due to environmental factors, the quantity of a particular item ordered is modified after it has already been allocated, machines break down, delivery trucks arrive late, and so on.

In this chapter we examine examples of such practical time-varying issues that arise in a specific scheduling problem, namely, what we will refer to as the Wine Bottling Problem. We show how an EA along with supporting software components can, by proper design, effectively deal with these issues.

The rest of this chapter is organized as follows. We start with a classification of time-varying issues into three categories, alongside a brief literature review of work in each category in Section 2, and of scheduling problems in Section 3. Following this, in Section 4, the main case study of this chapter, an industrial wine bottling problem, is elaborated in sufficient detail to enable the reader to more easily visualize the kinds of problems being addressed. We go on to describe the real-world business issues that had to be considered and resolved in order to build a solution around an EA core in Section 5. The design and construction of the EA itself is fully specified in Section 6 and alongside this we illustrate how the various time-varying issues were resolved in relation to EA components.

2 Dynamic Optimization Problems

The dynamic nature of real-world optimization problems is well known, but a closer examination reveals a few different aspects of the problem that can described as *dynamic*. In this section we introduce a classification of dynamic optimization problems into three categories, characterized respectively by:

1. Time-varying objective functions.
2. Time-varying input variables.
3. Time-varying constraints.

There is a large body of EA research literature that addresses this dynamic property of such optimization problems, and we will undertake a brief review of work in each of these areas. It will be noticed that, while there is an abundance of work on problems fitting into the categories of time-varying objective functions and time-varying input variables, there are relatively few published reports dealing with dynamic optimization problems of the third kind, that is one that deal with time-varying constraints.

2.1 Time-Varying Objective Functions

These are problems in which the shape of the fitness landscape varies with time, that is, the location of the optimum value is continually shifting. The literature shows that there has been a considerable amount of research done on this kind of problem from the early days of evolutionary methods, to the present [5, 6, 7, 8, 9]. Some are based on artificially constructed objective functions and others on real-world applications. In many published reports, situations were set up wherein during a single run of an EA, the location of the optimum would move, and researchers were particularly interested in developing EAs that could detect that there has been a change in the optimum and continue to alter their search to find the new optimum.

Some examples of the earliest papers related to dealing with time-varying objective functions using EAs are [10] and [11]. These papers report on attempts to track optima in fluctuating, non-stationary environments (objective functions). Pettit and Swigger studied environments that fluctuated stochastically at different rates and they envisioned applications to areas such as voice recognition and synthesis.

Grefenstette extended the work of Krisnakumar [12], by pursuing ideas related to maintaining diversity. That work considered an abstract problem that used an optimization surface that would change randomly every 20 generations of the genetic algorithm, with new peaks appearing and previous optima being replaced by new ones. Although not addressed as an application in that paper, reference was made to real-world dynamic problems such as optimizing the fuel mixture for a jet engine based on the measurement of engine output. This problem is dynamic because of both slow and rapid changes. The components of the engine slowly degrade over time and hence, slowly alter the function being optimized, and it is also possible to experience a rapid change due to the sudden failure of a component. It was clear that the intent of the work was to serve as a foundation for dealing with these types of problems.

Different authors have worked to create standard test suites that can be used to study time-varying objective functions. For instance, Branke [13] has published a test function called the moving peaks benchmark (MPB) problem. This is an abstract function that represents a landscape with several peaks. Each peak is defined by a height, a width and a position in the xy-plane. The function is dynamic because these three characteristics of each peak are altered slightly as time progresses. It is easy to imagine this function surface slowly changing with time, with the xy-coordinates of the peak representing the global optimum slowly moving, and also possibly, the global optimum shifting from one peak to another.

The case study that will be considered in this chapter does not fall into the category of dynamic problems discussed in this section. When considered from a broad perspective, the wine bottling problem can be thought of as having a time-varying objective function, but it is not one that changes during the course of an EA run, rather the objective function changes with each run of the application as business conditions change.

2.2 Time-Varying Input Variables

These are problems in which the input data being processed in the optimization scenario changes from day to day, or in principle from run to run of the EA. For example, if we are interested in optimizing the production schedule of a factory, then each day, or possibly each minute, the set of customer orders to be allocated and sequenced on machines will change. This is because in a live business environment, as time passes, old orders are completed, and new ones arrive.

In [14], Johnston and Adorf describe how artificial neural networks can be used to solve complex scheduling problems with many constraints and report the real-world problem of scheduling the assigned usage of the NASA/ESA Hubble Space Telescope was solved using an application based on these techniques.

Chryssolouris and Subramaniam [15] explore the dynamic job shop problem where multiple criteria are considered and jobs may be processed by alternate machines. Although they recognized the lack of work on dynamic constraints such as random job arrival, in terms of input to the EA, their work primarily focused on multiple job routes.

Madureira, Ramos and Carmo Silva [16] investigated using a genetic algorithm to study how to produce schedules in a highly dynamic environment where new input variables require continual re-scheduling. Their work was based on an extended job-shop in a dynamic environment with simple products and such that require several stages of assembly. Another example of recent work done in this area include [17].

2.3 Time-Varying Constraints

These are problems in which the constraints of the environment within which a solution must be found change from day to day. These varying constraints add an additional level of complexity to the problem because a good EA approach must be able to operate equally well regardless of how many of these constraints are in place, and in what particular combination they occur. For an example of a time-varying constraint, consider a manufacturing environment in which the problem is to generate a production schedule for a set of jobs. After a schedule has been produced and passed to the factory floor for execution, it is customary to *freeze* a number of days of the schedule. This means that it is mandated that for the specified number of frozen days, the schedule will remain fixed. Any subsequent revised schedules should not alter the contents of the current schedule for these days. In this context, whenever a new schedule needs to be produced, it is a requirement that it mesh seamlessly with the frozen period of the existing schedule. In general, we do not know what the current schedule might be at any given moment in time, yet we must create an EA that generates a new solution that matches up with the existing one, and still produces an optimal result for the long term. As new schedules are produced every day, the contents of the frozen period varies, yielding a time varying constraint.

Good examples of the types of problems that can be categorized as time-varying constraints can be found in [18], wherein Jain and Elmaraghy describe circumstances that require the regular generation of new production schedules due to uncertainties (both expected and unexpected) in the production environment. They touch on typical examples such as machine breakdowns, increased order priority, rush orders arrival, and cancellations. All of these issues are also considered in this chapter, from the perspective of an integrated EA-based software solution. There are many other examples of works regarding this type of problem, for instance in [19, 20, 21].

2.4 Constraints Encountered in Practice

Based on our experience in solving real-world optimization problems for commercial organizations, we have found that the type of problems commonly experienced in industry belong to the second and third categories. Interestingly we have found that the vast majority of published research in EAs addresses the first category and to a lesser extent the second category. However, dynamic optimization of the third kind, i.e., where the problem involves time-varying constraints, although well-recognized in other domains, has been the subject of relatively few investigations in the EA literature. This observation is especially true when we extend our search to fully-fledged application of an EA to a dynamic real-world problem. Figure 1 illustrates some examples of issues that typically arise in real-world problems.

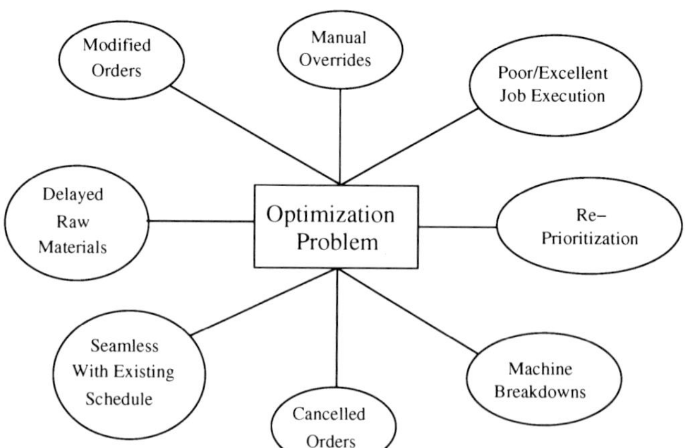

Fig. 1. Typical Issues in Real-World Problems

3 Scheduling Problems

The case study examined in this chapter, the wine bottling problem, is presented as a scheduling problem. It is a textbook example of a real-world industrial problem whose solution requires finding a production schedule to run a factory, or multiple factories. Essentially, given a number of tasks that need to be performed, we need to determine the best way of executing the required work, in terms of deciding where to assign work, and at exactly what time, and in what sequence the individual tasks should be carried out. Before getting into the wine bottling problem properly, in this section, we first take a brief look at some papers that give the reader a cross-section of examples of the application of EAs to scheduling applications, both classical academic problems, as well as real-world problems.

The Job-Shop Scheduling problem is a canonical example of a scheduling problem. It is one of the most difficult combinatorial optimization problems, and as is the case with most scheduling problems, it is NP-complete [4]. A very good survey of job-shop scheduling problems and different solution representations is given in [22]. [1] discusses the application of genetic algorithms to the job-shop scheduling problem. Wang and Zheng [3] solve it using a modified genetic algorithm. In [2], a simulated annealing approach to tackle the job-shop scheduling problem is discussed.

Yamada and Reeves [23] proposed an EA to solve a scheduling problem called the permutation flow shop problem. They combined stochastic sampling and best descent methods into one unified method to reach effective results. Their work was based on a method described by Nowicki and Smutnicki in their tabu search algorithm for the flow shop problem [24].

There are also many published reports of the application of EAs to solve scheduling problems in various domains of application. Marchiori and Steenbeek in [25] developed an EA for the real-world airline crew scheduling problem. Results of the real-world benchmark instances were compared with the results produced by the commercial systems and produced effective competitive results.

Ponnambalam and Reddy [26] developed a multiobjective hybrid search algorithm for lot sizing and sequencing in flow-line scheduling. The main idea is in the memetic type of algorithm that combines genetic algorithm with local search procedure.

A practical problem of optimal oil production planning was discussed by Ray and Sarker [27]. Production is based on several oil wells, from which a crude oil is extracted by means of injection of a high pressure gas. The goal of the problem was to optimize amount of gas needed to be used in each of wells to maximize output of oil taking in account a limited amount of gas each day. Single objective and multiple objective versions of the problems were considered.

Burke and Smith [28] investigated a real-world problem that addressed the maintenance problem of a British regional electricity grid. They compared the performance of a proposed memetic algorithm with other methods.

Martinelli shows the optimization of time-varying objective functions using a stochastic comparison algorithm in [29]. Values of the time-varying functions are known through estimates. Noise filtering was introduced to decrease probability of wrong moves.

Tinós and Yang introduce a self-organizing random immigrants scheme for algorithms in dynamic environments in [30]. Newly immigrated individuals are held in the subpopulation for some time until they develop good fitness values.

Yang discusses in his chapter [31] a memory scheme approach as a method of improving the performance of EAs for dynamic environments. Two kinds of memory schemes are described: direct and associative. These schemes are applied to genetic algorithms and on univariate marginal distribution algorithms in dynamic environments.

Schönemann [32] considers the application of evolutionary strategies for numerical optimization problems in dynamic environments. The main parameters of evolutionary strategies for problems in dynamic environments are presented and performance measures are discussed with advantages and disadvantages of each of them.

4 Scheduling Case-Study: Wine Bottling

Wine manufacturing and EAs go particularly well together. From the very starting point of planting grape vines and reaping the mature fruit, all the way through the crushing of the grapes, management of bulk tank movements during the fermentation process, to bottling of the finished product and sales, the wine manufacturing industry is a rich source of real-world application areas. Many of these problems are based on classical optimization problems and on that basis alone are quite difficult to solve. EAs, of course, by their very nature are natural takers for these kinds of challenges. In this section, we examine what is involved in one of the wine manufacturing steps just described, namely the bottling of the finished product.

Before getting to the point where wine is bottled into a finished product, the liquid would have gone through a series of fermentation and other processing steps. We will assume that we are at the point where the liquid is in a finished, consumable state, and is residing in a bulk storage tank, which could be anywhere from several tens of thousands of liters, up to more than one million liters in volume. This bulk liquid remains in storage awaiting the bottling process wherein it is pumped into a bottling factory and put into consumer size bottles, with a typical volume of less than one liter each.

A bottling factory houses several bottling *lines*. These are machines that are connected to intermediate feeding tanks that contain finished wine, and are used to transfer the wine into glass bottles and hence produce the items that everyday consumers are accustomed to purchasing. The bottling lines also take care of related tasks such as capping the bottle with a screw cap or a cork, applying labels to the bottle, and packaging the bottles into cartons. Each line is capable of bottling a subset of the types of finished wine products manufactured by the wine company.

These two elements, the bulk wine liquid, and the bottling lines, constitute the basic working elements of the wine bottling problem. The bottling process is illustrated in Figure 2. Customer orders determine which bulk wines are put into which bottles and the times at which that is done. The wine company receives orders for particular finished goods from their clients and it is those orders that must be carefully considered in order to determine the best way of running the bottling plant. In an ideal situation, customers place their orders with sufficient notice to ensure timely bottling of their goods.

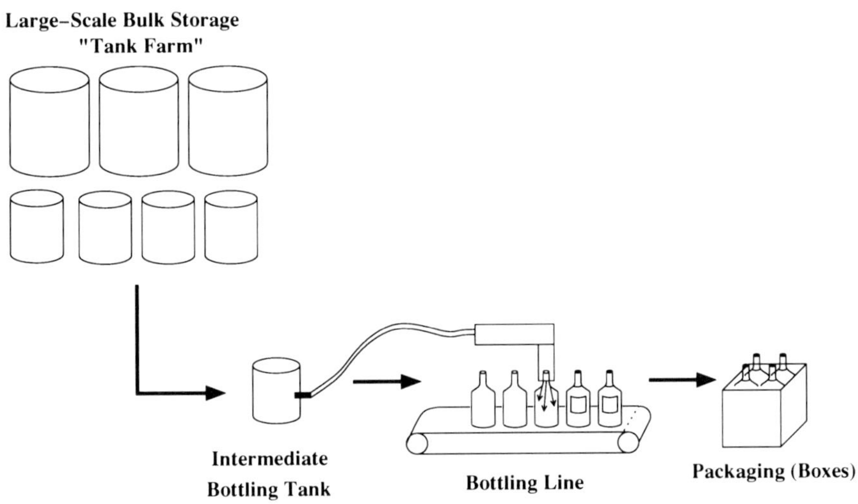

Fig. 2. The Wine Bottling Process

The problem is to determine a sequence of orders to be carried out on each bottling machine (hence classifying it as a scheduling problem) such that optimal use is made of the company's resources, from the point of view of making maximum profit, and also maximizing customer satisfaction. Hence a good schedule will minimize production costs and at the same time ensure that orders are produced in full, sufficiently before their due dates. This is the fundamental part of the wine bottling problem: deciding how to schedule production so as to make the best use of limited resources.

Wine making, as is the case in a multitude of other real-world applications, is a highly variable and complex business, and what, from the point of view of using an EA, would otherwise be considered a straightforward scheduling application, gives rise to a series of significant challenges. These practical day-to-day business considerations are what make this problem highly dynamic, and we are led to classify the problem as a dynamic optimization problem of the third kind, due to the fact that and many of the issues are based on time-varying constraints.

4.1 Basic Constraints in Wine Bottling

Before getting into the business intricacies that give rise to the time-varying constraints, we will first consider some more basic issues that affect the scheduling problem:

Due dates: When a customer places an order, the sales department will assign to it a due date that is acceptable to the customer and which should also be realistic taking into account the size of the order and available resources at the bottling plant. Sometimes, for various reasons, due dates are unrealistic, but nevertheless, the scheduling application needs to employ techniques that strive to maximize the number of orders that are *delivered in full on time* (this expression is sometimes abbreviated as DIFOT). In cases where it is impossible to have orders delivered on time, the algorithm must strive to minimize the delays incurred on each order. This is a very important objective in the application, since it is of utmost importance that customers should not be displeased due to late delivery of orders.

Bulk wine availability: Some orders may need to be inevitably delayed due to the fact that the bulk wine needed to fill the bottles may not yet be ready. This could easily happen since the process of creating wine is quite variable, and batches may not have responded to the fermentation process as expected, and could require additional processes to get the wine to the required specification and taste. There are a number of other processes involved in the preparation of the bulk wine that can lead to delays, such as filtration and temperature stabilization.

Dry Materials: In addition to the liquid wine, there are a few dry goods that are required to produce a finished bottle of wine. First there is the glass bottle, then there is its covering which may be a screw cap or a cork, and several other items such as labels, of which there may be several (for the front, back and possibly neck of the bottle), and foils and wire hoods for sparkling wines. If an order is scheduled at a particular time, then to proceed, the appropriate amounts of each dry good must be ready for installation into the bottling machine. Hence it is a requirement that the optimization software checks the availability of these materials in order to produce feasible schedules.

Job run lengths: It is inefficient to have machines frequently changing from one type of bottling job to another because this incurs set-up and take-down time and reduces the overall utilization of the machine. Hence, the scheduling algorithm must attempt to group similar orders for sequential execution so this type of inefficiency is avoided.

Wine changeovers: Wines are categorized broadly in terms of their color, there is red, white, and rose (pink). Below this level of classification, there are more detailed distinctions such as sweet red, dry red, aromatic white, full-bodied white, sparkling red, sparkling white, fortified wine, and others. Even for a very large wine company, it may not be the case that there is sufficient production volume to justify dedicating individual machines exclusively to the production of one type of wine. Hence the reality is that the same machine must be used at different times to bottle different types of wine. When a bottling line finishes working with one type of wine and switches over to another type, this is referred to as a *wine changeover*. Bottling lines must be cleaned during changeovers, to varying degrees, depending on the nature of the change. Certain types of wine changeovers are undesirable and need to be avoided where possible by the optimization algorithm. For example, if we are changing over from a run of white wine to a run of red wine, then a relatively brief cleaning is required, since residual amounts of white wine entering into red is not much of a problem. However, the reverse situation where we go from a run of red to a run of white requires a very extensive cleaning process, including sterilization. Minute amounts of red wine entering into a run of white wine are likely to cause a lot of damage.

Other changeovers: Although wine changeovers incur the most time, there are a variety of other changeovers that could happen, even within a run of the same color wine. Each different finished good that is being produced requires a particular size and type of glass bottle, as well as a particular type of covering, which may be a screw cap or a cork, of which there are several varieties, and unique labels for each brand of wine. The physical re-configuration of the bottling line hardware to accommodate these items requires a strip-down time to remove the items used by the previous run, and a set-up time to insert the new items required for the next run. The optimization algorithm needs to try to minimize these changeovers as far as possible by appropriately grouping orders.

Bottling line availability: Some industries use machines that are kept in operation continuously. This is sometimes the case in large-scale wine companies, but for only limited periods of time. In most scenarios, bottling lines have typical hours of operations that correspond to an average workday, which may be for example 8:00am to 6:00pm. The optimization algorithm needs to consider the availability of each machine in order to make appropriate scheduling decisions since one of the critical factors affecting the evaluation of a proposed schedule is the number of orders that are satisfied on time. In computing the amount of time that a job will take to perform and making a decision about where to assign it, machine availability must be factored in.

Routings: Each product can be bottled on a number of different lines. The choices are usually a proper subset of all of the machines. Although the same product could be bottled on a few different machines, the performance characteristics on each machine is likely to be different, sometimes significantly so. Set-up and strip-down time could vary, and also the speed at which the bottles are processed (called the *run rate*) could vary significantly. Each possible assignment of a finished good to a bottling line, together with its associated performance data, is referred to as a *routing*. The optimization algorithm must strive to choose the best routing to use under the circumstances. If all else is equal, then naturally the fastest routing would be chosen. However, this is often not the case, as some lines may be heavily loaded thus preventing such a choice from being made. In order to achieve better load balancing between the machines, alternate routings may have to be used.

4.2 The Software Solution

Our work on the wine bottling problem resulted in a full-featured piece of software, built around a core EA that deals with all of the above listed issues and creates feasible, optimal schedules for satisfying customer orders for wine. The application was launched for a major global wine company and it experiences heavy daily usage at international sites. It is a cornerstone of their scheduling department, and we think it is a good example of an integrated EA serving robustly in prime time. The main screen of the application is shown in Figure 3 with all confidential client information removed. The area of horizontally adjacent rectangles shown in the upper portion of the picture is a graphical representation of the schedule produced by the software. Each rectangle corresponds to a customer order that has been assigned to a bottling line at a particular time. The length of the rectangle is indicative of the amount of time required to execute the job and its color coding lets the user know at a glance the type of wine being produced. The graphical interface also allow the user to drag and drop the rectangles to make manual adjustments to the schedule produced by the optimization algorithm.

Figure 4 shows the configuration interface of the software application that allows the user to enter parameters that define the availability of each bottling line. This interface needed to be completely flexible in terms of allowing the user to specify any possible set of available and unavailable times for each line. In this application, the user is allowed to specify a typical weekly timetable, for example a machine may be available from Monday to Friday from 8:00am to 6:00pm, and unavailable otherwise. It also allows the user to specify periods of time in which there is a deviation from this typical schedule. For example, ahead of a festive season, it might be desirable to make the machines work longer hours, and possibly on weekends. This interface also allows the user to allocate periods of time for regularly scheduled downtime for maintenance of the machines.

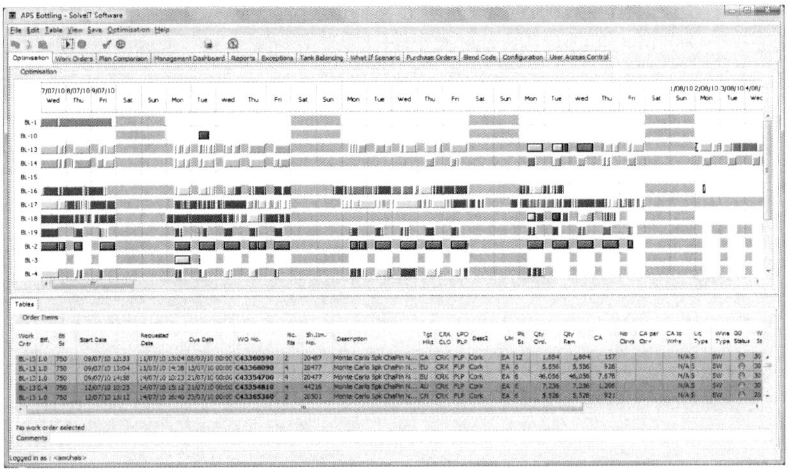

Fig. 3. Main Screen of Software Application

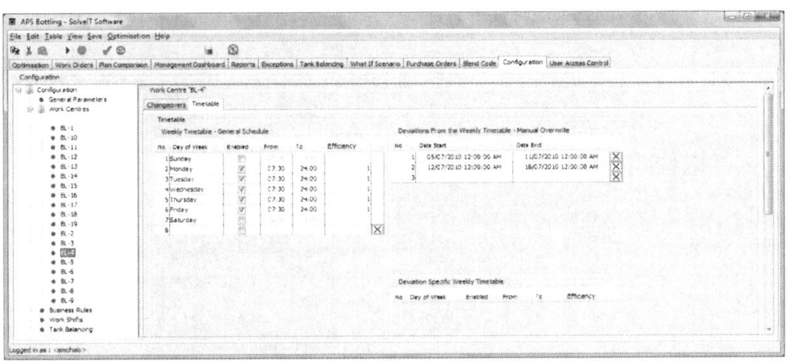

Fig. 4. Machine Availability Interface

Figure 5 shows a material requirements dialog that is part of the software application. For any given order, the user may view the materials required to produce the required number of bottles of wine, such as caps, corks, labels, and so on. The application provides the user with information on the availability of each material based on knowledge of opening inventory stock that it uses for a running simulation over the time frame of the production schedule. Additional knowledge of still-open purchase orders for additional materials, and also knowledge of the lead times of various suppliers of these materials allows the software's underlying algorithm to make appropriate scheduling decisions so as to best ensure that orders are sequenced in a feasible manner. If scheduled too early, at a time when the materials would not be available,

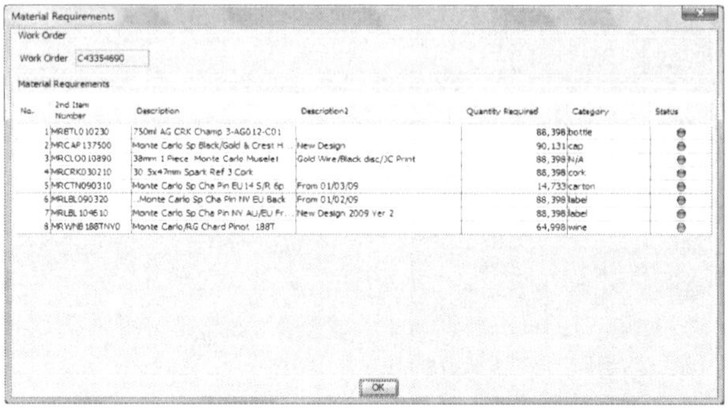

Fig. 5. Material Requirements Interface

an order would have to be passed over, possibly leading to other disruptions in the schedule and hence in the factory itself.

5 Time-Varying Challenges in Wine Bottling

In this section, we will look at some business requirements that lead to time-varying constraints that had to be addressed in the software. Here we will only consider the issues. Their actual solution, including algorithmic details, will be covered in a later section.

Manual assignments: There are various scenarios in which the human scheduler would need the ability to override the schedule produced by the EA. For instance, it might be that some of the information used as input by the algorithm such as the availability of dry goods and expected delivery times of raw materials, which are all imported automatically from the company's Enterprise Resource Planning (ERP) system, may for some reason be inaccurate. Hence the schedule that is created may not be feasible and would need to be corrected by the human scheduler. Another typical scenario is that a very important customer makes a late, but urgent request for a large quantity of wine. Even if this causes severe disruptions to the smooth running of the bottling plant, this type of request is usually accommodated due to the high value placed on some customers, and the importance maintaining a good business relationship with them. In this kind of situation the unusual placement of that urgent order in the production sequence would create a different optimization problem – a much more constrained one that would be pose greater difficulty for the EA to solve. The level of difficulty would increase even more dramatically when there are multiple such constraints. The software application that we developed had to be flexible enough to allow its built-in EA to work in such a way that it could actively seek out an optimal solution that satisfies the constraints described before, but at the same time allow inefficient manual overrides dictated by a human operator to co-exist with the otherwise optimal solution. Another way of looking at this is that the structure of the search algorithm had to be flexible enough to find locally optimal solutions in restricted neighborhoods of the overall search space, with those neighborhoods being defined by user-specified manual assignments.

The application presents the schedule with both graphical and tabular representations, each of which may be intuitively manipulated by the user, simply by dragging and dropping graphical rectangles, or rows of a table. This provides the capability of altering the sequencing of orders to suit a human user's preference. Additionally, as shown in Figure 6, there is a manual assignment interface that is accessed on a per-order basis and allows the user to manually specify all details about when and where a particular job should be scheduled.

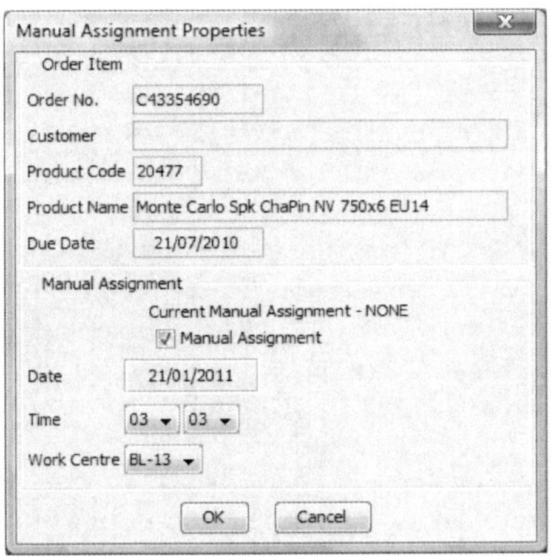

Fig. 6. Manual Assignment Interface

Machine breakdowns: From time to time, a bottling line will break down and become unavailable for use. It may be that a solution found by the optimizer previously would have been planned around that machine being available during a period of time that has now become unavailable due to the breakdown. The software must be flexible enough to repair the previous solution to take into account the breakdown. A very simple initial solution to this could be to merely shift the sequence of orders previously assigned to a machine, starting at the point where the machine breakdown start. All subsequence orders are scheduled later in time, by an amount that roughly works out to be the duration of the breakdown period. In some cases, a more sophisticated approach is required where a complete re-optimization is performed to repair damage that was done to the percentage of orders that would be delivered on time. This kind of optimization must attempt to keep as much of the existing schedule intact while repairing the placement of the affected orders.

Freeze periods: Every time the EA is run, the possibility exists that a very different schedule could be produced when compared to the previously generated one. This happens simply as a result of the EA doing its job of finding an optimal solution as dictated by its objective function (which changes each time new orders and constraints appear in the system, which is virtually always). In a real-world application, it is highly unlikely that this kind of approach could be tolerated. The preferred mode of operation is the following. The software is used to initially create a schedule for a fairly long period of time, for example 2-4 months. However, once that schedule is accepted and saved to an internal database, subsequent daily use of the optimizer to schedule newly arriving customer orders must be done in a controlled manner so that there is a buffer period at the beginning of the schedule that remains the same as if it was yesterday. This unchanging portion of the schedule is referred to as a *freeze period*. It is usually defined in terms of a number of days, for example 7 days. In order to achieve this effect, the EA has to perform what we refer to as seamless *meshing* with an existing schedule. The first 7 days of the existing schedule are frozen, and some of the previously existing orders, along with new ones are re-shuffled by the optimization algorithm and attached in a neat continuous way with the orders in the freeze period. The contents of the freeze period vary with time, as each day some orders are executed and so are removed completely from the schedule, and others that were once outside the freeze period gradually move into it. This is referred to as a *rolling time window*.

On a more granular level, freeze periods are affected by live update feedback data coming from the factory floor. The *current order* on a given bottling line is the first order showing up on the schedule. At any given moment, it usually corresponds to an order that physically is in the process of being executed in the factory. Our system was designed to receive updates in near-real-time (every 5 minutes) letting it know how much of that order has been carried out. That in turn affects the orders in the freeze period because the size of the current order needs to be gradually reduced to reflect what remains to be done, and consequently at the right-hand end of the freeze period, more orders will come in. This sometimes leads to a situation in which there is an order that straddles the freeze period and the open-optimization period.

Figure 7 shows a screen capture from our software application that illustrates how the frozen period is managed. In the graphical area of the window, where orders are represented by rectangles, on the left-hand side of the screen there are a number of orders that are surrounded by a thick bold border, indicating that they have been frozen. On the right-hand side, are the orders that the optimizer is free to move around to find an optimal schedule.

Modified orders: Once a schedule has been created and saved, there might be a situation in which the next time the software package is opened, it

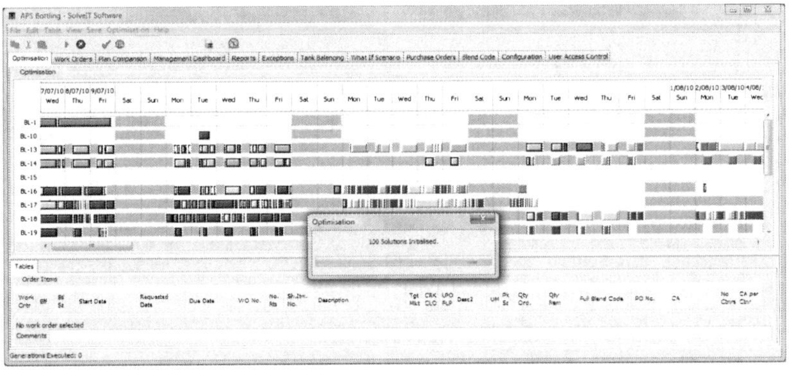

Fig. 7. Freeze Period Illustration

realizes that there was a modification made to one of the scheduled orders in the database. This might be, for example, a change in quantity. More or fewer bottles of wine may be required and by adjusting the scheduled order, the start and end times of all subsequent orders on the same bottling line become affected.

Poor/Excellent Job Execution: From historical performance data, any given bottling machine has an expected production rate. This rate is used by the software to estimate the time required to run a job of a particular size. However, due to a number of variables, the efficiency of a bottling machine may be better or worse than the expected rate on any given day, and the factory manager would expect the scheduling optimizer to take this into account when creating a new schedule, or when adjusting an existing one. A machine that is performing poorly will delay jobs scheduled further down in time, and likewise a machine that is performing unusually well, will bring jobs up earlier, which could, in some cases, have a detrimental effects when we take into consideration the availability of dry goods. The optimization software must be able to deal with this time-varying issue by re-adjusting the solution, or by re-optimizing as necessary.

6 The Solution Using an EA

In this section, we will look at the structural, algorithmic, and programmatic details required to solve the problem using an EA. We begin by examining the representation of a candidate solution and then look at how that abstract representation is converted (decoded) into an actual schedule with dates and times. Next we look at the key operators that were employed to alter candidate individuals. Finally we re-visit the time-varying problems described above and show how various elements of the EA had to be modified to accommodate those issues.

6.1 Representation

One of the most important aspects of the solution of a problem using an EA is the representation used to encode a candidate solution. Forcing the use of a particular representation may significantly impact on the quality of the solutions found since many useful operators may be overlooked, or may be cumbersome to program, thereby slowing down the execution of the algorithm. For this application, we chose a representation that closely matches the actual assignment of orders to machines. For reasons that will be described below, we chose to use both a genotypic and phenotypic representation of a solution.

For the wine bottling scheduling problem, the core of the problem was conceptualized as having a number of orders that must be placed on a fixed number of bottling machines in an efficient sequence. Hence the natural representation to use is a mapping of lists of orders to machines. This can be visualized as in Figure 8. The representation illustrated in this diagram is quite similar to the final schedule presented visually in Figure 3 above. The difference is that the actual decoding of the individual into a real-world schedule also takes into consideration several other time-varying factors such as machine availability, splitting of single orders into several sub-jobs, meshing with an existing schedule, and so on.

The representation shown in Figure 8 is stored programmatically as a map of machines to variable-length lists of orders. The list for any given machine is sorted chronologically in terms of which orders will be carried out first. The individual is constructed in such a way that the assignment of orders to machines is always valid, in other words, the assigned orders always respect product routings.

From a formalised perspective, the search space can be thought of as the set

$$S = ((m, o))^n \tag{1}$$

$m \in M$, where M is the set of bottling machines, and $o \in O$, where O is the set of customer orders to be scheduled, and n is the number of such orders, i.e., $n = |O|$. Hence an individual (as illustrated in Figure 8) can be represented mathematically as

$$I \subseteq S \tag{2}$$

If we wanted to use a mathematical structure that emphasizes the ease of accessing the jobs assigned to each machine, then instead, we could think of each individual as being represented as

$$I = (a_1, a_2, \ldots, a_k) \tag{3}$$

where, k is the number of bottling machines, and $a_i = (o_{i1}, o_{i2}, \ldots, o_{i\alpha(i)})$ is a sequence of customer orders assigned to machine i, and $\alpha(i)$ is the number of orders assigned to that machine.

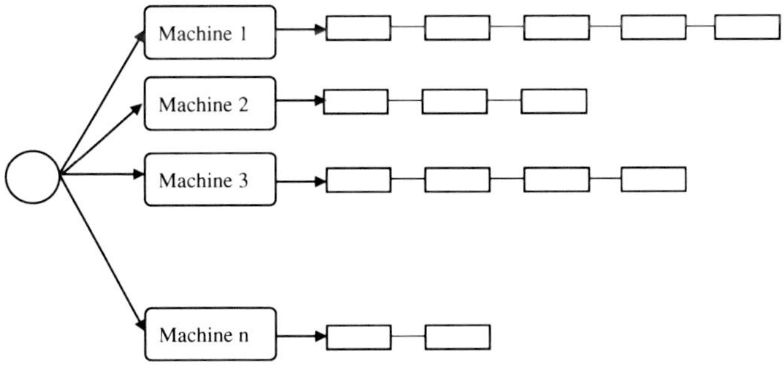

Fig. 8. Scheduling Individual Representation

6.2 Decoding

In order to manage the decoding process, we employed a concept we referred to as *time blocks*. A time block, illustrated in Figure 9 is a data structure that keeps track of a contiguous period of time. Each such block keeps track of a start time, an end time and an activity that is performed during that time. The time block data structure allows for the possibility that nothing is actually done during the period of time, in which case the time block is referred to as an *available* time block. The link to the activity performed during a time block may point to an external data structure that contains any information on any level of detail required to accurately model a scenario.

Decoding begins with a series of multiply-linked-lists of time block nodes associated with each machine. Each machine has a list of available time blocks, and occupied time blocks. At the outset, before anything is placed on a

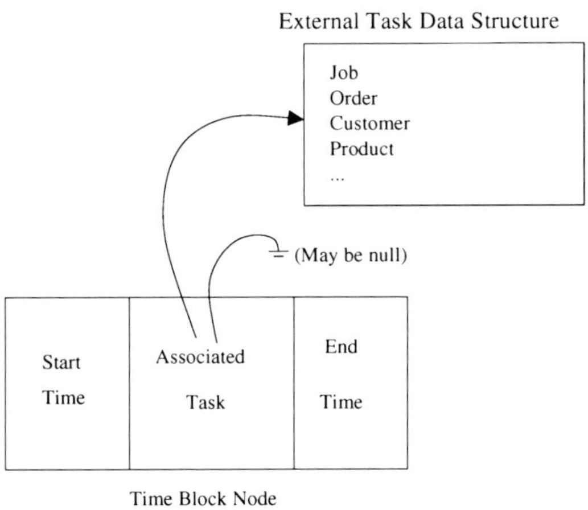

Fig. 9. A Time Block Node

machine, it would only contain a list of available time blocks, each representing a chunk of time during which the machine is available for use. This is what would happen in the nominal scenario. If there are manual decisions that were already made by a human operator, then these would be reflected by the existence of some occupied time blocks. More detail on this issue is given in section 6.5 below.

Decoding proceeds by going through all machine-job pairs found in the individual representation and proceeding to the corresponding machine, finding an available time block node, and marking it as occupied for the corresponding job. Some jobs may not fit into the first available time block and may need to be split into multiple parts. How this is done, and if it is permissible at all, depends on the policies of the business for which the scheduling application is being created. In the case of the wine bottling application that we are considering, orders were split across adjacent available time blocks. This process is illustrated in Figure 10.

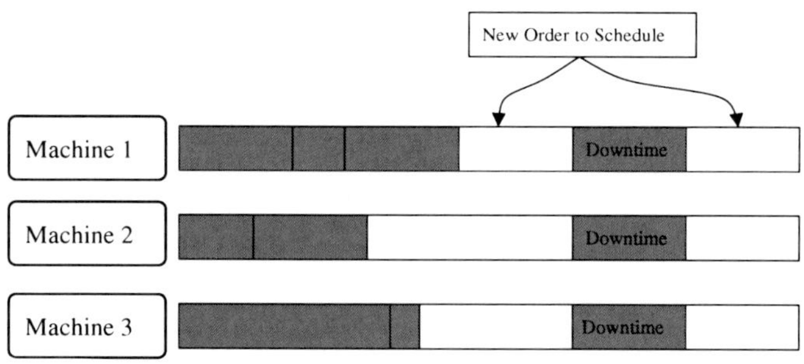

Fig. 10. Illustration of the Decoding Process

An important issue that arises when decoding is performed in this way for a scheduling problem is the question of in what order should the machine-job pairs be considered. Should we select Machine-A first and process all of the orders in its list, and then proceed to Machine-B and so on? Alternatively, should we process one job from Machine-A's list, then one job from Machine-B's list and so on and eventually cycle back to Machine-A, and keep on in that manner until all jobs have been decoded? The answer to this question depends on the industry being considered. For the wine bottling problem there is no dependency or constraints between jobs running on different machines and hence, the former approach was used.

However, there are cases in which the second approach (or yet others) may be needed. For example in some industries there might be a limit to the number of changeovers that are permitted to occur across an entire plant in one day. In those cases, it is important not to bias the decoding (or alternatively to deliberately and carefully bias it) to favor one machine, or a group of machines.

6.3 Operators

A number of operators were used to manipulate the representation given above. To avoid the problem of having to perform extensive repairs based on invalid representation states, crossover-type operators were avoided. The operators listed below, which are typical examples from the set used, may all be considered as mutation operators.

- *Routing Mutation*: This operator modifies the machine that was selected to execute a job. An alternative is randomly chosen from the set of possible options.
- *Load Balancing Mutation*: This is a variation to the routing mutation operator which, instead of merely randomly choosing an alternative machine for an order, chooses from a subset of machines that are under-loaded. Such an approach would help with load-balancing of the machines.
- *Grouping*: This operator groups orders based on some common characteristic, such as wine color, bottle type, or destination export country. There are several variants of the grouping operators. Some operate quite randomly, looking for a group of orders based on some characteristic and then looking left or right for a similar group and then merging the two. Other examples are given below.
- *Recursive Grouping*: A more directed variation on grouping is recursive grouping. This operator seeks out an existing group of jobs based on wine color, then within that group performs random grouping based on some other characteristic, such as bottle size. This process may then be repeated inside one of the subgroups.
- *Outward Grouping*: Outward grouping is a term we use to describe the process of identifying groups in a list of jobs based on a primary characteristic, then randomly selecting one of them and from that location looking left and right for another group with a common secondary characteristic and finally bringing together the two. As a concrete example, we may first identify groups based on a primary attribute, say closure type, that is, whether a screw cap or a cork is used to close the bottle, and then randomly select a group that uses screw caps as a starting point. Next a secondary attribute such as wine color is looked at. Suppose the group of screw caps involves white wines, then we look randomly to the left or to the right for a group based on the primary attribute that also involved white wines. In the end we may end up merging with a group of jobs that involves corks, but which happens to be all white wines. The end result is that we have a larger continuous group of white wine, with two sub-groups, one with screw caps and the other with corks. The process of finding a starting point and then looking outwards for subgroups to merge with gave rise to the name *outward grouping*.
- *Order Prioritization*: Orders that may be showing up as being produced late after decoding an individual are stochastically prioritized by moving them left in the decoding queue. This type of operator could be made more

intelligent by moving groups of jobs along with the one that is identified as being late. That way the job gets prioritized, but at the same time disruption to grouping is minimized. This operator works on one of the most important objectives of the solution, to maximize DIFOT.

6.4 A Pseudocode View of the EA

The coding and implementation of the algorithms used to solve the problem of this chapter are very lengthy, but from the point of view of core principles, the key parts of the EA can be succinctly summarized as in Algorithms 1, 2 and 3.

Algorithm 1: Main Loop of Algorithm

```
initialize()
while numIterations < targetIterationCount do
    individual = selectRandomIndividual()
    operator = rouletteSelectionOfOperator()
    newIndividual = operator.execute(individual)
    if evaluation(newIndividual) > evaluation(individual) then
        individual = newIndividual;
```

Algorithm 2: Population Initialization

```
for i = 1 ... populationSize do
    individual[i] = newBlankIndividual()
    for each order in customerOrdersList do
        capableMachine = order.getCapableMachines()
        selectedMachine = uniformRandomSelection(capableMachines)
        individual[i].assign(order, selectedMachine);
```

Algorithm 3: Individual Evaluation

```
evaluation = 0
evaluation := evaluation - delayedOrdersPenalty()
evaluation := evaluation - colourChangeoverTimePenalty()
evaluation := evaluation - toolChangeoverTimePenalty()
return evaluation
```

6.5 Solving the Dynamic Issues

We will now look at how we dealt with the time-varying issues that were identified in Section 5. As will be observed, the problems were addressed by a combination of modifications made to the initial time block node linked-lists, to the decoding process, by altering the input variables to the optimizer, and by introducing a step between the optimizer and the human user, called *solution re-alignment*.

Solving Manual Overrides: This problem was solved by applying a constraint to the decoding process, and indirectly affecting the fitness function. The software application allows the user to select a particular order, and specify which machine it should be done on, as well as the date and time of assignment. This becomes a timetable constraint for the genotype decoder. When a candidate individual is being decoded, the initial state of the machine time block usages includes the manually assigned orders as part of the list of occupied time blocks. Subsequent decoding of non-manually-assigned jobs would take place as usual using the remaining available time blocks.

The fitness function would then evaluate how well the individual was decoded into the partially occupied initial machine state. As in the case where there are no manual assignments, it is up to the evolutionary operators to modify the individual in such a way that its decoded form meshes well with the fixed orders to reduce changeovers and so on. The initial state of available time blocks used for decoding is illustrated in Figure 11.

With this approach, it is primarily the fitness function that would guide the search to a relatively good solution built around the manual assignment. However it is also important to de-emphasize some of the more structured and aggressive operators such as recursive grouping and outward grouping, which were designed with a clean slate in mind. They would still contribute toward the search for a good solution in parts of the time period that do not contain manual assignments, but the main EA loop should keep track of the performance of these operators and adjust their probability of application accordingly.

Solving Freeze Periods: The application allows the user to select a freeze date and time on each machine used in the bottling plant. During an optimization run, all assignments in the existing schedule that are before the freeze period cut-off are internally marked as manual assignments, and therefore behave in exactly the same way as described above for user-defined manual assignments.

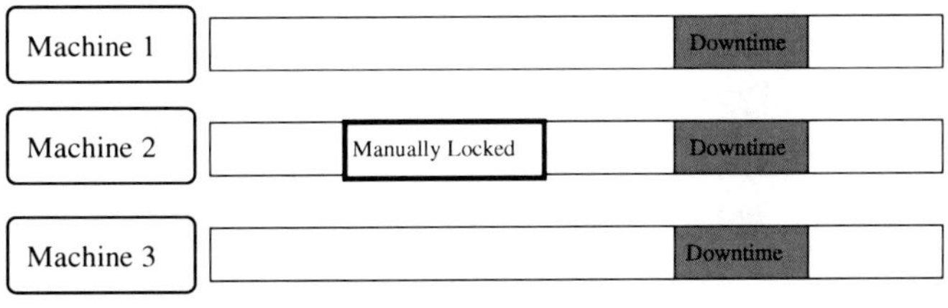

Fig. 11. Modified Initial Time Blocks for Manual Locks

It is also important to pay attention to jobs that might straddle the freeze period cut-off date. Such orders need to be fully frozen to ensure correct results. Another situation that needs attention is where a job may be split into several pieces, for example due to relatively low machine availability on consecutive days, and one some of those split pieces fall into the frozen period. Care must be taken to also freeze the parts that are outside of the freeze period, but which belong to the same order.

Solving Machine Breakdowns, Modified Orders, and Poor / Excellent Job Execution: These three problems were solved using a similar approach. Take Modified Orders for example. When the application re-loads and realizes that an existing order has been modified, for example its quantity has been increased or decreased, then the necessary action to be taken is at the level of modifying the existing solution, prior to it being fed back into the next run of the EA. This was accomplished using a process called *solution re-alignment.*

Essentially, on each machine, all assigned orders, sorted by starting date, are examined for any changes to the underlying orders. A new instance of an available time block nodes linked-list is created and the assigned orders are then re-inserted into it. By comparing each order as it last existed in the schedule with the order record coming from the company's ERP system allows assignments being shortened or lengthened as needed. Following the re-insertion of a modified assignment, all subsequent assignments are modified to fit in the resulting changed available space. This process has a direct effect on the number of orders ending up in the freeze period and therefore on the amount of available time that the optimizer has at its disposal for the next run. The process of solution re-alignment is illustrated in Figure 12.

Similarly, if unexpected machine breakdowns were encountered, then the linked list of available time block nodes into which the existing schedule is re-aligned is adjusted to reflect the new pattern of available operating times, with the breakdown periods marked as unavailable.

Solution re-alignment is a step that links the internal decoded representation of an individual with the dynamic world of the human operator in which multiple constraints can vary from one use of the application to another. It allows an existing solution to be adapted to match constraints that may have

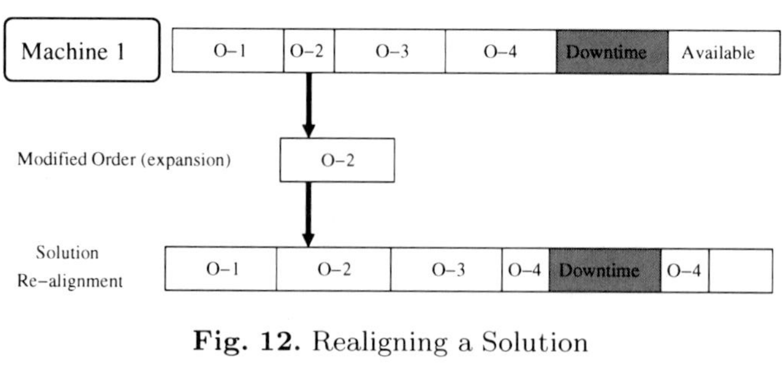

Fig. 12. Realigning a Solution

changed, and possibly subsequently, allow the EA to run with a freeze period that now accurately reflects the effect of the changed constraints. Re-alignment can be thought of as a second-level decoding process that works with an already decoded representation as opposed to working with a genotypic representation.

7 Conclusions

In this chapter, we have looked at issues that arise when dealing with dynamic optimization problems in real-world applications. A literature review was presented in line with our classification of dynamic problems according to time-varying objective functions, input variables and constraints. An important consideration that was emphasized throughout the chapter was that complexities arising due to practical, time-varying constraints had to be dealt with adequately in order to ensure that the solution produced is usable in an actual business environment. Given that a large real-world software application was presented as a main outcome of the research presented here, the need to ensure that those issues were resolved in a practical way that still allowed the power of EAs to be applied was a main theme.

The issues raised were exemplified by a case study based on an EA scheduling optimizer that was implemented for use in the wine bottling industry. The software, having been deployed and put into daily use in a live production environment, serves as empirical evidence that the approaches put forward in this chapter have passed the litmus test of suitability for real-world applications.

Numerical experiments were not included as part of this chapter, for the simple reason that they would be almost meaningless, given the objective of the research, which was not to compare one EA method against any other, but rather to demonstrate a way of marrying an EA with supporting software structures to enable the final software application to deal with difficult, practical day-to-day business realities.

One of the key issues that were dealt with was rooted in the need to be able to allow a human operator to manually decide certain parts of the eventual solution. In effect what this does is to modify the search space explored by the EA. Although there are undoubtedly better solutions that exist in the unconstrained search space, when we take into account these so-called manual assignments, the optimal solution based on those restrictions must be found in order to satisfy the immediate business needs that gave rise to the need for human intervention into the search process.

Another key issue that was explored was the need to ensure that future schedules mesh seamlessly with existing ones. This was handled by extending the concept of manual-assignments to the concept of a freeze period in which existing scheduled assignments are treated as unchanging, and therefore tantamount to numerous manual assignments. This has the benefit of allowing factory operators to continue working with the assurance that preparations made for the next few upcoming days would not be disrupted, and planning could proceed in a smooth and normal manner.

Overall, what has been illustrated is that in implementing a scheduling EA for use in a practical commercial application, it is necessary to design almost everything, from the representation, the decoding process, the operators and the solution structure, in such a way that maximum flexibility is maintained with respect to allowing time-varying constraints to be easily considered by the core algorithm that finds an optimal schedule, and with respect to allowing an easy and natural flow between data structures used directly by the optimization algorithm and the user interface that is manipulated by the human operator.

Acknowledgements. This work was partially funded by the ARC Discovery Grant DP0985723 and by grant N516 384734 from the Polish Ministry of Science and Higher Education (MNiSW).

References

1. Davis, L.: Job shop scheduling with genetic algorithms. In: Proceedings of the 1st International Conference on Genetic Algorithms, pp. 136–140. L. Erlbaum Associates Inc., Mahwah (1985)
2. Van Laarhoven, P.J.M.: Job shop scheduling by simulated annealing. Operations Research 40(1), 113–125 (1992)
3. Wang, L., Zheng, D.-Z.: A modified genetic algorithm for job-shop scheduling. International Journal of Advanced Manufacturing Technology 20(1), 72–76 (2002)
4. Garey, M.R., Johnson, D.S., Sethi, R.: The complexity of flowshop and jobshop scheduling. Mathematics of Operations Research 1(2), 117–129 (1976)
5. Branke, J.: Memory enhanced evolutionary algorithms for changing optimization problems. In: Proc. of the 1999 Congress on Evolutionary Computation, CEC 1999, pp. 1875–1882 (1999)
6. Branke, J.: Evolutionary approaches to dynamic optimization problems - a survey. In: GECCO Workshop on Evolutionary Algorithms for Dynamic Optimization Problems, pp. 134–137 (1999)
7. Morrison, R.W., De Jong, K.A.: A test problem generator for non-stationary environments. In: Proceedings of the 1999 Congress on Evolutionary Computation, CEC 1999, pp. 2047–2053 (1999)
8. Carlisle, A., Dozier, G.: Adapting particle swarm optimization to dynamic environments. In: International Conference on Artificial Intelligence, Las Vegas, NV, USA, pp. 429–434 (2000)
9. Yu, X., Tang, K., Yao, X.: An immigrants scheme based on environmental information for genetic algorithms in changing environments. In: Proceedings of the 2008 Congress on Evolutionary Computation, CEC 2008, pp. 1141–1147 (2008)
10. Pettit, E., Swigger, K.M.: An analysis of genetic-based pattern tracking and cognitive-based component tracking models of adaptation. In: Proceedings of the National Conference on Artificial Intelligence, pp. 327–332. AAAI Press, Menlo Park (1983)

11. Krishnakumar, K.: Micro-genetic algorithms for stationary and non-stationary function optimization. In: Proc. of the SPIE, Intelligent Control and Adaptive Systems, pp. 289–296 (1989)
12. John, J.: Grefenstette. Genetic algorithms for changing environments. In: Parallel Problem Solving from Nature, vol. 2, pp. 137–144. Elsevier, Amsterdam (1992)
13. Branke, J.: The moving peaks benchmark, `http://www.aifb.uni-karlsruhe.de/~jbr/MovPeaks/movpeaks/`
14. Johnston, M.D., Adorf, H.-M.: Scheduling with neural networks – the case of the hubble space telescope. Computers & Operations Research 19(3-4), 209–240 (1992)
15. Chryssolouris, G., Subramaniam, V.: Dynamic scheduling of manufacturing job shops using genetic algorithms. Journal of Intelligent Manufacturing 12(3), 281–293 (2001)
16. Madureira, A.M., Ramos, C., Silva, S.C.: Madureira, Carlos Ramos, and Slvio C. Silva. Using genetic algorithms for dynamic scheduling. In: 14th Annual Production and Operations Management Society Conference, POMS 2003 (2003)
17. Emperador, J.M., González, B., Winter, G., Galván, B.: Minimum-cost planning of the multimodal transport of pipes with evolutionary computation. Int. J. Simul. Multidisci. Des. Optim. 3(3), 401–405 (2009)
18. Jain, A.K., Elmaraghy, H.A.: Production scheduling/rescheduling in flexible manufacturing. International Journal of Production Research 35(1), 281–309 (1997)
19. Petrovic, D., Alejandra, D.: A fuzzy logic based production scheduling/rescheduling in the presence of uncertain disruptions. Fuzzy sets and systems 157(16), 2273–2285 (2006)
20. Kutanoglu, E., Sabuncuoglu, I.: Routing-based reactive scheduling policies for machine failures in dynamic job shops. International Journal of Production Research 39(14), 3141–3158 (2001)
21. Holthaus, O.: Scheduling in job shops with machine breakdowns: an experimental study. Computers & Industrial Engineering 36(1), 137–162 (1999)
22. Cheng, R., Gen, M., Tsujimura, Y.: A tutorial survey of job-shop scheduling problems using genetic algorithms—i: representation. Computers & Industrial Engineering 30(4), 983–997 (1996)
23. Yamada, T., Reeves, C.R.: Solving the c_{sum} permutation flowshop scheduling problem by genetic local search. In: Proceedings of the 1998 Congress on Evolutionary Computation, CEC 1998, pp. 230–234 (1998)
24. Nowicki, E., Smutnicki, C.: A fast tabu search algorithm for the permutation flow-shop problem. European Journal of Operational Research 91(1), 160–175 (1996)
25. Marchiori, E., Steenbeek, A.: An evolutionary algorithm for large scale set covering problems with application to airline crew scheduling. In: Oates, M.J., Lanzi, P.L., Li, Y., Cagnoni, S., Corne, D.W., Fogarty, T.C., Poli, R., Smith, G.D. (eds.) EvoIASP 2000, EvoWorkshops 2000, EvoFlight 2000, EvoSCONDI 2000, EvoSTIM 2000, EvoTEL 2000, and EvoROB/EvoRobot 2000. LNCS, vol. 1803, pp. 367–381. Springer, Heidelberg (2000)
26. Ponnambalam, S.G., Mohan Reddy, M.: A ga-sa multiobjective hybrid search algorithm for integrating lot sizing and sequencing in flow-line scheduling. International Journal of Advanced Manufacturing Technology 21(2), 126–137 (2003)

27. Ray, T., Sarker, R.A.: Optimum oil production planning using an evolutionary approach. In: Evolutionary Scheduling, pp. 273–292. Springer, Heidelberg (2007)
28. Burke, E.K., Smith, A.J.: A memetic algorithm to schedule planned maintenance for the national grid. Journal of Experimental Algorithmics 4, 1 (1999)
29. Martinelli, F.: Stochastic comparison algorithm for discrete optimization with estimation of time-varying objective functions. Journal of Optimization Theory and Applications 103(1), 137–159 (1999)
30. Tinós, R., Yang, S.: Genetic algorithms with self-organizing behaviour in dynamic environments. In: Evolutionary Computation in Dynamic and Uncertain Environments, pp. 105–127. Springer, Heidelberg (2007)
31. Yang, S.: Explicit memory schemes for evolutionary algorithms in dynamic environments. In: Evolutionary Computation in Dynamic and Uncertain Environments, pp. 3–28. Springer, Heidelberg (2007)
32. Schönemann, L.: Evolution strategies in dynamic environments. Evolutionary Computation in Dynamic and Uncertain Environments, pp. 51–77. Springer, Heidelberg (2007)

A Memetic Framework for Solving the Lot Sizing and Scheduling Problem in Soft Drink Plants

Claudio F.M. Toledo, Marcio S. Arantes, Paulo M. França, and Reinaldo Morabito

Abstract. This chapter presents a memetic framework for solving the Synchronized and Integrated Two-level Lot Sizing and Scheduling Problem (SITLSP). A set of algorithms from this framework is thoroughly evaluated. The SITLSP is a real-world problem typically found in soft drink plants, but its presence can also be seen in many other multi-level production processes. The SITLSP involves a two-level production process where lot sizing and scheduling decisions have to be made for raw material storage in tanks and soft drink bottling in various production lines. The work presented here extends a previously proposed memetic computing approach that combines a multi-population genetic algorithm with a threshold accepting heuristic. The novelty and its main contribution is the use of tabu search combined with the multi-population genetic algorithm as a method to solve the SITLSP. Two real-world problem sets, both provided by a leading market soft drink company, have been used for the computational experiments. The results show that the memetic algorithms proposed significantly outperform the previously reported solutions used for comparison.

Claudio F.M. Toledo
Institute of Mathematics and Computer Science, University of Sao Paulo,
Av. Trabalhador Sao-Carlense, 400, 13566-590, Sao Carlos, SP, Brazil
e-mail: `claudio@icmc.usp.br`

Marcio S. Arantes
Department of Computer Science, University of Lavras,
C.P. 3037, 37200-000, Lavras, M.G., Brazil
e-mail: `marcio@comp.ufla.br`

Paulo M. Franca
Department of Mathematics, Statistics and Computing, UNESP,
C.P. 266, 19060-900, P. Prudente, SP, Brazil
e-mail: `paulo.morelato@fct.unesp.br`

Reinaldo Morabito
Production Engineering Department, Federal University of Sao Carlos,
C.P. 676, 13565-905, Sao Carlos, SP, Brazil
e-mail: `morabito@ufscar.br`

1 Introduction

The present chapter is motivated by a problem commonly found in soft drink plants and it is established by a two-level production process. In the first level, raw materials are preprocessed and stored in tanks, and in the second level, they are spread to bottling lines responsible for producing several types of soft drinks. In order to compound a final product (raw material plus bottle type) a proper raw material and a right bottle type must be assigned to the tanks and bottling lines, respectively.

This synchronization aspect is a crucial issue when defining the lot sizes and scheduling for raw materials and final products. Therefore, a problem solution that integrates lot sizing and scheduling decisions for both levels in this problem is mandatory.

The objective of this work is to present an academic and applied contribution for researchers or practitioners dealing with real-world multi-level lot sizing and scheduling problems. This research was conducted using data provided by a soft drink company in Brazil.

The Brazilian soft drink consumption increased by 4.0% in 2008 and by 2.2% in 2009 with more than 14 billions of liters consumed. These numbers make the Brazilian soft drink market the third largest in the world with more than 800 plants [1]. These figures show a promising and competitive market, where applied research can take place.

The production processes in soft drink plants have several lines and tanks that can be seen as parallel machines in the context of two interdependent lot sizing and scheduling problems. It is possible to simplify some problem issues if each tank is previously assigned to a line. The situation of the problem with and without tanks already assigned to lines will be described through this chapter.

The memetic algorithms (MAs) presented in this chapter are relatively simple. However, our major concern is to show some advances from previous methods that have already found good results when applied to the same problem. The methods proposed here extend previous studies on the same problem reported in [2], [3], and [4]. A multi-population genetic algorithm was presented in [2] to solve artificial and real-world problem instances. These instances were solved without previous tank assignments, which means that one tank can feed several production lines at the same time. An MA that combines a threshold accepting (TA) local search with the multi-population genetic algorithm was applied to solve a set of 15 real-world instances in [3]. Subsequently, results evaluating relax-and-fix approaches for the same 15 real-world instances were given in [4]. All the instances solved in [3] and [4] take into account only the case where tanks are previously assigned to lines.

In this chapter, we use the MA presented in [3] as a starting point. Instead of the TA local search we now use a tabu search (TS) as the local search procedure embedded in the multi-population genetic algorithm. In order to deal with larger real-world instances while maintaining the good accuracy attained in prior tests, we propose to strengthen the method by adding a TS phase where the best individuals of each sub-population are selected for further enhancement. Next, a variant of this memetic approach is introduced. This method is defined by a different strategy to explore the neighborhood of solutions, as well as a different way to carry out the local search procedure.

The rest of this chapter is organized as follows. The soft drink industry problem is described in Section 2 and the MAs are introduced in Section 3. Computational tests are reported in Section 4 and conclusions follow in Section 5.

2 The Soft Drink Industry Problem

The soft drink industry problem studied in this chapter presents two synchronized and integrated production levels. The first level has storage tanks with limited capacity that store several soft drink flavors (raw materials). The second level has bottling lines (production lines) with also limited capacities within each week that cannot be violated. The line capacity is usually related to the number of hours available for production by week. Figure 1 illustrates this production process.

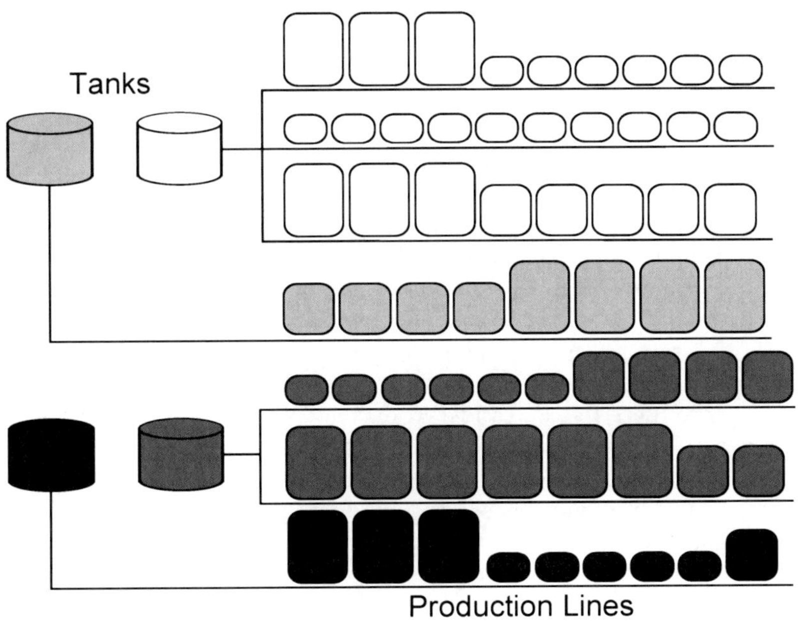

Fig. 1. Two-level production process without tanks assigned to line

The tanks are filled using one soft drink flavor at a time and they have to be filled by a minimum amount. The process of refilling a tank demands a cleaning phase. The tank is only filled up again after it becomes empty. It does not matter if a tank is filled by the same or by a different raw material, the tank should always be cleaned. The amount of time spent for cleaning a tank before being able to fill it depends on the raw material previously stored. Therefore, sequence-dependent setup times and costs take place. It is always faster and cheaper to clean a tank when filled by the same raw material.

Next, the bottling lines receive raw materials (soft drink flavors) from tanks producing the final product. This final product is defined as the soft drink flavor and bottle type. The lines can bottle raw materials in several bottle types: cans, bag-in-boxes, glass bottles and plastic bottles with different sizes. The production lines can have different processing times for each final product and each line usually produces a set of these products.

There are sequence-dependent setup times for lines. This setup occurs if a bottle type is changed in line or if a tank that provides raw material stopped to feed the line to be refilled. The sequence-dependent setup time will also generate a sequence-dependent setup cost for final products. Moreover, the sequence-dependent setup costs in both levels are proportional to their sequence-dependent setup times.

The soft drink industry usually has a time horizon of several months with weekly demands to be met. Inventory costs can occur if an excessive number of final products can be stored for the next week. However, a lack of products is not allowed as the demands for each week have to be fulfilled. A tank can store raw materials for several periods, but inventory costs incur when raw materials are kept stored.

Decisions about schedule and lot sizing have to be made for raw materials in tanks and products in the lines, so there are two lot sizing and scheduling problems. However, the two-level production process can happen appropriately only if there is a synchronization between the two lot size and scheduling problems. The lines cannot produce any lot sizes of soft drinks without the necessary amount of raw material scheduled in the tanks.

Let us illustrate these ideas based on the schedule presented in Fig. 2. One line, one tank, and three raw materials RM1, RM2, and RM3 are required to produce products P1, P2, and P3, respectively. A time horizon with two periods is shown, as well as the time spent taking raw materials from the tank and producing products in the line. The setup time to refill tanks (white boxes) and the setup time to change products in the line (black boxes) are also represented.

It is possible to think about the schedule for this production process without taking into account synchronization between the two production levels. In this case, Fig. 2 shows that it is allowed to try to produce P1 without available raw material RM1 in the tank in the first period. This situation can happen because some time is spent filling the tank with RM1 at the beginning of the production process. In the second period, the setup time to

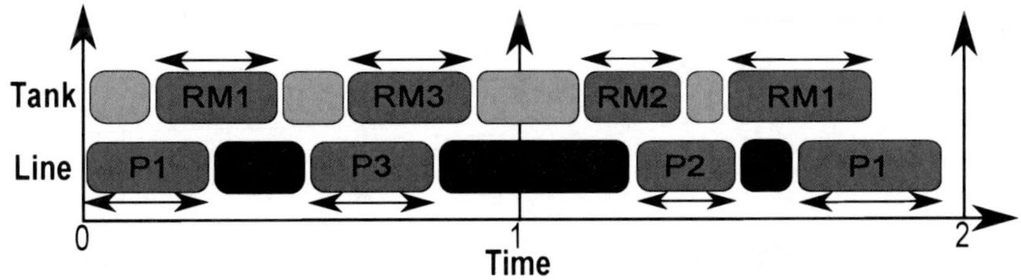

Fig. 2. Schedule without synchronization

change from RM3 to RM2 is shorter than the setup time to change from P3 to P2. Thus, it is not possible to refill the tank from RM2 to RM1 as shown as the line is still producing P2. Fig. 3 introduces the same schedule with synchronization.

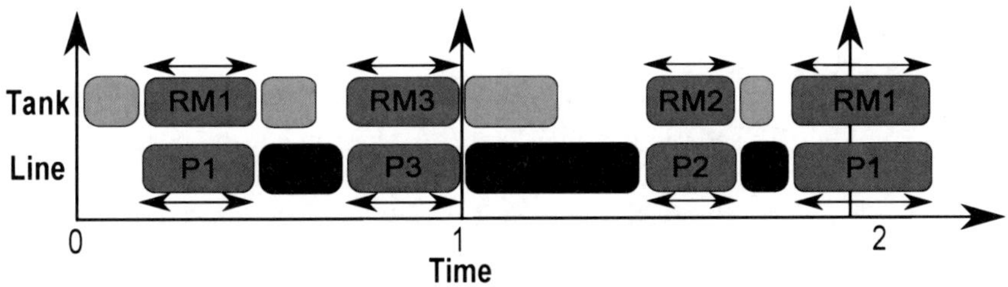

Fig. 3. Schedule with synchronization

Each raw material is being consumed in the tank level at the same time as the respective product is produced by the line. The problem related with this synchronization is the available capacity. At the end of the second time period, the synchronization shows that not enough time is left to produce P1 in the second period. Thus, the demand of P1 is not met following at least the previous line and tank sequencing.

This soft drink industry problem was called Synchronized and Integrated Two-level Lot Sizing and Scheduling Problem (SITLSP) in [5] and [2]. A simplification of the SITLSP was proposed in [3] and [4], where each tank is dedicated to only one line. However, the dedicated tank can be filled with any raw material demanded by this line (Fig. 4). This small simplification of the SITLSP reduces its complexity. The authors in [3] and [4] are able to describe a more compact mathematical model for the problem.

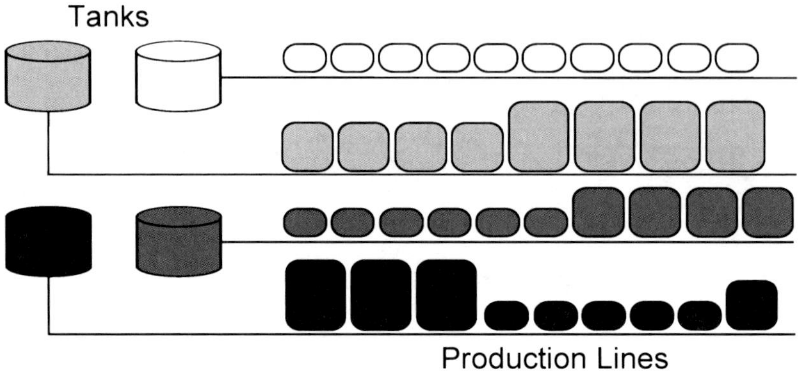

Fig. 4. Two-level production process with tanks assigned to the line

3 Related Works

The SITLSP covers various lot-sizing and scheduling issues such as capacity constraints, setup cost and setup time, multi-items and parallel machines. The capacitated lot sizing problem is proven to be an NP-hard optimization problem [6]. The SITLSP is a multi-item problem because there are several raw materials and products whose schedule and lot size have to be defined in lines and tanks, respectively. The multi-item capacitated lot-sizing problem is strongly NP-hard [7]. There are sequence-dependent setup times and costs in both production levels. It was showed in [8] that even finding feasible solutions for the capacitated lot sizing problem with setup times is an NP-complete problem.

There are various reviews which discuss models and solution approaches for the lot sizing and scheduling problem, including multi-level problems, such as [9], [10], [11], and [12]. An overview on modeling for industrial extensions of single-level dynamic lot sizing problems is provided by [13]. Meta-heuristics applied to solving the lot sizing problem are evaluated in [14], where several issues are discussed such as computational experiments, solution representation, evaluation functions, neighborhoods and operators.

The discrete lot sizing and scheduling problem with sequence dependent setup costs is introduced in [15], where the time horizon is subdivided into several micro-periods with identical capacity. It is a small bucket problem as the micro-period capacity is fully used or nothing is produced. The problem is formulated as a travelling salesman problem and lower bounds are found using Lagrangian relaxation.

A large bucket formulation is presented in [16] for the capacitated lot sizing problem with sequence dependent setup costs. The time horizon is subdivided into macro-periods with the same length. Demands for products are met at the end of some macro-periods. The model proposed in [16] differs from [15] because it allows continuous lot sizes and preserves the setup state over idle time. In big bucket problems, many items can be produced in the same time period without sequencing issues. A review on big bucket problems is presented in [17].

The general lot sizing and scheduling problem with sequence dependent setup costs is presented by [18], where a small bucket formulation is used to describe it. The time horizon is divided into several micro-periods, however each micro-period length is now defined as the time spent to produce the product assigned to it. A TA heuristic associated with a greedy method is used to solve problem instances. The heuristic assigns products to micro-periods and the greedy method determines lot-sizes.

A lot sizing and scheduling problem with setup costs and setup times is formulated as a general lot sizing and scheduling problem by [19]. Simulated annealing and TA heuristics are combined with a dual re-optimization algorithm. This approach is extended by [20] to solve the same problem with parallel machines.

In [21], one of the few optimal procedures for solving lot sizing and scheduling problems with sequence-dependent setup times is proposed. They define an enumeration method based on the branch and bound procedure which was able to optimally solve several instances ranging from 3 to 10 products and 3 to 15 periods. A model partially based on that proposed by Haase and Kimms [21] is presented in [22]. The production in some period is modeled using binary variables and a heuristic defines model parameters. The compact model obtained by the authors can solve large instances in a reasonable time.

A model is proposed in [23] for the single machine capacitated lot sizing and scheduling problem with sequence-dependent setup costs and non-zero setup times. A solver with an exact method was used to find optimal solutions for small-sized instances. Row aggregation and variable relaxation can determine lower bounds for large-scale instances. A heuristic with three steps (initialization, sequencing and improvement) is also used to solve problem instances. In [24] stronger models for the same problem that is solved by specific-purpose heuristics are proposed.

Rolling-horizon and relax-and-fix heuristics are proposed in [25] to solve the identical parallel machine lot sizing and scheduling problem with sequence dependent setup costs. The relax-and-fix method outperforms rolling-horizons with cost values closer to instances' lower bounds. Mixed integer programming (MIP) models are presented in [26] and [27] for lot sizing and scheduling in the presence of setup times for problems found in electrofused grain and animal nutrition industries. Toso et al. [28] studied a lot sizing and scheduling problem from a manufacturing plant for animal feed compounds. Model formulations and relax-and-fix heuristics are proposed.

A multi-level capacitated lot sizing problem (MLCLSP) is presented by Sahling et al. [29] as an extension of a single-level capacitated lot sizing problem. An uncapacitated MLLSP is solved in [30] using a particle swarm optimization algorithm. The results found are compared with those achieved with genetic algorithms (GAs). Mathematical programming and an ant colony algorithm are combined in [31] to solve the MLCLSP, where the heuristic fixes values for binary variables and a mixed-integer programming (MIP) software found values for continuous variables.

A variable neighborhood search (VNS) is used to solve the uncapacitated MLLSP in [32]. A setup shifting rule is proposed considering the interdependence among various interdependence items. In [33], Stadler described a multi-level model for a pharmaceutical production problem which is an extension from the single-level and single machine proportional lot sizing and scheduling problem. Computational tests are reported based on data provided by a pharmaceutical company.

A TS is a local search technique with a short-term memory structure that avoids visiting the same solutions or neighborhoods several times [34]. This metaheuristic has also been used to solve several lot sizing and scheduling problems. TS is used by Hung et al. [35] to solve a production planning problem with setups (time and cost) and multiple products, resources and periods. The potential solutions (neighbors) are evaluated solving LP problems whose information can define the ranks of the neighbors. This rank is used to evaluate different strategy approaches to explore neighborhoods.

TS is proposed as an approach to solve a production planning problem in a flexible production system by Al-Fawzan [36]. A lot sizing model with backorder and a set of random test problems are presented. The TS outperforms a random sequencing procedure within a short computational time. TS is applied by Buscher and Shen [37] to the job shop scheduling problem taking into account lot streaming. The heuristic is used in scheduling tasks where moves are defined to generate new schedules from an initial one. Moves that can return to previous schedules are kept tabu for several iterations.

GAs are evolutionary computation methods that simulate the evolution of biological processes. They were formally introduced by Holland [38] and studies about the implementation of GAs can be found in [39] and [40]. An MA usually combines the population approach of GAs with local search methods [41].

The MA proposed in this chapter will evolve various populations with individuals hierarchically structured in ternary trees. Prior works have demonstrated that evolutionary approaches with individuals hierarchically structured in trees, when dealing with large scheduling problems, have outperformed their corresponding unstructured ones. A hybrid population approach for an MA with individuals hierarchically structured in ternary trees is presented in [42]. Instances randomly created are solved for the total tardiness single machine scheduling problem, where the MA outperforms a GA.

The same approach was applied by França et al. [43] to solve the problem of scheduling a flowshop manufacturing cell with sequence dependent family setups. The use of hierarchically structured individuals improved the results found by the GA and the MA. A group scheduling problem for manufacturing cells is solved in [44] using a GA and a MA. The population has 13 individuals structured in ternary trees where an individual encodes two solutions: the better fitness solution and the current solution. The MA outperforms GA in a set of 15 instances.

Despite the economical relevance of the SITLSP, only a few solution methods have been proposed so far. A drink manufacturer problem with a single canning line was solved by Clark [45]. Sequence-independent setup times and a mixed-integer programming (MIP) model were proposed. A hybrid heuristic combining local searches and mathematical programming is reported as the best approach for this problem.

The first mixed-integer mathematical model for the SITLSP is proposed in [5] combining issues from the General Lot-sizing and Scheduling Problem (GLSP) and the Continuous Setup Lot-sizing Problem (CSLP). The time horizon is divided into macro-periods with the same length. Each macro-period has a maximum number of slots and each slot can determine for which product or raw material a particular slot in some line or tank is assigned. The synchronization among slots in the two-levels of the problem is done using micro-periods as each macro-period is divided into micro-periods with the same length. The authors were able to solve small-to-moderate sized instances using GAMS/CPLEX. A multi-population genetic algorithm with a hierarchically structured population was applied in [2] to solve small-to-moderate and moderate-to-large sized instances. The authors also solved a set of real-world instances provided by a soft drink industry. It is reported that the GA found better results for these problem instances.

As aforementioned, a simplifying formulation for the SITLSP is presented in [3] and [4] called P2SMM – Two-stage multi-machine lot-scheduling. The bottling lines here have dedicated tanks and these tanks can be filled with the liquid flavors needed by the assigned lines. In [3], results found using MAs and a relax-and-fix approach are presented. Relax-and-fix heuristics are proposed and evaluated in [4]. The computational results in this work are obtained using problems created based on real data from a soft drink company. A MIP model for small-scale soft drink plants is proposed by Ferreira et al. [46]. There is only one production line where the bottleneck is in the production line.

4 Memetic Algorithm Approaches

The MAs will be described in this section, where issues related with neighborhood exploration and local search execution are explained. To make this text self-contained, details about the multi-population genetic algorithm used by Toledo et al. [2] are also presented. For the same reason, the TA local search built-in the MA introduced by Ferreira et al. [3] is also described.

4.1 Memetic Algorithm Pseudocode

Algorithm 1 describes the MA introduced in [3]. This pseudo code comes from the multi-population genetic algorithm, where the MA in Fig. 5 has only added a local search procedure at line 18. Several overlapping populations

Algorithm 1. MultiPopulationMemeticAlg

```
1  begin
2      repeat
3          for i ← 1 to numberOfPopulations do
4              initializePopulation(pop(i));
5              evaluatePopulationFitness(pop(i));
6              structurePopulation(pop(i));
7              repeat
8                  for j ← 1 to numberOfCrossovers do
9                      selectedParents(ind1,ind2);
10                     newInd← crossover(ind1,ind2);
11                     if executedMutation then
12                         newInd ← mutation(newInd);
13                     evaluatedFitnessIndividual(newInd);
14                     insertPopulation(newInd, pop(i));
15                 structurePopulation(pop(i));
16             until populationConvergence(pop(i)) ;
17         for i ← 1 to numberOfPopulations do
18             localSearch(bestId(pop(i)));
19             executeMigration(pop(i));
20     until StopCriterion ;
21 end
```

evolve until a convergence criterion has been satisfied. A population stores individuals structured in a ternary tree with four clusters. A single cluster is constituted by individuals distributed in two levels: the fittest individual (node) is the cluster "leader" always positioned in the upper level and three other supporters in the lower level (Figure 5).

The population illustrated in Figure 5 has 13 individuals hierarchically structured in ternary trees. The best individual is the one with lowest cost value associated. Notice that the root node always keeps the best known individual (here: the node with value 100) of the whole population. First, the initial population is generated using initializePopulation(pop(i)) and its individuals are evaluated using a fitness function – evaluatePopulationFitness (pop(i)). The individual hierarchy is then built by procedure structure Population(pop(i)).

The genetic operator crossover(ind1,ind2) selects two individuals from the randomly chosen cluster in such a way that one of them is always the leader

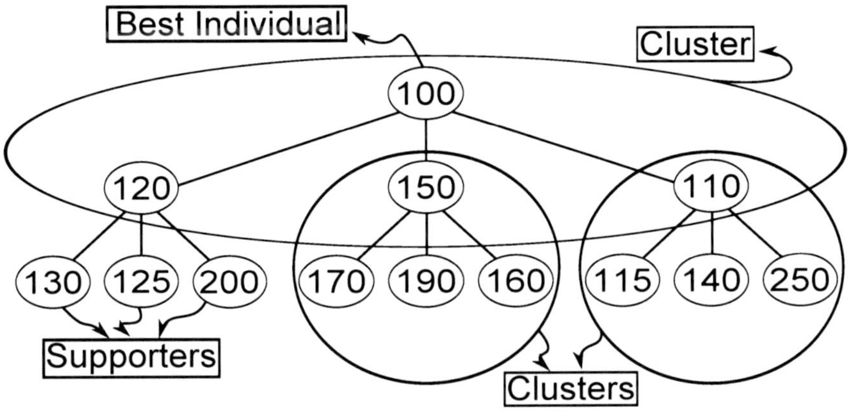

Fig. 5. Population hierachically structured in ternary tree

and the other is also randomly selected among their supporters. If the new individual (newInd) is better than the worst parent, it will be inserted into the population and the parent is discarded (insertPopulation(newInd, pop(i))).

The hierarchical structure must be kept, so structurePopulation(pop(i)) performs an update, rearranging individuals in each cluster so that the fittest individual becomes the leader of its cluster. Notice that after the updating process, the root node is always occupied by the fittest individual of the whole population. New individuals may also be created by means of a mutation operator mutation(newInd) applied to new offspring. The number of new individuals generated is defined by the crossoverRate parameter:

$$numberOfCrossover = crossoverRate * numberOfIndividuals \tag{1}$$

Convergence is reached when there is no individual insertion. After the last population has been converged, a local search is made for the best individual of each population. Finally, a migration operator exchanges the best individuals among the populations.

The local search procedure proposed in [3] is the TA method while here, TS is used. Other novelties in the MA introduced in this chapter are:

1. The local search will execute only a random selection of neighborhood moves.
2. The local search is executed only on the best individual from the whole population.

Thus, an MA with another strategy to select moves has been proposed with a local search execution which is more constrained. The local search will be executed after the population convergence in Algorithm 1 and only over the best individual found so far. There is no local search execution for the best individual of each population. The random choice aims to avoid an exhaustive use of the same type of move over and over again. The constrained use of local search aims to spend more time only looking for solutions in the neighborhood of the best individual.

4.2 Individual Codification/Decodification

The individual represents a problem solution. It is represented by a $T \times N$ matrix whose entry (t, n), $t \in \{1...T\}$ (periods) and $n \in \{1...N\}$, is shown in Figure 6.

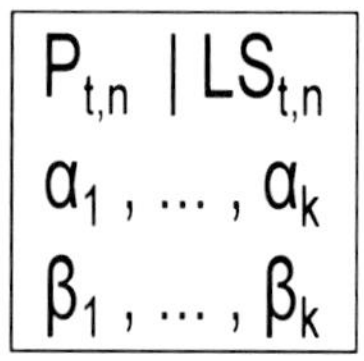

Fig. 6. Entry (t, n) in the solution representation

There are four problem data encoded in each entry (t, n): $P_{t,n}$ is the product in position n to be produced in period t, $LS_{t,n}$ is the lot size of product $P_{t,n}$, $SL_{t,n} = (\alpha_1, \ldots, \alpha_k)$ is the sequence of lines where $P_{t,n}$ can be produced and $STk_{t,n} = (\beta_1, \ldots, \beta_k)$ is the sequence of tanks where raw material for $P_{t,n}$ can be stored.

Individuals are initialized by splitting the product demands through these entries. The demand $D_{i,t}$ of some product i in period t is split into several lots $(LS_{t,n})$ and randomly distributed among entries in periods t, $t-1$, $t-2$,..., 1. The parameter $\alpha_i \in \{\alpha_1, \ldots, \alpha_L\}$ is a line number in $SL_{t,n}$ and L is the number of lines. The parameter $\beta_i \in \{\beta_1, \beta_2, \ldots, \beta_{2\bar{L}}\}$ tells us where and how the raw material of $P_{t,n}$ will be stored. The tank number j is defined as:

$$j = \begin{cases} \beta_i, & 1 \leq \beta_i \leq \bar{L}; \\ \beta_i - \bar{L}, & \bar{L} < \beta_i \leq 2\bar{L}. \end{cases} \tag{2}$$

Parameter $\bar{L}$ is the number of tanks. If $1 \leq \beta_i \leq \bar{L}$, the tank $j = \beta_i$ will be occupied after the previously stored raw material has been used. If the same raw material is replaced, this criterion can postpone tank occupation. If $\bar{L} < \beta_i \leq 2\bar{L}$, the tank $j = \beta_i - \bar{L}$ will be immediately filled. This criterion can use the available tank capacity.

There are exceptions for these rules. If tank j stores a raw material different from the raw material used by product $P_{t,n}$, it can be stored only after this different raw material has been pumped to the lines. Moreover, tank j which is already empty has to be filled.

Let us suppose two products ($P1$ and $P2$) with demands in periods $T1$ and $T2$, where $P1$ have demands of 150 units in $T1$ and 100 units in $T2$. Product $P2$ has demands of 110 units in $T1$ and 120 units in $T2$. Different raw materials are necessary for each product. There are two lines and two tanks able to store any raw material and to produce any product. Possible representations of solutions for this situation are illustrated in Figure 7.

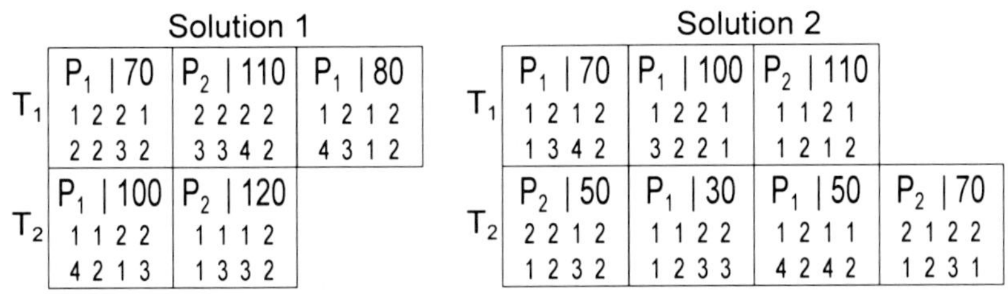

Fig. 7. Possible representation of solutions

The sequence of lines ($SL_{t,n}$) and tanks ($STk_{t,n}$) can repeat values of $\alpha_i \in \{1, 2\}$ and $\beta_i \in \{1, 2, 3, 4\}$ for $L = \bar{L} = 2$ and $k = 4$ (sequence length). Solution 1 has all the demands in their respective periods, but the $P1$ demand in $T1$ was split into two entries. In solution 2, product $P1$ has 20 units of demand in $T1$ assigned to the lot size of the first entry in period $T1$.

The problem data encoded in each individual is decoded to a problem solution. This decoding process starts by the first entry in the last period of the representation of the solution. The procedure is repeated until the data in the last entry of the first period has been decoded. A backward selection of entries aiming at postponing setups and processing time was chosen, but there is no guarantee that all demands will be produced at the end. Figure 8 shows this procedure.

If (α_i, β_i), with $\alpha_i \in SL_{mn}$ and $\beta_i \in STk_{mn}$, returns a pair of lines and tanks able to produce and store the lot size $LS_{t,n}$, then $LS_{t,n}$ is scheduled by the Execute(α_i, β_i) step. If there is no available capacity or it is sufficient to schedule only a part of $LS_{t,n}$, the next pair $(\alpha_{i+1}, \beta_{i+1})$ is selected for the remainder of the lot size. The complete decoding of the individual entries returns a synchronized lot size and schedule for lines and tanks.

This schedule is evaluated by the fitness function, which is determined by adding up all the problem costs: production, setup, and inventory costs for products and raw materials in lines and tanks, respectively. A high penalty cost for demands not satisfied is also added to the fitness value.

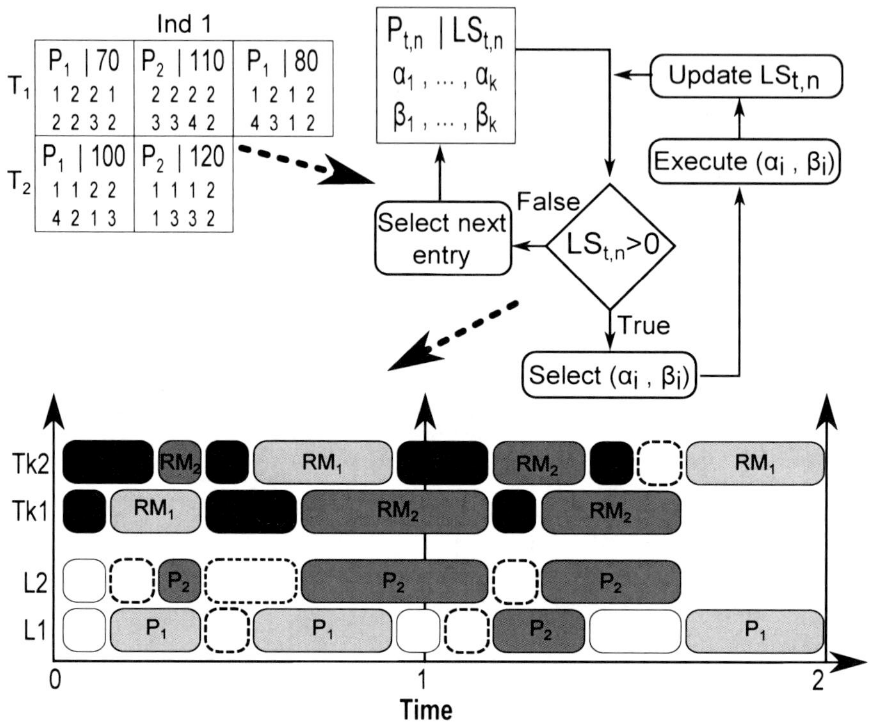

Fig. 8. Decoding procedure

4.3 Crossover, Mutation and Migration

New individuals are created using a crossover operator and a uniform crossover was designed for this representation of a solution in [2]. The authors reported that, among others, uniform crossover performed better in previous computational experiments. Figure 9 illustrates how the uniform crossover works taking into account the two solution representations (Ind1 and Ind2) from Figure 7.

A random value $\lambda \in [0, 1]$ is generated for each entry in the same position of Ind1 and Ind2. If $\lambda < 0.5$, the entry of Ind1 is inherited by Child. Otherwise, Child inherits the entry from Ind2. If one period of Ind2 has more entries than the same period of Ind1, Child will inherit them from Ind2 for $\lambda \geq 0.5$. Demand surplus or shortage is not allowed in Child. The circled entry of Ind1 is not inherited because it would exceed the total demand of $P2$ (100 units in $T2$). Those demands that are not met in any period are filled through a repair procedure.

Crossover and mutation operators are carried out over a cluster that is selected at random, where two individuals are taken. The first individual is a

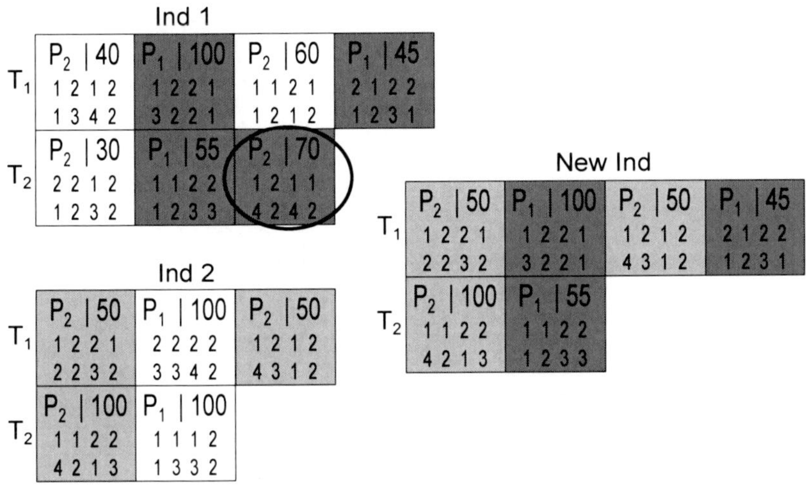

Fig. 9. Uniform crossover

supporter node that is randomly chosen in this cluster. The second individual is always the cluster leader. The mutation operator can be applied to this new individual according to the mutation rate. Next, this new individual is evaluated and, if it is better, it will replace the parent with the worst fitness value. Otherwise, the new individual is not inserted into this population. After these genetic operators, the population is rearranged aiming to keep the cluster hierarchical structure. Figures 10 and 11 illustrate this process.

The new individual in Figure 10a has a fitness value of 95, so it replaces the worst parent with value 130 (Figure 10b). The population is reorganized in such a way that a new individual becomes a cluster leader (Figure 11a) and then the best individual of the population (Figure 11b).

A population converges when new individuals are not inserted for numberOfCrossovers iterations. At this point, the next population evolves following the same procedures described in Algorithm 1.

Migration and a new initialization take place when all populations have converged. However, if the local search has to be applied to the best individual of each population, this is done before migration as illustrated in 12.

The restart will re-initialize the individuals, but it keeps the best individual and the migrated individual. If the local search has to be applied to the best individual from all the populations, this is also done before migration and new initialization, but only for the best individual as illustrated by Figure 13.

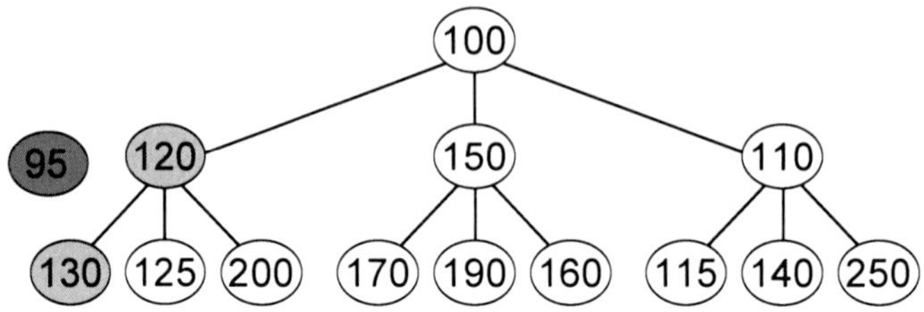

(a) New individual is created.

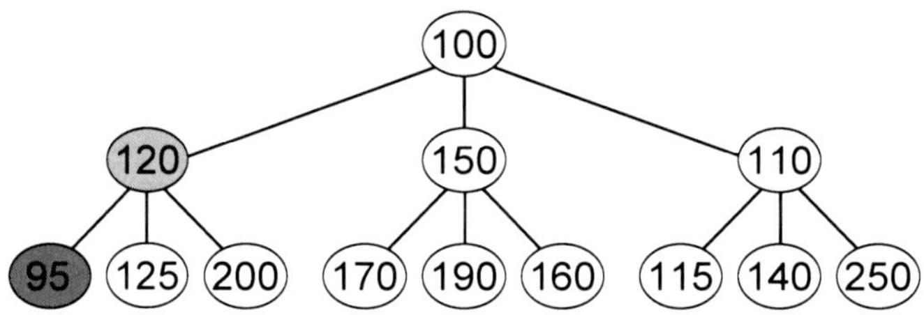

(b) New individual is inserted.

Fig. 10. Inserting a new individual.

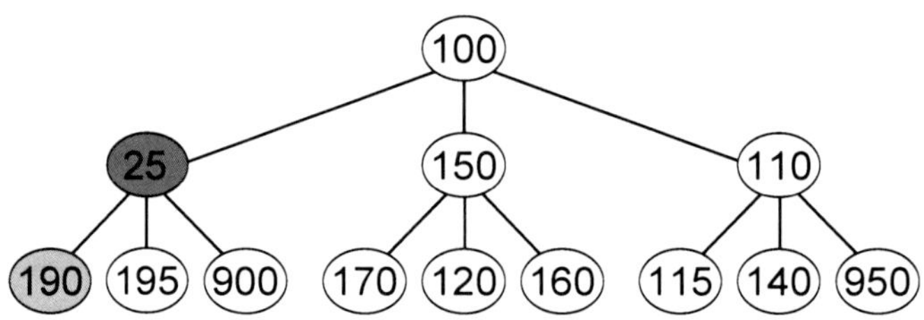

(a) Cluster is rearranged.

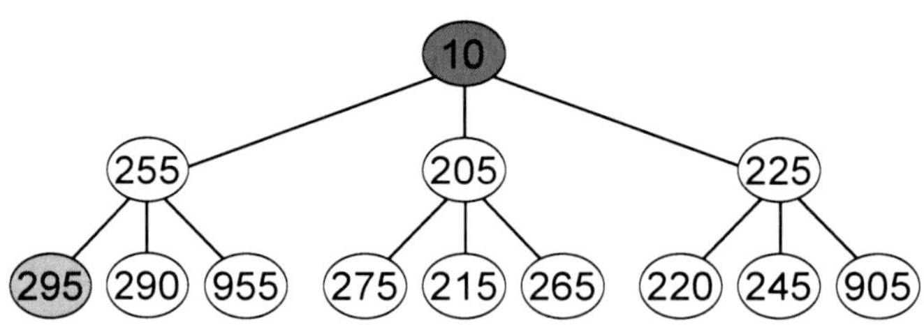

(b) Best individual is updated.

Fig. 11. Updating the hierarchical structure.

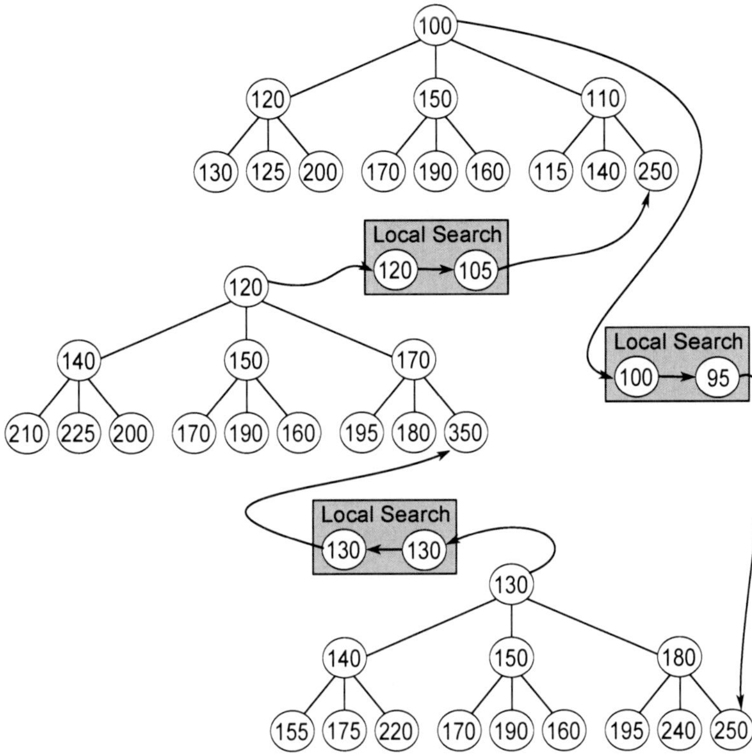

Fig. 12. Migration and local search execution

4.4 Local Search Procedures

A total of five neighborhood moves are defined using the representation of the solution (individual) previously described (Figure 7):

- Swap: two selected entries of the individual representation are exchanged.
- Insertion: an entry is randomly inserted into another position.
- Merge: take two entries with the same product, select one of them to merge the lot sizes and the other one to be removed.
- Split: the lot size of some entry is split and it is relocated as a new entry.
- Tank and Line sequences: the sequence of tanks and lines in a selected entry is reinitialized.

All these moves select entries or macro-periods randomly, but some problem constraints are checked. For example, it is not possible to insert an entry into the second period if its lot needs to meet the demand in the first period. Figure 14 illustrates these moves from an individual.

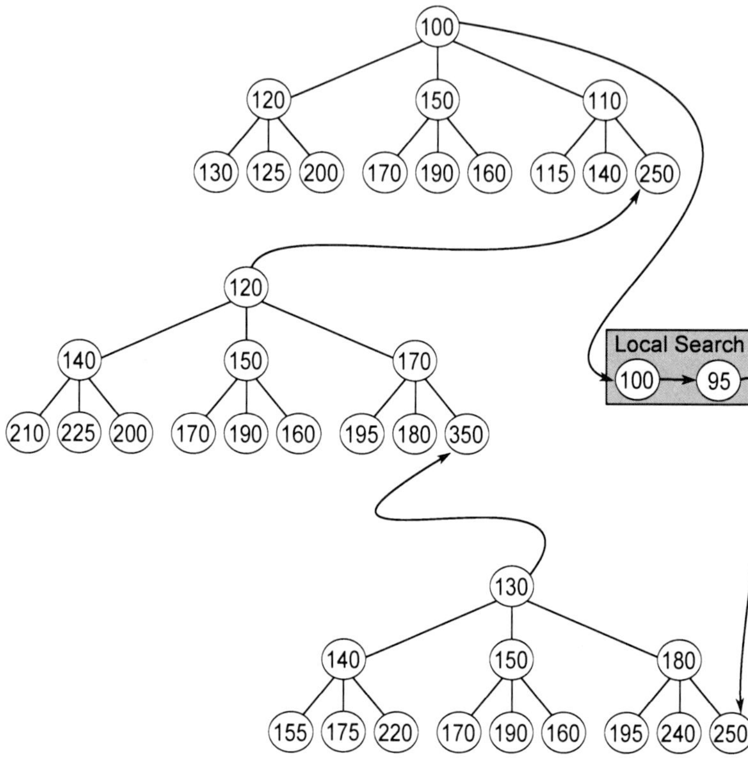

Fig. 13. Migration and local search execution about the best individual

Two strategies for neighborhood exploration are evaluated. The first strategy selects a move m randomly among all available moves, which are the moves previously defined (Figure 14). A selected move m is applied to the current solution which initially is the best individual of the all the populations. The move is applied until a maximum number of iterations have been reached. This procedure creates several neighbors. The neighbors are evaluated and the best neighbor found is returned.

The second strategy is the same as defined in [3] for the local search TA. This strategy is executed in three steps. First, a swap or an insertion move are randomly chosen and executed by the local search. After a maximum number of iterations, the local search (TA or TS) returns the best individual found in the neighborhood of these moves.

Next, the local search is executed selecting randomly between a merge or split move. After the same maximum number of iterations, the local search returns the best individual found with these moves. Finally, the local search is executed again randomly selecting a move among swap, insertion, merge or split moves to be applied to the best individual previously found. If better,

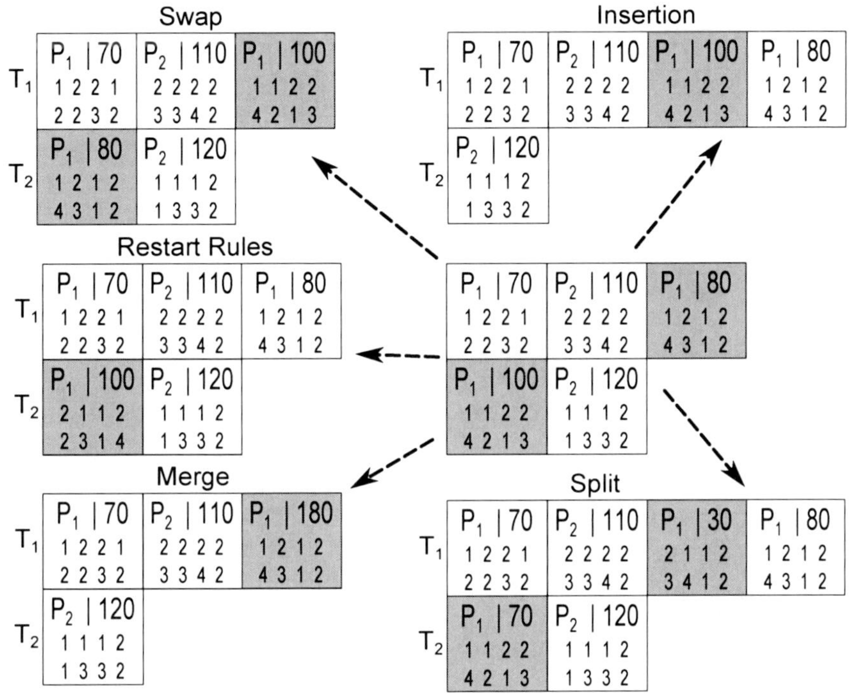

Fig. 14. Neighborhood moves

the best individual found at the end is inserted as the new best individual of the population.

The second strategy attempts to repeat the same subset of moves several times, where moves that change lot sizes (split and merge) are initially separated from those moves that change gene positions (swap and insertion). It is worth mentioning that this strategy for neighborhood exploration is executed for the best individual of each population. Thus, it is given a chance to improve the best individual found in each population.

The pseudo code of TA is presented in Algorithm 2. It is a local search method that accepts worse individuals when their fitness value remains within a threshold. Reduction of the threshold leads to method convergence. This reduction can follow different values considering whether an individual was or was not accepted.

Algorithm 3 shows the pseudo code for TS, where the method is executed until tsMax iterations have been reached. First, the current solution S is initialized as the best individual of the population. Then, S is changed by procedure executeNeiborhoodMoves(S) that returns the best neighbor (bestNeighbor) and the respective move that led to this neighbor (bestMove).

The current solution is updated by bestNeigbor and the bestMove is inserted into the tabu list. The best solution S^* can also be updated. The tabu

Algorithm 2. LocalSearch TA

```
begin
    individual ← bestIndividual;
    repeat
        newIndividual← moveExecution(individual);
        △f ← fitness(individual)-fitness(newIndividual);
        if (△f > −Th· fitness(individual)) then
            individual ← newIndividual;
        reduce(Th);
    until (maxNumberOfIterations) ;
end
```

list is dynamic, so its length changes after a certain number of iterations. If the size of the tabu list is reduced, older moves will be freed earlier. Otherwise, they will stay tabu longer. The size of the tabu list is a random value tabuListLength $\in \{1, \ldots, numberOfMoves - 1\}$.

Algorithm 3. TabuSearch

```
begin
    tabuList ← ∅;
    initialize(S);
    S* ← S;
    f(S*) ← f(S);
    count ← 0;
    stop ← 0;
    while (stop < tsMax) do
        (bestNeighbor, bestMove)← executeNeighborhoodMoves(S);
        S ← bestNeighbor;
        insertTabuMove(TabuList,mBest);
        if (f(S) < f(S*)) then
            S* ← S;
        count ← count +1;
        if (count > maxCount) then
            tabuListLength ∈ {1, ..., numberOfMoves − 1};
            update(tabuList,tabuListLength);
            count ← 0;
        stop ← stop + 1;
    return (S*, f(S*));
end
```

To sum up, four MA variants will be evaluated:

1. MA-TS1: A variant with the first strategy for neighborhood exploration and with a local search executed only of the best individual of the three populations.

2. MA-TA1: Identical to MA-TS1, but TA is the local search built-in method.
3. MA-TS2: A variant with the second strategy for neighborhood exploration and with local search executed before migration of the best individual of each population.
4. MA-TA2: Identical to MA-TS2, but TA is the local search built-in method.

5 Computational Results

Computational tests are carried out in problem instances defined from data provided by a large soft drink plant. Two sets of real-world instances will be solved. As mentioned before, a multi-population genetic algorithm solves artificial and real-world problem instances in [2]. However, the present chapter will make comparisons only with the real-world instances from that paper. This is the first set of instances to be evaluated.

The second set consists of one real-world instance and four instances derived from it as proposed in [3] and [4]. These instances stand for the case where tanks are previously assigned to lines. The computational tests were executed using an Intel Core 2 duo processor with 2.66 GHz and 2GB RAM.

5.1 Comparisons with the First Set of Real-World Instances

The first set has instances based on production plans executed by the soft drink company in several time periods. Table 1 presents the parameters used to define each instance.

Table 1. Parameters for instances.

Inst.	Lines	Tanks	Products	R.M.	Periods
A1	5	9	33	11	1
A2	6	9	49	14	2
A3	6	9	58	15	3
B1	6	10	52	19	1
B2	6	10	56	19	2
B3	6	10	65	21	3

There are six instances in total, separated into the types A and B. Instances of type A represent production planning where demands must be met at the end of seven days. Instances B cover a larger period with demands that have to be met at the end of ten days. There are nine tanks available in instances

A and ten tanks available in instances B. The number of lines is the same, except for instance A1 which has five production lines.

Instance A1 is the least complex one with one period (one week), 33 final products and eleven different raw materials that can be produced using five lines and nine tanks. Instance B3 is the most complex one with three time periods (30 days altogether) and 65 products that need 21 raw materials. This instance has six lines and ten tanks available for production.

The two levels of the production process work 24 hours per day, which means that tanks and lines are always available for production. Minimum and maximum tank capacities are 1,000 liters and 24,000 liters, respectively. The demand in these instances ranges from 47 to 180,000 units and processing times can vary between 50 to 2,000 units/hour. The inventory and production costs for products and raw materials are set as 1($/u). All products assigned to any line spend 0.5 hour for setup time and the setup cost is set as 3,000 ($/u) for all products.

There are no setup times and setup costs, if two consecutive products are the same. This situation changes at tank level. If a raw material is replaced by the same one, the setup time is 1 hour and it is 2 hours if they are different. Therefore, there are two different setup costs for the tanks. The values are estimated at 6,000($/u) if they are the same and at 12,000($/u) if different. These cost estimates are intended to provide an appropriate trade-off between costs present in the objective function of the problem.

Table 2 presents computational results obtained by the four MA approaches introduced in Section 4. The comparisons use the estimated cost for production plans elaborated by company's schedulers (Ind) as a benchmark, shown in the second column. Comparisons also include the results found by the GA proposed for the same problem in [2].

Both MAs and GA evolved three populations hierarchically structured in ternary trees. Each population consists of 13 individuals, the crossover and mutation rates are set to 1.5 and 0.7, respectively. This means that 19 numberOfCrossovers are executed, as well as there being 70% chance of performing a mutation.

The TS in MA-TS1 was executed 120 times to find the best individual in each population. It was adjusted to create 40 neighbors (tsMax) for each selected move. Furthermore, the size of the tabu list is updated after parameter count reaches value 40.

TA in MA-TA1 was executed 30 times also to find the best individual in each population with $Th = 0.2$. The reduce(Th) method reduces the Th value by 0.005 when the individual is not accepted. This value is reduced by 0.01 when the individual is accepted without improving the best individual. Finally, Th decreases by 0.02 when the new individual is accepted and it is better than the best individual found so far.

The TS in MA-TS2 was executed ten times to find the best individual in each population with $tsMax = 40$ neighbor evaluations. This reduction was necessary because TS is now executed to find the best individual in each population when populations have converged. This means that the local search is called more times during the evolving process of the GA.

The same happens with TA in MA-TA2, but this local search is computationally less expensive than TS. Therefore, TA is executed 30 times over the best individual in each population, evaluating 40 neighbors per iteration. The initial value of Th and the reductions applied to it are the same as defined for MA-TA1.

The GA and MAs ran 10 times with an execution time of 1hour per execution in each instance. It is worth mentioning that the soft drink company decision maker usually spends 4 hours defining a production plan for each period. Table 2 shows the average values of the best solutions found after 10 executions.

Table 2. Average values of final solutions for instances.

Inst.	Ind	GA	MA-TS1	MA-TA1.	MA-TS2	MA-TA2
A1	1692,1	1666,6	1672,4	1664,8	1705,3	1669,3
A2	3511,9	3405,5	3349,0	3359,6	3536,0	3387,3
A3	5002,7	5199,1	4878,5	4882,3	5100,9	5436,1
B1	3378,2	3271,0	3232,5	3240,7	3353,5	3285,8
B2	4278,5	4231,5	4055,3	4104,4	4256,7	4187,7
B3	7943,4	8056,7	7528,2	7548,5	8664,0	8133,8

The MA-TS1 outperforms the GA in 5 out of 6 instances. In instance A1, MA-TA1 found the best average values for the final solutions. In the most complex instances of each type, A3 and B3, the GA was not able to improve the industry solution. However MA-TS1 and MA-TA1 achieved better results with a relevant improvement in the industry solution for both instances.

MA-TS2 and MA-TA2 were not able to outperform MA-TS1 and MA-TA1 in almost all instances, where the only exception occurred with MAT-TA2 that found better results than MA-TS1 in A1. The GA outperformed MA-TS2 in 5 out of 6 instances and MA-TA2 in 4 out of 6 instances.

MA-TA2 is related to the MA proposed in [3]. The results in Table 2 indicate that this approach did not perform better solving the industrial instances proposed in [2]. Indeed, the solutions found by MA-TS2 reveal that there is no improvement even when the MA executes TS (MA-TS2) instead of TA (MA-TA2).

The proposed MAs, MA-TS1 and MA-TA1, use a random selection of moves. MA-TA2 and MA-TS2 always apply the same solution first to a position (swap and insertion) move, then a lot size (split and merge) move and finally another move randomly selected. The strategy for neighborhood exploration of MA-TS2 and MA-TA2 seems not to be efficient in these real-world instances evaluated.

The local search applied to the best individual found so far returns better results for these instances. The MAs MA-TS1 and MA-TA1 execute the local search procedure only of the best individual found, instead of many times of the best individual found in each population. Table 3 shows the solution deviation from the industrial solution called Ind.

Table 3. Average deviation values for instances.

Inst.	GA	MA-TS1	MA-TA1.	MA-TS2	MA-TA2
A1	-1.50	-1.17	-1.61	0.78	-1.35
A2	-3.03	-4.64	-4.34	0.69	-3.55
A3	3.93	-2.48	-2.41	1.96	8.66
B1	-3.17	-4.31	-4.07	-0.73	-2.73
B2	-1.10	-5.22	-4.07	-0.51	-2.12
B3	1.43	-5.23	-4.97	9.07	2.40

This deviation is defined as $Dev(\%) = 100(Z - Ind)/Ind$, where Z is the average value of the best solutions of GA or MAs. It is possible to see that MA-TS1 and MA-TA1 are able to reduce costs in the original production planning from the industry. This reduction is more relevant in the complex instances of type B, where MA-TS1 and MA-TA1 reached values greater than 4% of reduction.

Figure 15 shows the average deviation found by methaheuristics taking into account all instances. It is possible to see the better performance of MA-TA1 and MA-TS1 in this set of instances.

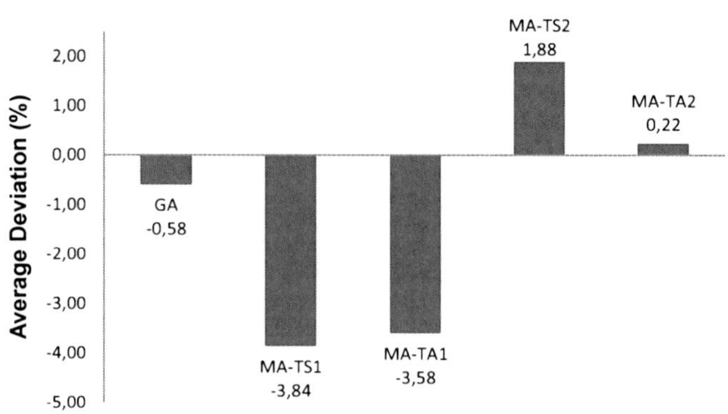

Fig. 15. Average deviation of methods for all instances

Figure 16 illustrates the outcomes returned by statistical comparisons using the two-tailed Mann-Whitney U test [47] as directed acyclic graphs (DAG) as recommended in [48, 49], where a directed edge from a node M_i to a node M_j indicates that M_j has better outcomes than M_i. Each node represents one algorithm configuration applied in the experiments. The significance level of 2% was used as threshold for the significance difference between the methods outcomes. Thus, this test allowed comparing the results returned by the 10 executions of each method in each problem instance.

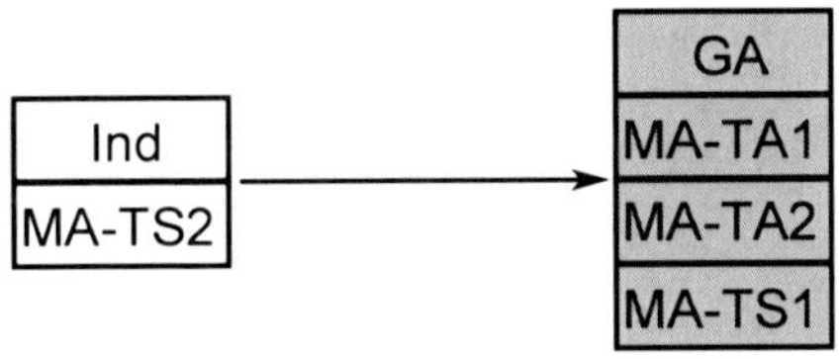

(a) DAG for A1.

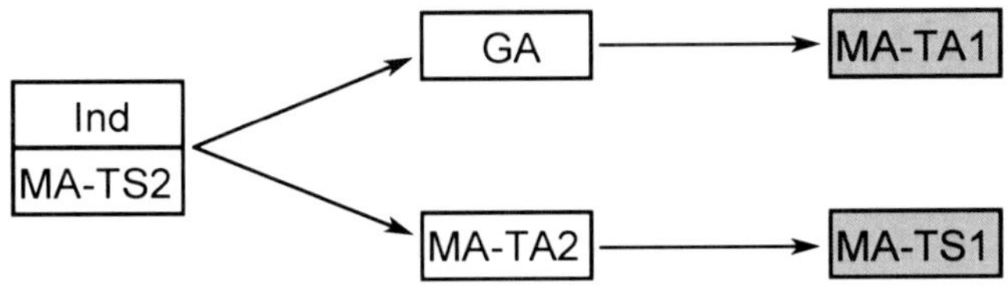

(b) DAG for A2.

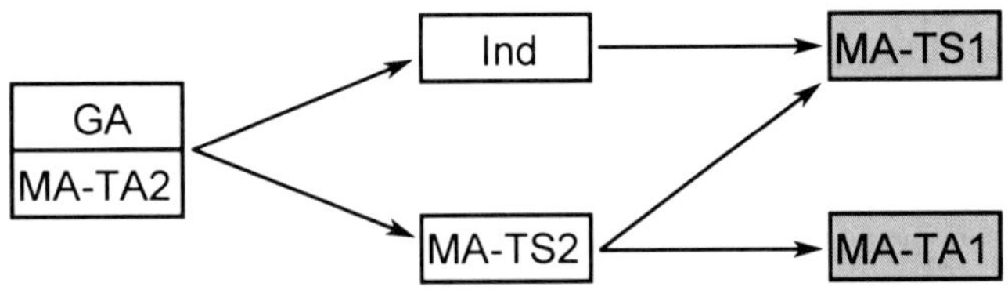

(c) DAG A3.

Fig. 16. Directed acyclic graphs (DAG) for A instances.

In Figure 16a, there is no significant difference among the results found by all methods, except the MA-TS2 and Ind whose results were outperformed. There is a similar result in Figure 16b, but now MA-TA1 is better than GA and MA-TS1 is better than MA-TA2. However, MA-TA1 and MA-TS1 are not significantly different. Figure 16c shows that MA-TS1 and MA-TA1 do not outperform each other, but they return better results than all the other methods.

Figure 17 illustrates the DAG-based presentation of the comparisons of the approaches based on the Mann-Whitney U test for the problem instances of type B. MA-TS1 and MA-TA1 outperform the other methods. There is no significant difference between MA-TS1 and MA-TA1 for instance B1 and B3 (Figures 17a and 17c), but MA-TS1 is better than MA-TA1 for instance B2 (Figure 17b).

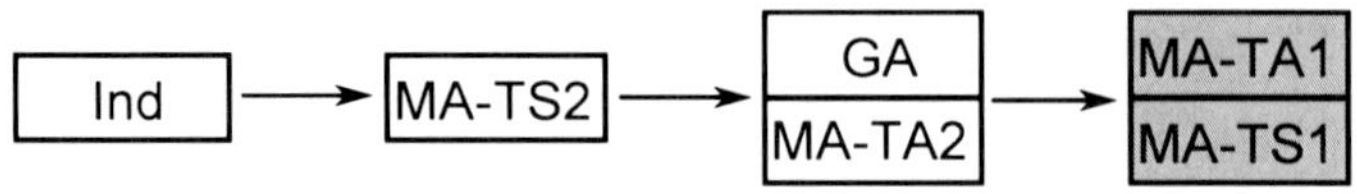

(a) DAG for B1.

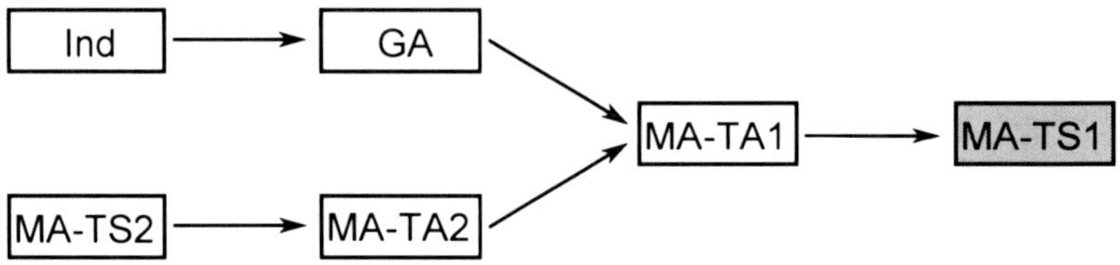

(b) DAG for B2.

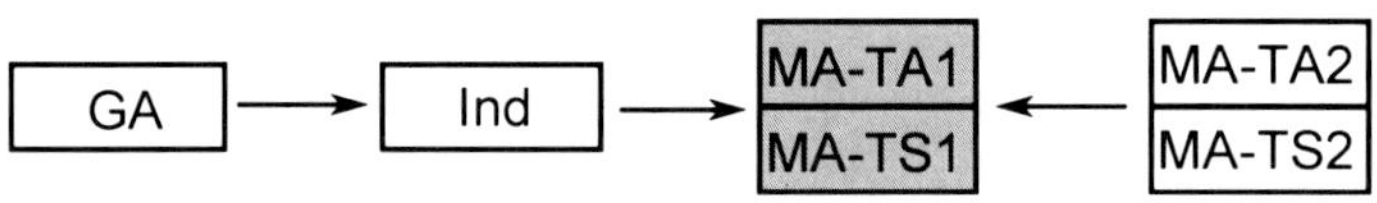

(c) DAG for B3.

Fig. 17. Directed acyclic graphs (DAG) for B instances.

To have an idea about the behavior of the methods in terms of execution time, the performance of the average solution values against the execution time is depicted by Figures 18 to 20. These values are those found by MA-TS1 which is compared with the GA. It must be mentioned that the first value plotted is already the best solution found after the first convergence of populations. This is because the methods have different starting points in some instances. It is not the value of the best individual after the populations have been initialized.

In Figure 18, MA-TS1 performed worse than the GA which returned the average best values since the beginning, for instance A1. The methods return initially similar values for the best individual found after the first population convergence. However, the next results show that the values reach their respective best values after only 500 seconds of execution time.

The GA did not repeat this performance for instance B1. MA-TS1 outperforms the GA during all the execution times for instance B1, where a better individual with a value above 3260 is found after the first convergence. The methods did not show relevant improvements in their average results after 2000 seconds of execution time.

In Figure 19, MA-TS1 was able to find better values since the beginning for both instances. Now the methods spend more time converging, where values near to their best average values arise only after 1500 seconds of execution time.

In Figure 20, the GA starts with very poor values in instance A3, but quickly improves the search without outperforming MA-TS1. Both methods start with high values for the first best individual found in B3, where MA-TS1 is

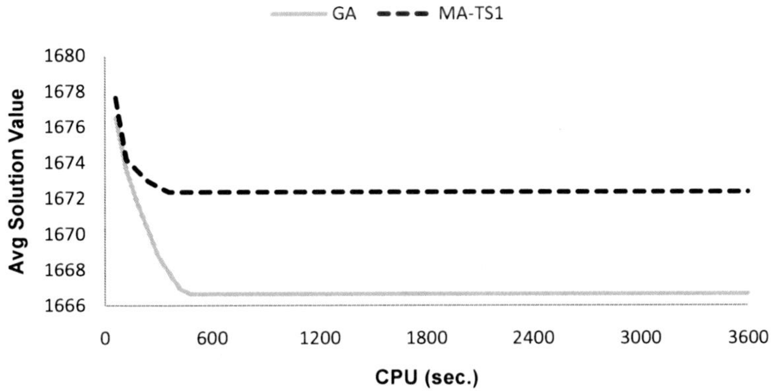

(a) Instance A1.

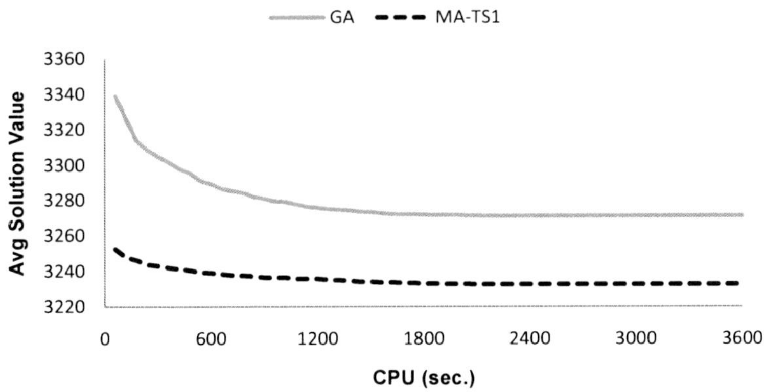

(b) Instance B1.

Fig. 18. Average solution values against the execution time for instances A2 and B2.

initially worse than the GA. They are then able to improve these values with MA-TS1 quickly, outperforming the GA from the beginning of execution.

5.2 Comparisons with the Second Set of Real-World Instances

A set of problem instances is presented in [4], where problem P1 represents the real-world instance taken from the soft drink company. This instance includes two machines (production lines) with one machine responsible for producing

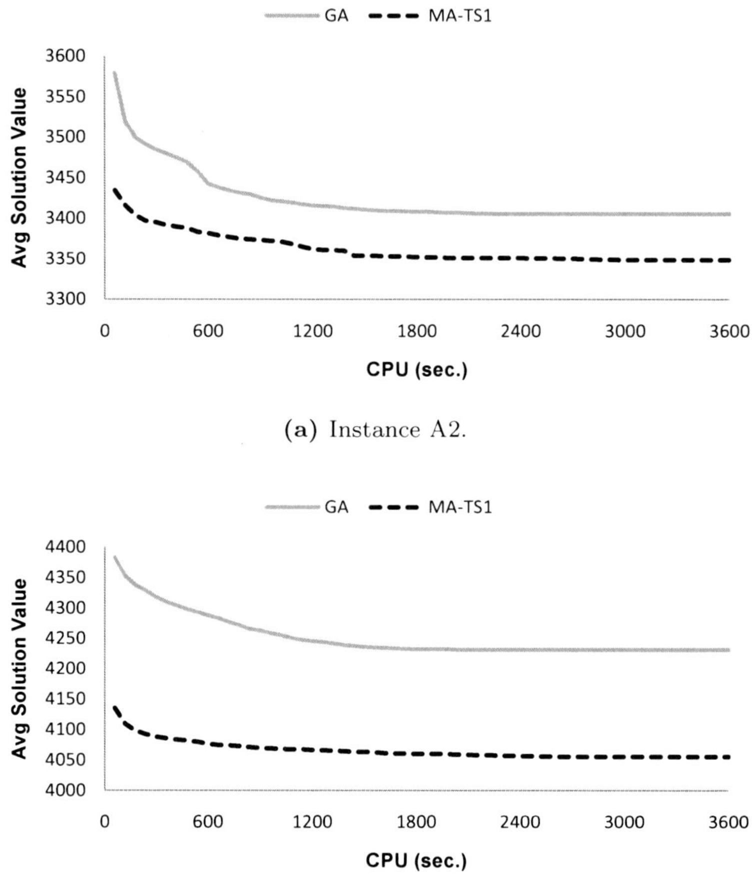

(a) Instance A2.

(b) Instance B2.

Fig. 19. Average solution values against the execution time for instances A2 and B2.

23 soft drinks (products) and the other machine responsible for producing 13 products. The machines need 18 soft drink flavors (raw materials) to produce these soft drinks.

The other problem instances P2, P3, P4, and P5 analyzed here are defined from problem P1. Problem P2 and P3 differ from P1, taking into account inventory and shortage costs, respectively. The inventory costs are duplicated in P2, as well as the shortage costs in P3. Problem P4 is identical to P1, except that the demands are randomly distributed among time periods. Problem P5 has the machine capacity reduced by 25%.

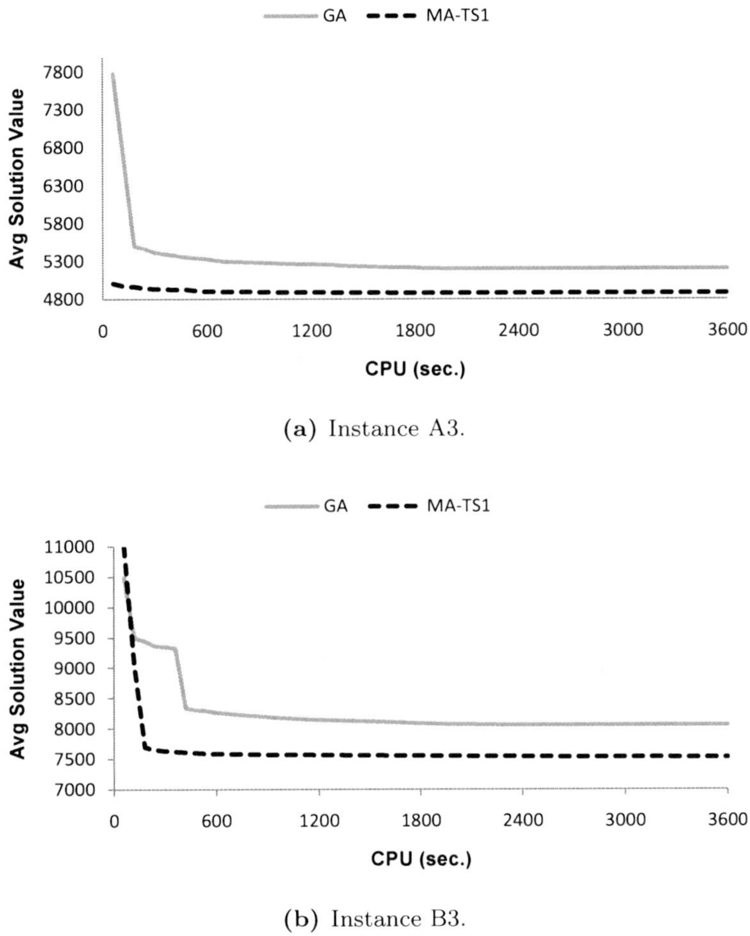

Fig. 20. Average solution values against the execution time for instances A3 and B3.

A time horizon of three weeks (periods) was considered and details of other problem parameters such as inventory costs, setup costs and processing time can be found in [4]. There, the authors proposed relax-and-fix approaches to solve these problems. They reported that best results were returned by the so called Relaxation Approach (RA). This method was executed within a time limit of 4 hours based on the time spent by decision-makers in the soft-drink company to plan the weekly production.

The MAs were again executed 10 times for each problem within 1 hour of execution time, where the average total costs are depicted in Table 5. The production cost, which is by far the largest component of total costs, was not considered in these values as it does not vary by period and all demands should be met. Thus, only inventory, shortage, and setup costs were considered. Methods

MA-TS1, MA-TA1, and MA-TS2 outperformed RA in all problem instances. The MA-TA2 was not better than RA in instances P2 and P5.

Table 4. Solution deviation values for instances.

Inst.	RA	MA-TS1	MA-TA1.	MA-TS2	MA-TA2
P1	306,8	234,2	258,2	233,0	278,0
P2	276,2	268,4	271,0	250,0	276,9
P3	290,8	221,7	260,5	261,5	288,1
P4	317,6	218,6	230,5	216,7	246,6
P5	379,5	272,6	316,9	364,1	423,5

The MA-TS2 outperformed the other methods in 3 out of 5 instances. MA-TS1 found the best results for P3 and P5. The MA previously designed for this set of instances (MA-TA2) seems to have its performance improved, when TS is executed as a local search (MA-TS2). Table 5 lists the deviation of MA from RA, where the best values are found by MA-TS1 for several instances.

Table 5. Average deviation values for instances.

Inst.	MA-TS1	MA-TA1.	MA-TS2	MA-TA2
P1	-23.68	-15.85	-24.06	-9.41
P2	-2.81	-1.88	-9.48	0.27
P3	-23.79	-10.44	-10.08	-0.94
P4	-31.16	-27.42	-31.77	-22.34
P5	-28.17	-16.51	-4.06	11.60
Average	-21.92	-14.42	-15.89	-4.16

The methods MA-TA1, MATS2 and MA-TA2 had more than 10% of improvement compared to the solutions found by RA, except in problem P2. It seems that the duplicated inventory cost in P2 reduced the possibility of improving solutions, as it becomes more expensive to create stocks for products. MA-TS1 and MA-TS2 performed better, taking into account the average deviations for the five problems.

The DAG obtained from the outcomes returned by the two-tailed Mann-Whitney U test in the P instances are depicted by Figure 21. There is no significant difference between MA-TS1 and MA-TS2 for instance P1 and the same happen between MA-TA1 and MA-TS1 for instance P5. MA-TS1 outperforms the other approaches in instance P3. There is no difference between the methods MA-TA1, MA-TS1, MA-TA2 and MA-TS2 for the results found in P2 and P4.

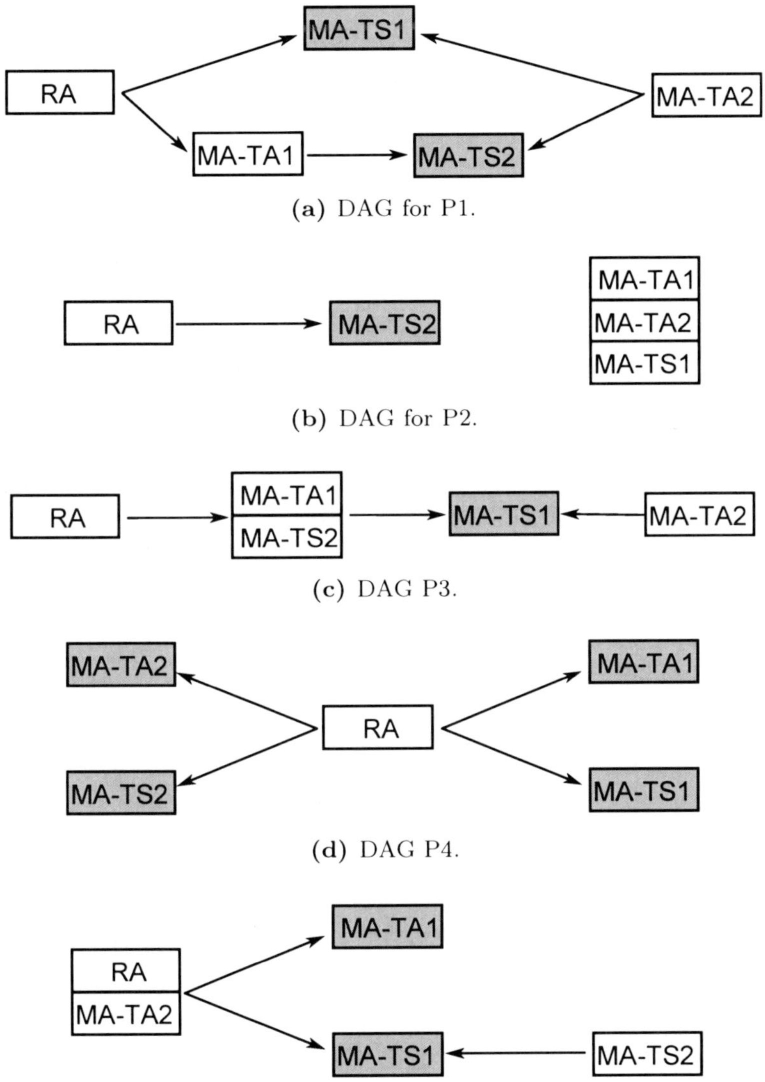

(a) DAG for P1.

(b) DAG for P2.

(c) DAG P3.

(d) DAG P4.

(e) DAG P5.

Fig. 21. Directed acyclic graphs (DAG) for P instances.

6 Conclusions

In this chapter, we evaluated a number of MA approaches to solve a soft drink industry problem – the SITLSP. It is a two-level production planning problem where lot-sizing and scheduling decisions have to be made in an interrelated way. It comprises parallel machines, capacity constraints and sequence-dependent setup costs and times.

The study was conducted by starting from an MA previously applied to the same problem. The objective was to propose variants of this MA able to provide improvements when compared with the previous results. A random selection of moves to explore the neighborhood of a solution representation was introduced, as well as applying the local search only to the best individual found so far. In addition, a TS with a dynamic tabu list was presented as an alternative to the original local search procedure.

The proposed algorithms were compared to solutions found by a GA and with the estimated costs of production plans elaborated by the soft drink company schedulers in the first set of benchmark instances. The MAs, MA-TS1 and MA-TA1, using the new ideas introduced in this chapter, outperformed the GA. The approach that uses the TS found the best results for most of the instances and it had the best deviation values.

In the second set of instances, the MAs were compared to results found by a relax approach (RA). This method had already outperformed the solution developed by decision makers in the soft drink company for one real-world problem instance. MA-TS1, MA-TA1, and MA-TS2 outperformed the RA solutions in all instances. MA-TS2 found the best solution values for most of the instances, but MA-TS1 returned the better average deviations taking into account all the problem instances.

Statistical tests (i.e., the Mann-Whitney U test) confirmed that MA-TS1 was always among the methods with the better results in the second set of instances. In several instances from the first set, there was no relevant difference between the results obtained with MA-TS1 and MA-TA1.

In summary, MA-TS1 shows up as a more robust approach, capable of solving the proposed large-scale real-world SITLSP instances. This approach uses a random selection of moves and applies the local search to the best individuals. Indeed, its fast convergence to minimum values demonstrates that MA-TS1 is quite valuable for commercial use. The results also indicate that our Memetic framework could be executed using other local search techniques or exploitation strategies.

Acknowledgements. This research was supported by Fundação de Amparo a Pesquissa do Estado de São Paulo (FAPESP). The results reported integrate the FAPESP project number 2010/10133-0.

References

1. ABIR. Brazilian Association of Soft Drink Industry (2011), `http://abir.org.br/2010/12/23/projecao0912/` (April 15, 2011)
2. Toledo, C.F.M., França, P.M., Kimms, A., Morabito, R.: A multi-population genetic algorithm approach to solve the synchronized and integrated two-level lot sizing and scheduling problem. International Journal of Production Research 47, 3097–3119 (2009)

3. Ferreira, D., França, P.M., Kimms, A., Morabito, R., Rangel, S., Toledo, C.F.M.: Heuristics and Metaheuristics for lot sizing and scheduling in the soft drinks industry: a comparison study. In: Xhafa, F., Abraham, A. (eds.) Metaheuristics for Scheduling in Industrial and Manufacturing Applications. SCI, ch. 8, vol. 128, pp. 169–210. Springer, Heidelberg (2008)
4. Ferreira, D., Morabito, R., Rangel, S.: Solution approaches for the soft drink integrated production lot sizing and scheduling problem. European Journal of Operational Research 196, 697–706 (2009)
5. Toledo, C.F.M., França, P.M., Morabito, R., Kimms, A.: An optimization model for the synchronized and integrated two-level lot sizing and scheduling problem in soft drink industries. Pesquisa Operacional 27(1), 155–186 (2007)
6. Bitran, G.R., Yanasse, H.H.: Computational complexity of the capacitated lot size problem. Management Science 28(10), 1174–1186 (1982)
7. Chen, W.H., Thizy, J.M.: Analysis of relaxations for the multi-item capacitated lot-sizing problem. Annals of Operations Research 26, 29–72 (1990)
8. Maes, J., McClain, J.O., Van Wassenhove, L.N.: Multilevel capacitated lot sizing complexity and LP-based heuristics. European Journal of Operational Research 53(2), 131–148 (1991)
9. Drex, A., Kimms, A.: Lot-sizing and scheduling - survey and extensions. European Journal of Operational Research 99, 221–235 (1997)
10. Kimms, A.: Multi-level lot sizing and scheduling: methods for capacitated, dynamic, and deterministic models. Physica-Verlag, Heidelberg (1997)
11. Karimi, B., Ghomi Fatemi, S.M.T., Wilson, J.M.: The capacitated lot sizing problem: a review of models and algorithms. Omega 31, 365–378 (2003)
12. Robinson, P., Narayananb, A., Sahinc, F.: Coordinated deterministic dynamic demand lot-sizing problem: A review of models and algorithms. Omega 37, 3–15 (2009)
13. Jans, R., Degraeve, Z.: Modeling industrial lot sizing problems: A review. International Journal of Production Research 46(6), 1619–1643 (2008)
14. Jans, R., Degraeve, Z.: Metaheuristics for dynamic lot sizing: A review and comparison of solution approaches. European Journal of Operational Research 177, 1855–1875 (2007)
15. Fleischmann, B.: The discrete lot sizing and scheduling problem with sequence-dependent setup costs. European Journal of Operational Research 75, 395–404 (1994)
16. Haase, K.: Capacitated lot-sizing with sequence dependent setup costs. OR Spektrum 9, 51–59 (1996)
17. Buschkuhl, L., Sahling, F., Helbeer, S., Tempelmeier, H.: Dynamic capacitated lot sizing problems: a classification and review of solution approaches. Operations Research Spectrum 32, 231–261 (2010)
18. Fleischmann, B., Meyr, H.: The general lot-sizing and scheduling problem. OR Spektrum 19, 11–21 (1997)
19. Meyr, H.: Simultaneous lotsizing and scheduling by combining local search with dual reoptimization. Europena Journal of Operational Research 120, 311–326 (2000)
20. Meyr, H.: Simultaneous lotsizing and scheduling on parallel machines. Europena Journal of Operational Research 139, 277–292 (2002)

21. Haase, K., Kimms, A.: Lot-sizing and scheduling with sequence dependent setup costs and times and efficient rescheduling opportunities. International Journal of Production Economics 66, 159–169 (2000)
22. Kovacs, A., Brown, K.N., Tarim, S.A.: An efficient MIP model for the capacitated lot-sizing and scheduling problem with sequence-dependent setups. International Journal of Production Economics 118, 282–291 (2009)
23. Gupta, D., Magnusson, T.: The capacitated lot-sizing and scheduling problem with sequence-dependent setup costs and setup times. Computers & Operations Research 32, 727–747 (2005)
24. Almada-Lobo, B., Klabjan, D., Carravilla, M.A., Oliveira, J.F.: Single machine multi-product capacitated lotsizing with sequence-dependent setups. International Journal of Production Research 45(20), 4873–4894 (2007)
25. Beraldi, P., Ghianib, G., Griecob, A., Guerriero, E.: Rolling-horizon and relax-and-fix heuristics for the parallel machine lot-sizing and scheduling problem with sequence-dependent set-up costs. Computers & Operations Research 35, 3644–3656 (2008)
26. Luche, J.R.D., Morabito, R., Pureza, V.: Combining process selection and lot sizing models for production scheduling of electrofused grains. Asia-Pacific Journal of Operational Research 26(3), 421–443 (2009)
27. Clark, A.R., Morabito, R., Toso, E.: Production setup-sequencing and lot-sizing at an animal nutrition plant through ATSP subtour elimination and patching. Journal of Scheduling 13(2), 111–121 (2009)
28. Toso, E., Morabito, R., Clark, A.R.: Lot sizing and sequencing optimisation at an animal-feed plant. Computers & Industrial Engineering 57, 813–821 (2009)
29. Sahling, F., Buschkuhl, L., Tempelmeir, H., Helber, S.: Solving a multi-level capacitated lot sizing problem with multi-period setup carry-over via a fix-and-optimize heuristic. Computers & Operations Research 37, 2546–2553 (2009)
30. Han, Y., Tang, J., Kaku, I., Mu, L.: Solving uncapacitated multilevel lot-sizing problems using a particle swarm optimization with flexible inertial weight. Computers and Mathematics with Applications 57, 1748–1755 (2009)
31. Almeder, C.: A hybrid optimization approach for multi-level capacitated lot-sizing problems. European Journal of Operational Research 200, 599–606 (2010)
32. Xiao, Y., Kaku, I., Zhao, Q., Zhang, R.: A variable neighborhood search based approach for uncapacitated multilevel lot-sizing problems. Computers & Industrial Engineering 60(2), 218–227 (2011)
33. Stadler, H.: Multi-level single machine lot-sizing and scheduling with zero lead times. European Journal of Operational Reserach 209, 241–252 (2011)
34. Glover, F., Laguna, M.: Tabu Search. Kluwer Academic Publishers, Norwell (1997)
35. Hung, Y.F., Chen, C.P., Shih, C.C.: Using tabu search with ranking candidate list to solve production planning problems with setups. Computers & Industrial Engineering 45, 615–634 (2003)
36. Al-Fawzan, M.A.: An algorithm for production planning in a flexible production system. Computers & Industrial Engineering 48(4), 681–691 (2003)
37. Buscher, U., Shen, L.: An integrated tabu search algorithm for the lot streaming problem in job shops. European Journal of Operational Research 199(2), 385–399 (2009)

38. Holland, J.H.: Adaptation in natural and artificial systems. MIT Press, Cambridge (1992)
39. Goldberg, D.E.: Genetic Algorithms in search, optimization, and machine learning. Addison-Wesley, Reading (1989)
40. Michalewicz, Z.: Genetic Algorithms + data structure = evolution programs. Springer, Heidelberg (1996)
41. Moscato, P.: On Evolution, Search, Optimization, Genetic Algorithms and Martial Arts: Towards Memetic Algorithms. Technical report, Caltech Concurrent Computation Program, report 826 (1989)
42. França, P.M., Mendes, A.S., Moscato, P.: A memetic algorithm for the total tardiness single machine scheduling problem. European Journal of Operational Research 132, 224–242 (2001)
43. França, P.M., Gupta, J.N.D., Mendes, A.S., Moscato, P., Veltinky, K.J.: Evolutionary algorithms for scheduling a flowshop manufacturing cell with sequence dependent family setups. Computers & Industrial Engineering 48(3), 491–506 (2005)
44. Moghaddam, R., Kanani, Y., Cheraghalizadeh, R.: A genetic algorithm and memetic algorithm to sequencing and scheduling of cellular manufacturing systems. International Journal of Management Science and Engineering Management 3(2), 119–130 (2008)
45. Clark, A.R.: Hybrid heuristics for planning lot setups and sizes. Computers & Industrial Engineering 45, 545–562 (2003)
46. Ferreira, D., Morabito, R., Rangel, S.: Relax and fix heuristics to solve one-stage one-machine lot-scheduling models for small-scale soft drink plants. Computers & Operations Research 37, 684–691 (2010)
47. Mann, H.B., Whitney, D.R.: On a Test of whether One of Two Random Variables is Stochastically Larger than the Other. The Annals of Mathematical Statistics 18(1), 50–60 (1947)
48. Weise, T.: Illustration of statistical test results for experiment evaluation. Technical report, www.it-weise.de (March 2, 2011), `http://www.it-weise.de/documents/files/W2011IOSTRFEE.pdf` (May 03, 2011)
49. Weise, T.: Global Optimization Algorithms – Theory and Application. Germany (2009), `http://www.it-weise.de`, `http://www.it-weise.de` (May 03, 2011)

Simulation-Based Evolutionary Optimization of Complex Multi-Location Inventory Models

Jörg Lässig, Christian A. Hochmuth, and Stefanie Thiem

Abstract. Real-world problems in economics and production often cannot be solved strictly mathematically, and no specific algorithms are known for these problems. A common strategy in such cases is a simulation-based optimization approach, which requires the complete simulation of the system to evaluate its configuration. This principle is introduced using the example of hub-and-spoke inventory systems. By comparing evolutionary approaches with Particle Swarm Optimization and ensemble-based Metropolis-type strategies, we show that evolutionary algorithms with elitism perform very competitively for the discussed optimization task. Hub-and-spoke systems serve as proof of concept for evolutionary simulation-based optimization of inventory models. Subsequently, we introduce the optimization of multi-location inventory models with lateral transshipments. As a consequence of recent economic developments, these very general models are in the spotlight of scientific discussions. In contrast to current approaches, we consider assumptions about the future behavior of the system for deciding how to reallocate and to increase assets. Our experiments show that this approach is suited for a broad class of systems and reveals interesting insights about their optimal structure.

Jörg Lässig
Algorithms Group, International Computer Science Institute,
Berkeley 94704, California, USA
e-mail: jla@icsi.berkeley.edu

Christian A. Hochmuth
Manufacturing Coordination and Technology, Bosch Rexroth AG,
97816 Lohr am Main, Germany
e-mail: christian.hochmuth@boschrexroth.de

Stefanie Thiem
Institute of Physics, Chemnitz University of Technology,
09107 Chemnitz, Germany
e-mail: stefanie.thiem@cs.tu-chemnitz.de

1 Introduction

In recent years, information technology has become a vital element of industrial engineering. Production processes are planned, controlled, and monitored by means of centralized and decentralized systems [1]. Today it is possible to track the complete production cycle from suppliers and subcontractors along the supply chain to the customers [2]. Machine data acquisition systems collect machine and facility data in order to enable fast reactions to failures and to determine the workload, efficiency, performance, and quality for a specific time period in a comparative way. Task management systems extract data from production planning and control systems to support the coordination of production tasks on the production layer [3]. All these achievements aim for an increase of efficiency and flexibility in the production process, for a reduction of cycle times as well as for better capacity utilization while reducing the costs [4].

Focusing on today's complex production and distribution networks, the availability of straightforward application tools for the control of the material flow through the different stations of the system has gained an increasing importance [5]. In the field of storage management and warehousing, shorter delivery times and reduced assets are desired. To achieve progress for both of these seemingly contradicting objectives, different strategies exist. Single-level storage policies apply decision rules whose parameter values have to be optimized. However, currently these rules consider only the current state of the inventory system. Assumptions about the future behavior of the system are not considered while deciding questions of asset reallocation and asset increase. Additional strategies for the control of transshipments between different locations are also the current questions of research. Compared to the increase of the assets in a single location, lateral transshipments between different locations have the advantage that surplus can balance shortages in other locations, which increases the availability, while the assets do not increase on average [6].

This chapter focuses on two major concerns. The first one is the modeling of complex inventory and logistics systems to evaluate different layouts and strategies and the special requirements for optimizers to find close to optimal setups and policies. We specifically emphasize on the question of how eligible evolutionary approaches are able to fulfill this task by first comparing them to other common iterative improvement schemes. Secondly, two different application classes within the inventory and logistics simulation domain are chosen, while the first application serves as a proof of concept.

The rest of this chapter is structured as follows. In Section 2 we first introduce the inventory models under investigation and related results from the literature as well as general approaches to the optimization of these inventory systems. Then, Section 3 describes global optimization methods which are eligible for this optimization task, focusing especially on Genetic Algorithms, Threshold Accepting and Particle Swarm Optimization for reasons of comparison. Readers who are more interested in the problems under investigation can skip this section and continue in Section 4, which considers the specific case of the simulation-based optimization of hub-and-spoke inventory systems, and Section 5, which considers the more general case of multi-location inventory systems with lateral transshipments. We conclude the chapter with Section 6 and give a prospect to future work in Section 7.

2 Related Work

This section consists of two parts. First, we describe how the modeling of inventory systems changed over the last decades and how the investigated problems became increasingly complex. Subsequently, we focus on the development of different strategies to optimize these systems.

2.1 Inventory Systems and Related Optimization Problems

Inventory studies discuss strategies that determine when to order what amount at which location from which source. The objective of these strategies is to optimize a given performance measure for a defined planning horizon. In the 50s of the past century, first theoretical investigations of systems with stochastic demand at one location and a planning horizon of one period were presented. Examples of this research originate from Arrow, Harris, and Marschak [7] as well as Dvoretzky, Kiefer, and Wolfowitz [8]. An expansion to several periods was obtained by Bellman's proposal for dynamic optimization [9].

The step to the consideration of systems consisting of several locations was made by Allen [10–12] and by Clark and Scarf [13]. In these studies, Allen discussed strategies for the redistribution of stock between the locations, and Clark and Scarf introduced so-called multi-echelon models, which can be distinguished by either a vertical or a hierarchical structure. The results of Clark and Scarf laid the foundation for a wide range of subsequent examinations of multi-echelon systems. Such systems consist of sections which are arranged to stages (echelons) representing physical locations or processing steps. In this representation, serial manufacturing systems are described by concatenated stages, where, e.g., assembly systems converge to a defined point and distribution systems diverge from a defined point. As an analytical tool, especially dynamic programming was used. The hub-and-spoke inventory system introduced in the forthcoming sections is an example of a model with vertical structure. Spokes have to meet a certain demand and control their stock by ordering product units from a central hub. In this case simulation-based optimization allows us to resolve restrictions of analytical models such as zero lead times.

In contrast to models with vertical structure, models distinguished by a horizontal structure consist of equal elements, which are not ordered in a predefined way. Consequently, a certain flow direction of product units is not defined in advance for such a model. Even though the first research by Allen [10] about inventory systems with horizontal structure dates back to 1958, only few results are available for these systems. This can be explained by the complexity of the analytical models which arise in the case that future orders have to be considered at the ordering time at a location [14]. These orders usually change the state of the system. Thus, for a long time, only single-period models were investigated, e.g., as described by Krishnan and Rao [15], Aggarwal [16], and Köchel [17]. As an analytical tool, Köchel proposed a periodic-stationary Markov decision model, which allows to derive first results based on a dynamic model [18]. The multi-location inventory system with lateral transshipments discussed in the second part of this chapter is a very general horizontal system. To surmount restrictions of analytical models simulation-based evolutionary optimization is proposed. As a consequence, completely arbitrary systems can be modeled, in contrast to analytical models, which rely heavily on strong assumptions.

Fewer results are known in the literature concerning models with mixed structure. Such models describe systems which define predecessor-successor rules for the flow of product units through a defined set of locations whereas lateral transshipments between the locations are allowed [14, 19–21]. Thus, these models combine aspects of models with vertical and horizontal structure.

Due to the complexity which arises by modeling the system as realistic as possible, an analytical approach is not advisable. Simulation-based evolutionary optimization represents a suitable way to find optimal solutions for systems with vertical, horizontal, or even mixed structure as models can be formulated in almost arbitrary detail.[1] Although this leads to a disadvantageously fast increase in computation time, the extent of that increase can be partly confined by the application of techniques to accelerate the algorithm and, especially, by the parallelization of the simulation task. To obtain an accurate estimate of the objective function, the simulated time period must be sufficient. However, the vast increase in computation speed during the last decades and the broad availability of multi-core processors support the adoption of simulation as an evaluation tool for real-world problems.

[1] Of course this is limited by the available resources for the optimization procedure afterwards.

2.2 Optimization of Complex Systems

In the context of systems, optimization can be part of the design process on the one hand in order to construct systems which behave optimally under certain fixed aspects, and on the other hand to control systems to achieve optimal parameter setups and characteristics. For an objective function g and the decision set Θ comprising all solutions $\vartheta \in \Theta$ an optimization problem can be formally described by

$$\min_{\vartheta \in \Theta}\{g(\vartheta)\} \,. \tag{1}$$

To denote all current solutions of an algorithm with N concurrent individuals as introduced later we use a vector notation $\boldsymbol{\vartheta} = (\vartheta_1, \ldots, \vartheta_N)$. The objective function $g : \Theta \longrightarrow \mathbb{R}$ serves for the performance evaluation of the system. Each given solution $\vartheta \in \Theta$ is mapped to a real value which describes the performance of the system under a certain aspect. We consider the set $\Theta^\star$ as the set of optimal decisions with regard to the objective function g. Formally $\Theta^\star$ is given by

$$\Theta^\star := \{\vartheta^\star \in \Theta \;:\; g(\vartheta^\star) = \min_{\vartheta \in \Theta}\{g(\vartheta)\}\} \,. \tag{2}$$

The objective function value $g(\vartheta^\star)$ of an optimal decision is called optimal objective value. Local optima are a problem in searching the optimal value. A decision ϑ' is called local optimal decision with regard to $g(\vartheta')$, if there is an environment U for each element in Θ such that

$$g(\vartheta') = \min_{\vartheta \in \Theta \cap U}\{g(\vartheta)\} \,. \tag{3}$$

An optimization problem $\mathcal{P}(g, \Theta)$ is solved if an optimal decision $\vartheta^\star$ and the optimal value $g^\star = g(\vartheta^\star)$ have been determined. A pair $(g^\star, \vartheta^\star)$ is called optimal solution of $\mathcal{P}(g, \Theta)$.

2.2.1 Complexity of Stochastic Optimization Problems

In case of stochastic systems it is often not possible to formulate an analytical objective function g. For the determination of the objective value of a decision $\vartheta \in \Theta$ the performance of a stochastic system is determined by an evaluation function $B(\vartheta, X(\vartheta))$. The random variable $X(\vartheta)$ describes the stochastic influences on the system. The evaluation function B is itself a random variable and, hence, difficult to compare. Therefore, the performance measure of a stochastic system is usually the expectation value $\langle B(\vartheta, X(\vartheta))\rangle$ of an evaluation function. This means that we have to consider the expected performance of the decision $\vartheta \in \Theta$ if a stationary state is reached, i.e., the performance of the system does not depend on the initial state. This is the case if the cost functions are approximately constant over time. Consequently, a stochastic optimization problem can be formally described by

$$\min_{\vartheta \in \Theta}\{g(\vartheta)\} = \min_{\vartheta \in \Theta}\{\langle B(\vartheta, X(\vartheta))\rangle\} . \tag{4}$$

Usually the solution of a stochastic optimization problem is more complex than the solution of a deterministic problem and, hence, in general specific solution methods are required.

2.2.2 Simulation-Based Optimization

In practice the objective function g is replaced by an approximate objective function g_{app}, which is determined by averaging the evaluation function $B(\vartheta, X_i(\vartheta))$ over several runs i, i.e.,

$$g_{\mathrm{app}} = \frac{1}{M}\sum_{i=1}^{M} B(\vartheta, X_i(\vartheta)) . \tag{5}$$

Simulation-based optimization suffers mainly from two problems. First, it is obviously not possible to calculate any gradient of the objective function, and secondly, there is a difference between the objective function g and the approximate objective function g_{app}, and any method can only optimize g_{app}. There are two basic approaches for simulation-based optimization – the non-iterative approach and the iterative approach, see Fu [22], Fu and Healy [23], and Pflug [24].

Non-iterative simulation-based optimization. For this approach two consecutive problems have to be solved, the simulation problem to estimate the objective values and the optimization problem to search for a minimum. In a first step, realizations of the relevant random variables are determined, which are either used as real values or which are used to calculate an estimation of the objective function. The resulting deterministic approximated objective function is then used as objective of the actual optimization problem, which has to be solved in a second step. This approach is also called the *retrospective approach* and was proposed by Healy and Schruben [25]. For increasing problem complexity the realization of this approach gets more difficult [23, 26] because for complex problems the necessary sample size becomes too large to approximate the function precisely enough. Additionally, it is hard to predict what sample size is necessary to get sufficiently good results. Often this motivates an additional analytical investigation of the objective function to determine its structure. There are also other non-iterative approaches, e.g., the SPO-Approach or the response-surface method described by Greenwood, Rees, and Siochi [27].

Iterative simulation-based optimization. In contrast to the non-iterative *retrospective* approach, iterative *prospective* methods can be applied, which are characterized by a cycle where simulation and optimization alternate. Each suggested parameter setup of the optimizer is evaluated by the simulator and the simulation result is then used by the optimizer to proceed with the next iteration. This strategy is repeated until an acceptable solution to the problem has been found. The procedure is visualized in Figure 1.

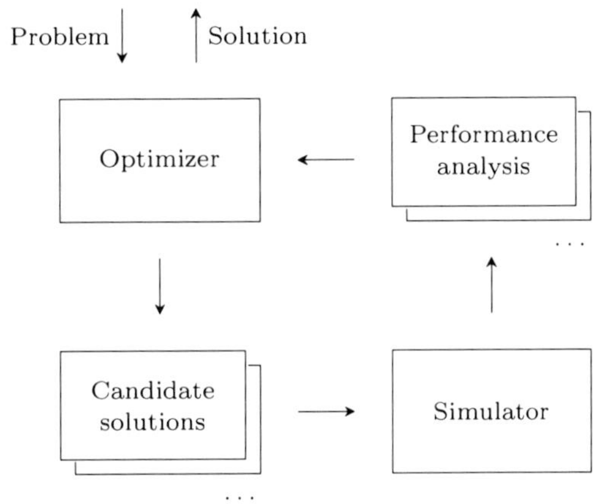

Fig. 1. Scheme for iterative simulation-based optimization.

Initially, some set of candidate solutions is chosen by a starter, usually uniformly at random. The complete system is simulated once for each candidate solution in this set with the corresponding parameter setup.[2] Then the solutions can be evaluated based on the results of the simulation to get the value $g_{\text{app}}(\vartheta)$ as approximation for the objective function value $g(\vartheta)$. Then either the algorithm terminates or a new search point is calculated, i.e., the optimizer suggests new candidate solutions based on the current solutions, which are then treated equivalently in a new iteration. The iterative approach is more generally applicable compared to the non-iterative approach. Resources are used more efficiently and information is only calculated if it is really required. Our investigations focus on the iterative approach to simulation-based optimization.

3 Applied Optimization Methods

The approach of iterative simulation-based optimization works only together with methods which use only search point and value pairs for the optimization. This covers relatively simple iterative improvement approaches such as

[2] It is possible to add further logic here, e.g., to simulate each system more than once to reduce stochastic influences.

Simulated Annealing or *Threshold Accepting* [28, 29] but also more elaborate techniques which deal with more than one candidate solution, e.g., classical *Evolutionary Strategies* [30], *Genetic Algorithms* [31] or newer paradigms like *Particle Swarm Optimization* [32]. Also ensemble-based techniques [33] can be competitive, as we will see [34, 35].

Besides the obvious quest for competitive and thorough optimization results, for practical applications also acceptable running times are important. Especially for the considered case of simulation-based optimization, the evaluation of a single solution is extremely expensive and, hence, fast converging methods with a relatively small number of required function evaluations are suited best for this case. Another pitfall, which does not exist for classical problems, is the fact that a simulation with the same parameter set can return different objective function values due to the stochastic nature of the problem as argued above, which results in the same behavior as adding noise when searching the parameter space. An additional complication is that the objective function takes on very similar values for most candidate solutions near the global optimum. Not all methods applicable in principle can manage these additional requirements equally well.

The chapter comprises results for the application of Ensemble-based Threshold Accepting, Particle Swarm Optimization, and Genetic Algorithms with elitism. The performance of the different methods is compared for different setups and suggests that the evolutionary approaches perform competitively, showing some advantages when compared to the other approaches. Independently of the algorithmic insights, also quite surprising results for the different logistics strategies themselves are described.

3.1 Ensemble-Based Threshold Accepting

Threshold Accepting (TA) is a simple iterative improvement scheme which has been introduced by Dueck and Scheuer in 1990 [36]. It fits into the scheme of Monte Carlo-type techniques, which is described by Algorithm 1. The degrees of freedom within this template are the *initial state*, the *temperature schedule*, the *move class*, the *acceptance condition* and the *stopping condition*. The initial state can be chosen uniformly at random, while the stopping condition is in our case given by a fixed number of unsuccessful iterations. For different designs of the acceptance probability $P_{\vartheta'\vartheta}$ depending on the current solution ϑ and a new solution ϑ' and of course on the objective function g, the algorithm shows a different behavior. The simplest approach is *random search* with $P^{(\mathrm{RA})}_{\vartheta'\vartheta} = 1$.

The parameter $\Delta g = g(\vartheta') - g(\vartheta)$ is the difference between the objective function values of the new and the current solution. For TA an elementary step function is applied:

Algorithm 1. Metropolis-type techniques

Data: problem $\mathcal{P}(g, \Theta)$ with a solution space Θ.
Result: solution $\vartheta_{\text{final}} \in \Theta$.

```
begin
    ϑ ⟵ generate_initial_state(Θ);
    for t ⟵ 1 to ∞ do
        T ⟵get_new_temperature(ϑ, t, T);
        ϑ' ⟵get_new_state(ϑ, t, Θ);
        acceptance_condition ⟵ true with probability P_ϑ'ϑ,
        false otherwise;
        if (acceptance_condition) then
            ϑ ← ϑ';
        if (stopping_condition) then
            return ϑ;
end
```

$$P_{\vartheta'\vartheta}^{(\mathrm{TA})} = \begin{cases} 1 & \text{if } \Delta g \leq T \\ 0 & \text{otherwise} \end{cases}, \tag{6}$$

where a temperature parameter T is introduced, which in general gets reduced during the algorithm execution. Franz et al. were able to prove that Equation (6) is an optimal decision rule for this kind of algorithm [37].[3]

An obvious generalization of these methods is to increase the number of solutions for which exactly the same process is done in parallel. This is called ensemble-based approach. One advantage of this procedure is that more information about the current situation is known which can be used, e.g., to establish adaptive temperature schedules. Considering Metropolis-type algorithms, several strategies for the temperature schedule have been investigated in the literature. Especially, adaptive schedules show excellent performance such as, e.g., *constant thermodynamic speed scheduling* by Salamon, Sibani, and Frost [38], which is applied for the comparison in this chapter.

Using the mean ensemble objective value[4] $\langle g \rangle_t$ at time t and its standard deviation in the equilibrium at temperature T, defined by $\sigma_g(T) = \sigma_g = \sqrt{\langle g^2 \rangle_t - \langle g \rangle_t^2}$, where this equation is evaluated at the end of a fixed temperature run, the rule is to reduce the temperature if

$$\langle g \rangle_{t-1} - \langle g \rangle_t > \frac{c_1 \cdot \sigma_g}{\sqrt{N}} \tag{7}$$

is satisfied, where c is a constant and N the ensemble size. The temperature is then reduced using

[3] Note that this depends on the choice of other parameters, especially on the choice of the temperature schedule, for which the optimal choice is in practice unknown.

[4] Using E for energy instead of g is more common in this context, but here we stick to g to denote the objective function.

$$T_{t+1} = T_t - \frac{c_2 \cdot T_t^2}{\sigma_g} \ . \tag{8}$$

In the application case of the optimization approach to hub-and-spoke inventory systems, the current solution is represented by a vector of real values. The move class creates a new solution by altering each element of the vector adding a perturbation to the current value which is chosen in the interval $[-c_{\text{step}}, c_{\text{step}}]$ uniformly at random. This is equivalent to choosing a new position in a small hypercube around the current solution.

3.2 Particle Swarm Optimization

Another approach which uses also more than one candidate solution at the same time is Particle Swarm Optimization (PSO), originally proposed by Kennedy and Eberhart [32]. PSO uses the dynamics of swarms to find solutions to optimization problems with continuous solution space. Meanwhile, many different versions and additional heuristics were introduced, where we restrict our considerations here to the Standard PSO 2007 algorithm [39].

PSO is, similar to Genetic Algorithms, an iterative population-based approach, i.e., PSO works with a set of feasible solutions, the swarm. Let N denote the number of swarm individuals (particles) or the swarm size, respectively. The basic idea of PSO is that all swarm individuals move partly randomly through the solution space Θ. Thereby individuals can share information about their so far best previous position $\vartheta^{\text{bsf}} \equiv \mathbf{r}^{\text{bsf}}$, where each particle has a number of K informants. Additionally, each individual i has an internal memory to store its best so far (locally best) solution $\vartheta_i^{\text{lbsf}} \equiv \mathbf{r}_i^{\text{lbsf}}$. In every iteration the movement of each individual beginning from its actual position $\vartheta_i \equiv \mathbf{r}_i$ is then given by a tradeoff between its current velocity $\mathbf{v}_i$, a movement in the direction of its locally best solution $\mathbf{r}_i^{\text{lbsf}}$ (cognitive component) and of its so far best known solution $\mathbf{r}_i^{\text{bsf}}$ of its informants in the swarm (social component) (cp. Figure 2). Thus, the equations of motion for one individual i and a discrete time parameter t are given by

$$\mathbf{v}_i^{t+1} = w \cdot \mathbf{v}_i^t + c_1 \cdot \boldsymbol{\mathcal{R}}_1 \cdot \left(\mathbf{r}_i^{\text{lbsf},t} - \mathbf{r}_i^t\right) + c_2 \cdot \boldsymbol{\mathcal{R}}_2 \cdot \left(\mathbf{r}_i^{\text{bsf},t} - \mathbf{r}_i^t\right) \tag{9}$$

$$\mathbf{r}_i^{t+1} = \mathbf{r}_i^t + \mathbf{v}_i^{t+1} \ . \tag{10}$$

The diagonal matrices $\boldsymbol{\mathcal{R}}_1$ and $\boldsymbol{\mathcal{R}}_2$ contain uniform random numbers in $[0, 1)$ and thus randomly weight each component of the connecting vector $(\mathbf{r}_i^{\text{lbsf}} - \mathbf{r}_i)$ from the current position $\mathbf{r}_i$ to the locally best solutions $\mathbf{r}_i^{\text{lbsf}}$. The vector $(\mathbf{r}_i^{\text{bsf}} - \mathbf{r}_i)$ is treated analogously. Since every component is multiplied with a different random number, the vectors are not only changed in length but also perturbed from their original direction. The new position follows then from the superposition of the three vectors as it is visualized in Figure 2. By choosing the cognitive parameter c_1 and the social parameter c_2, the influence of these two components can be adjusted.

The Standard PSO 2007 setup uses $c_1 = c_2 = 0.5 + \ln(2) \simeq 1.193$ and $w = 1/(2\ln 2) \simeq 0.721$ as proposed by the stability analysis by Clerc and Kennedy [40]. Depending on the problem dimension d, a number of $N = \lfloor 10 + 2\sqrt{d} \rfloor$ particles are chosen with a number of $K = 3$ informants.

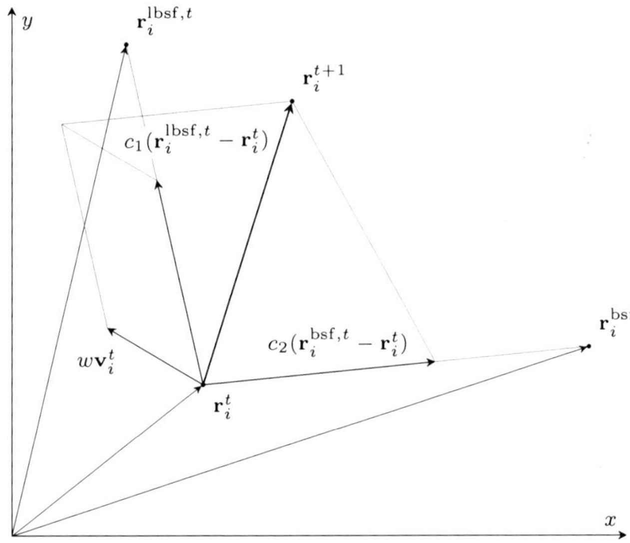

Fig. 2. Iterative solution update in PSO. From the current particle position $\mathbf{r}_i^t$ the new position $\mathbf{r}_i^{t+1}$ is obtained by vector addition of the velocity, the cognitive and the social component. Here, the random influence by the matrices $\mathcal{R}$ is omitted, which usually further scales the length and perturbs the direction of the vectors.

The pseudocode is shown in Algorithm 2. The position and the velocity components for the different particles i and dimension d are written as subscripts, i.e., v_{id} is the dth component of the velocity vector $\mathbf{v}_i$ of particle i.

3.3 Genetic Algorithm

Finally, we introduce Genetic Algorithms (GAs) as an optimization method which can be used to optimize complex inventory systems. Because GAs are not only applied to the optimization of hub-and-spoke inventory systems

Algorithm 2. Position and velocity update rule in PSO

Data: position $\vartheta_i \equiv \mathbf{r}_i$, velocity $\delta_i \equiv \mathbf{v}_i$, locally best position $\vartheta_i^{\mathrm{lbsf}} \equiv \mathbf{r}_i^{\mathrm{lbsf}}$ and globally best position $\vartheta_i^{\mathrm{bsf}} \equiv \mathbf{r}_i^{\mathrm{bsf}}$ for each particle i, cognitive parameter c_1 and social parameter c_2.

Result: $\boldsymbol{\vartheta}$, with $\vartheta_i \in \Theta$, the solution space, and updated velocities $\boldsymbol{\delta}$.

begin
    forall *particles i* **do**
        forall *dimensions d* **do**
            $r_1 \leftarrow get\_uniform\_random\_number(0,1)$;
            $r_2 \leftarrow get\_uniform\_random\_number(0,1)$;
            $v_{id} \leftarrow w \cdot v_{id} + c_1 \cdot r_1 \cdot \left(x_{id}^{\mathrm{lbsf}} - x_{id}\right) + c_2 \cdot r_2 \cdot \left(x_{id}^{\mathrm{bsf}} - x_{id}\right)$;
            $x_{id} \leftarrow x_{id} + v_{id}$;
    return $\boldsymbol{\vartheta} \equiv \{\mathbf{r}_i\}, \boldsymbol{\delta} \equiv \{\mathbf{v}_i\}$;
end

but also for the more complex case of multi-location inventory systems with lateral transshipments, we introduce them in more detail here.

GAs are heuristic optimization methods which are predestined for problems whose solution is not closed or cannot be efficiently evaluated. Further, GAs can be applied without specific knowledge about the application domain, and they are thus appropriate for very general optimization problems [14]. Since less information about the objective function is necessary, they are easily transferable to new optimization tasks. With respect to the initial solutions of the optimization process, GAs show a robust behavior. In the context of simulation-based optimization they have the positive property of processing the seemingly random objective values of simulation experiments well. However, in general it is not guaranteed that the best candidate solution found represents in fact the global optimum. The consideration of a high number of candidate solutions, randomly distributed initial states and a sufficiently large number of optimization cycles increase the quality of the optimization result. Also, the implementation of appropriate genetic operators, which are described in the following paragraphs, is not trivial. Their parameter values strongly depend on the problem under consideration and should therefore be adapted to obtain reasonably good results. Furthermore, the use of GAs is difficult if the evaluation of candidate solutions leads to a categorization of right and wrong only. Because the cost criteria are real-valued objective functions and due to the properties described, GAs are suitable to find adequate solutions to the presented systems.

For our application, a candidate solution, i.e., an individual, is represented by a vector of parameters to be optimized, the alleles of the genome. In the first step, the *initialization*, a sufficiently large set of individuals with randomly distributed values is generated, constituting the initial population. Subsequently, the *evaluation* assigns a numerical value $g(\vartheta_i)$ to each individual i. On this basis the fitness function may be analytically calculated or, in our case, estimated by a simulation experiment. In the following, a fraction of the population is extracted by *selection*, whereas the selection probability for each individual corresponds to its fitness value. Next, in the *recombination* step, new individuals are created by merging randomly advantageous partial candidate solutions of selected individuals to new candidate solutions. In the last step, the *mutation* step, the values of individuals selected according to a defined probability are altered slightly. Thus, the population of the next generation is represented by individuals selected and newly created and, depending on the implementation, by the individuals of the previous generation. The cycle is completed by re-evaluating the population. If a defined *stopping condition* is satisfied after the evaluation, the current optimum and, depending on the design, the best solutions of the last generation are returned, otherwise the cycle is continued. The procedure is formally given in Algorithm 3.

3.3.1 Initialization

The initial population is represented by individuals whose values have been chosen uniformly at random. To ensure a wide coverage of the search space and to prevent early convergence to a local optimum, the interval limits of the uniform distribution should be chosen sufficiently large. Thus, less advantageous candidate solutions are created intentionally. The interval limits correspond to the reference values of the inventory model and will be specified in Sections 4 and 5.

3.3.2 Evaluation

Each new individual has to be evaluated. Therefore, discrete simulation is used. Thus, for large population sizes N it is crucial that the simulator is implemented as efficient as possible and, if possible, parallelized to take advantage of current multi-core architectures. For the evaluation of a given candidate solution during the simulation run, values of stochastic variables are generated by a mixed-linear congruential random number generator. Its seed is initialized in the first generation, and the sequence of random numbers is inductively passed on to the next generations. This can be interpreted as fluctuating environmental conditions and, hence, prevents overfitting, i.e., the emergence of candidate solutions which perform well only under specific circumstances – represented here by sequences of random numbers – but not necessarily on average. However, in a fixed generation, all simulation runs are conducted on the basis of the same seed so that for individuals within one generation comparability is guaranteed using the same sequence of random numbers. The final optima of all generations are eventually re-evaluated, applying the initial seed to guarantee comparability. The evaluation function $B(\vartheta, X(\vartheta))$ assigns a specific value to each candidate solution $\vartheta \in \Theta$, whereas

$X(\vartheta)$ denotes the observed stochastic influences on the system as described above, i.e., the specific sequence of random numbers generated during the simulation run. In our case, the value $B(\vartheta, X(\vartheta))$ that is assigned to each evaluated candidate solution ϑ corresponds to the sum of the total costs $c_{\text{total},j}(X(\vartheta))$ over all locations j. Thus, we obtain

$$B(\vartheta, X(\vartheta)) = \sum_{j=1}^{n} c_{\text{total},j}(X(\vartheta)). \tag{11}$$

3.3.3 Selection

After evaluating the individuals of a population, their fitness values are estimated based on the simulation results. These fitness values $F(\vartheta)$ describe the performance of each candidate solution ϑ comparatively. As the simulation returns cost function values over a defined planning horizon, a minimization problem is defined. However, a candidate solution is preferable by definition if it displays a high fitness value, and in order to achieve this, the sign of the objective values is alternated. Furthermore, negative fitness values have to be excluded, and in case that all values are positive, comparability of large positive values has to be ensured. Thus, all values are rescaled so that a fitness value of zero is assigned to the least performing candidate solution. Consequently, the fitness value $F(\vartheta)$ of all candidate solutions $\vartheta \in \Theta$ is given by the sum of its negated objective value $-B(\vartheta, X(\vartheta))$ and an upper bound[5] on $B(\vartheta', X(\vartheta'))$ such as $\max\{B(\vartheta', X(\vartheta')) : \vartheta' \in \Theta\}$, and the fitness for all candidate solutions $\vartheta' \in \Theta$ is given by

$$F(\vartheta) = -B(\vartheta, X(\vartheta)) + \max\{B(\vartheta', X(\vartheta')) : \vartheta' \in \Theta\} \tag{12}$$

with

$$F(\vartheta) \in [0, -\min\{B(\vartheta', X(\vartheta')) : \vartheta' \in \Theta\} + \max\{B(\vartheta', X(\vartheta')) : \vartheta' \in \Theta\}] . \tag{13}$$

The fitness values measure the performance of the individuals relative to the current population rather than representing their absolute performance. Thus, slight deviations in the cost function values become more important as the interval limits of the population converge, which is not the case for an analysis in absolute terms. Although the sequence of the candidate solutions according to their fitness values is not affected by rescaling, selection probabilities are fitness-proportional, and therefore, the selection pressure is higher if fitness values are rescaled to the interval limits of the population. Assigning fitness values without rescaling would probably lead to the optimum, too, but slower due to lower selection pressure.

[5] This upper bound corresponds to the objective values of the worst performing individual.

Selection now describes the process to choose a set of individuals from the current generation in order to generate new offspring. In our implementation the next generation of size M consists of three groups of individuals:

- The best individual of the current population is always selected and unconditionally passed to the next generation without any further changes by recombination or mutation (see below). This approach is called *elitism* [41], constituting the first group of individuals in the new generation.
- Further $m-1$ individuals are chosen according to their (rescaled) fitness values to constitute a mating pool $\mathcal{M}$. The individuals of the mating pool are used twice. First, they are mutated (see below) and then unconditionally added to the new generation. This constitutes the second group of individuals in the new generation.
- In order to generate the third group of individuals which are added to the new generation, the mating pool $\mathcal{M}$ is utilized again, i.e., $M-m$ individuals are generated by choosing individuals from $\mathcal{M}$, uniformly at random, and subsequently recombining and mutating them (see below).

Thereby, the parameter m is prescribed by the recombination probability P_{C}, which is determined by Equation (14) for a fixed population size M. The recombination probability describes the probability that an arbitrary but fixed individual of the mating pool $\mathcal{M}$ is chosen at least once for recombination to generate $M-m$ new individuals. Equation (14) can be explained as follows: The probability that an arbitrary but fixed individual ϑ' of $\mathcal{M}$ is *not* chosen for recombination in any of the $M-m$ selection steps is $(\frac{m-2}{m-1})^{M-m}$ because one of $m-2$ individuals has to be chosen $M-m$ times from $m-1=|\mathcal{M}|$ available individuals. Hence, the probability that ϑ' is chosen at least once is given by the complementary event, and the recombination probability P_{C} for an individual is given by

$$P_{\mathrm{C}} = 1 - \left(\frac{m-2}{m-1}\right)^{M-m} . \tag{14}$$

Note that this equation is utilized to determine the parameter m iteratively for given values of P_{C} and M. Here $P_{\mathrm{C}} = 0.6$ has been applied.

For the random, fitness-proportional selection of $m-1$ individuals out of all M individuals we use *stochastic universal sampling* [42]. It offers the advantage of using a single random value uniformly distributed in $[0,1)$ to select a fixed number of individuals out of a population, while it is optimal in terms of the absolute difference between the normalized fitness value of an individual and its selection probability (*bias*) as well as regarding the range of the possible number of offspring (*spread*). Conceptually, this procedure corresponds to mapping the fitness onto a disc, whereas a sector is assigned to each individual. The angle fraction of such a sector is equivalent to the fitness value of an individual related to the overall fitness of the population. Around the disc $m-1$ pointers are arranged equally spaced, each pointing to an individual to be selected. The disc is then figuratively put in motion by the random number, whereby the selected individuals are indicated. Consequently, an individual may be selected several times, given that its fitness value is high and thus the angle fraction of its sector is wide. Note that also the elitist individual may be among the further selected $m - 1$ individuals. In fact this individual is probably represented several times, due to its fitness value.

3.3.4 Recombination

As described, the recombination operation produces in each generation $M-m$ new offspring from $m - 1$ individuals previously selected, so that the next population contains again a number of M individuals including the elitist individual. The objective is in general to combine good parts of different individuals in order to create a new potentially better one. The resulting steps in the search space are therefore large, especially in comparison to the mutation operation (see below). For each recombination step, two parents are chosen uniformly at random from $\mathcal{M}$ to generate two new individuals. However, since elitism ensures that the best candidate solution is passed on besides the new individuals, the recombination probability for this individual should not be disproportionately high and, hence, the elitist individual is not considered as a potential parent. After copying the chosen individuals, for each value of the first child it is checked whether to swap that value with the corresponding one of the other child according to the recombination probability per value $P_{\mathrm{C},x}$, a parameter set to $P_{\mathrm{C},x} = 0.5$ in our case by observing the convergence of the algorithm. This *crossover* operation is similar to uniform crossover [43]. However, not single bits are swapped, but complete values, as bitwise operations on floating point values lead to chaotic results.

3.3.5 Mutation

The mutation operation randomly alters slightly the values of individuals to perform a kind of local search on the objective function. Thus, it complements the recombination operation, which works on a larger scale. As described in the selection step, $M - 1$ individuals have to be mutated to generate a new population. In accordance with elitist selection the best candidate solution of the previous generation is not altered to let a good solution persist. Also, this individual is probably present in the remaining $M-1$ individuals several times corresponding to its high fitness value. Therefore, it is also possible to evolve this candidate solution further by a slight change. An individual is mutated according to the mutation probability P_{M}. With respect to the convergence behavior we here set $P_{\mathrm{M}} = 0.2$. However, mutation is constrained to a single, randomly chosen location (in the simulated model) of this candidate solution, which has been proven empirically to be advantageous for our application. Each value in the representation of this location, e.g., reorder level s_j and order-up-to level S_j, is then changed according to the mutation probability per value $P_{\mathrm{M},x}$, fixed in this context to $P_{\mathrm{M},x} = 0.2$. The extent of this change depends on the relative mutation variance $\sigma^2_{\mathrm{M}}(X/x_{\mathrm{ref}})$, in our case empirically set to $\sigma^2_{\mathrm{M}}(X/x_{\mathrm{ref}}) = 0.0004$ for the reference value x_{ref} of the parameter. Similarly to the initialization, this reference value x_{ref} equals the capacity $y^+_{\mathrm{max},j}$ of each location j. Thus, the local search is performed on a smaller scale for smaller reference values. The actual change X/x_{ref} is a uniformly distributed random variable with limits determined by the variance $\sigma^2_{\mathrm{M}}(X/x_{\mathrm{ref}})$ and the expectation value $\langle X/x_{\mathrm{ref}}\rangle_{\mathrm{M}} = 0$.

Algorithm 3. Genetic Algorithm

```
Data: Problem 𝒫(g,Θ) with a solution space Θ.
Result: Solutions ϑ_final ∈ Θ.
begin
    ϑ ← generate_initial_states(Θ);
    m ← calculate_m(M, P_C)                    /* according to
        Equation (14)                                          */
    for t ← 1 to ∞ do
        ϑ_elitist ← get_best_individual(ϑ, 𝒫);
        G ← ∅;
        for i ← 1 to m − 1 do
            ϑ ← get_individual_by_stochastic_universal_sampling(ϑ, 𝒫);
            G ← G ∪ {ϑ};
        H ← ∅;
        for i ← 1 to ⌈(M − m)/2⌉ do
            ϑ1 ← choose_individual_for_recombination(G);
            ϑ2 ← choose_individual_for_recombination(G);
            ϑ'1, ϑ'2 ← recombine_individuals(ϑ1, ϑ2, P_C,x);
            H ← H ∪ {ϑ'1};
            if (i ≤ ⌊(M − m)/2⌋) then
                H ← H ∪ {ϑ'2};
        I ← ∅ ∪ {ϑ_elitist};
        forall ϑ ∈ G ∪ H do
            if (get_uniform_random_number(0, 1) ≤ P_M) then
                ϑ ← mutate_individual(ϑ, P_M,x);
            I ← I ∪ {ϑ};
        ϑ ← assign_new_population(I);
        if (stopping_condition) then
            return ϑ;
end
```

3.3.6 Stopping Condition

The mutation operation finally returns the individuals of the next generation. According to the cycle of evolutionary optimization these individuals are again evaluated and eventually selected, recombined, and mutated, unless a certain stopping condition is satisfied after the evaluation. A minimum cycle count is suitable to increase the possibility to find new optima. Further, a minimum cycle count after the last cycle which returned a new optimum has been found to adapt the overall cycle count to the convergence speed of a specific optimization task. For the optimization of multi-location inventory systems with lateral transshipments we propose a combination of both, i.e., at least 3000 cycles under all circumstances, but additionally at least 1000 cycles after the last optimum has been found.

4 Simulation-Based Optimization of Hub-and-Spoke Inventory Systems

The first simulation-based optimization task to investigate is the optimization of the transport resources and order policies in a hub-and-spoke inventory system. Such systems are of special interest because they can often be found in real-world applications which deal with the problem how to control the flow of goods from the production locations to the points of demand, i.e., to define optimal transshipment resources and using them efficiently to serve demands with a given number of transport units as described by Köchel, Kunze, and Nieländer [44] and Köchel and Nieländer [14]. Hub-and-spoke systems, which consist of one central hub that is connected to all spokes, arranged similarly to a bicycle wheel, have several benefits including the fact that for a network of n spokes, only n routes are necessary to connect all nodes and this generally leads to a more efficient use of scare transport resources. Thus, the model is commonly used in the industry, in particular in transport, telecommunications and freight, as well as in distributed computing. Additionally, complicated operations can be consolidated at the hub, rather than maintained separately at each node and, therefore, the spokes are simpler and new spokes can be connected easily.

4.1 General System Description

Inventory systems with hub-and-spoke structure have been investigated before by Köchel et al. [44] and Zhou, Min, and Gen [45]. In this section results are shown which combine approaches from inventory theory and logistics to investigate the following problem as described by Köchel and Thiem [46, 47]:

- The system consists of one single warehouse, the *hub*, and a number of n retailers, the *spokes* (cp. 3).
- The *hub* owns an infinite amount of one specific product, which can be ordered by the retailers, and can rent as many transport units as needed from an external rental service up to a maximum number, which is optimized. Transportation orders from retailers, which cannot be served immediately due to a stock-out or no available transport units, are queued in an order queue with given capacity. Ordering and storing of products as well as holding and leasing of transport units and realizing transshipments generates costs.
- The systems allows the realization of different classes of *transport units* (trucks, pickups, ships, etc.).
- Each *spoke* has a storage for product units with the current inventory position r, and there is stochastic demand according to certain probability distributions for these product units, which is specified in Section 4.2. Depending on the order policy each retailer can order new product units from the hub. Arriving demand is handled corresponding to an acceptance-rejection policy.

It is worth noting that a comparison to previous results from literature is not reasonable because previous models only include certain aspects of our model. To find optimal values for the fleet size allocation problem, an analytical queuing model can be used. However, such a model would define limiting restrictions, e.g., a Poisson demand process, or would only allow approximative solutions. The purpose of our approach is to assess the behavior of the system and to surmount these restrictions.

The whole system is modeled in a simulation tool developed by Hochmuth [48], which we will use as an integrated part in the optimization process. A visualization of this system is shown in Figure 3. Concretizing the setup and corresponding criteria to be optimized allows us to model and to optimize a large variety of real-world hub-and-spoke inventory systems.

In our examples the optimization problem is to find an optimal order policy. This includes the optimization of the reorder levels s_j for each retailer j and the maximum number of available transportation units for each type. We have the choice between two different options for the order policy:

- The (s_j, S_j)*-order policy* is the restocking to $r_j = S_j$ product units when the current inventory position r_j of the retailer j has fallen below the reorder point s_j.
- The (s_j, nQ_j)*-order policy* is the ordering of an integer multiple n of the order quantity Q_j of product units (lot size) when the inventory level r_j of the retailer j has fallen below the reorder point s_j. For the current inventory position r the parameter n is set according to $n = \lceil (s - r)/Q \rceil$.

For our inventory system, the solution ϑ_i of an individual i encodes the variables for the order policy, i.e., the reorder levels s_j for all retailers j and, if necessary, the number of transport units t_k for each class k as a real-valued

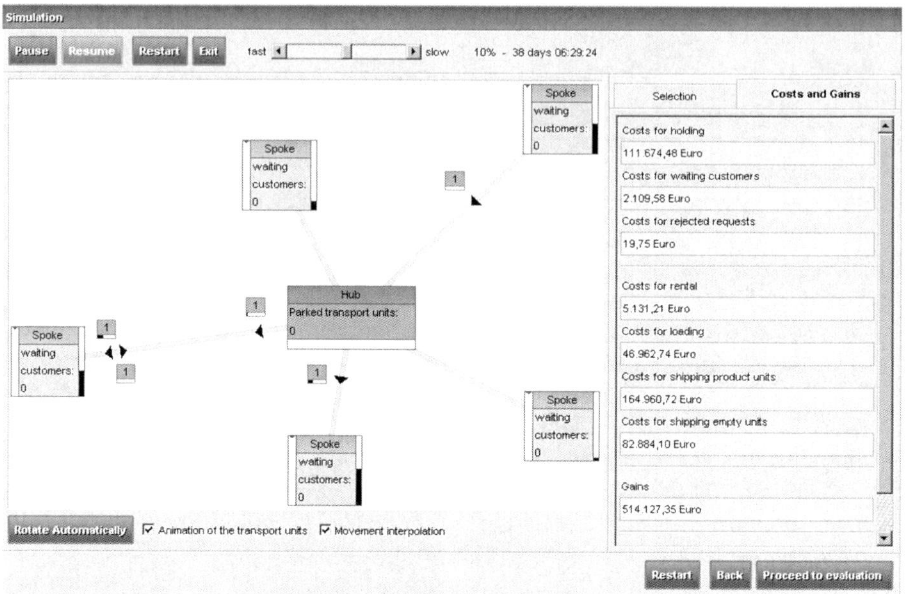

Fig. 3. Simulation for a hub-and-spoke inventory system with five spokes.

vector, which is necessary for continuous methods such as classical PSO algorithms. Since a feasible solution can only contain integer numbers, we simply round the real-valued entries of ϑ_i to obtain the input variables for the simulation of the hub-and-spoke inventory system. In our view this approach leads to the most natural movement of the swarm individuals in an integer solution space and further allows the unchanged use of the common PSO variants originally designed for continuous optimization problems. Also the move classes of the Monte-Carlo type techniques and the mutation operator of GA are based on this representation of the solution.

Another problem we faced for these systems is the stochastic noise in the objective function, which adds non-static local maxima and minima to the solution space. To reduce this effect a sufficiently long time period for the simulation has to be defined and it is also advisable to average the results over several optimization runs.

Since an evaluation of the objective function requires the complete simulation of the whole system, we are dealing with a very time-consuming optimization problem. Therefore, it is desirable to use optimization algorithms with powerful operators in order to minimize the number of objective function evaluations. In the forthcoming examples we compare the three different optimization algorithms as introduced in Section 3.

4.2 Experimental Settings for the Hub-and-Spoke System

In the following we present the optimization results for two different test cases. The first one deals with a system of identical retailers and the second case with retailers varying either in their distances to the hub or in their demand structure. We first define the general setup for all model elements. For the warehouse we assume

- a FIFO service policy, and
- an allocation policy with priority sequence according to maximum velocity, maximum loading, minimum empty units, and minimum transport costs.

The retailers are specified by

- 200 km distance from the warehouse,
- a demand with exponentially distributed inter-arrival times with mean 1 hour and deterministic demand of 1 product unit per client,
- impatient clients with zero waiting time (i.e., the lost-sales case),
- a storage capacity of 500 product units,
- cost parameters: inventory costs 50€ per day and product unit, rejection costs 50€ per rejected client and reward 200€ per satisfied demand unit, and
- all retailers j use an (s_j, nQ_j) order policy with an order quantity Q_j, which is equal to the capacity of the used transport units.

For the transport units we distinguish between two different classes:

- fast, small transport units (delivery vans) and not quite as fast but larger transport units (trucks),
- their average velocity is equal to 80 km/h and 50 km/h, respectively,
- the capacities are 1 product unit or 4 product units, respectively, and
- the corresponding cost parameters: using a transport unit costs 50€ and 100€ per day, 0.10€ and 0.20€ per km and transported product unit, and 10€ and 30€ for loading/unloading, respectively.

The optimization criterion is the maximum total profit over the planning period, which is chosen to equal 3 years. The starting inventory is 10 product units for each retailer.

Based on few pre-considerations, we obtain that the average demand per day is about 120 product units. One fast transport unit can deliver at maximum 4.8 product units per day, whereas a slow transport unit can deliver 12 product units. This corresponds to either the usage of 25 fast or 10 slow transport units a day and also motivates the first example, where we will investigate the influence of different numbers of available transport units.

4.3 Experimental Results

4.3.1 Example 1 – 5 Identical Retailers

In this example we use the setup for the model components as specified above and the numbers of transport units are chosen according to the following cases:

a) 20 transport units with velocity 80 km/h and capacity 1 product unit,
b) 8 transport units with velocity 50 km/h and capacity 4 product units,
c) 10 fast transport units and 4 slow transport units with capacities according to a) and b),
d) optimization of the number of fast and slow transport units.

Additionally, we consider two setups for the reorder levels s_j. Because we assume identical retailers, we first choose identical s_j for all retailers j. Subsequently, we optimize s_j individually for each retailer. The results are expected to be nearly identical for both cases and are shown in the format $\vartheta^\star = (\mathbf{s}^\star, \mathbf{t}^\star)$ with the optimal reorder levels $\mathbf{s}^\star = (s_1^\star, \ldots, s_5^\star)$ and the number of transport units $\mathbf{t}^\star = (t_{\text{fast}}^\star, t_{\text{slow}}^\star)$.

For our test cases the initial values for the reorder levels are chosen randomly in a region of the solution space with $s_j \leq 100$, and for PSO the initial velocities are set to zero. For TA, the initial temperature is chosen to be $T = 5 \cdot 10^6$ and we apply thermodynamic speed scheduling with $c_1 = 0.5$ and $c_2 = 0.05$ and a move class with $c_{\text{step}} = 10$. The number of individuals in one generation has been adapted to $N = 20$ for the GA and $N = 10$ for TA. The swarm size in PSO is set to the standard value $N = 15$ for seven degrees of freedom in the considered examples. For reasons of fair comparison we allow each algorithm only a fixed number of objective function evaluations, which are 500 for the simple cases with identical reorder levels (first four lines in Tables 1 through 3) and 3000 for the cases with individual reorder levels (lines 5 to 8 in Tables 1 through 3). Tables 1 through 3 present for each of the three algorithms the best obtained solution with the corresponding total profit as well as the average objective value over 100 optimization runs. The results show that the GA performs very competitive compared to the other algorithms in any case. For identical reorder levels (first four lines of the tables) PSO performs sometimes slightly better, e.g. in the case a) (first line) if the average total profit is considered as measure. These differences are not significant as statistical tests show. For the case of individual reorder levels the results of the statistical tests are presented in Table 7 and also visualized in Figure 4. This shows that the GA never performs significantly worse compared to the other algorithms for individual reorder levels, which correspond to the bottom lines in the Tables 1 through 3 and is denoted as setup 1a to 1d in Table 7.

In many cases the GA performs significantly better, which is denoted by a + sign in the Table 7. For identical reorder levels, all algorithms find solutions with relatively high order levels for the setups b) and c), due to a lack of transport units in these cases. For the case in which the reorder levels s_j are optimized for each retailer j individually, we expect to obtain similar results. This is not the case for all setups. In a few cases, the average total profit for individual reorder levels is better compared to the results for identical reorder levels, e.g. for PSO in the setups 1b) and 1c). Interestingly, all three algorithms perform worse in setup a) for individual reorder levels compared to the setup a) for identical reorder levels. This indicates that the solution space for individual reorder levels is already too complex to find the global optimum. This can be seen also if the reorder levels themselves are compared.

For all cases of this example it can be seen that the best solutions found by the different algorithms are very different in some cases, where the total profit is not too different. Also if the cases of identical reorder levels and the cases with individual reorder levels are compared for the same algorithm, the reorder levels can be very different. This shows that the search space is very shallow, so that many good solutions exist. This makes it hard for the algorithms to converge to the global optimum. Another reason for the differences is the statistical nature of the problem: even for the same parameters the simulation returns different results for the total profit. In particular in case b), i.e., when 8 large transport units are used, there are noticeable fluctuations in the reorder levels, which seems to be caused by only small differences between different order policies in this area of the solution space. However, for some setups, e.g. setup d), the solutions are very close to that of identical s_j, which shows that the optimization algorithms are able to adequately optimize inventory systems with more degrees of freedom.

Comparing the different setups it can be seen that when optimizing also the number of transport units, we can increase the solution quality considerably. The results show that the operation of fast transport units yields in this example much better results, and hence, the optimized solution uses hardly slow transport units, where the optimal number of fast transport units is

Table 1. PSO optimization results for the reorder levels $\mathbf{s}^\star$ and the transportation units $\mathbf{t}^\star$ for the hub-and-spoke inventory system with 5 identical spokes of Example 1. The upper (lower) results are obtained by assuming identical (different) reorder levels for each spoke.

Case		Order policy		Best total profit in €	Average total profit in €
identical reorder levels s for each retailer	a)	$\mathbf{s}^\star = 6$		14,802,081	14.750.390
	b)	$\mathbf{s}^\star = 60$		15,107,527	15,079,495
	c)	$\mathbf{s}^\star = 94$		15,022,905	15,005,687
	d)	$\mathbf{s}^\star = 7$	$\mathbf{t}^\star = (50, 4)$	19,615,544	19,099,452
individual reorder levels s for each retailer	a)	$\mathbf{s}^\star = (65, 80, 64, 80, 84)$		14,695,688	14,685,395
	b)	$\mathbf{s}^\star = (141, 46, 148, 129, 226)$		15,105,961	15,088,160
	c)	$\mathbf{s}^\star = (145, 222, 74, 117, 55)$		15,025,249	15,011,521
	d)	$\mathbf{s}^\star = (7, 8, 7, 7, 7)$	$\mathbf{t}^\star = (57, 4)$	19,639,569	18,040,788

Table 2. GA optimization results for the same system as in Table 1.

Case	Order policy		Best total profit in €	Average total profit in €
identical reorder levels s for each retailer	a) $\mathbf{s}^\star = 6$		14,805,119	14,743,772
	b) $\mathbf{s}^\star = 91$		15,108,690	15,080,055
	c) $\mathbf{s}^\star = 53$		15,037,343	15,005,075
	d) $\mathbf{s}^\star = 7$	$\mathbf{t}^\star = (72, 32)$	19,647,434	19,494,856
individual reorder levels s for each retailer	a) $\mathbf{s}^\star = (104, 83, 31, 67, 44)$		14,700,701	14,686,095
	b) $\mathbf{s}^\star = (79, 88, 93, 61, 47)$		15,115,583	15,043,077
	c) $\mathbf{s}^\star = (97, 82, 91, 95, 30)$		15,025,800	15,011,041
	d) $\mathbf{s}^\star = (7, 7, 8, 8, 7)$	$\mathbf{t}^\star = (52, 19)$	19,657,910	18,801,212

Table 3. TA optimization results for the same system as in Table 1.

Case	Order policy		Best total profit in €	Average total profit in €
identical reorder levels s for each retailer	a) $\mathbf{s}^\star = 6$		14,798,201	14,710,025
	b) $\mathbf{s}^\star = 115$		15,102,836	15,073,917
	c) $\mathbf{s}^\star = 99$		15,013,344	14,992,152
	d) $\mathbf{s}^\star = 7$	$\mathbf{t}^\star = (32, 0)$	19,532,798	19,087,389
individual reorder levels s for each retailer	a) $\mathbf{s}^\star = (102, 37, 23, 31, 44)$		14,697,488	14,671,210
	b) $\mathbf{s}^\star = (111, 91, 64, 131, 38)$		15,107,413	14,930,035
	c) $\mathbf{s}^\star = (99, 40, 36, 74, 96)$		15,016,770	14,996,140
	d) $\mathbf{s}^\star = (8, 7, 9, 8, 9)$	$\mathbf{t}^\star = (40, 0)$	19,382,452	18,345,893

greater than necessary for the satisfaction of the average daily demand, given by 25 fast TUs. The reason might be that it is optimal for the system to have enough products on-hand in order to avoid rejection costs.

Comparing the three different algorithms, GAs seem to be the best suited approach for the given problem. The particles in PSO contract usually relatively fast and once the swarm converged the system seems to be frozen and often no further improvement is possible. Also for TA, even if the adaptive temperature schedule is used, the algorithm does not accept worse solutions after some time. Our GA approach, as described in the previous section, retains some explorative behavior until it is finally stopped. Especially, mutations help to maintain this explorative character. Probably some diversity preserving mechanisms such as fitness sharing or deterministic crowding could increase the performance further.

4.3.2 Example 2 – 5 Different Retailers, 2 Types of TUs

In this example we use the setup for the model components as specified above and optimize the reorder levels of the retailers as well as the number of fast and slow TUs for the following cases:

a) *Different demand at the retailers:* The demand at the retailers is given by an exponential distribution with average inter-arrival times $\langle T_1 \rangle = 1.5$h, $\langle T_2 \rangle = 1.2$h, $\langle T_3 \rangle = 1$h, $\langle T_4 \rangle = 6/7$h and $\langle T_5 \rangle = 3/4$h. This results in an average daily demand of 16, 20, 24, 28 and 32 product units respectively, which also sums up to the former 120 product units.
b) *Different distances from the warehouse:* Again this is based on Example 1, but with different retailer-warehouse distances: $d_1 = 50$km, $d_2 = 100$km, $d_3 = 200$km, $d_4 = 300$km, $d_5 = 400$km.
c) Same as setup of b) but with distances: $d_1 = 100$km, $d_2 = 150$km, $d_3 = 200$km, $d_4 = 250$km, $d_5 = 300$km.

For the setup in a) we expect a general tendency towards an increase of the reorder levels from retailer 1 to 5 since the average daily demand also increases in this order. The results are summarized in Tables 4 through 6. For case a) we clearly see the expected increase in the reorder levels. Further, the results are almost identical to the total profit of about 19 Mio. € as obtained in Example 1.

In case b) we again obtain for both variants the expected increase in the reorder levels. This has been expected because the retailers with larger distance have to hold more product units on-hand to compensate the longer delivery times. Since variant c) has a shorter average distance than variant b), the total profit of variant c) is slightly better. Further, comparing the results of setup c) with the results for identical reorder levels in Example 1, which has the same average setup, we obtain very similar results for the total profit as expected.

Also for the more complicated cases, where the distances of the spokes to the hub are not identical, GAs perform significantly better than the other two approaches, see also Table 7 and Figure 4. The reason may again be that the applied GA is more explorative and, hence, able to return good results on a very bumpy and shallow search landscape.

The three algorithms have been compared using statistical tests for the final best objective values of each run, based on 100 runs for each setup. We use Mann-Whitney tests for the comparison of two such distributions. Separate tests are made for each setup. Table 7 shows the resulting exact significance probability of the test statistic. Low values indicate that the two algorithms have different underlying distributions of the final best objective values. In case of large values no conclusion can be made.

Table 4. PSO optimization results for the reorder levels $\mathbf{s}^\star$ and the transportation units $\mathbf{s}^\star$ for the hub-and-spoke inventory system with *5 different spokes* of Example 2.

Case	Order policy		Best total profit in €	Average total profit in €
a) diff. demand	$\mathbf{s}^\star = (6,6,8,8,10)$	$\mathbf{t}^\star = (33,0)$	19.691.730	19.307.838
b) diff. distan. var. 1	$\mathbf{s}^\star = (4,5,7,9,11)$	$\mathbf{t}^\star = (48,3)$	19.621.243	17.464.169
c) diff. distan. var. 2	$\mathbf{s}^\star = (5,6,7,8,9)$	$\mathbf{t}^\star = (46,3)$	19.648.951	17.850.843

Table 5. GA optimization results for the same system as in Table 4.

Case	Order policy		Best total profit in €	Average total profit in €
a) diff. demand	$\mathbf{s}^\star = (5,7,7,7,9)$	$\mathbf{t}^\star = (61,1)$	19.628.163	19.469.835
b) diff. distan. var. 1	$\mathbf{s}^\star = (4,5,7,9,10)$	$\mathbf{t}^\star = (50,6)$	19.597.861	18.436.872
c) diff. distan. var. 2	$\mathbf{s}^\star = (5,6,7,9,9)$	$\mathbf{t}^\star = (42,15)$	19.628.663	18.820.892

Table 6. TA optimization results for the same system as in Table 4.

Case	Order policy		Best total profit in €	Average total profit in €
a) diff. demand	$\mathbf{s}^\star = (5,6,7,8,11)$	$\mathbf{t}^\star = (45,0)$	19.398.392	19.046.810
b) diff. distan. var. 1	$\mathbf{s}^\star = (4,4,7,11,11)$	$\mathbf{t}^\star = (48,0)$	19.256.950	18.182.436
c) diff. distan. var. 2	$\mathbf{s}^\star = (7,6,8,8,11)$	$\mathbf{t}^\star = (35,0)$	19.466.977	18.282.670

Table 7. Significance levels from the Mann-Whitney test for a hub-and-spoke inventory system. For the + (−) sign the first algorithm is significantly better (worse) than the second one and for the ∘ sign the results are non-significant. For Example 1 (setup 1a–1d) only results for different reorder levels are shown. Setups 2a–2c belong to Example 2.

	Setup 1a	Setup 1b	Setup 1c	Setup 1d	Setup 2a	Setup 2b	Setup 2c
GA⇔PSO	∘ 0.2058	∘ 0.2439	∘ 0.8189	+ 0.0001	+ 0.0014	+ 1.3e-5	− 7.4e-7
GA⇔TA	+ 0.0	+ 1.3e-6	+ 0.0	+ 6.2e-9	+ 2.9e-5	+ 0.0	+ 2.2e-41
PSO⇔TA	+ 0.0	+ 3.0e-6	+ 2.4e-29	∘ 0.9764	− 0.0002	− 0.9788	∘ 2.9e-11

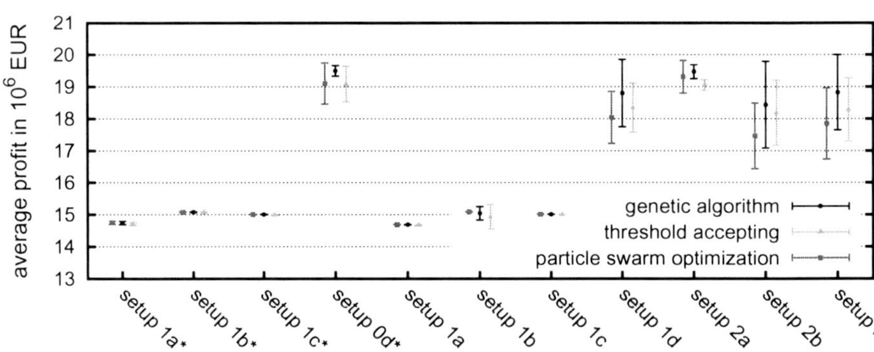

Fig. 4. Comparison of the solution quality of a hub-and-spoke inventory system for TA, PSO and the GA. For each of the problems stated above we show the average value of 100 optimization runs. The error bars represent the corresponding standard deviations. The first four cases marked by a star share identical reorder levels.

The examples of the simulation-based optimization of hub-and-spoke inventory systems show that GAs perform competitive compared to the other approaches. For PSO we obtain sometimes the contraction of the swarm to local optima and it is almost impossible to revert this. Also ensemble-based TA performs well but the probability to accept worse individuals gets lower and lower during the optimization, which sometimes also causes convergence to some local optimum. GAs keep different explorative components all the time, i.e., worse individuals from crossover are accepted without specific conditions and also mutated individuals manage to be passed on to the next generation unconditionally. Especially, for the often bumpy search landscape of problems in simulation-based optimization with additional stochastic influences, this seems to be a good choice. This can be seen in particular in the second example, which is more difficult to optimize. Hence, for the second part of the chapter we restrict ourselves to the GA to optimize more complex multi-location inventory systems with lateral transshipments. But it is worth to remark that the other methods showed competitive results as well.

5 The Hard Case: Simulation-Based Optimization of Multi-Location Inventory Systems with Lateral Transshipments

A multi-location inventory system with lateral transshipments is far more general compared to the systems investigated in the first part of the chapter. A given number of locations faces customer demands and each of them manages its stock individually. In contrast to a hub-and-spoke system as discussed above, the locations can receive new product units in different ways, i.e., either by ordering from an outside supplier or by transshipments from other locations. The problem of defining order and transshipment policies, which optimize the performance measures for the whole system in the long run, is a very demanding optimization task. Often the optimizer prefers certain flows of transshipments if customer demands and costs are defined differently for each location throughout the model. Also, different pitfalls have to be avoided here. Especially, undesirable forth-and-back transshipments have to be ruled out. As a consequence, suitable policies increase the set of parameters and the optimization task becomes more complicated. These very complex systems have been especially optimized by the application of evolutionary strategies. The second part of the chapter covers the optimization of these generalized systems in detail.

5.1 Introduction

To be competitive, companies are challenged to improve service levels and to reduce costs at the same time. Despite these objectives seem to be contradictory, they may be reached. Spreading of service locations is expected to improve customer service while lateral transshipments between the locations help to balance stock levels to decrease costs. For the design and control of such multi-location systems, suitable mathematical models are needed. Multi-location inventory models with lateral transshipments (MLIMT) model the following situation. A given number of locations have to meet a demand for some products during a defined planning horizon. Each location can get new product units either by ordering from an outside supplier or by transshipments from other locations. Figure 5 illustrates the system topology of a general MLIMT. The problem arises to define such ordering and transshipment decisions (OD and TD) that optimize given performance measures for the whole system.

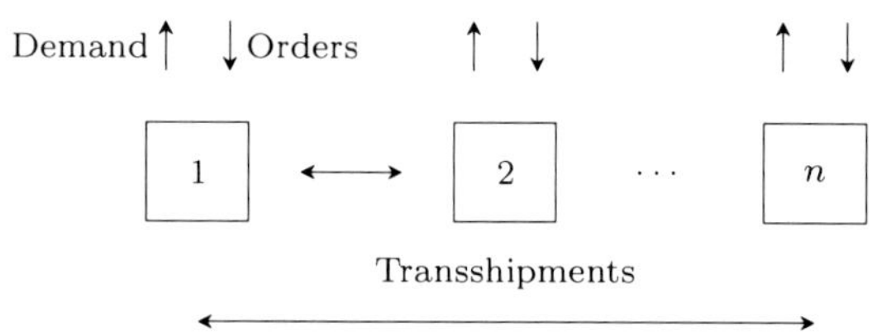

Fig. 5. Topology of a general MLIMT. Each of the n locations can refill its stock by ordering from an outside supplier or by transshipments from other locations.

Up to date a great variety of models and approaches exist to solve this problem. The most common and broadly studied class of models assumes a single product, discrete review, independent and identically distributed demand, backlogging, i.e., allowing clients to wait, complete pooling of product units between locations, emergency transshipments at the end of an order period, zero lead times, linear cost functions, and the total expected costs as performance measure (see Köchel [49] for a review). In this case analytical examination is an option, but in the general case, analytical solutions are not feasible for MLIMTs due to the transshipments. TDs change the state of the system and thus influence the OD. Hence the total consequences of an OD cannot be calculated directly. Approximate models and simulation are alternatives [49–51]. Further difficulties connected with TDs arise for continuous review models, e.g., the prevention of forth-and-back transshipments. This is closely connected with the task to forecast demand. Therefore, continuous MLIMTs are usually investigated under several simplifying assumptions, such as two locations [52, 53], Poisson demand [6], a fixed order policy not considering future transshipments [54], restriction to basic strategies such as a one-for-one ordering policy [6], and an all-or-nothing transshipment policy [52], or the constraint that at most one transshipment with insignificant time and a single shipping point is possible over an order period [53]. In none of these approaches the question for optimal order and transshipment policies has been answered. All models assume certain order policies and heuristic transshipments. In a few cases simulation is employed either to test approximate analytical models [53, 54] or for the definition of the best reorder point s_j for an (s_j, S_j)-order policy [55] by linear search and simulation. Consequently, the results are restricted to models with small system sizes.

Herer, Tzur, and Yücesan [56] find optimal order-up-to quantities S by sample-path-based optimization. Under the assumption of instantaneous transshipments, optimal transshipment quantities are calculated using simulation-based optimization combining a network flow formulation with infinitesimal perturbation analysis (IPA) to estimate the gradient. Extensions include finite transportation capacities [57] and positive replenishment lead times [58]. We introduce positive lead times for ordering as well as for lateral transshipments. Furthermore, we investigate the effect of non-stationary transshipment policies under continuous review. Thus, the complexity of this general model motivates the application of evolutionary simulation-based optimization instead of gradient-based methods. In this regard, we follow an approach similar to Arnold and Köchel [59], and recently Belgasmi, Saïd, and Ghédira [60], who analyze the effect of different parameters using evolutionary optimization. The MLIMT presented here has been developed with respect to three aspects:

- connection of discrete review for ordering and continuous review for transshipments,
- creating a simulator as general as possible to abandon restrictions of existing studies and to ensure broad applicability, and
- iteratively connecting an MLIMT simulator with an evolutionary optimization algorithm to investigate the search space by simulation-based evolutionary optimization [61].

Most advantageous is that various performance measures can be optimized for a large class of MLIMTs. Compound renewal demand processes as well as arbitrary ordering, demand satisfaction and transshipment modes can be investigated thanks to this approach (see [62, 63] for details). Because computing time increases fast for a higher dimensionality of the problem, parallelization is useful.

5.2 Experimental Settings

Despite a simulation model can represent any real system with arbitrary accuracy, our objective is not to design an oversized simulator for all multi-location inventory systems but to describe a simulator which can be used to find efficient solutions for the optimal control of an important class of MLIMTs in reasonable time. In addition that class of MLIMTs should go beyond models which can be analytically investigated [62]. Consequently the application domain is extended considerably.

The most common class of models defines emergency lateral transshipments at the end of a period [49]. However, even for simple models of this type, an analytical solution cannot be formulated in general, because potential transshipments have to be considered for the ordering decision at the beginning of a period. But after the demand has been realized at the end of a period, the optimal transshipment decision is the solution of an open (linear) transportation problem. Such problems do not have closed form solutions. Thus, prior to the demand realization, no expression is available for cost savings as a result of transshipments. Hence, the overall costs of an order decision cannot be defined.

Approximate models and simulation are approaches to solve this dilemma [49–51]. Here we propose the latter one. The essential elements of a multi-location inventory model with lateral transshipments are listed below.

5.2.1 Number of Locations

With respect to the analytical tractability for the number of locations n the cases $n = 2$ and $n > 2$ are distinguished. The simulator is suited for MLIMTs with an arbitrary number of independent non-homogeneous locations, but the optimization task becomes more difficult if they increase in number. A longer simulation time is necessary to ensure sufficient accuracy for a higher number of locations, and a higher optimization cycle count is essential to converge to the global optimum. To limit this increasing complexity, potential transshipments between locations may be excluded by defining pooling groups (cp. 5.2.6 Pooling Mode).

5.2.2 Number of Products

There may be a single product or a finite number of different products, whereas a substitution order between products may be defined in the latter case. Most approaches as well as the MLIMT simulator assume a single product. For multiple products, sequential simulation and optimization is viable in general, provided that shared fixed costs are insignificant relative to total fixed and variable costs, and capacities for storage and transportation are considered to be infinite.

5.2.3 Ordering Mode

The ordering mode defines when to order, i.e., the *review scheme*, and what *order policy* to use. The review scheme defines the time moments for ordering, where discrete or continuous review is possible. Under the discrete review scheme, the planning horizon is divided into periods, which is the case for most analytical models. In most practical applications, order requests are sent in certain intervals. Therefore, we define a periodic review scheme with fixed length $t_{\mathrm{P},j}$ of the review period for orders at location j. Arbitrary order policies based on the inventory position at the end of the order period are allowed. So far, (s_j, S_j) and (s_j, nQ_j)-order policies are available, and in the experimentation we concentrate on the first one as visualized in Figure 6. To balance shortage and excess in the system, a continuous review scheme is defined for transshipments. Thus, a transshipment request may be released at any time.

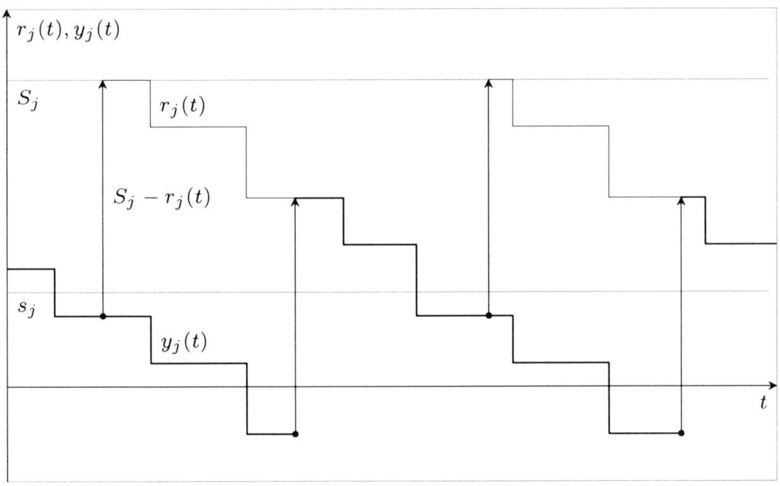

Fig. 6. (s_j, S_j)-order policy.

5.2.4 Demand Process

Demand may be deterministic or random, identical or different for all locations, stationary or non-stationary in time, independent of or dependent on locations and time as well as with complete or incomplete information. The most common class of models assumes demand independently and identically distributed over time. We define a *compound renewal demand process*, which is distinguished by two independent random variables T_j for the inter-arrival time of clients at location j and B_j for their demand, $j = 1, 2, \ldots, n$. Thus, exact holding and penalty costs can be calculated, which is an extension of analytical models with discrete review, where the whole demand of a period is transformed to the end of a period. That disadvantage does not exist for models with continuous review, but in almost all such models, a Poisson demand process is assumed, which is a limiting restriction as well.

5.2.5 Demand Satisfaction Mode

The mode for demand satisfaction describes how to process arriving demand for each location. It is common to assume a queue for backlogged demand. In dependence on an infinite, finite, or zero queuing capacity there exist the backlogging, intermediate, and lost-sales cases. Waiting demand is served according to a service policy and may finally have a random impatient time, after which clients leave unserved. The MLIMT presented here defines several service policies, such as FIFO (first in, first out), LIFO (last in, first out), SAN (smallest amount next), BAN (biggest amount next) and EDF (earliest deadline first). In the experimentation we focus on FIFO for demonstration purposes. The impatient time W_j is a random variable which is sampled according to an arbitrary (e.g. uniform, normal, exponential, etc.) distribution for each location j.

5.2.6 Pooling Mode

The pooling mode comprises all rules by which the on-hand stock is used to respond to shortages. Pooling may be complete or partial, defining which location and what quantity of product units are pooled. The MLIMT simulator includes all pooling modes from complete pooling to time-dependent partial pooling. A symmetric $n \times n$ matrix $\mathbf{P} = (p_{jj'})$ defines pooling groups in such a way that two locations j and j' belong to the same group if on only if $p_{jj'} = 1$ (otherwise $p_{jj'} = 0$). Transshipments are useful to balance shortages within an order cycle, but may be counterproductive near the end of such a cycle. Therefore, the pooling time $t_{\text{pool},j} \in [0, t_{\text{P},j}]$ is defined for each location j. After the kth order request, location j may receive transshipments from all other locations as long as for the actual time $t \leq kt_{\text{P},j} + t_{\text{pool},j}$ holds. Subsequently, location j may obtain product units only from locations within the same pooling group. Note that this parameter suppresses the effect of other transshipment parameters and thus increases the requirements on the evolutionary optimization. In the experiments we assume that all locations belong to the same pooling group.

5.2.7 Transshipment Mode

The transshipment mode describes when to transship and which transshipment policy to apply. The objective is to use a finite quantity of product units as efficiently as possible for the whole system. Therefore, shortage and excess at different locations are balanced. There may be *preventive* lateral transshipments to anticipate a stock-out or *emergency* lateral transshipments after a stock-out is observed. Most existing models define the latter ones at the end of a period. According to continuous review, the MLIMT simulator allows transshipments at any time over an order period as well as multiple shipping points and partial deliveries to release a TD. To control the flow of transshipments, a great variety of rules can be defined. Three basic ideas ensure broad applicability: *priorities*, the generalization of common transshipment *rules* and the introduction of a *state function.*

Difficulties arise in calculating the effects of a TD. Therefore, TDs should be based on appropriate forecasts for the dynamics of the model, especially the stock levels. For each location, transshipment orders (TO) and product offers (PO) are distinguished. Moments for TOs or POs are the arrival events of clients or transshipments and order deliveries, respectively. *Priorities* define a certain sequence of transshipments in one-to-many and many-to-one situations. Because of the continuous time, only these occur. The priorities result from arbitrary combination of the following three *rules*: BAN (biggest amount next), MTC (minimum transshipment costs per unit), and MTT (minimum transshipment time). *State functions* are observed to decide if a TO or a PO should be released. The following notations for each location j and time $t \geq 0$ are useful for further statements.

$y_j(t)$	Inventory level at time t
$y_j^{\pm}(t) = \max(\pm y_j(t), 0)$	On-hand stock $(+)$ respectively shortage $(-)$ at time t
$b_{\text{ord},j}(t)$	Product units ordered but not yet delivered at time t
$b_{\text{ord},k,j}$	Product units ordered in the k-th request
$b_{\text{tr},j}(t)$	Transshipments on the way to location j at time t
$r_j(t) = y_j(t) + b_{\text{ord},j}(t) + b_{\text{tr},j}(t)$	Inventory position at time t
$t_{\text{P},j}$	Order period time
$t_{\text{A},j}$	Delivery lead time of an order
$n_{ord,j} = \lfloor t_{\text{A},j}/t_{\text{P},j} \rfloor$	Number of periods to deliver an order

To decide at time t and location j about a TO or PO, the state functions $f_{\text{TO},j}(t)$ and $f_{\text{PO},j}(t)$ are defined based on the available stock plus expected transshipments $f_{\text{TO},j}(t) = y_j(t) + b_{\text{tr},j}(t)$ and the on-hand stock $f_{\text{PO},j}(t) = y_j^+(t)$ respectively. Since fixed cost components for transshipments are feasible, a heuristic (h_j, H_j)-rule for TOs is proposed in the following way, which is inspired by the (s_j, S_j)-rule for order requests $(h_j \leq H_j)$:

$$\text{If } f_{\text{TO},j}(t) < h_j \text{ then release a TO for } H_j - f_{\text{TO},j}(t) \text{ product units.} \quad (15)$$

However, in case of positive transshipment times, it might be advantageous to consider expected demand, i.e., a TO is released on the basis of a forecast of the state function $f_j(t')$ for a time moment $t' \geq t$. The MLIMT simulator offers three such moments in time, the current time (i.e., no forecast) $t' = t$, the next order review time $t' = t_1$, and the next time of a potential order delivery $t' = t_2$. In the experimentation we focus on the comparison of time t and t_1. For example the state function $f_{\mathrm{TO},j}(t) = y_j(t) + b_{\mathrm{tr},j}(t)$, $t \geq 0$ is considered. Let $kt_{\mathrm{P},j} \leq t < (k+1)t_{\mathrm{P},j}$, i.e., we assume that we are in the review period after the kth order request. Then t_1 is defined as follows:

$$t_1 = (k+1)t_{\mathrm{P},j} \; . \tag{16}$$

For t_2 we introduce two events:

$$\begin{aligned} &ev(t) \leftrightarrow \{\text{in the actual period was no order delivery until } t\} \qquad \text{and} \\ &\overline{ev}(t) \leftrightarrow \{\text{there was an order delivery until } t\} \; . \end{aligned}$$

Then the following holds:

$$t_2 = (k - n_{\mathrm{ord},j})t_{\mathrm{P},j} + t_{\mathrm{A},j} + \begin{cases} 0 & \text{if } ev(t) \leftrightarrow t < (k - n_{\mathrm{ord},j})t_{\mathrm{P},j} + t_{\mathrm{A},j} \\ t_{\mathrm{P},j} & \text{if } \overline{ev}(t) \leftrightarrow t \geq (k - n_{\mathrm{ord},j})t_{\mathrm{P},j} + t_{\mathrm{A},j} \end{cases} . \tag{17}$$

Considering $m_j = \langle B_j \rangle / \langle T_j \rangle$, the expected demand per time unit at location j, the following forecasts are used:

$$\hat{f}_{\mathrm{TO},j}(t) = f_{\mathrm{TO},j}(t) = y_j(t) + b_{\mathrm{tr},j}(t) \; , \tag{18}$$

$$\hat{f}_{\mathrm{TO},j}(t_1) = f_{\mathrm{TO},j}(t) - m_j(t_1 - t) + \begin{cases} b_{\mathrm{ord},k',j}, \; k' = k - n_{\mathrm{ord},j} & \text{if } ev(t) \\ 0 & \text{if } \overline{ev}(t) \end{cases} , \tag{19}$$

$$\hat{f}_{\mathrm{TO},j}(t_2) = f_{\mathrm{TO},j}(t) - m_j(t_2 - t) \; . \tag{20}$$

Thus, by replacing function $f_{\mathrm{TO},j}(t)$ in (15) by various forecast functions a great variety of transshipment modes can be described, which control the release of TOs by several locations. Figures 7 and 8 illustrate general forecast functions for both events. We point out that in case of linear transshipment cost functions without set-up component, the (h_j, H_j)-rule degenerates to the (H_j, H_j)-rule. As a well-designed evolutionary optimization algorithm is expected to approximate that solution, we generally work with the (h_j, H_j) rule.

A condition to serve a TO is that at least one location offers product units. To decide what quantity is offered at time t, an additional control parameter is introduced, the offering level o_j, corresponding to the hold-back level in Xu et al. [53]. Since only physically available stock is offered, the state function $f_{\mathrm{PO},j}(t) = y_j^+(t)$ is defined. To prevent undesirably small and frequent transshipments, the offered quantity $y_j^+(t) - o_j$ must not be smaller

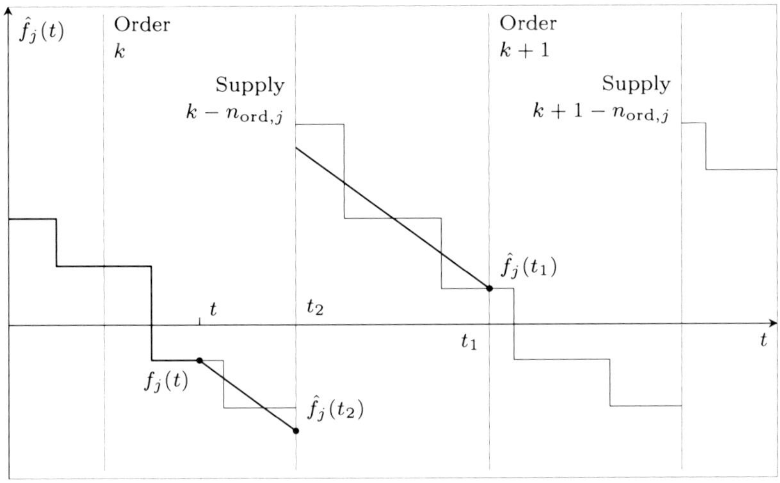

Fig. 7. Forecast functions for $ev(t) \leftrightarrow t < (k - n_{\text{ord},j})t_{\text{P},j} + t_{\text{A},j}$.

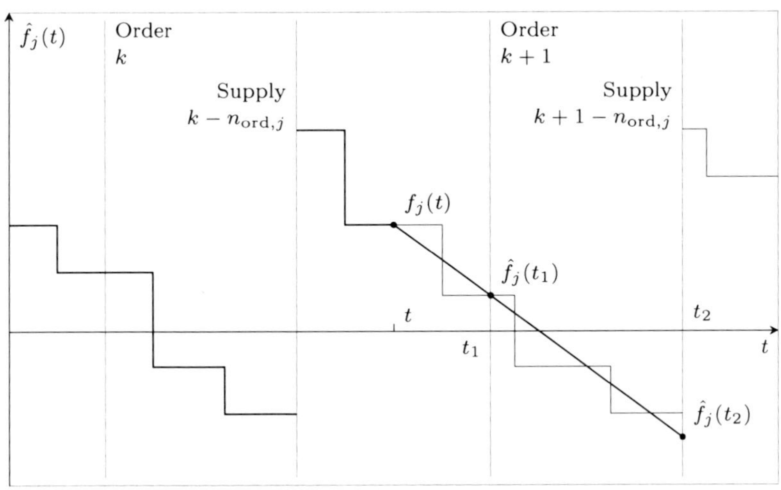

Fig. 8. Forecast functions for $\overline{ev}(t) \leftrightarrow t \geq (k - n_{\text{ord},j})t_{\text{P},j} + t_{\text{A},j}$.

than a certain value $\Delta o_{\min,j}$. Similar forecasts are applied to take future demand into account with forecast times t, t_1 and t_2 (see [62] for details). Thus, the PO rule is as follows:

$$\text{If } \hat{f}_{\text{PO},j}(t) - o_j \geq \Delta o_{\min,j} \text{ then release a PO for } \hat{f}_{\text{PO},j}(t) - o_j \text{ product units .} \tag{21}$$

Hence, the most important transshipment mode generalizations are as follows: First, introducing various control parameters we have extended the set of available transshipment policies considerably, including all commonly used policies. Second, multiple shipping points with partial deliveries are allowed. Finally, introducing time-dependent forecasts for the state function, the proposed decision rules become non-stationary in time.

5.2.8 Lead Times

Lead times for order deliveries and transshipments of product units may be negligible, positive constants, or random. Most analytic models assume zero lead times, while the MLIMT simulator defines location-specific order lead times and a not necessarily symmetric distance matrix for all locations. In conjunction with the transport velocity, transshipment lead times are determined.

5.2.9 Cost and Gain Functions

Costs may occur for ordering, storing and transshipping product units as well as for waiting and lost demand. These functions may be linear, linear with set-up part, or generally non-linear. A location may also receive gain from product units sold to clients. To solve models analytically, often linear cost functions are assumed. With respect to the MLIMT simulator order cost functions, holding cost functions, shortage cost functions and transshipment cost functions contain components which are fixed, linear in time and quantity as well as components which are linear in time. Fixed costs arise from each non-served demand unit. All cost parameters are location-specific and the gain from a unit sold is a constant.

5.2.10 Planning Horizon

The planning horizon may be finite or infinite. In case of periodic review, it may consist of a single period. Simulating infinite planning horizons by a finite simulation is of course not possible. Thus, appropriate approximations are used. An adequate estimation of the stationary properties can be achieved by a sufficiently long simulation time. The only difficulty is the fast increase of necessary computation time.

5.2.11 Objective Function

As objective function, various cost criteria can be formulated, such as total expected costs, total expected discounted costs, long-run average costs, and non-cost criteria like service rates or expected waiting times. Both criteria types may form a multi-objective problem. Alternatively, one criterion can be optimized while given restrictions have to be satisfied. Existing models commonly observe the total expected costs criterion. For the MLIMT simulator, different objectives can be optimized by choosing corresponding cost functions. For example the service rate or average waiting times can be expressed by corresponding shortage cost functions. Thus several criteria are applicable.

5.3 Experimental Results

To get an impression of simulation-based evolutionary optimization applied to multi-location inventory models with lateral transshipments, the results of the GA for two experiments on a four-location model are discussed in this section. The system topology is visualized in Figure 9. In both experiments we assume identical location-dependent model characteristics, excluding the client demand distributions, which are uniform distributions in example 1 and exponential distributions in example 2. However, equal mean values are assumed. Although simulation-based evolutionary optimization is applicable to in fact general models, the following assumptions are made.

1. All locations j use an (s_j, S_j)-order policy. For all locations j, an order period is equal to 10 days, i.e., $t_{\mathrm{P},j} = 10$ days. The order lead times are $t_{\mathrm{A},1} = 48\,\mathrm{h}$, $t_{\mathrm{A},2} = 60\,\mathrm{h}$, $t_{\mathrm{A},3} = 72\,\mathrm{h}$ and $t_{\mathrm{A},3} = 84\,\mathrm{h}$ for locations 1 to 4, respectively.
2. Transshipment times are 8.66 h between the outer locations 1–3 and 5 h between an outer location and the central location 4. The priority sequence for transshipment orders and product offers is MTC, BAN, and MTT. Stocks of all locations j are completely pooled, i.e., $t_{\mathrm{pool},j} = t_{\mathrm{P},j}$, and transshipments are therefore not constrained.
3. The inter-arrival time of customers to each location j is an exponentially distributed random variable with $\langle T_j \rangle = 2\,\mathrm{h}$. The customer demand D_j is uniformly distributed in the interval $(0, 5(j+1))$ in example 1 and exponentially distributed with mean $\langle D_j \rangle = 2.5(j+1)$ in example 2 for each location j. The service policy is FIFO, and the impatient time is triangularly distributed in the interval $(0\,\mathrm{h}, 8\,\mathrm{h})$, thus, $\langle W_j \rangle = 4\,\mathrm{h}$.
4. The inventory costs are 1€ per unit and day, whereas the linear order and transshipment costs are 1€ per unit and per day transportation time. The fixed transshipment costs equal 500€ for each location and the gain per unit sold is 100€. Out-of-stock costs are fixed to 50€ per lost client and 1€ per hour. The objective function is total costs over the simulation time of 260 weeks excluding the transition time of 52 weeks.

The state function chosen for transshipment orders and product offers is $f_{\mathrm{TO},j}(t) = y_j(t) + b_{\mathrm{tr},j}(t)$ and $f_{\mathrm{PO},j}(t) = y_j^+(t)$, respectively. To analyze the effect of forecasting, all four combinations of the current time t and the forecast moment t_1 are compared for both state functions. For optimization the GA as introduced in Section 3.3 is used with a population of 50 individuals, where an individual is a candidate solution, i.e., a vector of the following policy parameters for each location j.

s_j	Periodic reorder level
S_j	Periodic order-up-to level
h_j	Transshipment reorder level
H_j	Transshipment order-up-to level
o_j	Offer level
$\Delta o_{\min,j}$	Minimum offer quantity

The optimization stops if a new optimum has not occurred for the last 1,000 cycles, but at least 3,000 cycles are realized. The cycle count to find the actual optimum and the total cycle count, respectively, are stated next to the results. The number of function evaluations is determined by the product of the total cycle count and the population size (50). In most cases, 3,000 cycles, i.e., 150,000 function evaluations, are calculated for each state function combination.

For the initialization of the GA, inventory-related values are set in accordance with the capacity $y^+_{\max,j}$ for each location j. The reorder level s_j is uniformly distributed in the interval $[-y^+_{\max,j}, y^+_{\max,j})$. In case of an (s_j, S_j)-order policy, the order-up-to level S_j is uniformly distributed in the interval $(s_j, s_j + 2y^+_{\max,j})$, and in case of an (s_j, nQ_j)-order policy, the order quantity Q_j is uniformly distributed in the interval $(0, 2y^+_{\max,j})$. Similarly the (h_j, H_j)-transshipment request policy is initialized for each location j. Thus, the request level h_j and the request-up-to level H_j are uniformly distributed in the intervals $[-y^+_{\max,j}, y^+_{\max,j})$ and $(h_j, h_j + 2y^+_{\max,j})$, respectively. The product offer policy defines an offer level o_j, uniformly distributed in the interval $[-y^+_{\max,j}, y^+_{\max,j})$, and a minimum offer quantity $\Delta o_{\min,j}$, uniformly distributed in the interval $(0, y^+_{\max,j}/4)$.

For all experiments, the total results, optimized parameter values and cost function values are discussed. Prohibitive parameter values are enclosed in brackets, i.e., values that prevent ordering or offering product units. In general, the GA may choose values arbitrarily if there is no difference in the objective function. This is the case for values below or above a certain limit. Thus, after the optimization, the minimum absolute values of all parameters not changing the cost function values are determined using binary search.

5.3.1 Example 1 – Uniformly Distributed Demand

Tables 8–10 show the results for Example 1. For uniformly distributed demand, solution 3 gives the least total costs, taking advantage of forecasting demand for transshipment order decisions. This solution also shows the least

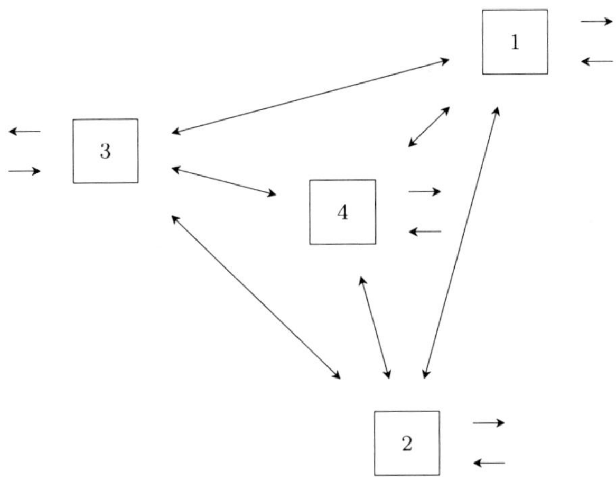

Fig. 9. Topology of a four-location model with lateral transshipments.

out-of-stock costs, closely followed by solution 2. Except for solution 2, all solutions use transshipments to compensate shortages. However, only solution 4 shows a directed flow of transshipments from location 4 to location 2, whereas transshipments in both directions are observed in case of solutions 1 and 2. Thus, the system has evolved an implicit structure as a result of evolutionary optimization, although all locations share identical parameters at the beginning. For all solutions displaying transshipments there is one location controlling its inventory level completely via transshipments instead of releasing periodic orders.

5.3.2 Example 2 – Exponentially Distributed Demand

Tables 11–13 show the results for Example 2. For exponentially distributed demand, solutions 2 and 3 have the lowest total costs and lowest out-of-stock costs. While the first one uses demand forecast for product offers and thus has less out-of-stock costs, the latter one applies demand forecast for transshipment orders resulting in less total costs. For all solutions transshipments are observed, but only solution 2 shows a directed flow. For the solutions 1, 3 and 4, alternating transshipments are released, i.e., flows in both directions occur. Again, the development of an implicit structure can be observed. In each case there is one location substituting periodic orders completely by transshipments from other locations.

Table 8. Overall results for Example 1.

	Transshipment order	Product offer	Result	Cycle	Rank
1	current time t	current time t	–54,885,174	1,784 (3,000)	3
2	current time t	time of next order t_1	–54,938,874	1,784 (3,000)	2
3	time of next order t_1	current time t	–55,149,067	1,784 (3,000)	1
4	time of next order t_1	time of next order t_1	–54,386,423	2,710 (3,710)	4

Table 9. Optimized parameter values for Example 1. Values that prevent ordering or offering of product units are enclosed in brackets.

		Periodic order		Transshipment order		Product offer	
	j	s_j	S_j	h_j	H_j	o_j	$\Delta o_{\min,j}$
1	1	356	795	[0]	[0]	[0]	[806]
	2	[0]	[0]	914	1,462	0	1,042
	3	1,200	1,701	1,250	1,250	[297]	[1,216]
	4	1,019	3,124	1,573	2,043	1,577	0
2	1	365	804	[0]	[0]	[0]	[317]
	2	578	1,257	[0]	[0]	[0]	[436]
	3	887	1,732	[0]	[0]	[0]	[623]
	4	1,093	2,244	[0]	[0]	[0]	[1,332]
3	1	379	818	[–48]	[0]	[135]	[663]
	2	576	1,255	[0]	[0]	[0]	[1,241]
	3	1,200	3,224	0	1,276	208	1,057
	4	[0]	[0]	948	2,439	1,526	0
4	1	367	806	[0]	[0]	[0]	[355]
	2	[0]	[0]	312	1,205	[0]	[0]
	3	885	1,730	[0]	[0]	[472]	[427]
	4	1,116	3,191	[0]	[0]	815	321

6 Conclusions

The major benefit of applying simulation-based optimization is that simulation models and the layout of the system under investigation can be changed or extended by further features in a relatively straightforward manner. For analytical approaches, this is often not the case. Especially for very complex systems that cannot be handled analytically, the simulation approach provides a powerful tool for practical applications. Of course, the optimization methods under consideration have to meet specific assumptions and have to be adjusted to avoid pitfalls of the specific problem under investigation. Nevertheless, the results for different meta-heuristics – in particular for evolutionary approaches – have been quite satisfactory and hence, a best practice example for the application of evolutionary methods for solving real-world planning problems has been presented, with a particular focus on the layout and system design in the logistics and transportation domain.

One of the findings is that GAs seem to be predestinated for shallow and bumpy objective functions, which is often the case for simulation-based

Table 10. Cost function values for Example 1.

	j	Inventory costs in €	Out-of-stock costs in €	Periodic order costs in €	Transshipment costs in €	Gain in €
1	1	557,741	7,155	244,099	0	8,613,278
	2	816,789	9,723	0	25,543	13,062,067
	3	1,136,491	4,016	573,325	0	17,483,570
	4	1,314,275	15,885	1,311,993	51,619	21,794,912
	$\sum$	3,825,296	36,779	2,129,417	77,162	60,953,827
2	1	570,168	5,684	244,321	0	8,624,377
	2	858,426	6,371	398,649	0	13,087,921
	3	1,124,607	9,541	593,731	0	17,422,405
	4	1,413,377	9,347	833,417	0	21,871,808
	$\sum$	3,966,578	30,942	2,070,118	0	61,006,512
3	1	588,592	3,762	244,560	0	8,636,330
	2	855,411	6,647	398,600	0	13,085,942
	3	891,662	6,703	1,246,027	169,650	17,460,357
	4	1,312,262	12,459	0	124,626	21,827,396
	$\sum$	3,647,927	29,570	1,889,186	294,276	61,010,025
4	1	572,078	5,272	244,349	0	8,625,792
	2	898,369	50,474	0	0	12,702,304
	3	1,121,349	9,763	593,651	0	17,419,721
	4	1,512,392	1,339	1,279,337	27,794	21,954,773
	$\sum$	4,104,188	66,848	2,117,337	27,794	60,702,591

Table 11. Overall results for Example 2.

	Transshipment order	Product offer	Result	Cycle	Rank
1	current time t	current time t	–54,299,368	1,784 (3,000)	3
2	current time t	time of next order t_1	–54,505,197	1,784 (3,000)	2
3	time of next order t_1	current time t	–54,820,380	1,592 (3,000)	1
4	time of next order t_1	time of next order t_1	–53,437,920	1,784 (3,000)	4

optimization problems. The algorithm applied in this work performed quite well for the described examples and has been adapted to the problem by using specific selection, recombination, and mutation operators which fit well to the problem and also the parameter setup has been adapted specifically. There is an ample number of other possibilities to apply evolutionary approaches, such as parallel evolutionary algorithms with certain communication structures or other global optimization heuristics to the problem. One specific idea would be the integration of further diversity preserving mechanisms.

Quite remarkable is also the optimization result for the simulation-based optimization of multi-location inventory systems with lateral transshipments as described in the chapter. It has been demonstrated that a four-location model with transshipments can be optimized by applying simulation-based optimization and the practitioner gets a detailed description of how to achieve this. For the future, it would be interesting to standardize the approach and

Table 12. Optimized parameter values for Example 2. Values that prevent ordering or offering of product units are enclosed in brackets.

		Periodic order		Transshipment order		Product offer	
	j	s_j	S_j	h_j	H_j	o_j	$\Delta o_{\min,j}$
1	1	428	832	[0]	[0]	[0]	[823]
	2	664	1,288	[0]	[0]	[0]	[1,286]
	3	[0]	[0]	826	2,237	0	1,386
	4	1,369	3,468	1,269	3,259	1,685	0
2	1	[0]	[0]	610	642	[0]	[523]
	2	654	1,278	[0]	[0]	[0]	[656]
	3	868	1,700	1,157	1,187	[0]	[889]
	4	1,131	3,003	[0]	[0]	-839	1,537
3	1	368	1,019	[0]	[0]	703	0
	2	632	1,256	[0]	[0]	[0]	[1,254]
	3	1,200	3,161	0	1,684	120	1,164
	4	[0]	[0]	991	2,773	1,386	0
4	1	421	825	[0]	[0]	[0]	[423]
	2	1,021	2,988	946	1,010	-675	1,432
	3	1,433	2,819	1,643	1,670	[1,082]	[931]
	4	[0]	[0]	0	2,153	643	68

Table 13. Cost function values for Example 2.

	j	Inventory costs in €	Out-of-stock costs in €	Periodic order costs in €	Transshipment costs in €	Gain in €
1	1	611,801	4,805	243,755	0	8,589,685
	2	900,869	7,816	399,531	0	13,123,242
	3	1,193,483	27,144	0	8,864	17,159,198
	4	1,452,823	9,624	1,429,314	45,857	21,762,930
	$\sum$	4,158,977	49,389	2,072,600	54,721	60,635,056
2	1	604,484	2,512	0	0	8,601,402
	2	886,560	8,859	399,252	0	13,112,049
	3	1,211,481	1,644	545,235	0	17,422,146
	4	1,539,422	6,566	1,187,311	30,139	21,793,064
	$\sum$	4,241,946	19,581	2,131,797	30,139	60,928,660
3	1	591,947	6,227	301,900	7,522	8,586,351
	2	858,075	11,975	398,600	0	13,085,992
	3	870,787	2,546	1,158,381	300,270	17,410,444
	4	1,331,795	2,571	0	261,276	21,841,466
	$\sum$	3,652,603	23,320	1,858,882	569,068	60,924,253
4	1	602,220	5,622	243,645	0	8,584,211
	2	1,057,626	49,876	951,486	86,419	12,735,925
	3	2,009,390	6,876	531,654	0	17,375,827
	4	1,355,478	15,394	0	41,249	21,698,890
	$\sum$	5,024,714	77,768	1,726,784	127,667	60,394,853

to offer services for the simulation-based optimization of given systems. This could be possible by sending the description of a specific inventory model to a web-service which calculates optimized parameters and strategies for the inventory system under consideration.

7 Future Work

Considering the simulation-based optimization of multi-location inventory systems with lateral transshipments, several extensions of the model are feasible. Functional extensions include new policies for periodic orders, transshipment orders, and product offers. Furthermore, we have shown that the demand distribution has a significant impact on the optimization result. As such, investigations of more complex and empirically derived distribution functions are expected to lead to parameters that are suited best for certain classes of models. Extensions of the parameter set itself comprise the capacity of the locations. Therefore, the introduction of costs for unused storage is necessary, constituting estate and energy costs. The analysis of such a system would result not only in optimal parameters to control the flows of product units, but also in the reallocation of capacities. In addition to these static aspects of the model, dynamic properties such as the location-specific order period time might be added to the parameter set. However, there certainly are restrictions in real-world applications.

Besides these extensions, there is another fundamental idea regarding orders from more than one location at a time. The examples show that under specific circumstances one location may take the role of the supplier, which orders product units periodically and then distributes these to other locations. This results in a directed flow of product units. The basic idea is that a periodic order or a transshipment order may be released by several locations. Thus, extended order policies referring not only to one location but to a group of locations are necessary. Moreover, for each order the Traveling Salesman Problem with minimal costs has to be solved. However, the examples suggest that such a transportation logic can already be approximated by applying heuristics, and thus, the inclusion of more elaborate policies is expected to lead to a higher complexity without considerable advantages. Further research may also focus on model characteristics that promote such a flow through a location network. The examples show that a vertical or mixed model structure may be advantageous under certain conditions, although the model shows a horizontal structure prior to the application of evolutionary optimization.

Acknowledgements. The authors would like to thank the German Academic Exchange Service (DAAD), the Robert Bosch doctoral program and the Foundation of German Business (SDW) for funding their research.

References

1. Shaw, M.J., Fulkerson, W.F.: Introduction to the Technology Strategy and Industrial Applications of Information-Based Manufacturing, pp. 920–6299. Springer, Heidelberg (2001)
2. Heinrich, S., Lässig, J., Dürr, H.: Generating, Planning and Control of Cross-Company Cooperation in Production and Supply Chain Networks. In: Proceedings of the 2nd International Conference on Changeable, Agile, Reconfigurable and Virtual Production, Toronto, Canada, pp. 1087–1096 (July 2007)
3. Lässig, J.: Algorithms and Models for the Generation and Control of Competence Networks, Mensch und Buch Verlag (2009); ISBN: 978-3-86664-648-3
4. Adeleye, E.O., Yusuf, Y.Y.: Towards Agile Manufacturing: Models of Competition and Performance Outcomes. International Journal of Agile Systems and Management 1(1), 93–110 (2006)
5. Lässig, J., Heinrich, S., Dürr, H.: An Online Solution for SME Service Enhancement with Short Term Cooperation Networks. In: Proceedings of the 24th International Manufacturing Conference, pp. 545–552. Waterford Institute of Technology, Ireland (2007)
6. Kukreja, A., Schmidt, C.P., Miller, D.M.: Stocking Decisions for Low-Usage Items in a Multilocation Inventory System. Management Science 47, 1371–1383 (2001)
7. Arrow, K.J., Harris, T., Marschak, J.: Optimal Inventory Policy. Econometrica 19, 250–272 (1951)
8. Dvoretzky, A., Kiefer, J., Wolfowitz, J.: The Inventory Problem. Econometrica 20, 187–222 (No. 2), 450–466 (No. 3) (1952)
9. Bellman, R.E.: On the Theory of Dynamic Programming. Proceedings of the National Academy of Sciences 38, 716–719 (1952)
10. Allen, S.G.: Redistribution of Total Stock over Several User Locations. Naval Research Logistics Quarterly 5, 337–345 (1958)
11. Allen, S.G.: A Redistribution Model with Set-up Charge. Management Science 8, 98–108 (1962)
12. Allen, S.G.: Computation for the Redistribution Model with Set-up Charge. Management Science 8, 482–489 (1962)
13. Clark, A.J., Scarf, H.: Optimal Policies for a Multi-Echelon Inventory Problem. Management Science 6, 475–490 (1960)
14. Köchel, P., Nieländer, U.: Simulation-based Optimisation of Multi-Echelon Inventory Systems. International Journal of Production Economics 93-94(1), 505–513 (2005)
15. Krishnan, K.S., Rao, V.R.K.: Inventory Control in N Warehouses. Journal of Industrial Engineering 16, 212–215 (1965)
16. P. Aggarwal, S.: Inventory Control Aspects in Warehouses. In: Symposium on Operational Research, Indian National Science Academy, New Delhi (1967)
17. Köchel, P.: A Stochastic Inventory Model for some Interconnected Locations. Mathematische Operationsforschung und Statistik 6(3), 413–426 (1975) (in German)
18. Köchel, P.: A Dynamic Multi-Location Inventory Model with Transshipments between Locations. Mathematische Operationsforschung und Statistik 13(2), 267–286 (1982) (in German)

19. Ekren, B.Y., Heragu, S.S.: Simulation Based Optimization of Multi-Location Transshipment Problem with Capacitated Transportation. In: WSC 2008: Proceedings of the 40th Conference on Winter Simulation, pp. 978–971 (2008); ISBN 978-1-4244-2708-6
20. Chiou, C.-C.: Transshipment Problems in Supply Chain Systems: Review and Extensions. In: Kordic, V. (ed.) Supply Chain, Theory and Applications, pp. 558–579. I-Tech Education and Publishing, Austria (2008)
21. Tlili, M., Moalia, M., Bahroun, Z., Campagne, J.P.: A Simulation-Optimization Model for Two-Echelon Multi-Location Inventory Systems with Transshipment and Lost Sales. In: ILS 2008: Proceedings of the International Conference on Information Systems, Logistics and Supply Chain, pp. 89–100 (2008)
22. Fu, M.C.: Sample Path Derivates for (s, S) Inventory Systems. Annals of Operations Research 53, 351–364 (1994)
23. Fu, M.C., Healy, K.J.: Techniques for Optimization via Simulation: An Experimental Study on an (s, S) Inventory System. IIE Transactions 29, 191–199 (1997)
24. Pflug, G.C.: Optimization of Stochastic Models: The Interface Between Simulation and Optimization. Kluwer Academic Publishers, Dordrecht (1996)
25. Healy, K.J., Schruben, L.W.: Retrospective Simulation Response Optimization. In: Nelson, B.L., Kelton, W.D., Clark, G.M. (eds.) Proceedings of the 1991 Winter Simulation Conference, Piscataway, NY, pp. 901–906 (1991)
26. Köchel, P.: Retrospective Optimization of a Two-Location Inventory Model with Lateral Transshipments. In: Proceedings of the 2nd International Conference on Traffic Science ICTS 1998, pp. 129–139 (1998)
27. Greenwood, A.G., Rees, L.P., Siochi, F.C.: An Investigation of the Behavior of Simulation Response Surfaces. European Journal of Operational Research 110, 282–313 (1998)
28. Kirkpatrick, S., Gelatt Jr., C.D., Vecchi, M.P.: Optimization by Simulated Annealing. Science 220(4598), 671–680 (1983); doi:10.1126/science.220.4598.671
29. Dueck, G., Scheuer, T.: Threshold Accepting: a General Purpose Optimization Algorithm Appearing Superior to Simulated Annealing. Journal of Computational Physics 90(1), 161–175 (1990); doi:10.1016/0021-9991(90)90201-B
30. Rechenberg, I.: Evolutionsstrategie 1994. k Frommann Holzboog, Stuttgart Bad Canstatt (1994); doi:10.1002/biuz.19950250620, ISBN 978-3-772-81642-0
31. Goldberg, D.E.: Genetic Algorithms in Search, Optimization and Machine Learning. Addison-Wesley, Reading (1989)
32. Kennedy, J., Eberhart, R.: Particle Swarm Optimization. In: Proceedings of IEEE International Conference on Neural Networks, vol. 4, pp. 1942–1948 (November-December 1995); doi:10.1109/ICNN.1995.488968
33. Ruppeiner, G., Pedersen, J.M., Salamon, P.: Ensemble Approach to Simulated Annealing. Jounal de Physique 1, 455–470 (1991); doi:10.1051/jp1:1991146
34. Thiem, S., Lässig, J.: Empirical Comparisons of Different Global Optimization Heuristics. Journal of the University of Applied Sciences Mittweida 7, 3–6 (2008); ISSN: 1437-7624
35. Lässig, J., Thiem, S.: An Integrated Framework for the Realtime Investigation of State Space Exploration. International Journal of Intelligent Systems and Technologies 3(1), 24–29 (2008)

36. Dueck, G., Scheuer, T.: Threshold accepting: a general purpose optimization algorithm appearing superior to simulated annealing. Journal of Computational Physics 90(1), 161–175 (1990)
37. Franz, A., Hoffmann, K.H., Salamon, P.: Best Possible Strategy for Finding Ground States. Physical Review Letters 86(23), 5219–5222 (2001); doi:10.1103/PhysRevLett.86.5219
38. Salamon, P., Sibani, P., Frost, R.: Facts, Conjectures and Improvements for Simulated Annealing. SIAM, Philadelphia (2002)
39. Clerc, M.: Standard PSO 2007, `http://www.particleswarm.info/Programs.html` (Online: accessed July 31, 2010)
40. Clerc, M., Kennedy, J.: The Particle Swarm – Explosion, Stability, and Convergence in a Multidimensional Complex Space. IEEE Transactions on Evolutionary Computation 6(1), 58–73 (2002)
41. De Jong, K. A.: An Analysis of the Behavior of a Class of Genetic Adaptive Systems. PhD thesis, University of Michigan, Ann Arbor (1975)
42. Baker, J.E.: Reducing Bias and Inefficiency in the Selection Algorithm. In: Proceedings of the Second International Conference on Genetic Algorithms and their Application, pp. 14–21 (1987)
43. Syswerda, G.: Uniform Crossover in Genetic Algorithms. In: Proceedings of the 3rd International Conference on Genetic Algorithms, pp. 2–9 (1989)
44. Köchel, P., Kunze, S., Nieländer, U.: Optimal Control of a Distributed Service System with Moving Resources: Application to the Fleet Sizing and Allocation Problem. International Journal of Production Economics 81(1), 443–459 (2003)
45. Zhou, G., Min, H., Gen, M.: The Balanced Allocation of Customers to Multiple Distribution Centers in the Supply Chain Network: a Genetic Algorithm Approach. Computers & Operations Research 43(1-2), 251–261 (2002)
46. Köchel, P.: Hub-and-Spoke-Systems: On Optimal Allocations of Transportation Resources in the Hub under Optimal (s, nQ) Ordering Policies at the Spokes. In: Proceedings of the 2nd German-Russian Logistics Workshop (May 2007)
47. Köchel, P., Thiem, S.: Search for Good Policies in a Single-Warehouse, Multi-Retailer System by Particle Swarm Optimisation. In: 15th International Symposium on Inventories, International Society for Inventory Research, Hungary (August 2008)
48. Hochmuth, C.A.: Modeling, Simulation and Visualization of a Hub-and-Spoke-System. Term Subject, Chemnitz University of Technology (2007) (in German)
49. Köchel, P.: A Survey on Multi-Location Inventory Models with Lateral Transshipments. In: Papachristos, S., Ganas, I. (eds.) Inventory Modelling in Production and Supply Chains, Research Papers of the 3rd ISIR Summer School, Ioannina, Greece, pp. 183–207 (1998)
50. Köchel, P.: About the Optimal Inventory Control in a System of Locations: An Approximate Solution. Mathematische Operationsforschung und Statistik, Serie Optimisation 8, 105–118 (1977)
51. Robinson, L.W.: Optimal and Approximate Policies in Multi-Period Multi-Location Inventory Models with Transshipments. Operations Research 38, 278–295 (1990)
52. Evers, P.T.: Heuristics for Assessing Emergency Transshipments. European Journal of Operational Research 129, 311–316 (2001)

53. Xu, K., Evers, P.T., Fu, M.C.: Estimating Customer Service in a Two-Location Continuous Review Inventory Model with Emergency Transshipments. European Journal of Operational Research 145, 569–584 (2003)
54. Minner, S., Silver, E.A., Robb, D.J.: Silver, and D.J. Robb. An Improved Heuristic for Deciding on Emergency Transshipments. European Journal of Operational Research 148, 384–400 (2003)
55. Kukreja, A., Schmidt, C.P.: A Model for Lumpy Parts in a Multi-Location Inventory System with Transshipments. Computers & Operations Research 32, 2059–2075 (2005)
56. Herer, Y.T., Tzur, M., Yücesan, E.: The Multilocation Transshipment Problem. IIE Transactions 38, 185–200 (2006)
57. Özdemir, D., Yücesan, E., Herer, Y.T.: Multi-Location Transshipment Problem with Capacitated Transportation. European Journal of Operational Research 175(1), 602–621 (2006); ISSN 0377-2217
58. Gong, Y., Yucesan, E.: The Multi-Location Transshipment Problem with Positive Replenishment Lead Times. Technical Report ERS-2006-048-LIS, Erasmus Research Institute of Management, ERIM (2006)
59. Arnold, J., Köchel, P.: Evolutionary Optimization of a Multi-Location Inventory Model with Lateral Transshipments. In: Proceedings of the 9th International Working Seminar on Production Economics, vol. 2, pp. 401–412 (1996)
60. Belgasmi, N., Saïd, L.B., Ghédira, K.: Evolutionary Multiobjective Optimization of the Multi-Location Transshipment Problem. Operational Research 8(2), 167–183 (2008)
61. Köchel, P.: Simulation Optimisation: Approaches, Examples, and Experiences. Technical Report CSR-09-03, Chemnitzer Informatik-Berichte (2009)
62. Hochmuth, C. A.: Design and Implementation of a Software Tool for Simulation Optimization of Multi-Location Inventory Systems with Transshipments. Master's thesis. Chemnitz University of Technology (2008) (in German)
63. Köchel, P., Hochmuth, C.A.: Optimization Experiments with a New Simulator for Inventory Systems with Multiple Locations and Lateral Transshipments. In: Ivanov, D., Meinberg, U. (eds.) Logistics and Supply Chain Management: Modern Trends in Germany and Russia, Proceedings of the 4th German-Russian Logistics Workshop, pp. 67–81. Cuvillier Verlag, Göttingen (2009)

A Fuzzy-Evolutionary Approach to the Problem of Optimisation and Decision-Support in Supply Chain Networks

Sven Schellenberg, Arvind Mohais, Maksud Ibrahimov,
Neal Wagner, and Zbigniew Michalewicz

Abstract. This chapter deals with the problem of balancing and optimising the multi-echelon supply chain network of an Australian ASX Top 50 company which specialises in the area of manufacturing agricultural chemicals. It takes into account sourcing of raw material, the processing of material, and the distribution of the final product. The difficulty of meeting order demand and balancing the plants' utilisation while adhering to capacity constraints is addressed as well as the distribution and transportation of the intermediate and final products. The aim of the presented system is to minimise the time it takes to generate a factory plan while providing better accuracy and visibility of the material flow within the supply chain. The generation of factory plans within a short period of time allows for what-if-scenario analysis and strategic planning which would not have been possible otherwise. We present two approaches that drive a simulation to determine the quality of the generated solutions: an event-based approach and a fuzzy rule-based approach. While both of them are able to generate valid plans, the rule-based approach substantially outperforms the event-based one with respect to convergence time and quality of the solution.

1 Introduction

Understanding and managing a company's supply chain is one of the hardest tasks procurement planners and supply chain managers face in today's

Sven Schellenberg · Arvind Mohais · Neal Wagner
SolveIT Software, Pty Ltd., 99 Frome Street, Adelaide, SA 5000 Australia
e-mail: {ss,am,nw}@solveitsoftware.com

Maksud Ibrahimov · Zbigniew Michalewicz
School of Computer Science, University of Adelaide,
South Australia 5005, Australia
e-mail: {maksud.ibrahimov,zbigniew.michalewicz}@adelaide.edu.au

business environment. Ideally, a supply chain is driven by demands generated from customers placing orders. An order may include one or many order items composed of processed and unprocessed raw materials and components. In addition to the customers' orders which, generally speaking, draw final products out of the supply chain network (i.e., pull factors), supply chain networks are often subject to push factors which are caused by suppliers feeding raw material into the supply chain [1]. In most cases, the supply of unfinished goods cannot be synchronised with the demand generated at the other end of the supply chain, or an adjustment is delayed. Suppliers constantly, periodically, or spontaneously deliver raw material or components into the supply chain network. An arbitrary imbalance is created by customers and suppliers which the supply chain network tries to even out or trade-off (see Figure 1).

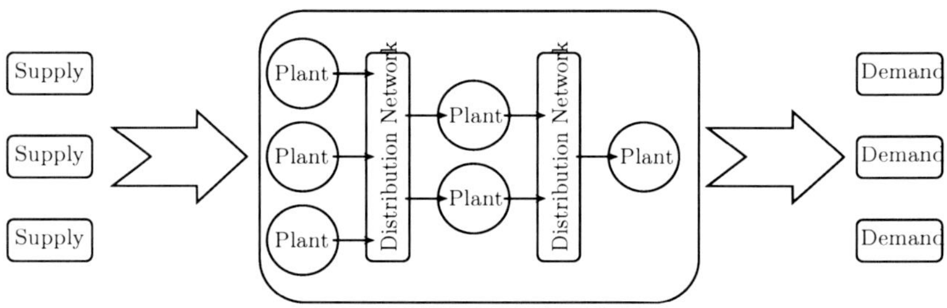

Fig. 1. Schematic view of multi-echelon supply chain network (3-echelon in this case)

Reconciling supply and demand by determining the amount of finished goods to produce becomes a labour-intense and time-consuming endeavour. It involves sourcing decisions when there are more than one internal or external suppliers of raw material, factory management decisions to define production plans and to determine maintenance outages, and decisions on how to effectively distribute the finished goods within the supply chain and to the end customer. Generating an optimal plan which takes all the previous considerations into account becomes virtually impossible for human operators who are in most cases only equipped with spreadsheet tools.

This chapter presents ways to synchronise and reconcile the drivers of multi-echelon supply chain networks demonstrated on a real-world example of an Australian manufacturer of agriculture chemicals. Although the model presented in this chapter aims to represent a specific supply chain model of a particular manufacturer, the components the network is based on represent supply chain entities as they can be found in many other businesses. We present two approaches which generate optimised production and material procurement plans. The system spawned off this project is currently evaluated and fine-tuned, and final completion and total business integration are expected to be finished in a few months. When deployed, the system will facilitate the planning process. The planning includes determining the production across the supply chain (type and quantity of final and intermediate products), scheduling trains and trucks, making sourcing decisions based on contractual obligations, internal supply, and potential supplement deliveries. The operators will have the opportunity to regenerate plans as soon as the environment changes, obtaining a response within a few minutes. The prompt generation of plans allows for strategic planning which could not be achieved by the previous mode of operation. The power of what-if-scenarios can be harnessed to benefit from early structural decisions of the supply chain, such as added production, storage or transportation capacities.

This chapter details two approaches on how to balance the output of producing entities such as plants and factories with storage facilities like tanks, stock piles, or silos. The aim is to generate an optimised production plan for each site including decisions such as sourcing of raw material and production rate while honouring storage capacity constraints.

Although both approaches employ a simulation as part of the solution evaluation, the kind of simulation differs. In the first approach, an Evolutionary Algorithm (EA) tries to find a sequence of events. An event is defined by the date it occurs and an impact it has to the simulation state. The events are stored in a priority queue and executed in chronological order. An event could for instance cause a material changeover or a wind down of the production rate of a plant. An actual simulation only occurs at these discrete events (Discrete Event Simulation, DES); in between any two adjacent sample points (events), the system's state is assumed to be linearly changing, which means a tank's level can be inferred at any given time between two sequential events. In this particular implementation, the reduction of plant's utilisation in order to comply with capacity constraints is performed deterministically, by taking the excess production of a storage and propagating back to its supplying nodes, reducing their production proportionally to their share of the excess amount.

The second approach presented in this chapter does not use events to change the state of the system. As opposed to the previous approach, it tries to find the underlying rules that will balance the supply chain's producers and product changeovers. During the optimisation process, an EA generates a rule base and compares its performance on the simulation run at a predefined interval.

The two above approaches are compared in terms of their overall performance comprising violation of hard constraints such as storage capacity, total final product yield at the end of the run, and satisfaction of demand.

This chapter is structured as follows: After this introduction, the description of the problem is presented in detail followed by the description of the approaches to tackle the stated problem. Experimental results are given in the following Section 4. Before finishing the chapter with a conclusion and outlook of future work, the result of a literature review on related problems is briefly summarised in Section 5.

2 The Problem

In many industries which base their operations on multi-echelon production systems, procurement planners and factory operators often face the same sort of problems when they create production plans for a short or medium term planning horizon:

- What is the best supplier for raw materials (in terms of reliability, contractual bindings, availability, costs)?
- How much of it should be sourced in a given period?
- How much finished or intermediate product should be produced?
- When is the best time to schedule maintenance outages?

There are many more questions that can be considered. The bottom line is that the problem is rather complex, involves trade-offs which cannot be made in isolation, that is disregarding processes that happen further down the supply chain, and all decisions are heavily interdependent. Applying rule-of-thumb reasoning will not lead to an optimal or not even to a feasible plan if hard-constraints such as capacity limits are violated.

In order to understand the complexity of the presented problem and the manual interaction involved, let us consider the following example: A factory plan is supposed to be developed that balances the production and storage capacities of the supply chain presented in Figure 2.

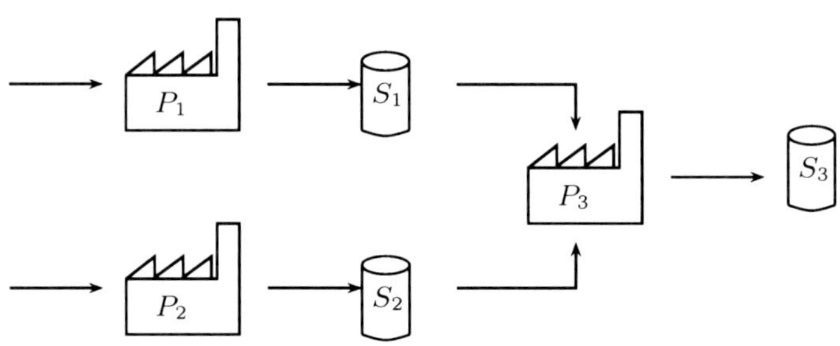

Fig. 2. Simple supply chain example to demonstrate complexity of balancing the components involved.

The network is composed of 3 plants ($P_1 \ldots P_3$) producing material M_1, M_2 and M_3 (P_1 produces M_1, P_2 M_2, and P_3 M_3) at a rate of 50 units per time unit t, and 3 storage tanks (S_1, S_2, and S_3). Assume that all the tanks have the same capacity of 100 units, and S_1 and S_3 are already filled to 80% of their maximum capacity. The planner tries to run the plants as hard as possible, that is, it is desired to run them on maximum production capacity. Running P_1 hard within a period of t means it produces 50 units of M_1 which are stored into S_1. Since there is a constant consumption of

M_1 by plant P_3 an overflowing of the tank does not occur. P_2 also produces its material M_2 and conveys it into its tank S_2. P_3 is sourcing its two input products and converting it at a ratio of 1:1 (which means we obtain 50 units of M_3 at the end of t from P_3). Since S_3 is already filled by 80 units (assume no consumption at this point), an excess amount of 30 units has been produced at the end of the time period. As the excess amount cannot be stored elsewhere, the production of the plant feeding into S_3 has to be reduced.

In this example, a utilisation of 40% would avoid the excess capacity to be created and storage S_3 to contain 100 units at the end of the period. Winding down the production in P_3 however reduces the demand of M_1 which is produced by P_1. The planner has to go back to the producer of M_1 and also reduce the production rate of this plant, as the material cannot be stored in S_1 (which is also filled to 80% of its capacity). This process of propagating back the plant utilisation has to be done for every storage facility that is exceeding its capacity. In this case, only two plants were affected, but for multi-echelon networks, it can be easily conceived that the amount of work that has to be done once a storage constraints are violated is enormous. The dynamics of the environment the plan is going to be implemented in often force the planners to re-create a plan multiple times (imagine unplanned outages of production plants as an example).

In essence, the process of creating such a plan can be very labour-intense, particularly in a real-world dynamically changing environment. Expanding the planning horizon to generate long-term plans (for instance yearly plans) is even more prone to change, as uncertainties and the level of their impact have implications on larger parts of the plan. A benefit of long term planning, the ability to plan strategically by evaluating what-if-scenarios, becomes virtually impossible when plans are generated manually. Analysing the impact of an event and generating a sufficient number of contingency plans is only possible if the time to generate those plans is very low and involves minimal amount of user interactions.

The proposed system automates the planning process to the extent that constraints are entered via an intuitive user interface and a plan is delivered within minutes after the optimisation process has been started. This allows for strategic planning as well as a prompt update as soon as new orders are placed which impact on the production schedule.

3 The Approach

In this section we describe two approaches that we developed to optimise the supply chain of the given business. Both approaches perform a simulation in order to determine the quality (or fitness) of the solution. The first approach generates a sequence of events and applies these to a discrete event simulation (DES). We call this one "Event-Based Optimisation" approach. A simulation is performed at any event occurring (next-event time advance [2]). In contrast, the second approach uses rules to make decisions for factory utilisation and sourcing. The "Rule-Based Optimisation" approach uses also a simulation to evaluate evolved solutions, but advances the time at a fixed interval (fixed-increment time advance [2]). In addition to the supply chain model being used and the common parts of the EA, this section describes both approaches in detail and lists advantages and disadvantages.

3.1 The Supply Chain Model

When modelling a real-world system, careful attention has to be paid to the level of abstraction of the model. On the one hand, a high level of details improves the fidelity of the examined properties of the system, but on the other hand, the simulation process becomes computational expensive which prolongs the run time of the actual optimisation. A trade-off has to be made between reflecting as much as possible to draw the required conclusions from the system and minimisation of details. The model we employ for both of our optimisation approaches is described as follows:

The supply chain network can be thought of as a directed graph $G = (V, E)$ with $V = \{0, \ldots, n\}$ as a set of vertices and $E = \{0, \ldots, m\}$ as the set of edges. Each vertex/node can be of any of the three following types:

1. Plant
2. Storage
3. Switch

Although nodes of different type process material differently, the three node types have common properties. Each node has a set of predecessors which they depend on with respect to their source products. These predecessors are defined by the incoming edges of the vertex. Figure 3 illustrates a very simple supply chain network. *Plant 1* produces two products, stores both of them in separate storages and *Plant 2* converts *Product A* into *Product C* which is finally stored into *Storage 3*. Note that in this simplified version, *Plant 1* has no predecessors, whereas *Plant 2*'s predecessor is *Storage 2*. The conversion from a raw material into an intermediate or final product can be thought of as a chemical reaction with arbitrary input and output products.

A switch, the third kind of supply chain node, is not a physical entity, but it serves to route the material through the supply chain network. Due to the internal design and the paradigm to keep functional units as simple as possible, plants and storages are only allowed to source a material from one predecessor node. In case there is more than one predecessor nodes which output the same product, a switch has to be inserted which controls the material flow (see Figure 4). A switch can either implement a local heuristic to determine which successor to deliver to or it may be controlled by the optimiser which evolves the routing as part of its usual individual reproduction process (the switch becomes part of an individual's genotype). An example for the former may be that *Switch A* is implemented in a way such that it always exhausts the capacity of a tank before it starts filling up another tank

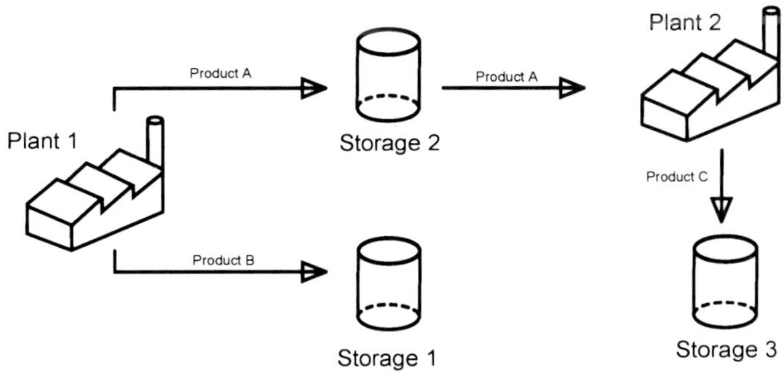

Fig. 3. Supply Chain model showing relationship between supply chain nodes

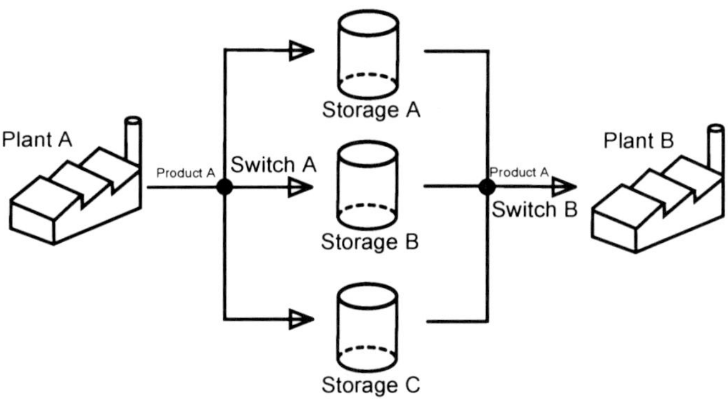

Fig. 4. Supply Chain illustrating material flow routed by Switches

(see Figure 4, *Switch A*). The local deterministic heuristic incorporated into *Switch B* may reduce the storage's level evenly draining all tanks at the same time.

In addition to nodes having predecessors, nodes also contain output buffers (one per product) in which the products they produce or store and their quantity are temporarily kept. Switches do not contain output buffers, but delegate requests for material to their predecessors depending on their implemented routing logic.

The previously described node-predecessor relationship only works for continuous material flow such as liquids that are pumped from a producing plant into a storage tank. In many supply chain networks, the distribution of material is done via transportation means with limited capacity or infrequent transportation times (trucks, ships, air planes or railway). Therefore, an additional property of the edge between two nodes is a schedule defining the availability of the transportation means and its capacity. Only when the transportation means is available at the plant or storage, it can empty the node's buffers and store the material temporarily in its own buffer. Plants either have to shut down or store their products into adjacent tanks at times at which the transportation means is not available.

Given the supply chain network of the particular business, a simulation process can be performed that is mostly similar for both optimisation approaches. The only difference is in the way the decisions are inferred that change the state of the supply chain network.

Before the simulation is started, the supply chain nodes have to be initialised and a list of the sample points (event queue) will be created. During the initialisation process of the nodes, the buffers of some storage entities are filled to simulate an opening stock. The next step is to determine the points of the planning horizon at which the system is sampled. These points are the set of all events that occur during the planning horizon, such as

- Change of availability of a plant (plants may run at reduced run rate due to partial maintenance or outages).
- Product changeovers.
- Arrival/departure of transportation means travelling between two nodes.

It is of paramount importance to the validity of the simulation to add all events that cause a discontinuous change of the supply chain network's state, that is, a change that causes nonlinearity which would prevent inference of nodes' properties in between two adjacent sample points, to this event queue. If desired, additional sample points can be added to verify the result of the simulation.

Once the event queue is filled, the nodes are sorted by precedence. Nodes without predecessors occur first whereas terminal nodes that base their buffer's material and quantity on the result of all previous components' production are located last in this list. The next step is to iterate through the list of predecessors (beginning with the least dependant node) and process the node. Processing a node breaks down into three steps:

1. Pull resources (raw material, intermediate material, final product, etc) from predecessor.
2. Apply the conversion rule.
3. Push the converted material into output buffer(s).

The first step is straightforward: Since each node knows about its predecessors and the material it required, it pulls the maximum amount of this material from each of its predecessor nodes. Switches forward the call to their pull method to the appropriate predecessor. Essentially, the pull step boils down to copying the content of the predecessor's output buffers and passing it on to the conversion step.

In the conversion step, the actual business logic for each plant is implemented. If we consider a chemical formula like $1A + 2B \rightarrow 2C + 1D$ with hypothetical elements A, B, C and D, the factory that processes this formula would source all the material it could get for product A and B and determine the quantities of the resulting product C and D (taking into account the ratio of A:B=1:2). The amount of input material, which has actually been used during the conversion process, is reduced from the predecessor's output buffers to account only for the actual consumption and to determine overproduction. While *switches* do not implement a conversion routine, they pass on the incoming materials to their output buffers.

Finally in the last step, the produced material is stored into its output buffers to be available for the successor node's processing procedure. At the end of the sample cycle, that is, once all nodes have been processed, the simulator captures the state of the system by storing the buffers which contain the quantity/products tuple for each node. The final buffer capacity allows determining the plants' utilisation or storages' remaining capacity. The simulation terminates once the planning horizon's end date is reached.

3.2 Features of the Optimiser

The meta-heuristic used to optimise the supply chain network is a steady-state EA, that is, only one individual alteration occurs at each generational step, replacing a parent. EAs are well understood, reliable and they can be customised to solve a variety of problems of different domains. They have been applied to other problems of similar level of difficulty and performed satisfactory. This section recapitulates the general working principle of the EA and elaborates on the specific implementation used here.

The EA operates as follows: The population contains a number of individuals (approximately 100 individuals) which encode a solution in their genotype. After initialising and evaluating the individuals of the population, the recombination process is started. As part of it, an operator from the set of available operators is chosen to alter the individual. These operators are application-specific as their operation strongly depends on the encoding chosen as well as the particular business problem they try to solve. Some operators may have the ability to consider certain business constraints to operate in the feasible search space which reduces the search space and leads to faster convergence. The specific operators used for both optimisation approaches are explained in the following sections.

After the operator changed the individual, the new individual is immediately evaluated. A self-tuning procedure compares the fitness value of the individual before and after the operation and adjusts operator weights according to whether the new individual yields a better solution or not. The weight is taken into account the next time an operator is chosen. In case an operator performed poorly, it is less likely to be selected for evolving the next individuals (similar to roulette-wheel selection). After a defined number of generations, the weights are reset, so dominating operators have to proof again their performance. This is particularly important if the optimiser prematurely converges and the optimisation process gets stuck at a local optimum. The previously preferred operator would not contribute to improving the solution's quality which means other (potentially more explorative) operators come into play.

The evaluation procedure uses application-specific measures to determine the fitness of the individual. For this particular optimisation problem, both optimisation approaches use an identical fitness evaluation function. Part of the evaluation is the simulation. Once finished, the system state is evaluated. The objectives of the optimisation process for supply chains include, but are not limited to, (a comprehensive list of measures of supply chains can be found at [3]):

- Maximisation of product yield.
- Minimisation of product changeovers.
- Adherence to a specified product ratio for the remaining planning period.
- Satisfaction of demand.
- Minimising transportation, inventory, backlogging costs.

The optimisation process is terminated once a pre-defined number of generations is reached or if the search stagnates over a certain period of generations. The best individual that was found throughout the optimisation process is returned (keep-the-best strategy [4]).

Both techniques used in this chapter leverage of the same optimisation engine described in this section (and in fact applications for other customers do as well). Different business rules are encoded within the individual and as operators which makes the EA reusable for other optimisation problems.

3.3 Event-Based Optimisation

The event-based optimisation approach tries to determine a sequence of events that, when applied to the simulation, results in an optimal solution. Events are defined by the date they occur and the action they perform, i.e., the state change they cause to the system. Examples are a changeover event in a factory which causes a factory to produce a different product, a change in the factory's utilisation, a change of the availability of the factory or a factory specific event such as a cleanout of storage tanks. The event-based approach has a very shallow hierarchy of representations. From the event sequence (which can be understood as the genotype) a conversion into the final solution is made by means of the simulation.

At the start of the optimisation run, each individual is initialised with a random sequence of events. The operators alter the event queue in different ways. They change the type of event (which changes the action they perform) or the date of their occurrence as illustrated in Figure 5. Some of them insert a delay at a specific time which causes all subsequent events to be delayed. A crossover operator randomly determines an event of the event queue of two individuals and swaps all of the following events with the other individual, similar to the classical crossover operation for genetic algorithms.

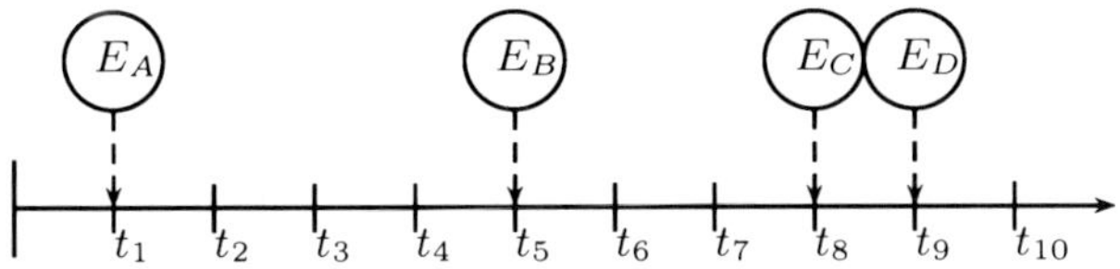

(a) Event queue before mutation

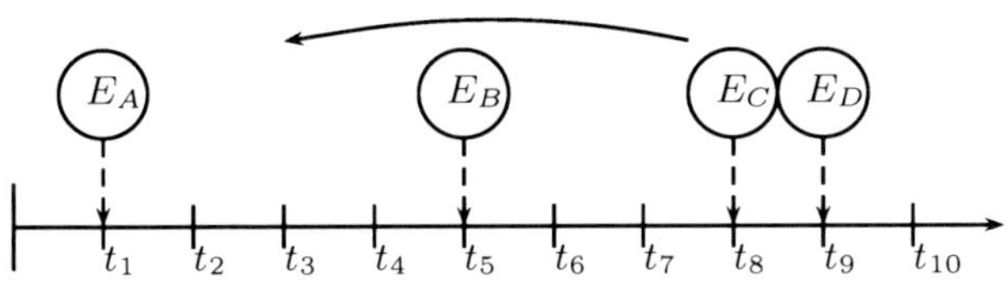

(b) Mutation operator in action changing date of E_C to t_3

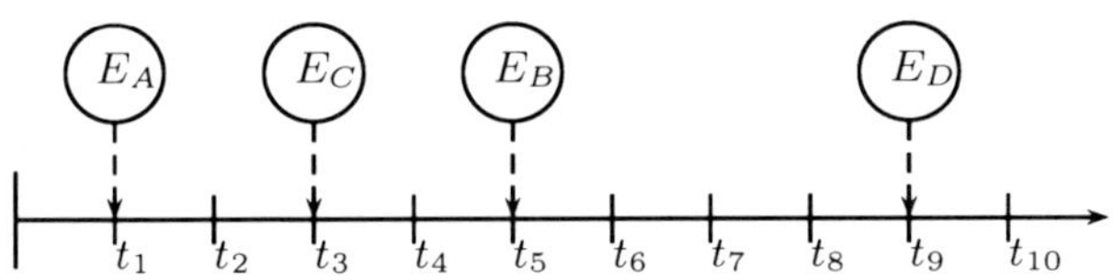

(c) Event queue after mutation. E_C now alters the state at t_3 according to its encoded type.

Fig. 5. Timeline of the simulated horizon illustrating the impact of the date mutation operator on the event queue (dashed lines indicate the time events occur, i.e. when they cause a change of the state of the simulated system).

One part of the evaluation process is to transform the genotype of the individual (the representation of the solution we apply the operators to) into a representation that can be evaluated (the other part is to aggregate the fitness value components). The transformation is done by running the simulation according to the encoded sequence of events. At each of the event dates, a simulation is executed which samples the state of the system (e.g., fill level of storage tanks, excess level, utilisation, amount of raw material sourced, etc). After the event date, the system state will change non-linearly (for instance, a different product is produced; a plant shuts down its operation, etc). Once the first simulation pass is completed a repair function is triggered. Before the repair, each plant's produced products are consumed by their subsequent storages, disregarding the current available storage capacity.

The repair function deterministically winds down the producers (the plants) in order to balance the production with the available storage capacity. This process starts at the last storage in the supply chain, i.e., the one that relies on all of the previous nodes, and works back to the first node in the supply chain (i.e., the one without any predecessor nodes). The excess amount of an overflowing storage is proportionally reduced from the previous producers in order to determine the appropriate production rate (planned utilisation). Each time a plant's utilisation is adjusted, a partial simulation has to be performed again as the minimised production has implications to other supply chain nodes downstream in the supply chain network. The advantage of this deterministic method to compute the utilisation is that these kinds of events do not fall into the search process and, therefore, decrease it.

On the other hand, this way of adjusting the plant's utilisation is rather expensive as the simulation is rerun every time the utilisation is changed. An optimisation process not honouring the storage constraints, that is, with disabled repair function, was able to evaluate about 1500 individuals per minute, whereas a normal optimisation process (repair-enabled) could only evaluate a maximum of 150 to 200 individuals on the same machine and the same supply chain network (this comparison of course is not exact as violated storage constraints result in more raw material being available which skews the production plan and the obtained product yield).

3.4 Rule-Based Optimisation

The approach proposed in this section is based on a combination of fuzzy logic and EAs. Rather than directly evolving decisions made during the simulation of the supply chain (in form of events), rules are generated which drive the decision making process.

Since this book is mainly about EAs, a very short introduction into Fuzzy Logic is given in this section. Thereafter, the application of fuzzy logic to the subject of this chapter, optimising supply chains, is demonstrated and the particular implementation is discussed by providing examples on encoding and decoding of the individual's genotype.

3.4.1 Fuzzy Logic

Fuzzy logic is based on the fuzzy set theory in which values are expressed as degree of membership rather than crisp inputs like discrete measurements. If we were to classify the age of a person having two categories (young and old), classical Boolean logic would either map a given age to young or old. Let's say the cut-off point separating old from young would be exactly 40 years of age. In Boolean logic, every person below 40 years would be young, whereas everyone with an age greater or equal to 40 is classified as old. Instead, in fuzzy logic, an age can belong to multiple categories (or degrees of membership). The degree of membership is denoted by a number between 0 and 1. Given two membership functions *young* and *old*, a person aged 30 years could be young to the degree of 0.8 and old to the degree of 0.3 (depending on the shape of the membership functions). Fuzzy terms such as "age is young" are called linguistic terms ("age" is a linguistic variable and "young" a linguistic value). Those terms can be combined by Fuzzy-And or Fuzzy-Or (or other fuzzy operators, or T-Norms [5]) and further extended to IF-THEN rules of the form "IF age IS old AND health IS bad THEN health-insurance-premium IS high" (The IF part is called "antecedent" and THEN part is named "consequent"). Many of these rules can make up a rule base.

The type of fuzzy logic system (FLS) as it is used for this approach is called Mamdani-type FSL [6]. The fuzzy inference process can be decomposed into three major steps. First of all, the crisp input data obtained by measurements is transformed into fuzzy sets (fuzzification step) by determining the degree of membership of the fuzzy terms specified in the fuzzy rule applied. The next step combines the degrees of the fuzzy sets by the given fuzzy-operators and summarises each term into a rule weight. This rule weight is used to determine the output of the rule by limiting the shape of the output function. The last steps, the defuzzification, combines all reshaped output functions and applies a defuzzification function in order to obtain a crisp output value. This function is usually the centre of mass or a weighted average function.

3.4.2 Application of FL to SC Optimisation

Translated to our actual problem of generating decisions to drive a supply chain simulation, the fuzzy rules determine in their consequent part when to schedule product changeovers (that is which products are produced at which time) or the utilisation of the factories (utilisation of the facility). The advantage for using fuzzy logic to encode the rules driving the supply chain decisions is that these natural language rules facilitate diagnostics and audit features of the system. A planner controlling the system can deduct the reasoning of the system by analysing the rule bases which is more intuitive than the previous event-based approach.

Another downside of the event-based approach is also that the EA has to find the correct event for each time, albeit the conditions may be similar to a previous point in the plan at which the correct decision was made. Say a product changeover has to be triggered every time a storage tank is about to overflow, as the product is consumed from the overflowing tank as soon as the changed-over product is produced. This changeover may not be triggered at another point in time as the EA has not yet generated such an event at this time. It may or may not randomly generate such an event at the particular time. The rule-based system however would have had developed a rule that triggers a changeover upon reaching of maximal storage capacity of the tank, which means every time the condition holds (storage tank is high), the changeover is triggered. The application of rules makes the generated plan become more predictive and coherent. In addition, the underlying logic can be investigated and manually fine-tuned.

The individuals that are evolved in the evolutionary process encode the rule base. We decided to generate the rules by the EA as they may differ for different settings and different products. Evolving the rule base by the EA allows adaptation to the current constraints and environment. There are many other means to combine EAs with fuzzy logic. They differ in the part of the fuzzy inference system that is subject to the optimisation. Genetic tuning for instance changes the database (shape or number membership functions, linguistic terms) of the fuzzy inference system. Other methods evolve the knowledge base or generate new knowledge base components. An elaborate taxonomy and survey on methods to combine genetic algorithms and fuzzy logic systems (GFS) can be found at [7].

A fuzzy rule (consider this example for explanatory purposes: `IF Level_Of_Tank IS high THEN Utilisation_Of_Plant is low`) of the structure consists of an antecedent part (`IF ...`) and a consequent part (`THEN ...`). The antecedent can contain multiple linguistic terms (`Level_Of_Tank IS high`) which can be combined by different operators (T-Norm). For our purposes, Fuzzy-And and Fuzzy-Or are sufficient. A linguistic term has two parts, the linguistic variable (`Level_Of_Tank`) and the linguistic value (`high`). The number of linguistic values can be arbitrary, but in order to reduce the complexity, we opted for 5 linguistic values and triangle membership functions (see Figure 7).

The possible linguistic variables are the level of each storage tank, the currently produced product of a plant and previous utilisation. A set of rules forms a rule base. For the proposed system, a number of rule bases is created, one rule base per possible consequent part. Since the linguistic variables for all the rules of a rule base are the same (that is, all rules in one rule base pertain to the utilisation of a specific plant), the only additional information

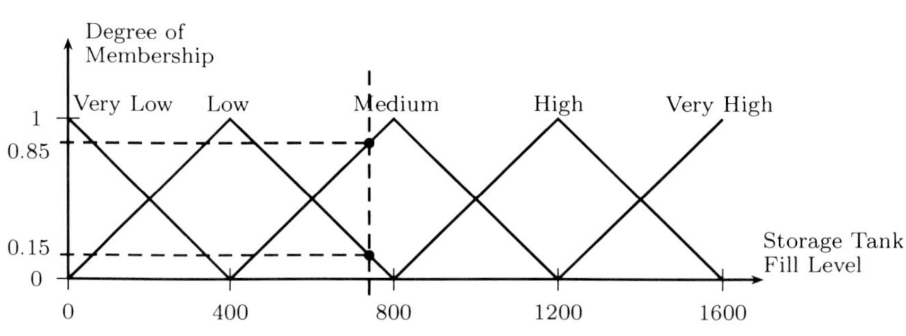

Fig. 6. Overlay of 5 membership functions denoting the fill level of a tank. At a level of 740l, the tank level is member of the "low" function by 0.15 and member of the "medium" function by 0.85. For each storage, the parameters for each membership function are different as the maximum capacity may differ (and hence "very high" would relate to a different maximal level).

that needs to be encoded in the chromosome is the linguistic value of the consequent part. Figure 6 displays the integer number vectors encoding each rule. A rule base to n rules encoded as sketched in Figure 9: The first index represents whether the rule is enabled, the second specifies the fuzzy operation which combines the linguistic terms (Fuzzy-And or Fuzzy-Or) and the following pairs of integer values contain a pointer to a look-up-table of linguistic variables and linguistic values. The last cell holds the linguistic value for the rule base's consequent part.

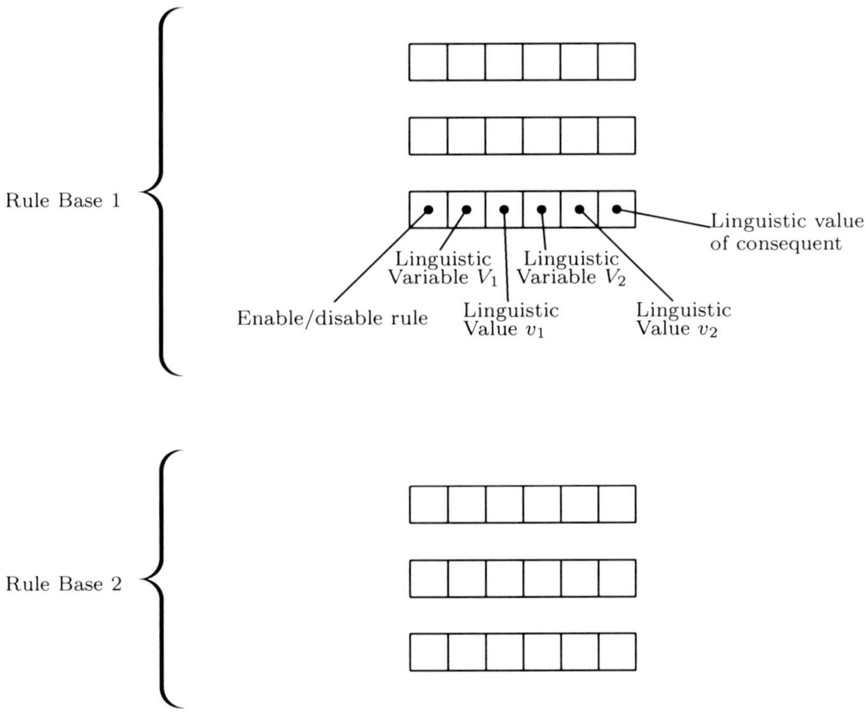

Fig. 7. Encoding the rule bases. Elements in squares are of type integer number and stand either for booleans or indices in look-up-tables that hold linguistic variables and values.

3.4.3 Generation of Rules

As in the previous section, the same EA with its self-calibrating capabilities is employed. The only differences lie in the method of evaluation, that is, how the simulation is carried out, and the set of operators used to evolve the individuals (as the encoding of the individuals). The evaluation function itself is exactly the same. It takes into account the maximal production yield, a penalty is deducted in case storage constraints are violated and also a penalty in case of a delay in providing enough final product to satisfy firmed orders is applied.

Unlike the event-based optimisation, the rule-based approach samples the system at pre-defined intervals. At these sample points, all properties of the simulated system are evaluated and actions are derived from the current state. These actions are triggered by the rules described above. A rule may pertain to the fill level of a storage shed and cause a reduction of the feeding plant upon reaching of a "high" fill level.

Different operators modify the genotype of the individual at each recombination step. A mutation operator randomly changes bits of the genes by honouring the feasible maximal possible integer number value at the position in the chromosome. The meaning of such a mutated chromosome changes in the decoding step which leads to different rules and thus different decisions in the simulation step (see Figure 8).

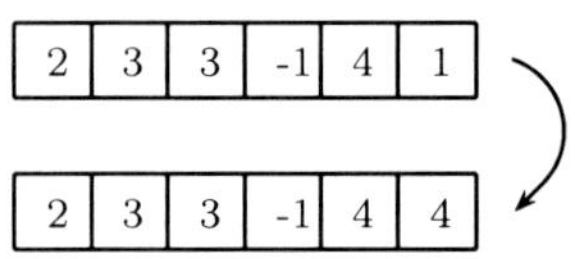

(a) Mutation of chromosome (before and after mutation)

```
Before: IF Level(TankC) IS high
THEN Util.(PlantA) IS low

After : IF Level(TankC) IS high
THEN Util.(PlantA) IS very high
```

Index	Ling. Variable	Ling. Term
0	Level(TankA)	very low
1	Level(TankB)	low
2	Level(TankC)	medium full
3	Level(TankD)	high
4	Util(PlantA)	very high
5	Util(PlantB)	
7		
...		

(b) Phenotype decoding

Fig. 8. Impact of mutation on decoded phenotype ('-1' means the term is not considered in the rule)

Another typical operator for genetic algorithms, the crossover operator, was also adapted and implemented. Two flavours of single point crossover are employed. One of them cuts individual rules of two parent individuals into two pieces (at a random crossover point) and swaps its right-hand part with the other parent. The other crossover operator swaps entire rules at once by determining again a crossover point and replacing one part with the parent's rule set.

4 Experimental Results

We have tested the two approaches on two identical supply chain networks. The aim was to maximise production while honouring storage constraints. The EA was configured to terminate its search after a maximum of 5000 generations, or prematurely if the search would not yield any improvement within 1000 generations. By virtue of the system, the event-based approach runs a simulation whenever an event occurs. The rule-based approach was configured to sample the system at a fixed interval of one day and change the system state by applying its rules.

Both algorithms were able to generate feasible solutions without violating constraints. The quality of the averaged solutions of each approach, however, differed significantly. While the event-based approach managed to fill up the final product storage shed to $172,000$ tonnes (see Figure 9), the rule-based approach exceeded this value by 37% ($233,000$ tonnes). In addition, the search procedure of the latter one terminated much earlier (on average at about 1300 generations) while the event-based algorithm used up the full span of available generational cycles. Another interesting observation is that the time it takes to evaluate an individual is much less for the rule-based approach (approx. 800 individuals per minute vs. 150 i/min). This and the premature termination caused the rule-based approach to find an (even better) solution after a few minutes of run time only.

As already stated in the introduction of this chapter, the system presented is applied to a real-world problem and, as such, it is difficult to compare it to synthetic problems as they are usually used as a baseline in academia. The only plausible baseline can be obtained by comparing the factory planner's schedule and the expected product yield with the schedule generated by the system. Preliminary test results indicate a high degree of similarity between human and system generated factory schedules with respect to the length of product runs (or in other words, the date of scheduled product changeovers) and accumulated product yield at the end of the planning horizon. Taking only these few measures into account, one can conclude that the model adequately represents the supply chain. The time it takes to generate a yearly plan by the planning personnel is tremendous. This means a what-if-scenario analysis becomes virtually impossible. Unless we deal with strategic what-if-scenarios, the result of such a scenario would become obsolete by the time it is obtained. Using the proposed system, a near-optimal plan for an entire year could be created in less than 10 minutes for the event-based approach and about 2 minutes for the rule-based approach respectively (carried out on a standard Dual Core 1.6GHz computer optimising a 5-echelon supply chain). This allows for what-if-scenario analysis especially for short term planning horizons.

Another observation worth mentioning is that in some instances, the optimiser made decisions that were, by a human operator, hard to justify. These decisions dealt with production trade-offs that were done early in the planning

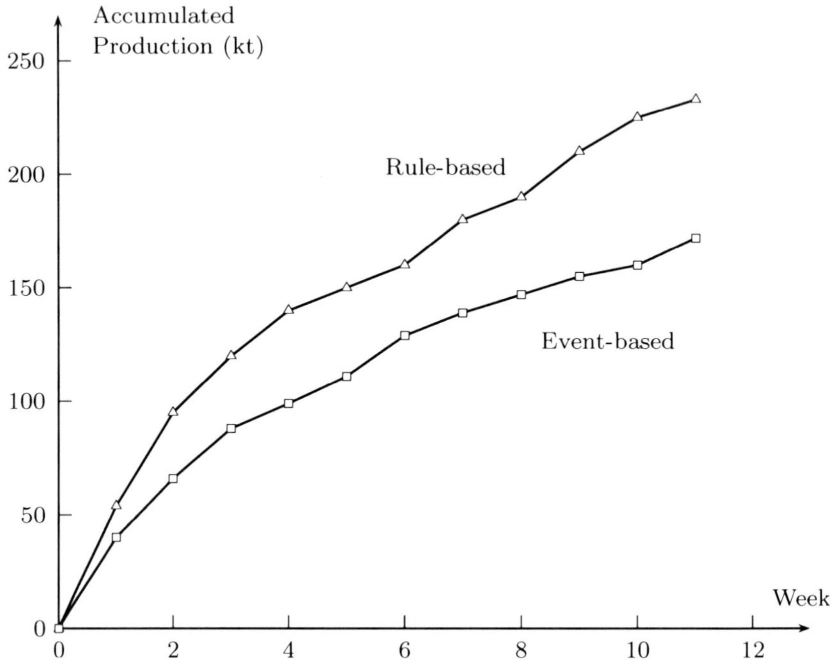

Fig. 9. Total product yield of event-based and rule-based approach (in kilotonnes)

period in order to benefit from an event that happened later with the aim to increase the overall production. This may be a valid decision with respect to the evaluation function (higher production means fitter individual), but, since we deal with a real-world environment, one has to consider the uncertainties that the future may bear (especially for long-term plans). The future benefit the optimiser was speculating for may in reality never materialise which would result in an overall inferior plan (compared to one that would not have made the trade-off decision in the first place). As a consequence, we introduced a staged optimisation that partitions the planning horizon in periods which are optimised in isolation. Once a period has been optimised, the resulting stock is carried over into the next period serving as opening stock.

We ran a trial to determine the impact of partitioning the optimisation. To obtain a normalised result we limited each period's runtime according to its share on the overall planning horizon. Optimising the whole period at once took about 8 minutes. For the test case with two periods, a maximum optimisation time of 4 minutes was allocated for each of the periods. Essentially, this means we allocated the processing time evenly, as opposed to allowing the EA to exhaust the maximum of 2000 generations for each period (in which case the result would be skewed as the search space is only half the size, but the same amount of computational resources are applied to optimise). The simulation was run four times over the whole period and the planning horizon split in 2, 4 and 8 periods. Figure 10 confirms our initial assumption. The total yield drops by about 25% when comparing the optimisation over the entire period to the 8-segments-run. The segmented solutions are of lower quality in terms of the individual's fitness, but when audited by human planners they are much more viable as they exploit short term opportunities while ignoring higher, yet more unlikely, long-term gains.

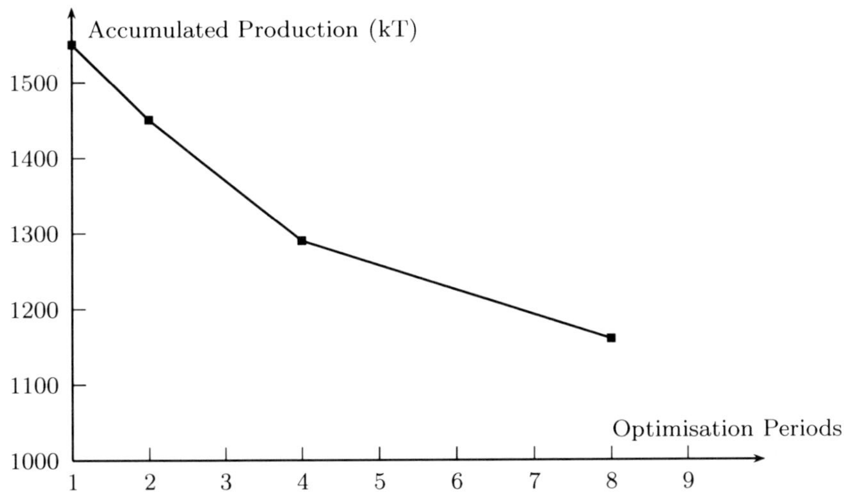

Fig. 10. Decline of total yield when optimising periods in isolation

5 Related Work

The optimisation of supply chain networks has been an ongoing research topic for many years. The research work can be categorised into two major streams. Either a supply chain has to be built from scratch with optimal location of production facilities, storage and distribution centres, or an existing supply chain has to be managed in order to make decisions that pertain sourcing of raw materials, amount of production, etc. A combination of both is also possible in which case an existing supply chain is supposed to be structurally altered (e.g., by adding new sites or negotiating new supply contracts). Structural or management decisions can also be considered with regards to their planning horizon. The former decisions are of strategic and the latter ones of tactical and operational nature. This section identifies and discusses a collection of related work undertaken in the area of optimising supply chains focussing on both streams: Operational/tactical as well as tactical planning.

An approach that addresses long and short term planning questions is presented in [8]. In this paper, the authors describe a system that uses a hybrid technique of mixed integer programming, a genetic algorithm, and discrete event simulation to make optimal decisions of where to produce (internally or externally), production planning, transportation as well as strategic decisions on location and capacity of facilities. While the genetic algorithm is employed to optimise sourcing policies and qualitative variables, mixed integer programming reduces computational costs by solving quantitative variables. The effectiveness of the obtained supply chain configuration is evaluated by a simulation.

In [9], an discrete-event simulation facilitates the evaluation of supply chain scenarios of a food supplier. The results obtained from the simulations suggest ways to improve the supply chain by changing inventory strategies. As a result, the stock level could be reduced which was beneficial to the freshness of the products, and additional products could be introduced as shelf space was freed.

Other researchers concentrate only on isolated parts of the supply chain. [10] describes a hybrid simulation-optimisation approach with the objective to select the best supplier of a supplier portfolio (in a strategic way rather than for daily sourcing: the supplier found is used throughout the planning horizon). A genetic algorithm is employed to search for possible configurations of suppliers. As with the previous approach, a discrete-event simulation determines the key performance indicators (KPIs) that form the input for the evaluation function.

Another approach concentrating on parts of the supply chain in isolation is discussed by Xie and Petrovic in [11]. Their approach defines a new decision-making system for stock allocation that is based on fuzzy IF-THEN rules. Initially, the rule base is generated by domain experts, but they allow for alteration by changing the rule's weights. A simulation on a 2-echelon supply chain (1 warehouse, and multiple retailers) is run to demonstrate the effectiveness of this approach.

Fuzzy set theory is also applied in [12]. Unlike [11], Wang and Shu use fuzzy logic to model uncertainty such as demand, processing time and delivery of supplies. The objective of their work is to develop a supply chain model that minimises the inventory costs by meeting the demands. A genetic algorithm tries to find the optimal order-up-to levels for all stock-keeping units. Again, this approach only looks at one objective (minimisation of inventory costs) and one method to achieve this objective (reduction of inventory).

The common denominator of the above papers is the application of a simulation (mostly DES) in order to obtain properties of the supply chain and evaluate the performance of the optimisation method. This observation is also backed by several surveys such as [13, 14, 15]. Methods used to obtain inputs for the simulation are mainly genetic algorithms (GA). However, even though GAs seem to dominate, recent publications indicate an advent of fuzzy logic systems as drivers for the simulation process.

Another aspect worth considering when building a solution that is meant to be reusable is the structure of the building blocks of the supply chain, that is, the model. Many attempts have been made to develop a unified supply chain model and terminology. Some of them like [16] represent common terminology or [17], which emphasises configurability and proposes event-discrete simulation as means to analyse supply chains, but this work has its emphasis on business processes rather than the definition of reusable component as they are desirable to create a programming model. Others like [18, 19, 20] use special modelling languages to express the complexity of business processes and automatically generate simulation models. Although all of them suggest employing simulations to analyse supply chain networks, they require expert knowledge of modelling languages like Rockwell Software's ARENA, one of the prevalent languages for modelling supply chain networks. The resulting models may be generated in a programming language that is incompatible to the rest of the system. We believe that a generic supply chain model can be developed that supports both, flexibility and ease of use when creating the model without the necessity to acquire expert knowledge on simulation languages. The ideas presented in this chapter are implemented in a framework which is employed to model the supply chain operations of a real-world business.

6 Conclusion and Future Work

In this chapter, we looked at the application of an EA to the problem of optimising the key decision points in the supply chain network of a major agricultural chemicals company. Modelling the highly complex nature of that company's operations was the first part of the challenge of successfully accomplishing this endeavour. A balanced mix of discrete and continuous event simulation had to be used. Furthermore, the model was designed in such a way as to be amenable to use within the framework of an EA.

Of key importance was developing the ability to represent the decision points in the supply chain network simulation as entities that could be manipulated by evolutionary operators. This was achieved in different ways. For the event-based approach, firstly by allowing the timing of decision events to be determined by values within the candidate individual representation, and also the types of those decision events. Thus, for example, a candidate individual could specify that the operation of "change from product A to product B" could be scheduled to happen in a particular part of the network on a given date and time. Secondly, the processing logic of "switching" nodes of the network could be manipulated by evolutionary operators. For example, the production of a particular chemical could involve a number of ingredients which have to be drawn from a variety of sources. In some cases, it might be easy to determine a set of rules to express the correct routing logic to use. In other situation, it may not be obvious and thus it would be preferable to let the individual represent decision making logic, as part of the encoding, and then evaluate the performance of evolved logic as a component of the overall fitness evaluation process.

The rule-based approach abandons the idea of having isolated nodes that handle the sourcing. In contrast to the event-based system, a decision is made at each sample point based on the current state of the system. The resulting plan produces consistent and traceable decisions. Despite the fact that this approach adds the process of balancing supply and storage capacities to the search space (as opposed to running a deterministic repair function), it is able to find a solution much faster while generating even better solutions. Reasons for this observation may be that the search space of the event-based approach, that is all combinations of type and date of events, seems to be larger than the permutation of rules used in the rule-based approach. In addition, many infeasible solutions seem to be generated which demand for repair by running a costly re-evaluation.

Using rules to make decisions constricts the search space. However, only those potential solutions seem to be neglected that are less viable, as indicated by the better solutions obtained. The reason is less surprising when considering the nature of the problem. Changeovers, reduced plant production and sourcing of material are based on rules. A changeover occurs once a tank is reaching its maximum capacity, the production is reduced upon downstream bottlenecks and a sourcing decision depends on minimal procurement costs. Trying to find these decisions without understanding their natural cause is more expensive and bears many lost opportunities compared to a supply chain that is driven by a condition-decision scheme as we presented it.

The results presented in this chapter also illustrate the trade-off that is frequently accepted by business managers in practice. By running the software in a global mode over a large time-frame, an excellent result could be achieved, but the validity of such a result could be doubted by human managers who would rightly point out that the further out into the future that assumptions are made about supply chain conditions, the less reliable those assumptions would be. Hence we chose to apply the simulation/evolution algorithm over the whole time-frame in phases, starting with a short term phase of a few months, and then looking further into the future. This gave our approach the benefit of seeking higher levels of optimisation in the short term, wherein knowledge of conditions is quite firm, and then freezing those results and progressively expanding the scope of inclusion to seek out optimisation further into the future.

As constructed, the software application arising from the rule-driven simulation model and the EA presented in this chapter was able to provide invaluable insight and speculative modelling ("what-if") capabilities to managers of the client company, allowing them to find ways to optimise their supply chain network, and of course maximise production. This is the litmus test of the value of this application, and it is a rewarding application of the power of evolutionary computation to a real-world business.

In addition to our work presented in this chapter, several promising ideas may improve the results obtained. As an example, we aim to expand the linguistic terms that can be taken into account by adding future availability of plants as well as past utilisation to smoothen out the plant's overall utilisation.

A promising method to improve the evolution of the rule bases is to employ a co-evolutionary approach in which rule bases would be developed in isolation. An instance of an EA would only concentrate on its designated rule base (i.e. one EA instance could be employed per plant to evolve rules for its utilisation, one EA instance for sourcing, etc.) passing on the best of its rule bases to form the overall solution rule bases.

This chapter has presented only the current state of our endeavour to find a common model and methodology for optimising supply chain networks which caters for many business cases and industries. We expect the current method to be fine-tuned and extended to allow maximal generalisation and applicability.

References

1. Hinkelman, E.G.: Dictionary of International Trade Handbook of the Global Trade Community, 6th edn. World Trade Press (2005)
2. Law, A.: Simulation Modeling and Analysis (McGraw-Hill Series in Industrial Engineering and Management). McGraw-Hill Science/Engineering/Math. (2006)
3. Gunasekaran, A., Patel, C., Tirtiroglu, E.: Performance measures and metrics in a supply chain environment. Internation Journal of Operations & Production Management 21(1), 71–87 (2001)
4. Michalewicz, Z. (ed.): Genetic algorithms + data structures = evolution programs, 2nd edn. Springer-Verlag New York, Inc., New York (1996)
5. Hájek, P.: Mathematics of Fuzzy Logic (Trends in Logic), 1st edn. Springer, Heidelberg (1998)
6. Mamdani, E.H., Assilian, S.: An experiment in linguistic synthesis with a fuzzy logic controller. International Journal of Man-Machine Studies 7(1), 1–13 (1975)
7. Herrera, F.: Genetic fuzzy systems: taxonomy, current research trends and prospects. Evolutionary Intelligence 1(1), 27–46 (2008)
8. Truong, T.H., Azadivar, F.: Simulation optimization in manufacturing analysis: simulation based optimization for supply chain configuration design. In: WSC 2003: Proceedings of the 35th Conference on Winter Simulation, pp. 1268–1275 (2003)
9. van der Vorst, J.G.A.J., Beulens, A.J.M., van Beek, P.: Modelling and simulating multi-echelon food systems. European Journal of Operational Research 122(2), 354–366 (2000)
10. Ding, H., Benyoucef, L., Xie, X.: Simulation optimization in manufacturing analysis: a simulation-optimization approach using genetic search for supplier selection. In: WSC 2003: Proceedings of the 35th Conference on Winter Simulation, pp. 1260–1267 (2003)

11. Xie, Y., Petrovic, D.: Fuzzy-logic-based decision-making system for stock allocation in a distribution supply chain. Intelligent Systems in Accounting, Finance and Management 14(1-2), 27–42 (2006)
12. Wang, J., Shu, Y.-F.: Fuzzy decision modeling for supply chain management. Fuzzy Sets and Systems 150(1), 107–127 (2005)
13. Semini, M., Fauske, H., Strandhagen, J.O.: Applications of discrete-event simulation to support manufacturing logistics decision-making: a survey. In: WSC 2006: Proceedings of the 38th Conference on Winter Simulation, pp. 1946–1953 (2006)
14. Simulation Study Group. Simulation in the uk manufacturing industry (1991)
15. Terzi, S., Cavalieri, S.: Simulation in the supply chain context: a survey. Computers in Industry 53(1), 3–16 (2004)
16. Supply Chain Council. Supply-Chain Operations Reference-model Version 9.0 (2008), `http://www.supply-chain.org/` (January 15, 2010)
17. Rabe, M., Jaekel, F.-W., Weinaug, H.: Reference models for supply chain design and configuration. In: WSC 2006: Proceedings of the 38th conference on Winter Simulation, pp. 1143–1150 (2006)
18. Mackulak, G.T., Lawrence, F.P., Colvin, T.: Effective simulation model reuse: a case study for amhs modeling. In: WSC 1998: Proceedings of the 30th Conference on Winter Simulation, pp. 979–984. IEEE Computer Society Press, Los Alamitos (1998)
19. Ding, H., Benyoucef, L., Xie, X., Hans, C., Schumacher, J.: One a new tool for supply chain network optimization and simulation. In: WSC 2004: Proceedings of the 36th Conference on Winter Simulation, pp. 1404–1411 (2004)
20. Ganapathy, S., Narayanan, S., Srinivasan, K.: Logistics: simulation based decision support for supply chain logistics. In: WSC 2003: Proceedings of the 35th Conference on Winter Simulation, pp. 1013–1020 (2003)

A Genetic-Based Solution to the Task-Based Sailor Assignment Problem

Dipankar Dasgupta, Deon Garrett, Fernando Nino,
Alex Banceanu, and David Becerra

Abstract. This chapter presents a study investigating a multi-objective formulation of the United States Navy's Task-based Sailor Assignment Problem and examines the performance of a widely used multi-objective evolutionary algorithm (MOEA), namely NSGA-II, on large instances of this problem. The performance of the evolutionary algorithm is examined with respect to both solution quality and diversity and has shown to provide inadequate diversity along the Pareto front. Domain-specific local improvement operators were introduced into the MOEA, producing significant performance increases over the evolutionary algorithm alone. Thus, hybrid MOEAs provided greater diversity along the Pareto front. Also a parallel version of the evolutionary algorithm was implemented. Particularly, an island model implementation was investigated. Exhaustive experimentations of the sequential and parallel implementations were carried out. The experimental results show that the genetic-based solution presented here is suitable for these types of problems.

1 Introduction

According to the United States Navy's personnel policies, roughly every three years sailors serving on active duty are reassigned to a different job. As a result, at any given time there exists a sizable population of sailors to be

Dipankar Dasgupta · Alex Banceanu
Department of Computer Science, University of Memphis, Memphis, TN, USA
e-mail: ddasgupt@memphis.edu, alexbanceanu@gmail.com

Deon Garrett
Icelandic Institute for Intelligent Machines/School of Computer Science,
Reykjavik University, Reykjavik, iceland
e-mail: deon@iiim.is

David Becerra · Fernando Nino
Department of Computer Science, National University of Colombia,
Bogota, Colombia
e-mail: {dcbecerrar,lfninov}@unal.edu.co

reassigned to available jobs. The Navy's goal is to identify sailor and job matches that maximize some overall criterion of satisfiability of sailors and commanders and is referred to as the Sailor Assignment Problem (SAP).

In this research, the SAP and a "fine-grained" and more complex version of SAP, called the Task-based Sailor Assignment Problem (TSAP), were studied. In the TSAP, sailors have to be assigned to different jobs (tasks) in different time slots. In this case, instead of assigning a single task to a sailor, a sailor is assigned to multiple tasks, which are distributed in a certain number of time frames. Thus, it may be considered that each day is divided into several shifts and sailors need to be assigned to particular tasks in each shift. Accordingly, in this TSAP, a sailor can be assigned to different tasks in different time shifts, whereas in the case of SAP, a sailor cannot be assigned to two different jobs at the same time.

In reality, the utility of a possible solution to the SAP/TSAP is not determined by a single measure. Instead, there are a number of attributes that go into deciding whether a solution is good. Typically, each of these attributes is represented by a distinct objective function. One approach to these types of problems is to attempt to formulate a single metric that encompasses all aspects of solution quality and optimization based on this metric. Such a single-objective formulation has the advantage that the resulting problem can be solved using any of a vast set of classical optimization methods. However, constructing the metric requires detailed knowledge of the structure of each objective function. Therefore, it can be difficult for a decision maker to specify an appropriate set of parameters to obtain a solution. To overcome the difficulties mentioned above, multi-objective evolutionary algorithms (MOEAs) can be applied to the multi-objective instance of the SAP/TSAP. In this chapter, the Non-dominated Sorting Genetic Algorithm II (NSGA-II) was run to obtain multiple diverse solutions to the SAP/TSAP in a single run of the algorithm. However, the results produced by this MOEA lack diversity when compared to the solutions produced by perfect solvers (in the case of SAP).

Multi-objective optimization has its roots in the late 19th century welfare economics, in the works of Edgeworth and Pareto. A mathematical formulation of a multi-objective optimization problem (MOP) is as follows. Find a vector $x = [x_1, x_2, \ldots, x_n]^T$ which:

1. satisfies the r equality constrains, $h_i(x) = 0, 1 \leq i \leq r$,
2. is subject to the s inequality constrains, $g_i(x) \geq 0, 1 \leq i \leq s$, and
3. which optimizes the vector function $z = f(x) = [f_1(x), f_2(x), \ldots, f_m(x)]^T$

It is clear that solving a MOP is focused on searching for the optimal values of the decision variables (vector x) that minimize/maximize the objective functions (vector $f(x)$) while satisfying the constraints. The vector x is an n-dimensional decision vector or solution and $\mathcal{X}$ is the decision space, i.e., the set of all expressible solutions. $z = f(x)$ is an objective vector that maps $\mathcal{X}$ into $\Re^m$, where $m \geq 2$ is the number of objectives. The image of $\mathcal{X}$ in objective space is the set of all attainable points $\mathcal{Z}$ (see Fig. 1).

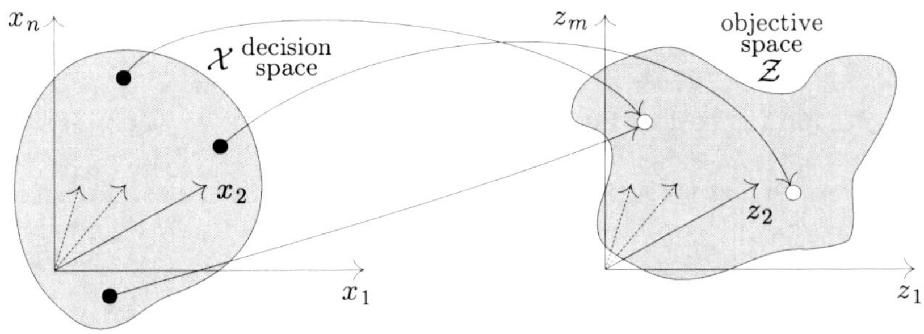

Fig. 1. The n-dimensional parameter space maps to the m-dimensional objective space.

The scalar concept of optimality does not apply directly in the multi-objective setting. A useful replacement is the notion of Pareto optimality. Essentially, a vector $x^* \in \mathcal{X}$ is said to be Pareto optimal for a given MOP if all other vectors $x \in \mathcal{X}$ have a higher value for at least one of the objective functions $z = f(x)$, and no smaller value for all objectives. Pareto optimal points are also known as efficient, non-dominated, or non-inferior points. We can also speak of locally Pareto optimal points, for which the definition is the same as the one just given, except that we restrict attention to a feasible neighborhood of x^*. Typically, there is an entire curve or surface of Pareto points, whose shape indicates the nature of the tradeoff between different objectives.

Over the past ten years, a large amount of research has been conducted on the use of evolutionary techniques to solve MOPs. The inherent use of a population of candidate solutions provides benefits unattained by other techniques when applied to problems with two or three objectives. In recent years, some attention has been devoted to the experimental study of the performance of evolutionary techniques in problems with many objectives. These are optimization problems with more than three conflicting objectives. The experimental studies conducted lead to the conclusion that the performance of evolutionary techniques is severely hindered when the dimension of the objective space grows. A recent study [1] shows that for problems with more than ten objectives, a purely random search may perform favorably when compared with an evolutionary technique. It is well-known that for single-objective problems the performance of evolutionary algorithms can often be improved through the introduction of a local search operator. These combinations of evolutionary and local searches are known by the names of Memetic Algorithms (MAs) [2, 3], Hybrid Evolutionary Algorithms [4, 5, 6], and Genetic Local Search [7]. These hybrids of evolutionary and local search algorithms have been shown to provide state-of-the-art performance on a wide range of hard single-objective combinatorial optimization problems [8, 9] and have also proven to provide an effective performance on MOPs [10, 11, 12].

Garrett et al. [13] used genetic algorithms (GAs) to solve single-objective SAPs and compared the results they obtained with an existing algorithm, the Gale-Shapley algorithm [14]. The Gale-Shapley algorithm is a quadratic time algorithm for finding an optimal set of stable matches (i.e., marriages) and produces a set of marriages that is stable and optimal with respect to the preferences of one group after $O(n^2)$ steps. If minimizing the total permanent change of cost is not one of the objectives in the SAP, the results produced by the Gale-Shapley algorithm are optimal sets of assignments with respect to the preferences of the sailors and commanders, and such a match may not necessarily involve all sailors. The work in [13] examined an alternative GA approach to the SAP in order to allow outside constraints to influence the quality of the match and also investigated the ability of a GA to generate multiple good solutions to the SAP through niching techniques. The solutions produced by the GA were significantly better with respect to coverage and fulfilling the objectives of SAP when compared to the Gale-Shapley algorithm. Due to the scalarization of the objective functions (to convert the multi-objective SAP to a single-objective problem), however, the impact of a change in a weight vector on the solution is unpredictable, thus making it necessary to run the algorithm several times with different parameter settings to get a broad view of the solution space. For this reason, diversity (coverage) of the solutions is highly dependent upon the ability of the algorithm to find Pareto optimal solutions as well as the proper selection of parameters employed by the user.

Therefore, to overcome such problems, Dasgupta et al. [4] extended the earlier work to incorporate each objective directly into the optimization problem. Two well-known MOEAs – NSGA-II [15] and SPEA2 [16] – were implemented to solve the multi-objective SAP. The results produced by these algorithms lacked diversity when compared to the CHC algorithm, a single-objective GA, repeated for many weight vectors. Then, a local search operator was constructed and integrated into the MOEAs in such a way as to emphasize the importance of finding diverse solutions across the Pareto set approximation. Specifically, a hybrid GA featuring an efficient SAP solver, the Kuhn-Munkres (KM) algorithm [17, 18], was used to further improve the performance. The KM algorithm solves linear assignment problems in $O(n^3)$ time and it was chosen due to the fact that the single-objective versions of SAP with small modifications are linear assignment problems. This approach, while yielding very good solutions, suffers from very long run times as the problem size increases. Therefore, a way to find more diversified solutions is to combine the solutions produced by the perfect solver KM with a MOEA in what was called an "informed initialization".

Dasgupta et al. in [19] presented an extension of the SAP (i.e., the TSAP). In the TSAP, the SAP is no longer considered as a static assignment but as a time-dependent multi-task SAP, making it a more complex problem; in fact, an NP-complete problem. Subsequently, a multi-objective formulation of the

United States Navy's TSAP was studied and the performance of NSGA-II was examined on large instances of this problem.

This chapter is a more comprehensive version of the work presented in [19]. The rest of the chapter is organized as follows. Section 2.5 presents the TSAP in detail. The proposed multi-objective evolutionary approach to solve the TSAP is then described in Section 3. In addition, Section 4 presents a parallel implementation of the evolutionary algorithm to solve the TSAP based on an island model. Section 5 describes the experiments carried out and a discussion of the results of the MOEA based approach is also presented. Finally, Section 6 presents the conclusions of this work as well as discusses possible improvements in future work.

2 Types of Assignment Problems

2.1 Assignment Problem

The assignment problem [20] is a well-studied and important combinatorial optimization problem. This problem can be stated as: Given a number of agents n and a number of tasks m to be performed by the agents, the goal is to find an assignment of the tasks to the agents in such a way that some objective measures (objectives) such as cost, training, and satisfaction, are optimized. In general, there are some restrictions that have to be satisfied; for example, workload limitations on the agents.

2.2 Linear Assignment Problem

The Linear Assignment Problem (LAP) [20] is the simplest assignment problem, in which the number of agents and number of tasks are equal and every agent can perform every task, but only one task can be assigned to each agent. In most cases, there is only one goal to be optimized, usually called cost. Then the problem is equivalent to the problem of finding an optimum weight vertex matching in an $n \times n$ cost-weighted complete bipartite graph.

Formally, the LAP can be stated as follows: given a set of agents $A = \{a_1, a_2, \ldots, a_n\}$ and a set with the same number of tasks $T = \{t_1, t_2, \ldots, t_n\}$ and the cost function $C : A \times T \rightarrow R$, find a bijection (matching) $m : A \rightarrow T$ such that the cost function:

$$\sum_{a \in A} C(a, m(a)) \tag{1}$$

is minimized (or **profit** maximized). Usually the cost function is also viewed as square real-valued matrix C with elements $C_{ij} = C(a_i, t_j)$. This problem can be expressed as an integer linear program with the objective function:

$$\sum_{i=1}^{n}\sum_{j=1}^{n} C_{ij}x_{ij} \tag{2}$$

subject to the constraints:

$$\sum_{i=1}^{n} x_{ij} = 1 \quad \forall j \in \{1, 2, \ldots, n\} \tag{3}$$

$$\sum_{j=1}^{n} x_{ij} = 1 \quad \forall i \in \{1, 2, \ldots, n\} \tag{4}$$

$$x_{ij} \in \{0, 1\} \quad \forall i, j \in \{1, 2, \ldots, n\} \tag{5}$$

$x_{ij} = 1$ if agent i is assigned to perform task j and it is 0 otherwise.

2.3 General Assignment Problem

The General Assignment Problem (GAP) is a generalization of the assignment problem that was originally studied as the problem of scheduling parallel machines with costs [9]. In this scenario, a number of agents n and a number of tasks m to be performed by the agents are given. Any agent (sailor) can perform any task, but each agent has a budget and the sum of resources required for tasks assigned to it cannot exceed this budget. When an agent is assigned to perform a task, it incurs some costs and resources associated with it. The solution to this problem is to find an assignment in which all agents do not exceed their budget and total cost of the assignment is minimized. Let b_i be the budget of agent i, let R_{ij} be the resources and C_{ij} be the cost incurred when agent i is assigned to perform task j, then the GAP can be expressed as the following integer linear program:

$$minimize \sum_{i=1}^{n}\sum_{j=1}^{m} C_{ij}x_{ij} \tag{6}$$

subject to the constraints:

$$sum_{j=1}^{m} x_{ij} R_{ij} \leq b_i \quad \forall\, i \in \{1, 2, \ldots, n\} \tag{7}$$

$$\sum_{j=1}^{m} x_{ij} \leq 1 \quad \forall\, i \in \{1, 2, \ldots, n\} \tag{8}$$

$$\sum_{i=1}^{n} x_{ij} = 1 \quad \forall\, j \in \{1, 2, \ldots, m\} \tag{9}$$

$$x_{ij} \in \{0, 1\} \quad \forall\, i \in \{1, 2, \ldots, n\}, j \in \{1, 2, \ldots, m\} \tag{10}$$

2.4 Sailor Assignment Problem

Every two years, each sailor in the Navy is required to change jobs. As a result, at any given time, there are many sailors who require assignment to new jobs. The problem the Navy faces is to find a set of assignments, also called a match, of sailors to jobs which keeps the sailors happy and maintains fleet readiness while minimizing the cost of implementing those assignments. Formally, the SAP can be formulated as follows:

$$Maximize \sum_{i=1}^{n} \sum_{j=1}^{m} \mathcal{F}_{i,j} d_{i,j} \tag{11}$$

subject to the constraints:

$$\sum_{i=1}^{n} d_{i,j} \leq 1 \quad \forall\, j \in \{1, 2, \ldots, m\} \tag{12}$$

$$\sum_{j=1}^{m} d_{i,j} \leq 1 \quad \forall\, i \in \{1, 2, \ldots, n\} \tag{13}$$

where $\mathcal{F}_{i,j}$ denotes the fitness of assigning sailor i to job j and D is an assignment matrix such that

$$d_{i,j} = \begin{cases} 1 & \text{sailor } i \text{ assigned to job } j \\ 0 & \text{otherwise} \end{cases} \tag{14}$$

2.5 The Problem: Task-Based Sailor Assignment Problem

Compared to the SAP, this problem is specifically designed for naval bases of the United States Navy, where there are a limited numbers of sailors and during one day, the same task may have to be done by different sailors. So depending upon the requirements, a day or a week is divided into different time shifts where different tasks have to be performed by sailors. However, the same task may not continue the next day or the next time shift requiring the sailor to be moved to another task.

This shows that a sailor may have to do different tasks in a day but, clearly the same sailor cannot perform more than one task in the same time shift. The goal here is to minimize the number of sailors working on the naval base. In short, instead of assigning a single task to a sailor, a sailor has to perform multiple tasks, which are distributed in a certain number of time frames. It may be considered that each day is divided into several shifts and sailors need to be assigned to particular tasks that are required to be completed in that shift. Accordingly, a sailor can be assigned different tasks in different shifts (see Figure 2).

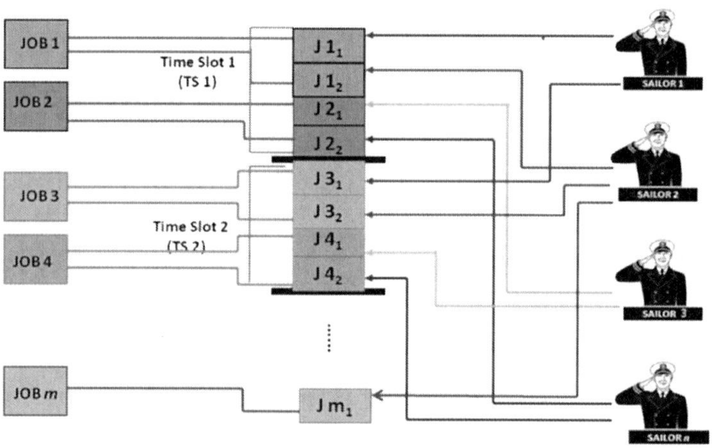

Fig. 2. Graphical description of the TSAP

Similar to the SAP, here also sailors are examined for a variety of qualifications and limitations to determine if each sailor is a valid match for one of the tasks, and a score is computed determining the extent to which the sailor is a valid match. The training score encompasses many factors, including the pay grades of the sailor and the proposed task, the amount of training required for the sailor to be able to perform the duties of the proposed job, among others. In addition, the monetary cost of assigning the sailor to the proposed task is computed.

After all candidates have been identified, the sailors are allowed to rate those tasks for which they are qualified. The final set of assignments must satisfy Navy regulations and must do so at a reasonable cost to the Navy. In this work, we will assume that sailors may be assigned only those tasks for which the sailors applied and were ranked.

After the command preview stage, the detailer must construct the set of assignments in accordance with the regulations and preferences of the Navy. The complexity of the detailing process limits the Navy's ability to make effective decisions. Therefore, an automated system for quickly finding good solutions to the problem is needed. Formally, the TSAP can be described as follows:

Let n be the number of sailors, m be the number of task classes and t be the number of time slots. Any eligible sailor can be assigned to any task in a time slot. Each sailor has his/her own capacity and consumes some resources for doing tasks assigned to him/her. The problem is to find a sailor-task assignment for each time slot in such a way that minimizes the number of sailors along with fulfilling the previous objectives of the SAP, namely, to maximize the total training score, the sailor preference, the commander preference, and to minimize the task sailor assignment cost. Additionally, the following constraints are involved in the TSAP:

1. the sum of the resources for a sailor-task assignment over given time should not exceed its capacity,
2. the same sailor cannot be assigned to multiple tasks in one time slot, and
3. the number of sailors working in one time slot should be equal to the number of tasks required to be done in that time slot.

Let cap_i be the capacity of sailor i, R_{ij} be the resources and C_{ijk} be the cost of assigning sailor i to perform task j on time slot k . Let y_{jk} be the requirement of the number of class j tasks in time slot k, and x_{ijk} be an indicator of whether the sailor i performs task class j in timeslot k or not.

Then the TSAP can be formally expressed as the following multidimensional integer linear program:

$$minimize\ \mathbf{F}(\mathbf{x}) = (f_1(\mathbf{x}), \ldots, f_N(\mathbf{x})) \tag{15}$$

where $\mathbf{x} = (x_{ijk})_{i,j,k}$; f_{obj} for $obj = 1, \ldots, N$ are the objectives to be minimized defined as

$$f_{obj}(\mathbf{x}) = \sum_{i=1}^{n} \sum_{j=1}^{m} \sum_{k=1}^{t} x_{ijk} C_{ijk}^{obj} \tag{16}$$

with C_{ijk}^{obj} the cost of assigning sailor i to task j on time slot k determined by objective obj; and $\mathbf{F}(\mathbf{x})$ is subject to the constraints:

$$\sum_{j=1}^{m} x_{ijk} R_{ij} \leq cap_i \quad \forall\, i \in \{1, 2, \ldots, n\}, \forall\, k \in \{1, 2, \ldots, t\} \tag{17}$$

$$\sum_{j=1}^{m} x_{ijk} \leq 1 \quad \forall\, i \in \{1, 2, \ldots, n\}, \forall\, k \in \{1, 2, \ldots, t\} \tag{18}$$

$$\sum_{i=1}^{n} x_{ijk} = y_{jk} \quad \forall\, j \in \{1, 2, \ldots, m\}, \forall\, k \in \{1, 2, \ldots, t\} \tag{19}$$

$$x_{ijk} \in \{0, 1\} \quad \forall\, i \in \{1, 2, \ldots, n\}, \forall\, j \in \{1, 2, \ldots, m\} \tag{20}$$

$$\forall\, k \in \{1, 2, \ldots, t\} \tag{21}$$

Without major difficulty, it can be shown that the TSAP is NP-Complete. Since there is no notion of time slot in the General Assignment Problem (GAP), for reducing the instances of GAP to TSAP, consider a time slot which has all tasks to be done by sailors listed in the GAP. As no sailor will be repeated in this time slot and the sum of the resources for sailor-task assignments will be less than or equal to the capacity and as there is only one time slot covering all the tasks, it is ensured that all tasks are assigned to different sailors. In this way all the constraints are fulfilled and a GAP instance is reduced to a TSAP instance. Hence by this reduction process, it can be shown that the solution for the TSAP and GAP will be identical.

3 The Approach: A MOEA for the TSAP

Some of the most promising approaches to many MOPs arise from evolutionary techniques. One of the primary benefits touted by practitioners is the ability of MOEAs to cover the Pareto front in a single run of the algorithm. Any multi-objective EA must push the initial population in the direction of Pareto optimal solutions and, ideally, the algorithm would terminate with a set of non-dominated solutions such that no possible solution could dominate any member of the Pareto front. This set is called the true Pareto optimal set, or true Pareto front. In practice, we often wish only to find solutions that are very good and non-dominated with respect to one another. This requires that the algorithm routinely improve the quality of the initial randomly generated solutions until some level of acceptability has been met. Several MOEAs have been proposed in the literature with varying degrees of success. Among these algorithms, the Non-dominated Sorting Genetic Algorithm by Deb and others (NSGA-II) [15] and the Strength Pareto Evolutionary Algorithm (SPEA2) of Zitzler et al. [16] have been widely studied and found to be effective across a range of common test functions as well as combinatorial optimization problems. In this work, NSGA-II is used to solve the TSAP.

3.1 Chromosome Representation Scheme

One of the main decisions to be made when adapting a GA to a particular problem is how to encode solutions to the problem in a manner amenable to genetic search. In this work, we utilize an integer encoding. A chromosome consists of a set of integers, each representing a sailor assigned to perform a particular task. Each sailor and job is given an identifying integer number. The length of the chromosome is equal to the total number of tasks that should be assigned to sailors along all the time slots. Each integer is chosen from a subset of possible numbers, precisely those representing valid sailors for that task. In this way, we try to ensure that only those sailors qualified to perform a task may be assigned. However, it may be possible that in the middle of the search process, no sailor can be assigned to a task, then a '-1' is assigned meaning that this task has not yet been assigned.

A linear chromosome representation is used as follows: the week is divided into a number of days d which are subdivided into k time slots. Each time slot contains variable number of tasks to be done by eligible sailors. Figure 3 shows the representation of a chromosome where each day is divided into multiple time slots and tasks contained in each time slot are done by the eligible sailors; here, $t_{m,j}$ represents the task class m and its jth instance to be done by sailor S_i. Decoding a chromosome is simply performed by assigning sailor S_i to the corresponding task on location i of the chromosome, which occurs at the corresponding time slot k.

It is important to note that tasks are associated with corresponding sailors but they are not explicitly specified in the chromosome, which contains sailors only. Therefore, it is necessary to map the sailors to tasks in the decoding process.

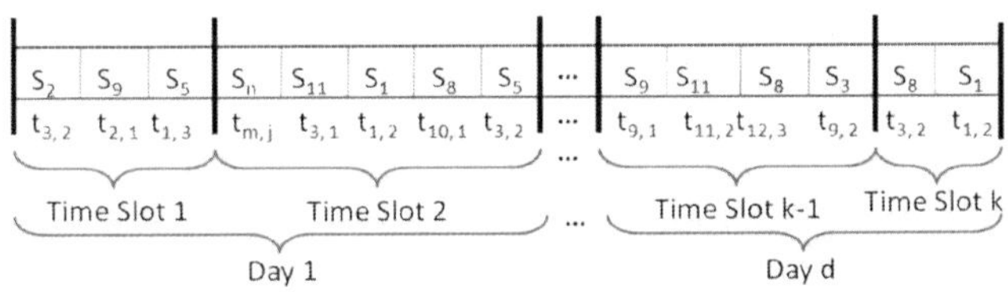

Fig. 3. Chromosome representation

Note that in Figure 3 a random crossover or mutation could easily result in a violation of the constraints. The second and third constrains of the TSAP (see Section 2.5) ensure that feasible solutions do not assign two different tasks to the same sailor at a particular time slot, the sailor's capacity should not be exceeded and that all the tasks are assigned a sailor to perform it. But, the proposed encoding cannot prevent violations of the constraints to occur. Notice that it can also happen that a sailor or task may be left unassigned.

3.2 Objectives and Fitness Measure

The objective functions of this entire problem are following:

1. Minimize the number of sailors assigned to navy tasks
2. Minimize the sailor-task assignment cost
3. Maximize the training score (TS) of a particular sailor
4. Maximize the sailor preference (SR) of specific tasks
5. Maximize the commander preference (CR)

The multi-objective optimization process used in this work consists of minimizing all of the five objectives defined above. Also the values of all the objectives are normalized in the [0, 1] interval. However, it should be noticed that while in principle, these values are scaled such that the objective function values range from 0 to 1, in practice the extreme values never occur. The reason is that multiple sailors are in contention for the same tasks. Also it is assumed that each sailor can be assigned to those tasks which maximize or minimize the values of each objective function. However, constraints make it very unlikely that all such situations can be simultaneously attained. Thus, the actual upper and lower bounds for each objective are unknown for any problem instance of any size. As a result, it is unlikely that a single solution (slate) reaches the value 1.0 in all or in any objectives.

A crucial issue here is the assurance that all necessary constraints (Section 2.5) need to be satisfied. Therefore, during the evolutionary optimization process, an infeasible solution is penalized for constraint violation as explained below. These necessary characteristics of the TSAP are taken into account by the inclusion of the following penalization functions. (These penalization functions were already used in [19]):

1. Unassigned Tasks Penalty (UTP): This value is used to penalize solutions which contain unassigned tasks and is defined as:

$$UTP = \frac{No.\ of\ unassigned\ tasks}{Total\ no.\ of\ tasks} \tag{22}$$

2. Redundant Sailors Penalty (RSP): This value penalizes solutions containing same sailors repeated in a single time slot and is computed as

$$RSP = \frac{No.\ of\ redundant\ sailors}{Total\ no.\ of\ tasks} \tag{23}$$

3. Capacity Exhaustion Penalty (CEP): CEP enforces penalty in a solution once it finds a sailor assigned to a task though his/her capacity gets exhausted already in that solution. It can be calculated as:

$$CEP = \frac{Capacity\ exhaustion}{Sum\ of\ max.\ resources\ to\ do\ all\ tasks} \tag{24}$$

A single penalization valued was computed as the sum of UTP, RSP, and CEP. Then this penalty value is divided and an equal "portion" of it is added to each objective. Therefore, the penalization is equally distributed among all the objectives.

After further analysis some additional constraints were introduced to take care of balancing the workload among all sailors, attempting to assign the sailor to consecutive tasks while guaranteeing a minimum of hours per day as follows:

1. Excessive fluctuations penalty (EFP): This value is used to penalize those solutions that assign the work to the sailors in such a way that they keep working for the whole day with very short breaks in between. It is computed as:
$$EFP = \frac{\sum_{i=0}^{n} TF_i}{n \times TS \times TD} \tag{25}$$
where n is the total number of sailors, TF_i is the total fluctuations for sailor i in a week, TS is the total number of time slots per day and TD is the total number of days in a week.
2. Uneven distribution of working hours penalty ($UDWP$): This constraint penalizes those solutions that allow timeslots very haphazardly, placing too much burden on some sailors and giving very little work to others.
$$UDWP = tanh\left(\frac{\sigma}{k \times \mu}\right) \tag{26}$$
where μ is the average number of timeslots worked by a sailor per week, σ is the standard deviation of the number of timeslots worked from the mean μ and k is a constant which has been set to the value 8 after rigorous experimentation.
3. Instability in 24 hour time frame penalty (IP): This constraint penalizes the solution to make sure that the sailor does not have to switch his state (i.e., work to rest and vice versa) too frequently
$$IP = \frac{\sum_{i=0}^{n} MF_k^x}{n \times TS^x} \tag{27}$$
where n is the number of sailors, MF is the maximum number of fluctuations in a 24 hour time frame for sailor k and TS is the number of time slots per day. All these latest constraints have been normalized and scaled to the range of 0 to 0.3.

3.3 Genetic Operators

One simple mutation operator is performed by choosing a particular task at random and then assigning a new qualified sailor to that task. However, after mutating a solution, the capacity of the newly assigned sailor could be exhausted. Also, a crossover operator is implemented by swapping the sailors assigned to do all the tasks in a particular time slot in two solutions. As in mutation, after performing crossover, it is likely that the offspring violate the capacity constraint. Therefore, a repair operator is implemented to try to maintain feasible solutions. These genetic operators are described in the following sections.

3.3.1 Mutation

The mutation operator works as follows: Initially, it chooses one random location from the chromosome. It finds out the sailor who has been assigned to that task. Then it determines the list of possible sailors who can perform that task and randomly picks a sailor out of them and substitutes the current sailor with the new one keeping two constraints in mind: the new sailor is not repeated in that time slot and has enough capacity to perform that particular task. This process continues until a new sailor can be assigned to that task or all the eligible sailors are checked, in which case, the sailor assigned to that task remains unchanged.

It is important to notice that given a specified mutation rate P_m, the effective mutation rate, the percentage of attempted mutations that actually effect some change in the individual, is less than P_m, since only some of the attempted mutations are completed successfully. As the proper setting of the mutation rate is a crucial aspect of any evolutionary algorithm, it is important to understand the relationship between the specified mutation rate and the effective mutation rate on the TSAP. Clearly, this relationship is intricately linked to the amount of contention for tasks, which in turn depends critically on the ratio of tasks to sailors in a particular instance of the problem. As this ratio increases, the likelihood of finding a sailor to perform a particular task which is not currently assigned to another sailor increases, thereby increasing the percentage of mutations that can be completed successfully.

Randomization of a newly created chromosome utilizes the same process as the mutation operator described above. Each task is assigned to a randomly selected qualified sailor if one is available. If no sailor is available, then the randomization procedure marks the task unassigned.

3.3.2 Crossover

The crossover operator picks a time slot at random from one parent and swaps it with the same time slot of another parent. This operator does not need to check for redundant sailors on the same time slot that is being swapped as the entire time slot is swapped instead of single sailor of the time slot. The crossover operator is depicted in Figure 4.

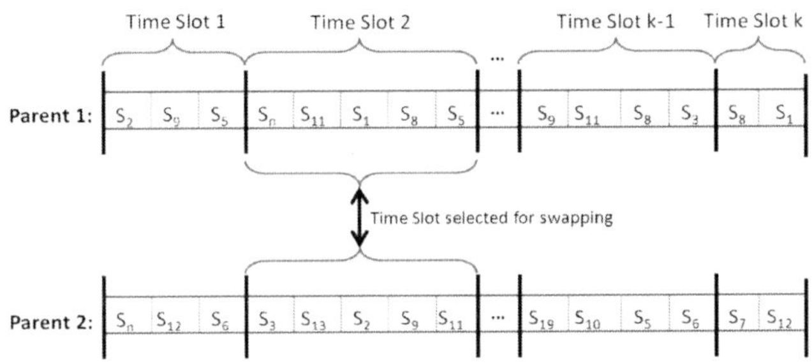

Fig. 4. The offspring produced by swapping of time slots between two parents

3.3.3 Repair Operator

A repair operator is implemented to attempt to recover from possible constraint violations after mutation and crossover and works as follows. First, the chromosome is traversed to look for unassigned tasks and for each unassigned task, a qualified sailor whose capacity has not been exhausted is assigned to this task. If no sailor is available to perform assigned to the task, it remains unassigned. Then each time slot is checked for redundant assigned sailors. If a solution contains the same sailor assigned to two different tasks in a single timeslot, then the list of eligible sailors for a randomly chosen task is examined and a new sailor is assigned. If after trying to assign a new sailor to that particular task this constraint is still violated, the task is marked as unassigned.

In order to perform a repair mechanism on the sailor capacity constraint, the sailor capacity exhaustion is defined as the number of extra resources (hours) a particular solution assigns to a particular sailor. Accordingly, for each sailor whose capacity has been exceeded, the repair operator attempts to assign a different sailor to the task he/she has been assigned until the capacity is no longer exhausted. Note that even after repair it is likely that the capacity of some sailors remains exhausted.

3.4 Hybrid Approach

The proposed genetic-based technique was combined with some local search operators as follows. Specifically, two local search operators were implemented: a sailor shift operator and a sailor swap operator. The sailor shift search operator performs a random shift of a sailor on a solution in the same way as the TSAP mutation operator. On the other hand, the sailor swap operator picks a particular sailor assigned to perform a task and tries to swap it with another sailor chosen at random, previously checking feasibility and constraint violations. This process is repeated for a pre-specified number of times. Several experiments were carried out for all four versions of GAs: one for standard NSGA-II and other three for hybrid NSGA-II.

4 Parallel Implementation of the Proposed Approach

The processing elements in parallel computation can be diverse and include resources such as a single computer with multiple processors, several networked computers, specialized hardware, or a combination of the above. Optimization problems are among the most interesting and challenging tasks suitable for the use of parallelization techniques.

Parallelization techniques have always been a focus of research in evolutionary algorithms for single-objective optimization because of their population-based nature. As a natural extension, parallelization techniques have also been used for evolutionary multi-objective optimization. Thus, different parallel approaches applied to single-objective optimization problems have been widely suited and reported, particularly, their modeling, performance and behavior have been investigated. However, a modest insight has been given to parallel approaches to solve MOPs. Different parallel models have been proposed in the design of multi-objective optimization algorithms, but only a few of the reported parallel approaches are accessible to compare the performance of those implementations. Then, determining the efficiency and effectiveness of parallel implementations is usually difficult.

The advances in the use of multi-objective evolutionary approaches for real-world applications, containing multiple objectives and high dimensionality, have led to the exploitation of the inherent parallelism of such algorithms. Since the United States Navy's TSAP is a complex NP-hard, CPU time consuming process, these features make it a perfect candidate to be solved using a parallel model. Then, the efficiency (i.e., how well it performs computationally) and effectiveness (i.e., how good its reported solutions are) of the model used to solve the TSAP could be improved by increasing the number of processors allocated to it.

Given that a MOEA approach for the TSAP was already implemented, the main idea of the proposed parallel implementation is to study how a parallelization model affects the MOEA performance. Specifically, the focus of the proposed approach in this work is to improve the effectiveness (i.e., how good its reported solutions are) of the algorithm to find the solutions to TSAP, discovering what solutions arise from simultaneous execution of multiple TSAP instances. Additionally, we are interested in investigating the effects of varied parallel algorithmic parameters, and in determining how the used policies affect the MOP solution process.

Accordingly, some groups or categories can be defined based on the shared features of the algorithms. Specifically, the most common parallel evolutionary algorithms can be grouped into one of three categories: master-slave, islands, and diffusion models. In the master-slave model the objective function evaluations are distributed among several slave processors while a master processor executes the MOEA, and other overhead functions. In the island model, the MOEA's populations are relatively isolated from each other but individuals within some particular island occasionally migrate to another one. As in the master-slave model, the diffusion model deals with one conceptual population, except that each processor holds only a few individuals [21]. Clearly, in all these models a different trade-off between exploration and exploitation of the search space is required.

4.1 Parallel Implementation Tools

The proposed parallel version of the MOEA for solving the TSAP was implemented using the JavaSpaces technology. JavaSpaces was the chosen high-level coordination tool for gluing processes together into the distributed application because it allowed the development of a simple design and implementation.

In addition, the proposed approach uses an island MOEA model as its parallel paradigm. Particularly, a master processor was used for managing tasks and several slave processors to execute the TSAP algorithm. Accordingly, the master processor generates tasks, writes them into a space and collects results from the space.

4.1.1 JavaSpaces

Building distributed applications with conventional network tools usually entails passing messages between processes or invoking methods on remote objects. Therefore, several technologies can be used to build these applications, including low-level sockets, message passing, and remote method invocation (RMI). In contrast, in JavaSpaces applications, the processes do not communicate directly, but their activities are coordinated by exchanging objects through persistent object exchange areas (i.e., space or shared memory).

Then, JavaSpaces provides a different programming model that views an application as a collection of processes cooperating via the flow of objects into and out of one or more spaces. This approach can simplify the design and implementation of sophisticated distributed applications. Hence, a process can write new objects into a space, take objects from a space, or read (make a copy of) objects in a space, as depicted in Figure 5 [22, 23].

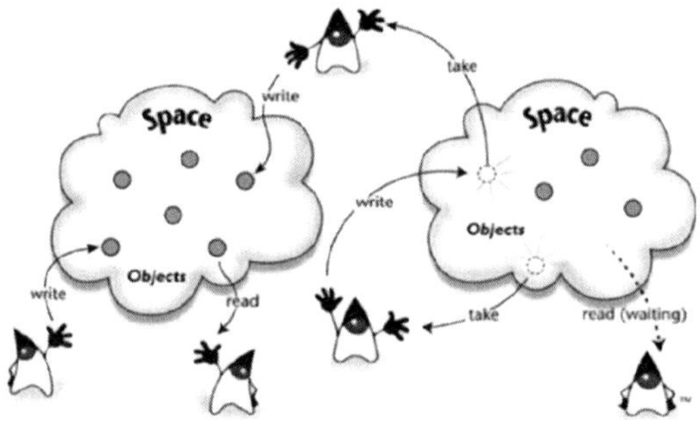

Fig. 5. Processes use spaces and simple operations to coordinate activitiesCopyright Sun Microsystems, Inc.

The JavaSpaces technology provides a significant number of utility interfaces and classes that provide a valuable set of tools for developers. One of these tools is the compute-server implementation, which is an implementation of an all-purpose computing engine using a JavaSpace. A compute-server provides a service that accepts tasks, computes them, and returns results. The server itself is responsible for computing the results and managing the resources that complete the job. Behind the scenes, the service might use multiple CPUs or special-purpose hardware to compute the tasks faster than a single-CPU machine. The typical space-based compute-server is depicted in Figure 6.

4.1.2 Island Models

The island paradigm is based on the biological evolution of natural populations in relative isolation, such as those that might occur within an ocean island chain with limited migration. Then, in the island model, every processor runs an independent MOEA using a separate sub-population. The processors might cooperate by regularly exchanging migrants which are good individuals in their sub-populations [24].

The island models can be categorized into two main groups, cooperating subpopulations and a multi-start approach. The cooperating subpopulation

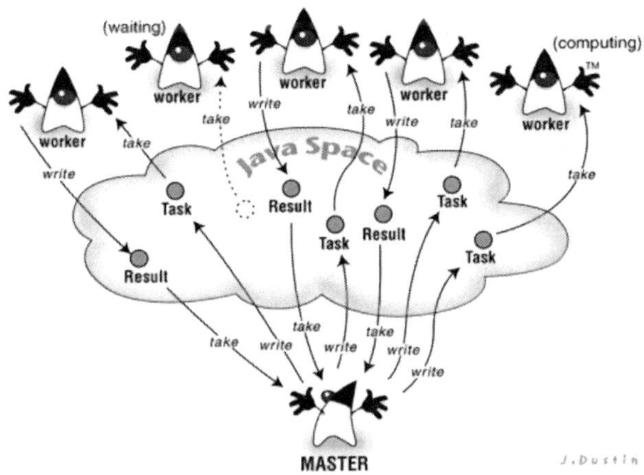

Fig. 6. A space-based compute-server(Illustration by James P. Dustin, Dustin Design)

methods are based on partitioning the objective/search space. In this group, the population is divided into subpopulations. These algorithms attempt to distribute the task of finding the entire Pareto optimal front among participating processors. This way each processor is destined to find a particular portion of the Pareto-optimal front. In the multi-start approach, each processor independently runs an optimization algorithm. The basic idea of using such a model is that running several optimization algorithms with different initial seeds is more valuable than executing only one single run for a very long time [21].

The island paradigm is termed coarse-grained parallelism because each island contains a large number of individual solutions. Communication backbones can connect processors in logical or physical geometric structures such as rings, meshes, toruses, triangles, and hypercubes. The communication backbone, the island model architecture (as the number of islands) and the migration policies (such as how often migration occurs, the number of solutions to migrate, how to select emigrating solutions, and which solutions are replaced by immigrants) are some of the key issues that should be faced when an island model is implemented.

4.2 The Proposed Parallel Implementation

As mentioned earlier, the proposed approach is based on an island model as parallel paradigm and on a compute-server model with regard to the

JavaSpace implementation. The implemented island model is categorized as a multi-start approach, and it used a ring geometric structure as its communication backbone. The proposed implementation performs three main tasks: initiating work and slave registration, handling migrations, and finishing work and sending and collecting results. Figure 7 depicts the main components of the JavaSpace technology. Note that the arrow represents one of the three actions (write, take, or read) that can be performed by the different clients (either master or slave) in order to communicate and synchronize through the shared JavaSpace area.

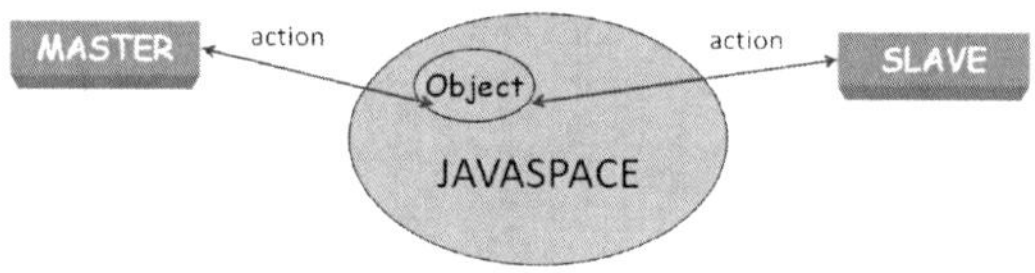

Fig. 7. JavaSpace technology components

4.2.1 Initiating Work and Slave Registration

The proposed implementation requires that the master keeps an updated list of the slaves working for it at any time so it can properly synchronize them. In order to accomplish this, all the clients must register with their master.

Once a master starts, the first step consists of writing (W) an entry (*InitEntry*) into the JavaSpace. This entry will then be taken by slaves that can perform the task. This entry is read (R) by the slaves. It is important to note that the entry must only be read (instead of being taken) by the slaves, this is because the entry "Init" must be kept in the JavaSpace during all the processes so that new slaves can be added at any time.

Subsequently, once a slave receives the *InitEntry*, it requests registration from the master. The slave then writes an unapproved *SlaveEntry*. Then, the master takes this entry. The master then writes the same entry that was taken before. The only difference is that this entry has been already approved by the master.

In the last step, the slave takes this entry, and writes it back. The reason for this rewriting is that the slave must be the author of the entry in the space in order to keep its lease alive. This entry lease is renewed in the background by the slave so that the master can know when the slave is not working on the task anymore, either because it completed the task successfully or due to an unexpected error, such as a network error or unavailability of the entry. At the end of this step, the master has a list of all the slaves that are working for it. The master will periodically check the status of its current slaves.

4.2.2 Handling Migrations

At a given point of the execution of the algorithm a migration will occur, and each of the slaves will send *EmigrantEntries* to the JavaSpace and then wait for an *ImmigrantEntry* to be sent to it by the master. The master will then collect all *EmigrantEntries* that are sent out by its slaves and will store them. After all slaves have sent their emigrants, the master creates an *ImmigrantEntry* for each one of its slaves based on the emigrant populations that it has collected in the previous step, and it will write all of them to the JavaSpace. Finally, each slave is able to take its corresponding immigrant and continue the execution of its task, i.e., the MOEA.

4.2.3 Finishing Work and Sending Results

When a slave client has reached the maximum number of evaluations in the TSAP algorithm, it writes a *ResultEntry* into the JavaSpace. These entries are then taken by the master, which adds them into its final result set. When the first *ResultEntry* is taken by the master, it also cancels the *InitEntry*, so that no new slaves begin working on it past this point in time. After this point the master waits for its slave list to empty, either because the slaves complete their work or due to an unexpected error that prevented a slave from finishing its computation. Once this occurs, the master presents its final result set, comprised of all the results received.

5 Experiments and Results

To test the evolutionary algorithm, a set of sample problems is produced which contain a reasonable ratio of sailors to tasks. Each problem was generated according to a specified number of sailors, tasks and time slots. Also, the mean and standard deviation of a normally distributed random variable determine number of tasks each sailor can perform.

The complexity of the problem and the time constraints prevented us from performing more rigorous statistical analysis of the results. Nonetheless, we believe that our experiments shed valuable light on the problem and can serve as a fruitful base for future research on the SAP and its variants.

In all the problems, 7 days with 12 time slots per day were considered which makes the problem harder since it increases the number of tasks for all the time slots. Also, a variable expected ratio of sailors per task was used. Table 1 shows the parameters of each sample problem used in this work.

5.1 Metrics Used for Evaluating the Solution

Many existing performance metrics in multi-objective optimization require knowing the set of Pareto-optimal solutions, and such solutions are used to compare against the approximate solutions obtained. It is also important to notice that many existing performance metrics for evaluating the distribution of solutions cannot be used in higher dimensions because the calculation of diversity measure is not straightforward and often computationally expensive [25].

In case of the SAP, to measure the performance of the algorithms, the *coverage* and *proximity* metrics [26] were used while on the other hand, for the TSAP, since Pareto-optimal solutions for the considered TSAP instances are not known, some of the existing performance metrics cannot be used to evaluate the results of the proposed approach. Therefore, the hypervolume was then used as a quality measure. Explanation about various performance metrics are as follows:

1. *Hypervolume.* The hypervolume indicator is a metric used in evolutionary multi-objective optimization that measures the volume of the dominated portion of the objective space. An important feature of this metric is that it is strict Pareto compliant.
2. *Proximity.* The distance between the approximation set S and the Pareto front P_F is defined as the average of the minimum distance between a solution and the Pareto optimal front over each solution in S, as in [26]:

$$D_{S \to P_F}(S) = \frac{1}{|S|} \sum_{Z^0 \in S} min_{Z^1 \in P_F} d(Z^0, Z^1), \tag{28}$$

   where $d(Z^0, Z^1)$ is the Euclidean distance between the two solutions. A small value for this indicator means that all the points in the approximation set are, on average, close to the true Pareto front. An ideal value of 0 is obtained when all the points are actually in the Pareto front.
3. *Diversity.* To measure the diversity of the approximation set the reverse of the proximity metric is used, as defined in [26]:

$$D_{P_F \to S}(S) = \frac{1}{|P_F|} \sum_{Z^1 \in P_F} min_{Z^0 \in S} d(Z^0, Z^1). \tag{29}$$

   In this indicator, for each solution in the Pareto front the distance to the closest solution is calculated, and the average is taken as the value for the indicator. A small value for this indicator means that all the points in the true Pareto front have, at least, one point in the approximation set which is very close. The ideal value of 0 is obtained when all the points in the Pareto front are also contained in the approximation set.

5.2 Experimental Results for the Serial Implementation

Simulated instances to test the MOEA contain values for all the objectives along with the resource requirements for performing tasks and capacities of the sailors. These values were generated assuming normal distributions. In previous works [5], it was found that one of the most important factors governing the difficulty of a problem instance was the contention for tasks. Given a fixed number of sailors, the more total tasks available for the sailors to choose from, the less difficult the problem is.

In a real world scenario, it is unlikely that there will exist many more available tasks in all time slots than sailors to fill them. Accordingly, a set of sample problems is produced which contain a reasonable ratio of sailors to tasks. Each problem was generated according to a specified number of sailors, tasks and time slots. Also, the mean and standard deviation of a normally distributed random variable determine number of task types each sailor can perform.

In all the problems, 7 days with 12 time slots per day were considered which makes the problem harder since it increases the number of tasks for all the time slots. Table 1 shows the parameters of each sample problem used in this work. The task-sailor ratio is defined here as the rate of task types a sailor can perform on average.

Table 1. Parameters of the randomly generated TSAP instances used in this work

	No. sailors	No. tasks	Task/sailor ratio
Problem 1	1,000	500	0.25
Problem 2	2,000	1,000	0.05
Problem 3	4,000	2,000	0.05
5 Problem 4	5,000	200	0.01

For each problem, 10 runs of the evolutionary algorithm were performed for three different population sizes: 100, 200 and 400 individuals where each run consists of 500,000 evaluations of the objectives. Binary tournament selection, where two individuals are chosen at random and the fittest one is selected, and mutation with probability (1.0/Total no. of tasks to be performed) were used. Also, all individuals in the population were chosen to be crossed over and the same amount of offspring was produced. The populations of parents and offspring were combined and the non-dominated solutions go to the mutation process. Additionally, after performing crossover and mutation, the repair operator function was executed.

Figure 8 depicts the change in the hypervolume in one run of NSGA-II for a 5,000 sailor instance of the TSAP with 100 population size. Notice that after a certain number of evaluations (about 250,000 evaluations) the hypervolume will remain almost constant. The reference point used to compute hypervolume was [1,1,1,1,1]). Note that the reference point has the same dimensionality as the objective space.

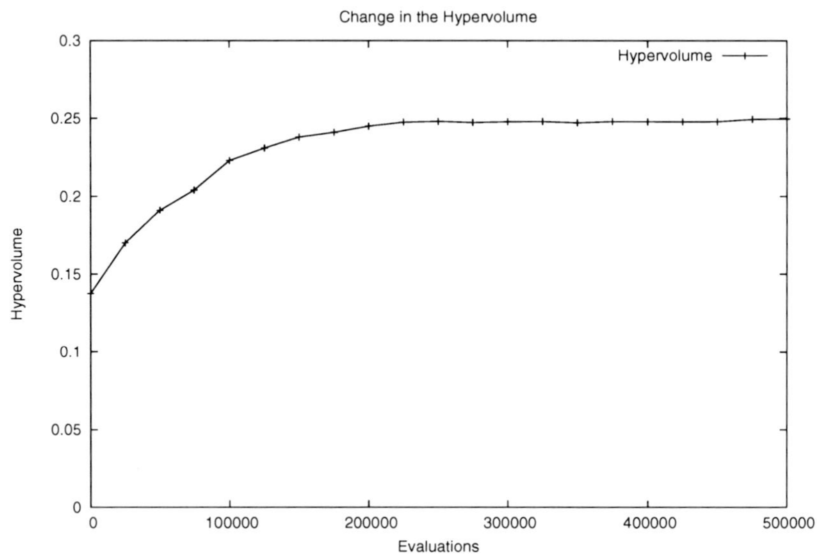

Fig. 8. Change in the hypervolume during 500,000 objective evaluations for a 5,000 sailor instance of the TSAP with 100 population size.

In this work, several approaches were implemented and thus the results obtained were also compared. Particularly, the results of the standard NSGA-II were compared against the hybrid NSGA-II. Accordingly, in each run, the final Pareto set approximation was recorded. The mean and standard deviation of the final hypervolumes over the specific number of trials are reported in Table 2 and Table 3. It can be seen from Table 2 that as the difficulty of the problem increases (see Table 1), correspondingly the standard deviation increases. As the size of the population increases, the number of non-dominated solutions also increases, which in turn gives more precise values of the hypervolume, which can be noticed from the value of the standard deviation for the corresponding population sizes.

Figure 9 shows a parallel plot for sample solution by applying NSGA-II for 1,000 sailors, while Figure 10 shows the results for the same problem instance obtained by the hybrid approach. In these figures, each objective is plotted along the x-axis, whereas ranges of these objectives are represented by y-axis and each line represents a complete solution, i.e., a particular assignment

Table 2. Mean and standard deviation of the hypervolume for the number of specified runs on each instance of the TSAP

	Population size		
Sailors	100	200	400
1,000	.1844±.0103	.1648±.0055	.1529±.0040
2,000	.2223±.0015	.1963±.0015	.1665±.0019
4,000	.2745±.0019	.2451±.0014	.2096±.0023
5,000	.2565±.0029	.2449±.0023	.2305±.0019

Table 3. Mean and standard deviation of the hypervolume for 10 runs of NSGA-II and the three different hybrid approaches based on local search (LS) operators (shift, swap and a combination of both) for different instances of the TSAP

	Approaches			
Sailors	1000	2000	4000	5000
NSGA-II	.1843±.0103	.2222±.0015	.274480±.00190	.25650±.0029
shift LS	.1864±.0106	.2228±.0017	.274500±.00190	.25590±.0028
swap LS	.1809±.0126	.2229±.0014	.274300±.00140	.25510±.0026
both LS	.1906±.0056	.2231±.0017	.275700±.00014	.25670±.0031

of sailors to tasks in all time slots. In general, the hybrid approach, including local search operators, provided better diversity in all the objectives as observed in the sample solutions shown in Figures 9 and 10. Particularly, notice in Figure 10, solutions with lower values for objectives 1, 3 and 5 are found. For instance, a specific solution that assigns only 201 sailors, which corresponds to only 20.1% of the total number of sailors, is provided.

Same as above, Figure 11 shows a parallel plot for a sample solution by applying NSGA-II for 5,000 sailors, while Figure 12 shows the results for the same problem obtained by the hybrid approach combining both local search operators. Specifically, the shift local search operator is performed in every $10{,}000^{th}$ evaluation whereas swap local search operator is executed in every $15{,}000^{th}$ evaluation. From Figure 10 it is observed that solutions with lower values are found for all objectives. Moreover, in the case of last objective (number of assigned sailors) the hybrid approach provides a solution with less number of sailors, such as 83 (0.0166) whereas NSGA-II comes with a solution with 87 sailors (0.0174). It is worth noticing that the diversity of the last objective depends entirely on the task-sailor ratio.

Table 3 summarizes the results of comparing 17 runs of NSGA-II to three different hybrid approaches based on local search operators (shift, swap and a combination of both) for the same problems defined in Table 1 and population size 100. It is observed that the hybrid approach that alternates both local search operators provides better solutions from a hypervolume point of view, i.e., better approximations towards the optimal Pareto front.

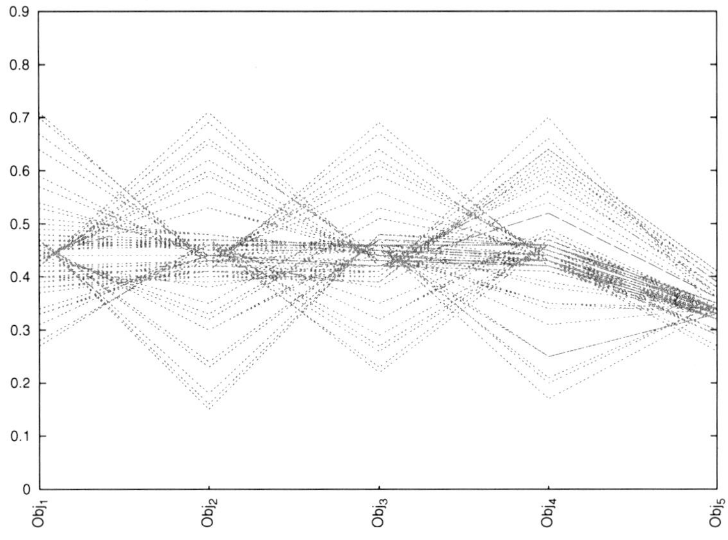

Fig. 9. Diversity of solutions found in the Pareto front using NSGA-II on a 1,000 sailor instance of the TSAP and population size equal to 100.

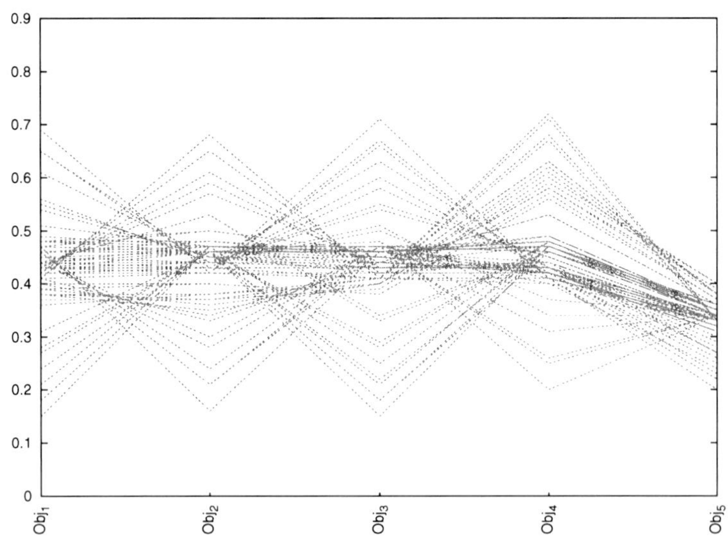

Fig. 10. Diversity of solutions found in the Pareto front using a hybrid approach alternating both local search operators on a 1,000 sailor instance of the TSAP and population size equal to 100.

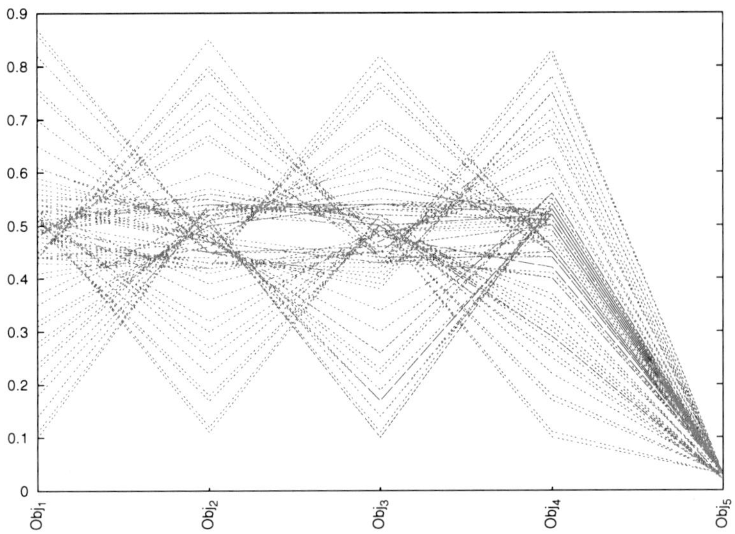

Fig. 11. Diversity of solutions found in the Pareto front using NSGA-II on a 5,000 sailor instance of the TSAP and population size equal to 100.

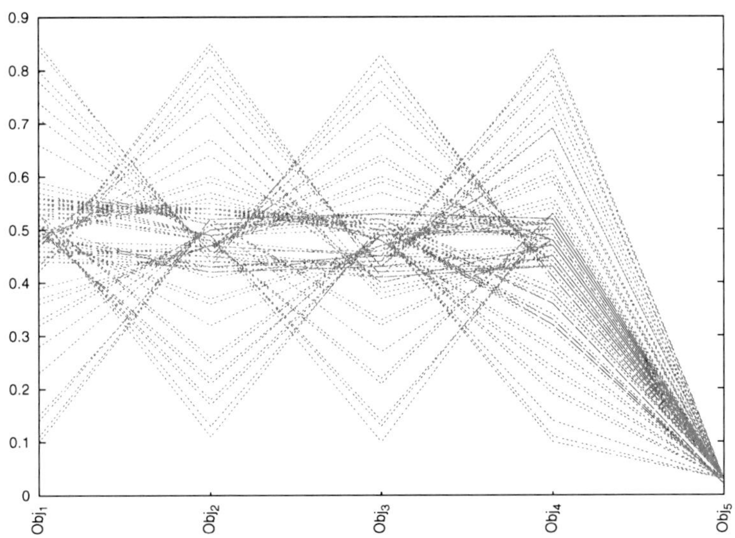

Fig. 12. Diversity of solutions found in the Pareto front using a hybrid approach alternating both local search operators on a 5,000 sailor instance of the TSAP and population size equal to 100.

Further experimentation was carried out after adding three new constraints to the EA. Experiments were performed on problems of variable sizes like 40 sailors and 400 job types problems or 1000 sailors and 500 job types problems etc. Following is the parallel plot comparison for the 1000 sailors problem, Figure 13 is the parallel plot of the GA run without the three new constraints and Figure 14 is the parallel plot of the GA run after adding the three new constraints

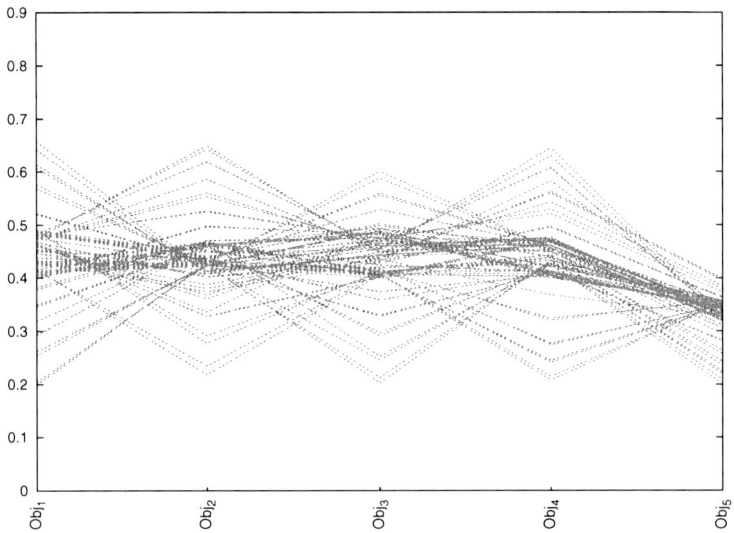

Fig. 13. Diversity found on the pareto front for a 1000 sailors instance and a population of size 100 using a GA without adding the new constraints.

Notice that the diversity of the Pareto front in Figure 13 is less as compared to Figure 14, one possible reason for this behavior is that, as we add more and more constraints, the size of the set of feasible solutions decreases as they have to fulfill all the constraints. Following is a comparison of the schedules of a 40 sailor instance. In Figure 15 and 16, the X-axis is the timeslots in the entire week and y-axis is the sailor number. Each dark block in the picture indicates that a job has been assigned to sailor y in timeslot x.

As is clearly shown in the two figures below, Figure 16 has a highest number of continuous horizontal dark blocks as compared to Figure 15 for each sailor. This indicates that due to the introduction of the new constraints, the GA selects the solutions which have continuous work periods and continuous rest periods alternatively for all sailors, rather than having constant fluctuation and switches between work and rest for the sailor. However, it also shows that the new GA is not a perfect one, the parameters need further fine tuning to get better results.

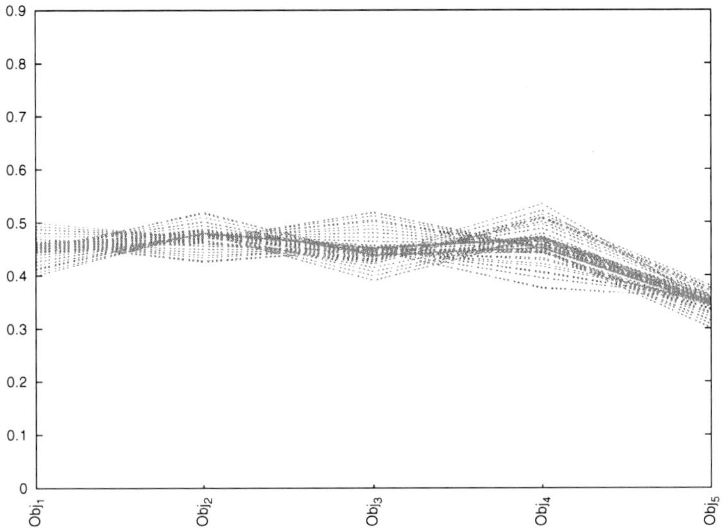

Fig. 14. Diversity found on the Pareto front for 1000 sailors instance and a population size of 100 using a GA with the newly added constraints.

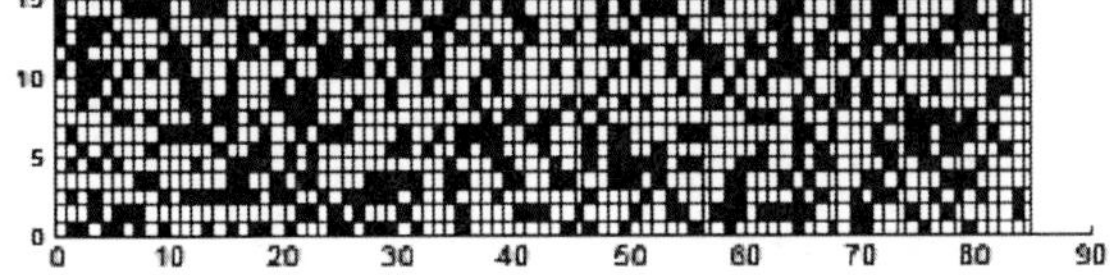

Fig. 15. A diagram of the schedule of the sailors for the entire week. This schedule was generated using the GA without the newly added constraints.

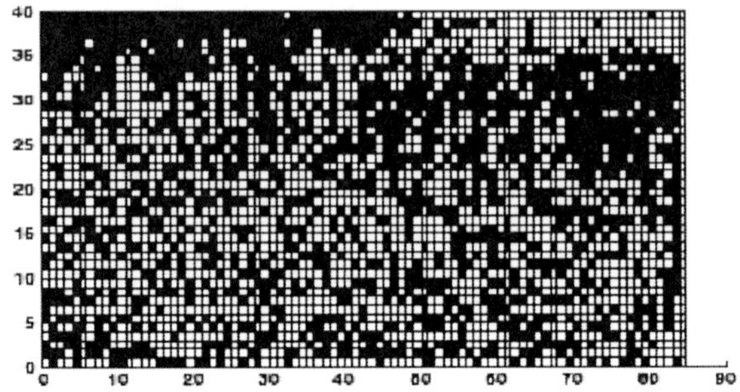

Fig. 16. A diagram of the schedule of the sailors for the entire week. This schedule was generated using the GA with the newly added constraints.

5.3 Parallel Implementation Results

5.3.1 Parallel Experimental Framework

An experimental framework was developed in order to study the impact of the proposed parallel model over the TSAP. A number of experiments were carried out to see if the parallel algorithm provides adequate results. Specifically, the experimental framework analyzes different parameters in the island model architecture and the migration policies; it also studies how those parameters affect the TSAP outcomes. Those experiments aim to determine the set of parameters that provide an adequate cover of the search space improving the precision of the model and the quality of the obtained Pareto fronts.

Although island models have been widely studied, there is still no full understanding of the role of the model's basic parameters. Since there are so many possible parameter settings for an island model, it is important to understand what influence each parameter has, how it interacts with the other parameters and how these interactions impact the results of a specific problem, for instance the TSAP.

5.3.2 Experimental Details

To thoroughly study the impact of the parallel implementation on the behavior of the TSAP algorithm, some parameters were fixed and several experiments for all combinations of those parameters were performed. Specifically, to evaluate the impact of the number of islands, a ring topology was used with three different numbers of islands (four, eight and sixteen islands). To study the impact of the migration policies three different migration rates (five, ten and twenty times) and three different portions of the population to be migrated (10%, 20%, and 30%) were used. The MOEAs parameters were fixed to perform 500.000 evaluations for 1000 sailors and 500 task types. The probability of mutation and crossover were set to 0.01 and 0.7, respectively. Those parameters are summarized in Table 4.

Table 4. Island architecture and migration policies.

Name	Values
Number of islands	4 and 8
Migration rates	5,10 and 20 times
Population to be migrated	10%, 20% and 30%
Number of evaluations	500000
Number of sailors	1000
Number of jobs	500
Crossover	70%
Mutation	1%

In order to compare the algorithm performance, the hypervolume metric has been used. The experimental framework allowed us to draw charts representing the average and standard deviation of the computed hypervolumes as a function of the parameters (island architecture and policies). Additionally, it allowed us to observe improvements and degradations of the algorithm with respect to its sequential version. The computational specifications of the implementation are as follows:

1. *Server.* Dual Intel Xeon Quad core 2.6 GHz CPU; 12 GB Memory; Debian Linux v4.3.2 Operating System; and 1 G Ethernet Communication Network;
2. *Workstations.* Dell Optiplex 755 Intel Core2 Quad 2.4 GHz CPU; 2 GB Memory; and Debian Linux v4.3.2 Operating System.

5.3.3 Parallelization Experimental Results

For each specific parameter configuration, five independent runs were performed. For each run, the improvement of the hypervolume with respect to the hypervolume obtained by the sequential algorithm was computed. Figure 17 and 18 report the average of the runs for a 4 island and an 8 island configuration. Figure 19 and Figure 20 depict the dispersion of the improvements. Note that a configuration is built as the combination of the parameters shown in Table 4. For example a configuration labeled as s4.p10.m5 means that a problem with 4 different islands, migrating 10 percent of the population, during 5 times in the 500,000 evaluations was executed.

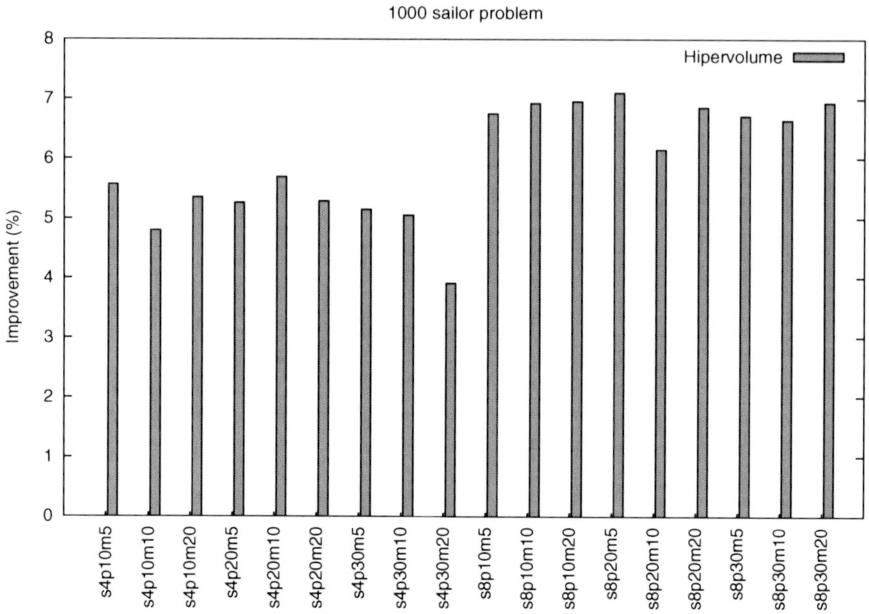

Fig. 17. Average of the improvement of the hypervolume of the parallel conformations with respect to its serial counterpart in the 1000 sailor problem.

From Figures 19, 20, 17 and 18, it is possible to derive the following conclusions:

1. When the number of evaluations is fixed at 500.000, the parallel implementation of TSAP provides better solutions than its sequential version, for all the tested configurations, i.e., the parallel implementation never obtains worse solutions than the sequential algorithm. The best performance is obtained by an island of 8 processors migrating 20 times for 20 percent of the population. Using this configuration an improvement of 7.11% in the hypervolume measure was obtained with respect to the hypervolume obtained by the sequential counterpart.
2. In general, the results obtained for all the configurations using 4 islands were outperformed by those using 8 islands.
3. The obtained improvements in hypervolume measures for the 2000 sailor problem are higher than those of the 1000 sailor problem (See Figures 17 and 18). It is clear that the 2000 sailor problem is a more difficult problem than the 1000 counterpart, and the parallel implementation seems to work better in this kind of harder problems. In the 1000 problem the hypervolume value is stabilized before the 500,000 thousand evaluations are exhausted. In the two thousand sailors problem this point is harder

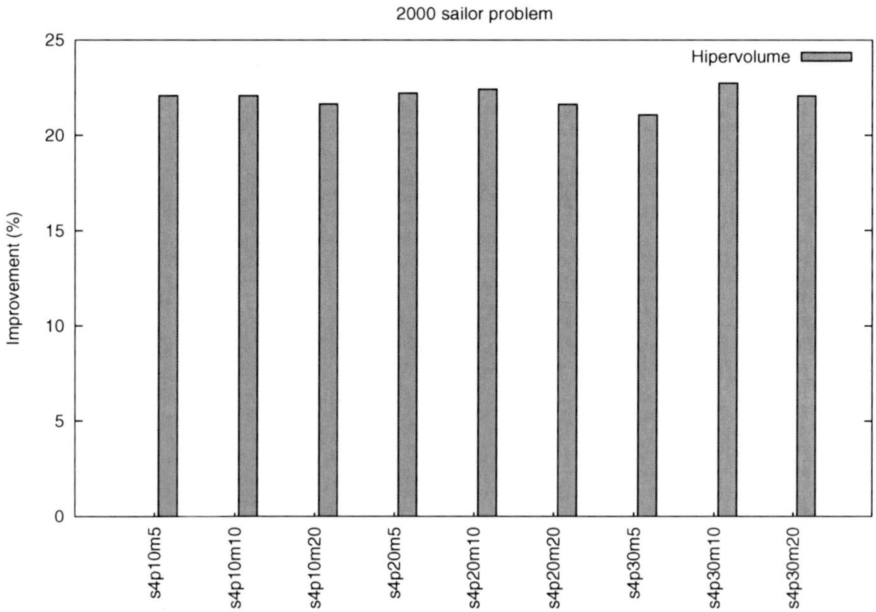

Fig. 18. Average of the improvement of the hypervolume of the parallel conformations with respect to its serial counterpart in the 2000 sailor problem.

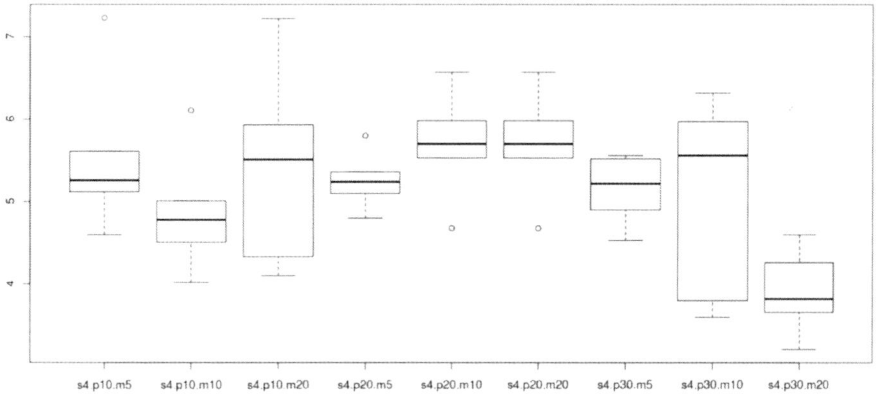

Fig. 19. Box plot diagrams for configurations using 4 islands.

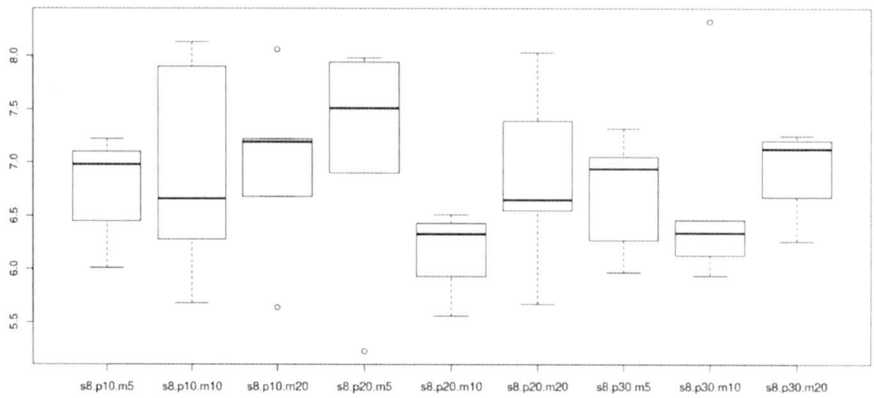

Fig. 20. Box plot diagrams for configurations using 8 islands.

to achieve. Then, the effectiveness of the two thousand problem benefits from the parallel implementation because more solutions arise from the simultaneous executions of the TSAP instances spreading the cover of the Pareto front.

4. However, in the TSAP, the iterations involving a higher number of processors working on it have a beneficial effect as they provide a way to improve the quality of the obtained Pareto fronts. It is not clear if this improvement is significant compared with the additional quantity of computational resources invested in a parallel approach. Serial iterations with the whole population seem to work well in the TSAP for the chosen number of evaluations (500.000), and that makes it unclear if this is the reason for the real improvement of the parallel approach. It has been reported in [4] that in the sequential algorithm for the TSAP, after a certain number of evaluations, the hypervolume value will remain almost constant. Thus, more tests should be performed in order to identify that point of stabilization to perform the comparison of convergence time.
5. It is not clear how the policies of migration impact the Pareto solutions. Thus, more experiments need to be carried out to get conclusions about how those parameters provide a beneficial effect to increment the diversity of the solutions to the TSAP.

6 Conclusions

In this chapter, a multi-objective evolutionary approach to solve the TSAP was presented. The proposed approach, based on the multi-objective NSGA-II, provided a good diversity of solutions along the Pareto set approximation.

An improvement of the TSAP results was obtained using a parallel approach. This has been verified by running several experiments over two TSAP instances (1000 and 2000 sailor problems) and changing the combinations of the parameters in the model. Although the standard island approach to the TSAP with migration seems to be quite robust and efficient, it is difficult to prove and compare the real improvement of the model. Specifically, it is not clear how the policies of migration impact the Pareto solutions. Therefore, more experiments need to be performed to conclusively determine how those parameters provide a beneficial effect to increment the diversity of the TSAP solutions.

Further work is needed to improve the MOEAs by including some local search operators or other hybridization mechanisms. In addition, implementing some domain specific genetic operators could potentially provide further improvements. In particular, a crossover operator which better maintains feasibility would reduce the need to repair individuals. Such an operator could potentially provide benefits both in solution quality as well as computation time.

Another potential improvement of the algorithm could be obtained through the injection of good solutions into the initial population. While the proposed approach currently incorporates five distinct objectives, there are several additional components which require further investigation (e.g., not all tasks are equally important to fill). An effective algorithm should bias solutions towards filling the most important tasks. Further work is needed to determine the effect on the performance of the algorithm. Also, a fair comparison with other approaches is necessary to better evaluate the proposed approach.

In this work, the degree of constraint violation was equally distributed among all the objectives. Thus, extensive experimentation is required to investigate other weighting schemes of the penalty functions. Also, a more exhaustive experimentation could be performed to fine-tune the parameters of the proposed MOEA.

Acknowledgements. The authors would like to thank James Simien for his contributions to this work. This research work is a better product due to his constructive help and comments.

This project is funded under the Scientific Services Program (TCN: 08179 Mod 1 and 07192), which is sponsored by the US Research Office and is managed by Battelle under Prime Contract No: W911NF-07-D-0001. The views, opinions, and/or findings contained in this report are those of the author(s) and should not be construed as an official Department of the Army position, policy, or decision, unless so designated by other documentation.

References

1. Knowles, J.D., Corne, D.W.: Quantifying the effects of objective space dimension in evolutionary multiobjective optimization. Multi-Criterion Optimization, 757–771 (2007)
2. Moscato, P.: On evolution, search, optimization, genetic algorithms, and martial arts: Towards memetic algorithms. Technical Report 826, Caltech Concurrent Computation Program, C3P (1989)
3. Moscato, P., Cotta, C.: A gentle introduction to memetic algorithms. In: Handbook of Metaheuristics, pp. 105–144 (2003)
4. Dasgupta, D., Hernandez, G., Garrett, D., Vejandla, P., Kaushal, A., Yerneni, R., Simien, J.: A comparison of multiobjective evolutionary algorithms with informed initialization and kuhn-munkres algorithm for the sailor assignment problem. In: Proceedings of the 2008 Genetic and Evolutionary Computation Conference (GECCO 2008). ACM Press, New York (2008)
5. Garrett, J.D., Vannucci, J., Silva, R., Dasgupta, D., Simien, J.: Genetic algorithms for the sailor assignment problem. In: Proceedings of the 2005 Genetic and Evolutionary Computation Conference, (GECCO 2005). ACM Press, New York (2005)
6. Knowles, J.D., Corne, D.W.: Memetic algorithms for multiobjective optimization: Issues, methods, and prospects. In: Krasnogor, N., Smith, J.E., Hart, W.E. (eds.) Recent Advances in Memetic Algorithms. Springer, Heidelberg (2004)
7. Jaszkiewicz, A.: On the performance of multiple objective genetic local search on the 0/1 knapsack problem: A comparative experiment. Technical Report RA-002/2000, Institute of Computing Science, Poznan University of Technology (July 2000)
8. López-Ibáñez, M., Paquete, L., Stützle, T.: Hybrid population-based algorithms for the bi-objective quadratic assignment problem. Technical Report 1, FG Intellektik, FB Informatik, TU Darmstadt, Journal of Mathematical Modelling and Algorithms (2004)
9. Shmoys, D.B., Tardos, É.: An approximation algorithm for the generalized assignment problem. Mathematical Programming 62(1), 461–474 (1993)
10. Knowles, J., Corne, D.: Towards Landscape Analyses to Inform the Design of Hybrid Local Search for the Multiobjective Quadratic Assignment Problem. In: Abraham, A., Ruiz del Solar, J., Koppen, M. (eds.) Soft Computing Systems: Design, Management and Applications, pp. 271–279. IOS Press, Amsterdam (2002)
11. Knowles, J.D., Corne, D.W.: M-PAES: A memetic algorithm for multiobjective optimization. In: Proceedings of the 2000 Congress on Evolutionary Computation (CEC 2000), pp. 325–332. IEEE Press, Los Alamitos (2000)
12. Krasnogor, N.: Towards robust memetic algorithms. In: Hart, W.E., Krasnogor, N., Smith, J.E. (eds.) Recent Advances in Memetic Algorithms, pp. 185–208. Springer, Heidelberg (2004)
13. Garrett, J.D., Vannucci, J., Dasgupta, D., Simien, J.: Applying hybrid multiobjective evolutionary algorithms to the sailor assignment problem. In: Jain, L., Palade, V., Srinivasan, D. (eds.) Advances in Evolutionary Computing for System Design, pp. 269–301. Springer, Heidelberg (2007)
14. Gale, D., Shapley, L.S.: College admissions and the stability of marriage. The American Mathematical Monthly 69, 9–15 (1962)

15. Deb, K., Pratap, A., Agarwal, S., Meyarivan, T.: A fast and elitist multiobjective genetic algorithm: NSGA-II. IEEE Transactions on Evolutionary Computation 6(2), 182–197 (1991)
16. Zitzler, E., Laumanns, M., Thiele, L.: SPEA2: Improving the strength pareto evolutionary algorithm. Technical Report 103, Computer Engineering and Networks Laboratory (TIK), Swiss Federal Institute of Technology (ETH), Zurich, Switzerland (May 2001)
17. Munkres, J.: Algorithms for the assignment and transportation problems. Journal of the Society of Industrial and Applied Mathematics 5(1), 32–38 (1957)
18. Kuhn, H.W.: The hungarian method for the assignment problem. Naval Research Logistic Quarterly 2, 83–97 (1955)
19. Dasgupta, D., Nino, F., Garrett, D., Chaudhuri, K., Medapati, S., Kaushal, A., Simien, J.: A multiobjective evolutionary algorithm for the task based sailor assignment problem. In: Proceedings of the 2009 Genetic and Evolutionary Computation Conference (GECCO 2009), pp. 1475–1482. ACM, New York (2009)
20. Akgul, M.: The linear assignment problem. Combinatorial Optimization, 85–122 (1992)
21. Talbi, E.-G., Mostaghim, S., Okabe, T., Ishibuchi, H., Rudolph, G., Coello Coello, C.A.: Parallel approaches for multiobjective optimization. In: Branke, J., Deb, K., Miettinen, K., Słowiński, R. (eds.) Multiobjective Optimization. LNCS, vol. 5252, pp. 349–372. Springer, Heidelberg (2008)
22. Sun. Javaspaces, `http://Java.sun.com/developer/technicalArticles/tools/JavaSpaces`
23. Freeman, E., Arnold, K., Hupfer, S.: JavaSpaces principles, patterns, and practice. E Addison-Wesley Longman Ltd., Amsterdam (1999)
24. Coello Coello, C.A., Lamont, G.B., Van Veldhuizen, D.A.: Evolutionary algorithms for solving multi-objective problems. Springer-Verlag New York Inc. (2007)
25. Deb, K., Thiele, L., Laumanns, M., Zitzler, E.: Scalable multi-objective optimization test problems. In: Proceedings of the 2002 Congress on Evolutionary Computation (CEC 2002), pp. 825–830. IEEE Press, Los Alamitos (2002)
26. Bosman, P.A.N.: Design and Application of Iterated Density-Estimation Evolutionary Algorithms. PhD thesis, Institute of Information and Computing Sciences, Universiteit Utrecht, Utrecht, The Netherlands (2003)

Genetic Algorithms for Manufacturing Process Planning

Guohua Ma and Fu Zhang

Abstract. Process planning has been defined as the systematic determination of the machining methods (operations, machine, tool, fixture) by which a product is to be manufactured economically and competitively. A process plan describes the manufacturing processes for transforming a raw material to a completed part, within the available machining resources. This chapter presents the application of genetic algorithms (GAs) in computer aided process planning (CAPP), and the development of a CAPP system based on a GA. The key to successfully applying GAs to a real-world application such as process planning is to model the problem from an optimization perspective, and to design a special representation mechanism, operators, and constraints to introduce the domain knowledge into the algorithms. These steps are discussed in great detail to show how evolutionary algorithms such as GAs can be used to solve a difficult real-world problem along with their advantages.

1 Introduction

Process planning is a production organization activity that transforms a product design into a set of instructions (machines, tools, set-ups, etc.) to manufacture the product. Over the last 30 years, there have been numerous attempts to develop computer-aided process planning (CAPP) systems in order to assist human planners. A process plan describes the manufacturing processes for transforming a raw material to a completed part, within the

Guohua Ma
Faculty of Department of Electronics & Mechanical,
Wentworth Institute of Technology, Boston, MA, USA
e-mail: mag@wit.edu

Fu Zhang
The MathWorks, Natick, MA, USA
e-mail: zhangfu@hotmail.com

available machining resources. The role of process planning in a manufacturing environment is illustrated in Figure 1 [1].

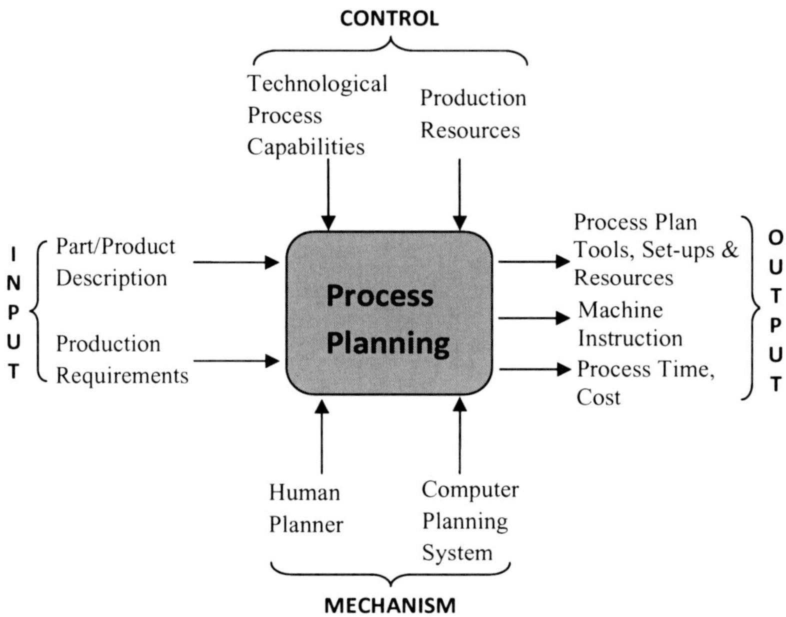

Fig. 1. The role of process planning in the manufacturing environment.

The use of CAPP systems has the following advantages [1]:

- Reduction of the demand on the skilled planner,
- Reduction of the process planning time,
- Reduction of both process planning and manufacturing costs,
- Creation of consistent plans,
- Production of accurate plans, and
- Increase of productivity.

Traditional CAPP approaches mainly aim at generating a single feasible plan for a given part. However, the introduction of new manufacturing technologies, such as design for manufacturing (DFM) and the integration of process planning and scheduling, have resulted in some new demands from CAPP. For example, to support DFM, the best process plan for a given part in a designated machining environment must be generated and fed back to the designer for evaluation. To support dynamic scheduling, a CAPP system must be able to generate plans with alternative routes and sequences to suit the variable status of the shop floor.

This chapter presents an optimization-centered CAPP model and search methods for prismatic parts machined in any given conventional job shop. For the traditional job shop, the approach aims to address the new roles of CAPP within the concept of concurrent engineering. The initial input is a solid model of a part. Machining features are then extracted, and each feature is mapped to all possible sets of operations, based on its geometrical and technological requirements and available machining resources in the job shop. Precedence relationships among all the operations are identified as constraints for operation sequencing. The process planning problem modeled in such a way contains the entire solution space constructed by the multiple choices of machines (M), tools (T), tool approach directions (TADs), and operation sequences. Several objective functions are provided for plan evaluation that minimizes the number of machines, set-ups, and tool changes. To find the optimal solution, genetic algorithms (GA) have been developed in which process planning heuristics are incorporated. A CAPP system based on the developed model and GAs has been implemented with Unigraphics, a commercial computer-aided design (CAD) software. The developed algorithm is able to achieve an optimal or near-optimal process plan by simultaneously considering operation sequences and selection of machines, tools, and TADs. Several case studies show that the developed CAPP system can generate multiple process plans to respond to dynamic events in the job shop, and also provide valuable information to support design for manufacturing. The main contributions of this chapter are:

- The process planning problem is modeled from an optimization perspective by considering all the possible combinations of M/T/TAD (operation selection) and operation sequencing concurrently.
- A GA has been successfully applied to solve this optimization problem by developing domain knowledge related GA operators.
- A CAPP system on top of a commercial CAD system is developed by embedding our GA into the system.

2 Related Work

CAPP has attracted much research interests for the past three decades. Numerical materials, including comprehensive reviews, have been reported in [2, 3]. The review in this chapter is unique in that it addresses the CAPP literature mainly from an optimization perspective, which reflects some trends of CAPP research in recent years.

2.1 Trends of CAPP: Towards Integration

The introduction of numeric control (NC), CAD, computer-aided manufacturing (CAM), and CAPP in manufacturing has resulted in enormous improvements in product quality, efficiency, and productivity. However, it also resulted in decreased flexibility, caused by the inability of the various departments to generate the required information adequately, accurately, and quickly. Departments have become isolated islands of automation. Integration is more than just communication and interfacing, such as the exchange of product, tool, and machine tool data. Co-operation is more important, i.e., how should different departments work together to come to joint solutions with regard to improved product quality, reduced cost and times-to-markets. New manufacturing philosophies, such as concurrent engineering (CE) and design for manufacturing (DFM), require CAPP to be fully integrated with other manufacturing activities. According to Bedworth et al. [4], the essence of CE is the integration of product and process planning into one common activity. In last three decades, the focus of the CAPP research has been on the partial integration of various planning tasks [3, 5, 6, 7, 8, 9, 10, 11, 12]. Some of the researchers focus on one manufacturing process in detail, i.e., boring of the turned component [13], or fix-axis mill-turn parts [14].

The content of integration of CAPP can be roughly categorized into two parts: integration of CAPP and scheduling, and integration of CAPP and product design.

2.1.1 Integration of CAPP and Scheduling

Scheduling is the allocation of resources over time to perform a collection of tasks in such a way that some relevant criteria are met. Traditionally, the output of process planning is used as the input to scheduling. These two parts are carried out as sequential, non-collaborative tasks. This situation leads to several drawbacks [2, 15]:

1. Process planning takes place without input from the job shop.
2. Scheduling follows process planning but only a fixed process plan is the input of the scheduling stage. Such a process plan gives unnecessary constraints to scheduling.

To overcome these shortcomings, the question of integration of process planning arose in the middle of the 1980s. Since then, this issue has been frequently addressed and has become a major issue in CAPP research [16, 17, 18]. Researchers have been attempting to build an integrated system that offers advanced production management by closely linking scheduling to process planning. Recently, more such works are reported [19, 20].

2.1.2 Integration of CAPP and Product Design

The integration of CAPP and product design has advanced from a concern with an interface to a state where the two systems can work in a shared information environment. It is a bi-directional activity: a) Design information must be transformed to CAPP for creating a process plan, and b)

Manufacturability information must be feedback from CAPP to design. In concurrent engineering, it is no longer sufficient to ensure a uni-directional flow of information from design to process planning. It is essential to feedback information from process planning to assist the designer in assessing the various features, not only from a functional point of view but also regarding manufacturability, assemblability, manufacturing time and cost, at an early stage [2]. One of the approaches to enable such an integration is called plan-based manufacturability analysis and redesign [8, 21]. An intelligent CAPP system to support early manufacturability analysis, cost estimation, and provide redesign suggestions will be most useful.

2.2 Integration Approach: An Optimization Perspective

One of the most widely used approaches to support the integration of process planning and scheduling is non-linear process planning [16, 17]. From an optimization perspective, integration in the context of non-linear process planning is to search for all the possible plans for each part before it enters the shop floor and a schedule is generated based on the best plans. However, some researchers [17] think this approach to be a kind of interface rather than "integration" because planning and scheduling are still carried out sequentially. The integration level between planning and scheduling of such an approach is low. They have tried to search a large combined space in order to generate optimal or near optimal plans and schedules concurrently. Nevertheless, in this case, it is necessary to perform deep search in a space that is usually too large in most real time settings [22]. This approach requires high capacity and capability from both hardware and software.

Process planning is a constrained optimization problem. Plans generated must meet various constraints imposed by the design specifications and the availability of manufacturing resources, as well as satisfy complex optimization criteria. Optimization in the context of process planning can be considered from the perspective of routing (machine tools and cutting tools assignment), sequencing (determination of set-ups and operations order), definition of tool path and machining parameters. Analytical models exist only for machining parameter selection. The other problems are dealt with largely by a heuristic or algorithmic approach [23].

During the last 20 years, process planning optimization has received significant research attention. Related research issues are: identify precedence constraints, plan evaluation and selection, and optimization strategy.

2.2.1 Identifying Precedence Constraints

For a given part, the machining operations cannot be performed in an arbitrary order. A number of geometric and technological constraints require some operations to be performed before or after certain other operations. Precedence constraints will highly influence operations sequencing and set-up planning. How to identify all these precedence constraints is essential for solving the plan optimization problem. Hayes and Wright [24] use feature interactions (mainly due to fixturing considerations) to guide the search for process plans. Feature interactions happen when the results of an operation destroy some of the requirement of others. Constraints (due to interaction) are used to cut down the search space. Gupta and Nau [25] present a systematic approach to find precedence constraints, which are used to generate all possible process plans. Chu and Gadh [6] use a rule-based approach to generate process plans with minimum set-ups. In their approach, features are classified according to the number of possible tool approach directions that can be used to machine them. Features are divided into two classes: single tool approach direction features and multiple tool approach direction features. Rules are applied to guide the search for proper ordering between operations and set-ups of a part. Generally, precedence relations are represented by a) precedence matrices [26], and b) precedence graphs [27, 28, 29], and precedence networks [30].

2.2.2 Plan Evaluation and Selection

In integration manufacturing systems, finding the most suitable process plan is a very complex task [31]. Ideally, the quality of a process plan should be evaluated from a technological as well as an economical standpoint. Meanwhile, the objectives in process plan selection might be imprecise and conflicting [32]. Therefore, some researchers use fuzzy approaches to deal with the process plan selection problem [31, 32]. Most of these approaches are useful if the micro-planning stage has been finished and the detailed process plans are available. In the macro planning stage, researchers always use a single value variable such as set-up change times, tool change times or some cost functions like weighted sum, to be the criterion [27, 33].

2.2.3 Optimization Strategy

As mentioned earlier, process planning optimization contains routing and sequencing tasks. How to deal with the relationship between the two tasks is a signature characterizing different optimization strategies. The first strategy is called the sequential decision making strategy. It treats routing and sequencing tasks as two separate, sequentially connected planning stages. This approach can certainly reduce the search space significantly, but the optimal plan may be lost. Search methods used in this approach are classical operations research (OR) algorithms such as A* [27], dynamic programming (DP) [34], and heuristic rules [35].

Typically, assignment of the machine, tool, and tool approach direction (TAD) has been performed before sequencing. In other words, the routing space is narrowed before search begins. For example, in [36], sequencing is carried out in three sequentially connected phases. In the initial phase, machining operations for a part are selected. Machines, fixtures, TADs, and sequence of the set-ups are determined in the second phase. The final planning phase determines all the detailed sequence of operations in each set-up, using some heuristic rules. Irani et al. [29] use a precedence graph (PG) representation for alternative process plans. Each Hamiltonian path in the PG that represents a feasible process plan is enumerated by the Latin multiplication method. However, the machining route is limited by the constraint that each feature is produced by only one feasible machine, tool, and TAD. This will reduce the flexibility to react to dynamic workshop disturbances. Sometimes, sequencing is performed before routing [34, 35]. Khoshnevis et al. [34] divide process planning into two stages. In the first stage, a sequence for processing the feature in such a way that the least amount of change of state with regards to machine, tool, and set-up is found. Search in this stage is based on "generation and best selection" rules. The specified machines, tools, and resting faces to be used are identified at the next stage by using dynamic programming, a form of OR search algorithm.

It is clear that neither of the above two approaches, adopting the sequential strategy, can obtain optimal results. Optimal plans can be obtained only by making these decisions concurrently. This is because the two tasks are interrelated; separating them would mean that decisions have to be taken in the absence of relevant information [27].

Some researchers have adopted the concurrent search strategy. They have tried to concurrently sequence operations and assign machining capacity to the operations. New search algorithms, such as simulated annealing (SA) [27] and GAs [37, 38], are often used in the concurrent approach. This approach is not widespread most likely due to the high complexity of the combined routing and sequencing tasks. This kind of approach could be extended to generate process plans and production schedules simultaneously [18, 27, 38]. Literature also shows that conventional operations research algorithms are not powerful enough to search for the large space if routing and sequencing are considered concurrently [27]. Thus, many newly developed optimization algorithms, especially evolutionary algorithms (EAs), have been used to find optimal process plans.

2.3 EAs for Process Planning

An EA is a generic population-based metaheuristic optimization algorithm. An EA uses some mechanisms inspired by biological evolution: selection, mutation, recombination [39]. Evolution algorithms (EAs) have been applied in many engineering fields [40, 41, 42]. EAs also have been applied for the process and scheduling problem in a multi-plant environment [43], and in forming machine cells and product families [44]. In [44], GAs are combined with a local search heuristic to minimize inter-cellular movement and maximize the utilization of the machines within a cell. GAs are well implemented in finding the optimal process plans, or integrating process planning and scheduling [20]. Dereli and Filiz [45] employed a GA for finding the optimal sequence of machining operations based on either minimum tool change or minimum tool traveling distance or safety. A comprehensive survey about evolutionary techniques for optimization problems in integrated manufacturing system is conducted by Gen et al. [46].

3 Modeling Process Planning from an Optimization Perspective

In this section, process planning is modeled as a constrained optimization problem. The modeling process contains the following steps:

1. Constructing the search space;
2. Finding constraints; and
3. Building the objective function.

3.1 Constructing the Search Space

The search space for process planning optimization is composed of all the feasible process plans. Several concepts are described as follows:

Feature, a concept used by the designer to describe the geometric shape of a design, such as a hole, a slot etc. The design of a part can be described as a set of features. A feature can be manufactured through several operations. For example, several operations such as drilling and milling are needed to finish a hole.

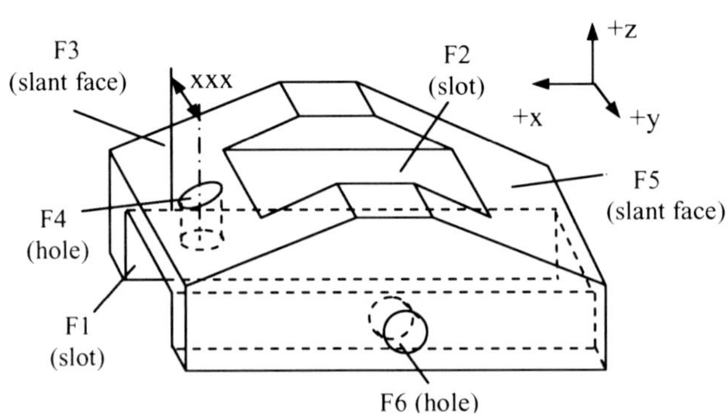

Fig. 2. Features and precedence relationships.

Operation, a part of a machining process such as milling, grinding or other treatment of a workpiece at a single workplace. This is the basic unit to compute the manufacturing cost and time. A process plan is composed of a set of operations.

Tool approach direction (TAD), the TADs of an operation are the unobstructed directions from which a cutting tool can access the feature to be manufactured by this operation. For example, in Figure 2, the feature F2 (slot) can be access from $-z$ with a vertical mill or $-y$, $+y$ with a horizontal mill.

3.1.1 Job Shop: The Environment of Process Planning Optimization

As mentioned earlier, one way to support integration is to increase the flexibility of process planning. In order to do so, the manufacturing environment that the CAPP system is developed for must be flexible enough to support alternate manufacturing capacities. Some traditional planning approaches are developed for handling planning tasks within a rather simple machining environment, such as a single CNC machine center. Such CAPP systems have little flexibility in terms of the ability to adjust to changes in their internal or external environment [28]. In the case of a dynamic event where a machine may break down, process plans generated as well as re-planning tasks cannot be executed.

On the other hand, a job shop is one of the basic forms of manufacturing, and its primary advantage is that it provides much flexibility. These include: 1) machine flexibility, 2) routing flexibility, 3) process flexibility, 4) product flexibility, and 5) volume flexibility [28]. A CAPP system developed for a job shop should have the following degrees of flexibility:

1. Route flexibility
   a. *Machine flexibility* refers to executing an operation on different machines. For example, drilling on a press drill or a CNC machine center.
   b. *Tool flexibility* refers to executing an operation with different cutting tools.
   c. *TAD flexibility* refers to executing an operation with different tool approach direction.
2. Sequence flexibility refers to executing an operation with different orders in the process plan.

Inside the job shop environment, the above-mentioned flexibility can be expressed by depicting the operation as a tree in Figure 3. The feasible M/T/-TAD and order of an operation are called operations attributes. An operation can be described as a collection of these attributes, i.e., [*order*, M/T/TAD].

Each unique, feasible combination of machine, tool, and TAD is called an *operation event (OpE)*. An operation can be treated as a collection of

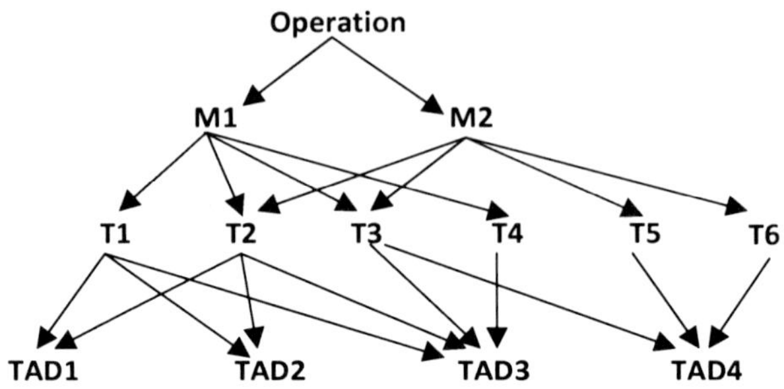

Fig. 3. Operation and its attributes.

operation events. A process planning problem in a job shop environment can be defined as follows:

1. Operation selection. For each feature, determine one or several operations required. This includes the selection of machines, tools, and TADs based on the feature geometry and available machining resources.
2. Operation sequencing. Determine the sequence of executing all the operations required for the part so that the precedence relationships among all the operations are maintained.

The decision-making tasks of operation selection and operation sequencing must be carried out simultaneously in order to achieve an optimal plan against a pre-determined criterion, such as minimum machine changes.

3.1.2 Tool Approach Directions (TADs)

For prismatic parts, six possible TAD's, i.e., the six normal directions of a prismatic block (±x, ±y, ±z), are assumed. For a tool acting on a feature alone, the theoretical TADs are fixed. However, interference may occur when considering the part and tool geometry. In the current approach, the solid model of a tool is moved from a pre-defined path towards the feature along its theoretical TADs. After reaching the feature, it is moved to cover the entire feature. During this simulation, any TAD that causes interference is discarded. For example in Figure 2, feature F4 can be machined through $+z$ and $-z$ direction. If the feature is a blind hole, it can only be approached from $-z$ direction.

3.1.3 Machine and Cutting Tool Selection

Selecting a machine capable of accommodating the individual processing needs of each operation can be a difficult job. Machine selection can be divided into two distinct aspects: technical and economical machine selection. Technical machine selection deals with the selection of machines, which can be used to produce a part in a technical sense. In the economical machine selection procedure, it is determined which machine is the preferable one with respect to availability and cost. In our approach, the machine is selected in a technical sense. All the possible machines that are capable of doing the job in the job shop are selected.

Due to the complexity of the cutting tool geometry, cutting tool selection heavily depends on the process planner's experience. Certain rules are summarized from the experienced process planners. Most rules specify cutting tool attributes in a range of values rather than a fixed value. According to the shape of each feature and the operation selected, the tool selection module searches in the cutting tool database, and analyzes sets of tool descriptions to find all possible cutting tools that match each operation.

Assuming that the part in Figure 2 is to be machined on a CNC vertical mill (CVM) and a drill press (Dril) with only one end-mill cutter (T1) and one drill (T2), the alternative OpE's for F1, F2, F4, and F6 are (CVM/T1/+z), (CVM/T1/-z), (CVM/T2/+z; CVM/T2/-z; Dril/T2/+z; Dril/T2/-z), and (CVM/T2/-y; Dril/T2/-y), respectively.

The final result of operation selection for a feature is expressed in a tree structure as shown in Figure 4. It has three levels: feature, Op sets, and OpE (M, T, TAD) sets. The OpE sets will construct the final search space.

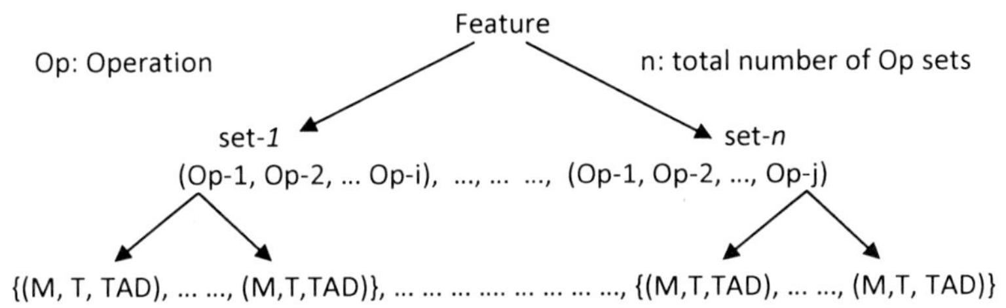

Fig. 4. Tree data structure of operation selection for a feature. The structure proceeds from features, to Op sets, to OpE sets (M, T, TAD).

3.2 Finding Constraints – Precedence Relationships Between Operations

The OpE sets form the basic elements of the plan solution space. This space is further enlarged by the possible sequences among the OpE in the final plan. However, not all possible sequences are valid. A valid sequence must satisfy the precedence relationships (PRs) between operations caused by geometrical and technological consideration. In the present approach, the PRs between operations are identified from the following:

1. Three types of constraints are considered to determine the PRs between features.

a. *Fixture constraint*: A PR between two features exists when machining one feature first may cause another to be unfixturable. An example is given in Figure 2, where F1 must be milled before F3 and F5 are milled. Otherwise, the supporting area for milling F1 is not sufficient.
b. *Datum dependency*: When two features have a dimensional or geometrical tolerance relationship, the feature containing the datum should be machined first. For example, F1 needs to be machined before F4 since the side face of F1 is the datum of F4 (see Figure 2).
c. *Good machining practice*: Good manufacturing practice or rules-of-thumb may also result in precedence relationships between features. Referring to Figure 2, if F4 is a blind-hole, F4 should be drilled before F3 in order to avoid cutter damage.

2. PRs among the set of operations of a feature: For every set of operations obtained through mapping from a feature, there exists a fixed PR among those operations, i.e., roughing operations come before finishing operations. Some examples are: drilling comes before reaming, milling comes before grinding, etc.

The PRs between two features can then be converted to the PRs between their OpEs according to the following rules.

If $F(X) \longrightarrow F(Y)$, $Op(i) \in F(X)$, and $Op(j) \in F(Y)$
Then $Op(i) \longrightarrow Op(j)$

The PRs between two operations are represented in the following manner.

If $Op(i) \longrightarrow Op(j)$
Then $Op(i)$ is the predecessor of $Op(j)$.

For a part needing n operations, all the PRs obtained can then be stored in an n by n matrix: $\{PR_{ij}, |i = 1, \ldots, n, j = 1, \ldots, n\}$. If operation i must be performed before operation j, $PR_{ij} = 1$; otherwise, $PR_{ij} = 0$.

3.3 The Process Planning Model – A Network Representation

Given the operation sets of the part, the process planning optimization problem can be conveniently described by a network constrained by their PRs. The network consists of parent node: the operation *Op*. Each *Op* consists of several child node: OpEs, the combinations of (M, T, TAD). There is a link between any two *Ops* that represents the PR between them, i.e., the one that the arrow points to must be performed after the other, while a link with double arrows means that there is no PR between the two *Ops* it connects. An example of OpE network is shown in Figure 5.

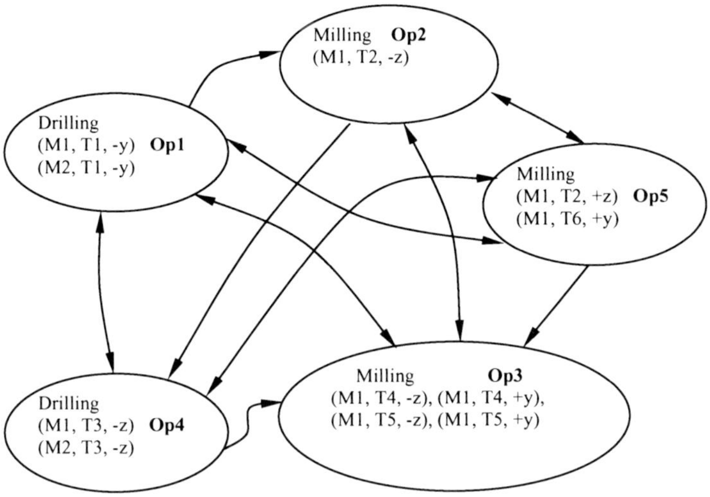

Fig. 5. The operation network representation.

3.4 Building the Objective Function: Plan Evaluation Criteria

Currently, the most commonly used criteria for evaluating process plans include the minimum number of set-ups, the shortest process time, and the minimum machining cost, etc. Since the detailed information on tool paths and machining parameters is not available at this stage, the total machining time and cost cannot be used for plan evaluation. Instead, the following five cost factors are identified as the plan evaluation criteria:

1. Machine cost (MC)

$$MC = \sum_{i=1}^{n} MCI_i \tag{1}$$

   where n is the total number of Ops and MCI_i is the machine cost index for using machine-i, a constant for a particular machine.
2. Tool cost (TC)

$$TC = \sum_{i=1}^{n} TCI_i \tag{2}$$

   where TCI_i is the tool cost index for using tool-i, a constant for a particular tool.
3. Machine change cost (MCC): a machine change is needed when two adjacent operations are performed on different machines.

$$MCC = MCCI \times \sum_{i=1}^{n-1} \Omega(M_{i+1} - M_i) \tag{3}$$

where $MCCI$ is the machine change cost index, a constant and M_i is the ID of the machine used for operation i.

$$\Omega(M_i - M_j) = \begin{cases} 1 & if \ M_i \neq M_j \\ 0 & if \ M_i = M_j \end{cases} \tag{4}$$

4. Setup change cost (SCC): a setup change is needed when two adjacent Ops performed on the same machine have different TADs.

$$SCC = SCCI \times \sum_{i=1}^{n-1} ((1 - \Omega(M_{i+1} - M_i)) \times \Omega(TAD_{i+1} - TAD_i)) \tag{5}$$

where $SCCI$ is the setup change cost index, a constant.

5. Tool change cost (TCC): a tool change is needed when two adjacent Ops performed on the same machine use different tools.

$$TCC = TCCI \times \sum_{i=1}^{n-1} ((1 - \Omega(M_{i+1} - M_i)) \times \Omega(T_{i+1} - T_i)) \tag{6}$$

where $TCCI$ is the tool change cost index, a constant.

These cost factors can be used either individually or collectively as a cost compound, based on the requirement and the data availability of the job shop.

Up to this point, the optimization process planning problem can be defined as follows: find the feasible process plan that traverses all the operations and the sequence among its *Ops* that satisfies their precedence relationships.

4 GA Implementation

To apply GAs to process planning problems, we model process planning as an optimization problem. The modeling process introduces the domain knowledge through *plan evaluation criteria*, GA *string representation* design, *operators* design, *constraints* design. The principles and procedures presented can be applied to many real-world problems.

4.1 Applying GAs to Process Planning

A GA is a stochastic search technique based on the mechanism of natural selection and natural genetics. The basic idea of GAs is that an initial population of candidate states of size n is chosen at random, and each state is evaluated according to the optimization function. Through crossover and mutation, good performing individuals have higher possibility go survive. After several generations, the algorithm converges to the best solution of the problem. The general algorithm is shown in Algorithm 1.

Algorithm 1. A General GA

```
begin
    n ⟵ 0;
    initialize population P(n);
    evaluate initial population P(n);
    while some convergence criteria is not satisfied do
        Perform competitive selection
        Apply genetic operators to generate new solutions P_new(n));
        evaluate solutions in the population P_new(n));
        n ⟵ n + 1
end
```

The key steps to modify a general GA to a special, domain specified GA in order to solve process planning problems are described in the following section. The same steps and principles can be applied to solve other real-world problems.

4.1.1 Knowledge-Based Representation of Process Plans

The first step in formulating a GA for process planning is to map the problem solutions (process plans) to string representations. Inspired by the works of Burns [42], we use a knowledge-dependent string to represent all possible elements in the solution space. For an n-operation problem, a string representing a process plan is composed of n gene segments. Each gene segment contains a child node OpE (M-ID, T-ID, TAD) from a unique parent node and its order number in the string. This representation is illustrated in Figure 6. It is clear that this string representation can cover all the solution space due to the selection of machine tools, cutting tools, TADs, and the sequence among operations.

Gene Segment

Op1	Op2	Op4	Op5	Op3	Op6
Machine 1	Machine 1	Machine 7	Machine 7	Machine 5	Machine 6
Tool 1	Tool 1	Tool 7	Tool 4	Tool 5	Tool 5
TAD +x	TAD +x	TAD +x	TAD +y	TAD +y	TAD -z

Fig. 6. A string representing a process plan with six operations.

4.1.2 Enforcing the Precedence Relationships

With the string representation, the first generation of gene strings can be generated randomly, and searching for a better solution can be carried out through the basic GA operators, i.e., selection, crossover, and mutation. However, for process planning problems, the precedence relationships (PRs) between operations must be maintained in the final solution. In other words, the challenge is how to enforce the PRs between operations throughout the GA operations.

Initially, we did not add any restriction on the generation of solution strings during the GA operations. In other words, the solution strings in each generation may include valid as well invalid sequences of operations. In order to eliminate the invalid strings, a penalty function was added to the compound cost function for every precedence violation. The invalid strings were expected to be eliminated during selection. This algorithm was tested on a range of process planning cases with the number of PRs in ascending order. The results showed that this algorithm was only able to find an optimal or near-optimal solution for cases with little precedence constraints. For cases having relatively large number of PRs, however, the solutions were often trapped at local minima formed by invalid solutions. This may be attributed to the limitation of the population of every generation. We are still investigating methods of solving this problem.

In the current approach, we use different methods to enforce the PRs in two stages of the GA, i.e., initialization and evolution. During initialization, we use a PRs filter to check every randomly generated solution string against all the non-zero PR_{ij}. If any precedence relationship is violated, this string is eliminated and a new solution string is generated. This process is repeated until a valid solution string is found. Enforcing PRs during the evolution stage, i.e., selection, crossover, and mutation, will be discussed with the introduction of these operators in the later sections.

4.1.3 Generation of Initial Population

Initial populations cannot be generated randomly. They must satisfy the PR constraints. A method similar to that reported in [47] has been adopted. The algorithm is described in Algorithm 2.

Algorithm 2. Generating initial Solution Based on Given PRs

```
begin
    Start with an empty string
    repeat
        Select (at random) an OpE (M, T, TAD) among those which have
        "no predecessors" and append it to the end of sequence;
        Delete the OpE just handled from the OpE network, as well as from
        PR matrix.
    until All OpEs are processed ;
end
```

4.1.4 Fitness Evaluation

Fitness is the measure of the quality of plans represented by the strings. The higher the fitness, the better the plan. Once all the solution strings in a generation are generated, the cost compound (CC) for the plan alternatives represented by these strings can be calculated using Eq.(1)–Eq.(6). CC is used as the fitness of the solution string. Since the objective is to search for a solution with minimum CC, the fitness of a solution string is calculated as:

$$fitness = UB - CC \tag{7}$$

where UB is an upper bound, is chosen to ensure a positive fitness value for all the solution strings. Therefore, a solution string possessing a high CC will have a low fitness value.

4.1.5 Domain-Related GA Operators

With the initial population of solution strings, our GA starts running by applying three operators – selection, crossover, and mutation. Specific domain knowledge must be incorporated into the design of GA operators.

1. **Selection.** In the selection process a new population is chosen with respect to the probability distribution based on fitness values. A widely known selection method is *proportionate selection* or *roulette wheel selection*. Here, the basic idea is to determine selection probability for each chromosome proportional to the fitness value. For chromosome k with fitness f_k, its selection probability P_k is calculated as follows:

$$p_k = \frac{f_k}{\sum_{j=1}^{Pop_{size}} f_j} \tag{8}$$

where Pop_{size} is the population size.

In order to enhance the selection pressure, the raw fitness function was scaled by a non-linear scaling: Fitness $= \frac{fitness^2}{a}$, where a is a constant dependent on the UB of the given problem. The scaled fitness is used instead of the raw fitness acquired by Equation 8. The aim of using scaling is to maintain a reasonable differential between relative fitness ratings of chromosomes. For example, for two stings, $s1$ and $s2$ with $fit_1 = 2000$, $fit_2 = 2050$, $fit_2/fit_1 = 1.025$, the competition between $s1$ and $s2$ is scaled to be more remarkable. We call this strategy "biased roulette wheel".

Meanwhile the "elite" mechanism developed by De Jong [41] is also incorporated. It works as follows: Let $p^*(t)$ be the best string generated up to generation t. If, after generating $P(t+1)$ in the usual fashion $p^*(t)$ is not in $P(t+1)$, then include $p^*(t)$ to $P(t+1)$ as the $(N+1)th$ number, where N is the population size. "Elitist" can be understood as the "best always survive" mechanism to enforce preserving the best chromosome. Numerical experiments show that this mechanism significantly increases the search speed. More importantly, it has been proven that simple GAs with standard genetic operators are not of global convergence [48] while GAs with "elitist" are of global convergence [49]. This mechanism is even more important in GAs for process planning problem since the mutation rate is very high.

2. **Crossover.** The strings obtained from selection are then mated at a given probability (crossover rate P_c). This operation is called crossover. To ensure that the crossover will not destroy the precedence constraint, and that each operation in the offspring must be executed once and only once, the cyclic crossover operator proposed by Dagli and Sittisathanchai [50] is adopted and modified to cross two strings (process plans). By applying this operator to two strings, the offspring generated is still a legal process plan that satisfies all the constraints. The algorithm is described in Algorithm 3.

Algorithm 3. Crossover of two process plan string-1 and string-2

```
begin
    Determine a cut point randomly from the all the positions of a string.
    Each string is then divided into two parts, the left side and the right
    side, according the cut point.
    Copy the left side of string-1 to form the left side of offspring-1. The
    operator constructs the right side of offspring-1 according to the order of
    operations in string-2.
    Copy the left side of string-2 to form the left side of offspring-2. The
    operator constructs the right side of offspring 2 according to the order of
    operations in string-1.
end
```

This process is illustrated with an example shown in Figure 7. A pair of strings, parent-1 and parent-2, undergoes the crossover operation in which the cut point is chosen between positions 3 and 4. The left side of parent- 1, Op4-Op1-Op2, is used to form the left side of offspring-1. The order of the right side of parent-1, Op5-Op7-Op8-Op3-Op6-Op9 is adjusted according to the order of parent-2 to form the right side of offspring-1. By doing so, the sequences among the operations in both parent-1 and parent-2 are maintained in offspring-1. A similar operation is applied to parent-2 and parent-1 to form offspring-2.

3. **Mutation.** So far, it can be seen that our GA traverses the solution space based on the alternative sequence between operations using crossover. The solution space, due to the assignment of machines, cutting tools, and TADs to each operation, must be also covered. This is achieved by using the

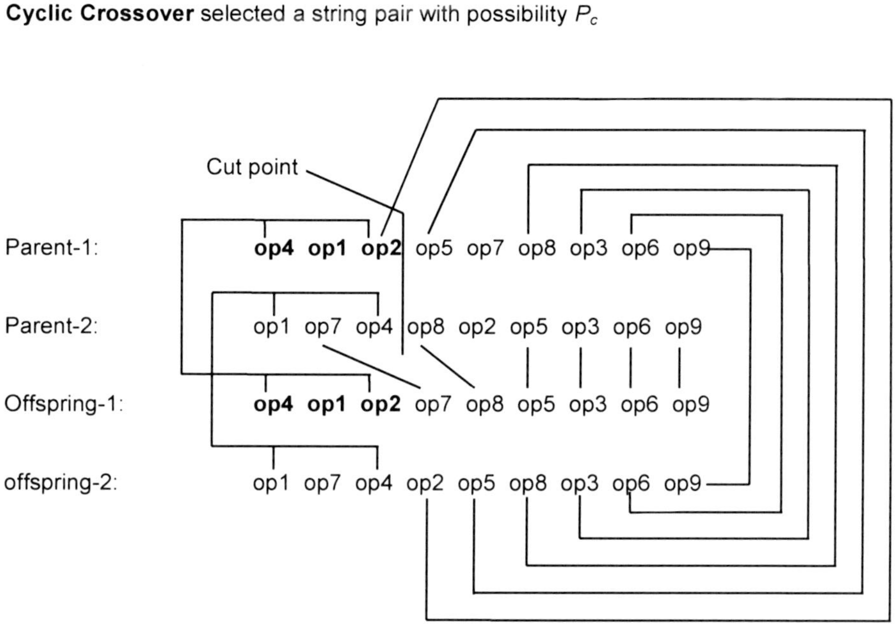

Fig. 7. An example of applying the cyclic crossover for changing operations sequence.

mutation operation. Mutation for a simple GA was described as the occasional (with small probability) random alternation of the value of a string position. For process planning, mutation can be defined as the random alternation of the child node OpE (M-ID, T-ID, TAD) of an operation. Therefore, we developed four advanced mutators to carry out mutation for process planning. They are described as follows:

a. *Machine mutator:* Machine mutator is used to change the machine to perform an operation if more than one machine can be applied. In order to reduce the total machine changes, machine mutation does not stop at the selected position. Instead, the "machine space" for every other operation is also checked to determine if a mutation can heuristically reduce machine changes. Algorithm 4 shows that the machine mutator operates on the solution strings in a generation.

   This mutation is illustrated with an example shown in Figure 8. It can be seen that Op3 (M1) is selected for mutation where M1 is the current machine. M3 is then assigned to Op3 to replace M1. It is also found that M1 is currently used by Op1, Op4, Op5, and Op2. Among them, Op1, Op4, and Op5 have M3 as one of their alternative machines.

Algorithm 4. Machine Mutation

begin

For every solution string, select an operation (a position in the string) randomly and use a predetermined probability (mutation rate) to determine if the machine needs to be changed. How the mutation rates are selected is determined in the following section.

Randomly choose a machine (M-b) from all the alternatives to replace the current machine (M-a).

Identify all the other operations in the same string whose current machine is M-a. If any one of these operations has M-b as an alternative, assign M-b to replace M-a.

end

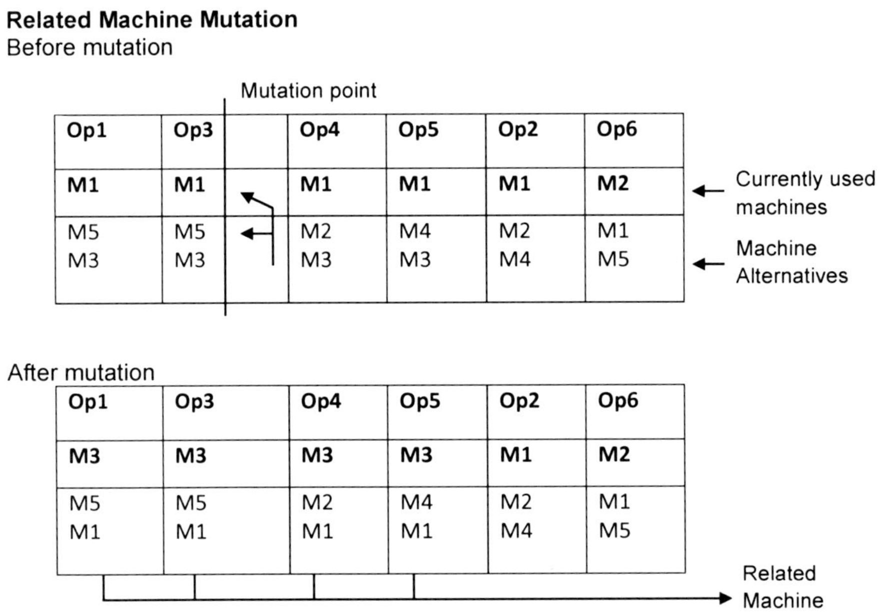

Fig. 8. An example of machine mutation.

Therefore, M3 is assigned to Op1, Op4, and Op5 to replace M1. The mutation rate of machine mutator is denoted as P_m^M.

b. *Tool Mutator:* The tool mutator operates on the solution strings after machine mutation is carried out. It has exactly the same mechanism as the machine mutator except the mutation rate is denoted as P_m^T.

c. *TAD Mutator:* The TAD mutator operates on the solution strings after machine mutation and tool mutation are carried out. It has exactly the same mechanism as the machine mutator except the mutation rate is denoted as P_m^{TAD}.

d. *Sequence Mutator:* The function of sequence mutator is used to introduce new sequence patterns. It was mentioned earlier that, by using cyclic crossover only, some of the sequence patterns will be lost. The sequence mutator is used to overcome this. It works as follows: select two gene segments randomly and then swap them, i.e., exchange their position in the whole string. The mutation rate is denoted as P_m^{Seq}.

By using the above four mutation operators, the machine, tool, and TAD mutators traverse the space of routing while the sequence mutator and cyclic crossover traverse all the space of sequencing. Therefore, the concurrent decision making mechanism can be realized, i.e., operations sequencing and routing are carried out concurrently.

4.1.6 Repairing Mechanism

By employing the sequence mutator, the shortcoming of cyclic crossover can be overcome. However, it is clear that such mutation may make a string infeasible because the precedence constraints may be violated by sequence mutation. A repairing mechanism is therefore developed to repair an infeasible string and make it feasible. This mechanism is called *Corrector*. The *Corrector* algorithm is shown in Algorithm 5, where $Oper_{num}$ is the number of operations to finish the part; i, j are the numbers to identify the operations, called $OpID$. $Sequence(i)$ is the order of the operation whose $ID = i$. For example, if a process plan is Op1-Op3-Op4-Op2, then $Sequence(4) = 2$.

Algorithm 5. Corrector

```
begin
    i ← 1;
    j ← 1;
    while (i, j < Oper_num) do
        if (PR_ij = 1 and Sequence(i) > Sequence(j)) then
            k ← j + 1;
            while k < i + 1 do
                Sequence(k) = Sequence(k) − 1
                k = k + 1
            Sequence(j) = Sequence(i) + 1
end
```

4.2 GA Parameters

The settings of GA parameters in our experiments are as follows:

1. ***Population size*** $n = 50$. In principle, the population size should be large enough such that the population associated with the solution domain can be

adequately represented. A larger population, however, needs higher computation costs, e.g., in terms of memory and runtime. Experimental results [51] have shown that $50 < Pop_{size} < 200$ works quite well for most problems.

2. ***Crossover probability*** In the present GA formulation, crossover is equivalent to a change of operation sequence that should therefore be vigorously performed to traverse more points in the solution space. Most GA literature suggests that $0.5 < P_c < 1.0$.
3. ***Mutation probability*** Generally, mutation acts as a background operator which provides a small amount of random search. Its purpose is to maintain diversity within the population, and inhibit premature convergence due to loss of information. Mutation alone induces a random walk through the search space. P_m is often in the range of [0.001, 0.1]. However, in process planning, the search space is constructed by two parts: sequencing space and routing space. Crossover cannot cover all the search space. The three group mutators play a similar role as crossover since the extended solution space due to the existence of alternative machines, tools, and TADs must be adequately traversed in the optimizing process. Meanwhile, sequencing will influence mutation since they interact, and the sequence mutator must re-introduce lost sequence pattern. Therefore, the four mutation probabilities should be similar to the crossover rate. Through numerical experiments by running GA with different P_m, the four mutation rates are determined as:

$$P_m^M = P_m^T = P_m^{TAD} = P_m^{Seq} = 0.6$$

Table 1 shows the comparison of GAs with different P_m, where "> 8000" means that after 8000 generations, the best fitness did not reach the best result ever found before; when computing the average value, if generation > 8000, let generation = 8000. It shows that the GA with $P_m \subset [0.5, 0.8]$ performs better.

Table 1. Influence of mutation rate P_m on the searching speed of GA.

P_m	0.2	0.4	0.6	0.8	1.0
1	>8000	1060	1480	>8000	940
2	>8000	>8000	6040	>3400	1940
3	>8000	3840	600	1300	4060
4	>8000	4240	380	560	>8000
5	>8000	1480	520	1340	5860
6		4920	3900	5860	>8000
7		1780	1180	1280	3560
8		6400	3520	2360	4380
9		>8000	460	360	4580
10		3420	180	2420	>8000
Average*		4314	1826	2688	4932

This value of P_m is much larger than the value suggested by literature. However, new research results in GA provided some bases for using such a large P_m. Firstly, some researchers [52] use mutation as the only genetic operator in solving combinational optimization problems. Secondly, they point out that the practical applications of GAs often favor larger or non-constant settings of the mutation rate. For example, in solving a scheduling problem [53], the mutation rate is set to 0.8 at the first generation and then gradually reduced by $P_m^{(t+1)} = aP_m^t$, where t is the generation number and $a = 0.99$. Thirdly, some studies by Fogel and Atmar [54] have shown that a GA with lower mutation rate is likely to outperform (slightly) a the GA with higher mutation rate in terms of average performance, but the GA with higher mutation rate has a higher variance and finds a better solution in a higher proportions of the trials. This indicates that, if only the best overall solution is sought, the mutation rate should be set higher than what has been normally perceived. The same phenomenon is found in [55]. Numerical experiments also show that in order to find a better solution with higher mutation rates, the "elitist" mechanism must be used to keep the best solution that has been found; otherwise higher mutation rates would soon cause it to be lost since the better strings are mutated. A GA with variable mutation rates was also tested and the results show that it performs as well as the constant mutation rate, $P_m = 0.6$. The mutation schedule used is:

$$P_m(t) = P_m^{max} * e^{-a\tau} \tag{9}$$

where $P_m^{max} = 0.8$, τ=t/8000, $a = 1 + \ln 10$, t is the generation number.

4. ***Stopping criterion*** When to stop the GA is a rather difficult and specific problem. At least three termination criteria are proposed in [56]. The most common practice is to terminate the GA after a pre-specified number of generations. A more intricate one is a number of generations in which no improvement is gained. More flexible termination criteria are based on a diversity test of the population. The degree of diversity is measured by entropy [56]. When the entropy drops below a given threshold, the GA is terminated. This is a convergence criterion. Since GA is a stochastic search method, it is quite difficult to formally specify the convergence criterion. As the fitness of a population may remain static for a number of generations before a superior string is found, the application of conventional termination criteria becomes problematic [41]. In this research, the mutation rate is rather high, so the strings keep high diversity for a large number of generations although no improvements are found. If the mutation rate is lowered, the convergence criterion will stop the GA in a locally optimal point. The stopping criterion sometimes reflects the goals of optimization. For some very complex problems, "the most important goal of optimization is improvement, to get some good, 'satisfying' level of performance quickly,... it is nice to be perfect: meanwhile, we can only strive

to improve." According to the above reasons and numerical experiments, all cases (number of operations less than 40) tested achieved very good results and showed no improvements after 8000 generations for a long time. Therefore, the stop criterion chosen is a static number of generations = 8000.

In summary, this section describes how domain specific knowledge is incorporated into the GAs. The overview of the GA-based process planner is shown in Figure 9. The ability of the proposed algorithms is demonstrated by examples in next section.

5 Case Study

5.1 Flexible Process Plan Generation

The chuck jaw shown in Figure 10b is used to test the developed algorithm. It consists of 9 features such as C_bores, slots, steps, pocket, and chamfer. The dimensions and tolerances of the part are shown in Figure 11. Since the feature F6 (C_bore) has partial intersection with F8 (Slot), in case of damaging the cutting tool, F6 should be machined before F8. The bottom surface of F1 (Slot) is the datum plane of F6, therefore, F1 should be processed before F6. The machining process starts from the raw stock shown in Figure 10a: a regular block. It is assumed that the block is pre-machined to size, 120mmX70mmX50mm. The features and PRs among them are extracted and shown in Table 2.

A job shop is assumed in this test, where there are 17 cutting tools, and 5 machines. The available machines and cutting tools are shown in Table 3 and Table 4. The MCI column in Table 3 lists the cost of each machine and the TCI column in Table 4 lists the cost of each cutting tool.

In order to test the capability of the developed GAs for generating optimal plans, as well as dealing with the dynamic characteristics of job shops, process planning for the part is carried out under two different settings. Setting-1 assumes that all the facilities are available while Setting-2 assumes that M2 (vertical mill) is down. For each setting, 60 trials (same set up, different initial plan) were conducted.

1. Setting-1: all machines and tools available Table 5 shows the OpE alternatives for machining each feature with all machines and tools available. The criterion used is the sum of the costs of setup and tool changes. The GA found the optimal plans (cost=833) 55 times out of the 60 trials. The parameters are set as follows: population size: 50, crossover rate: 0.7, mutation rate: 0.6 and the stopping criterion is 8000 generations. One of the best plans found is shown in Table 6.

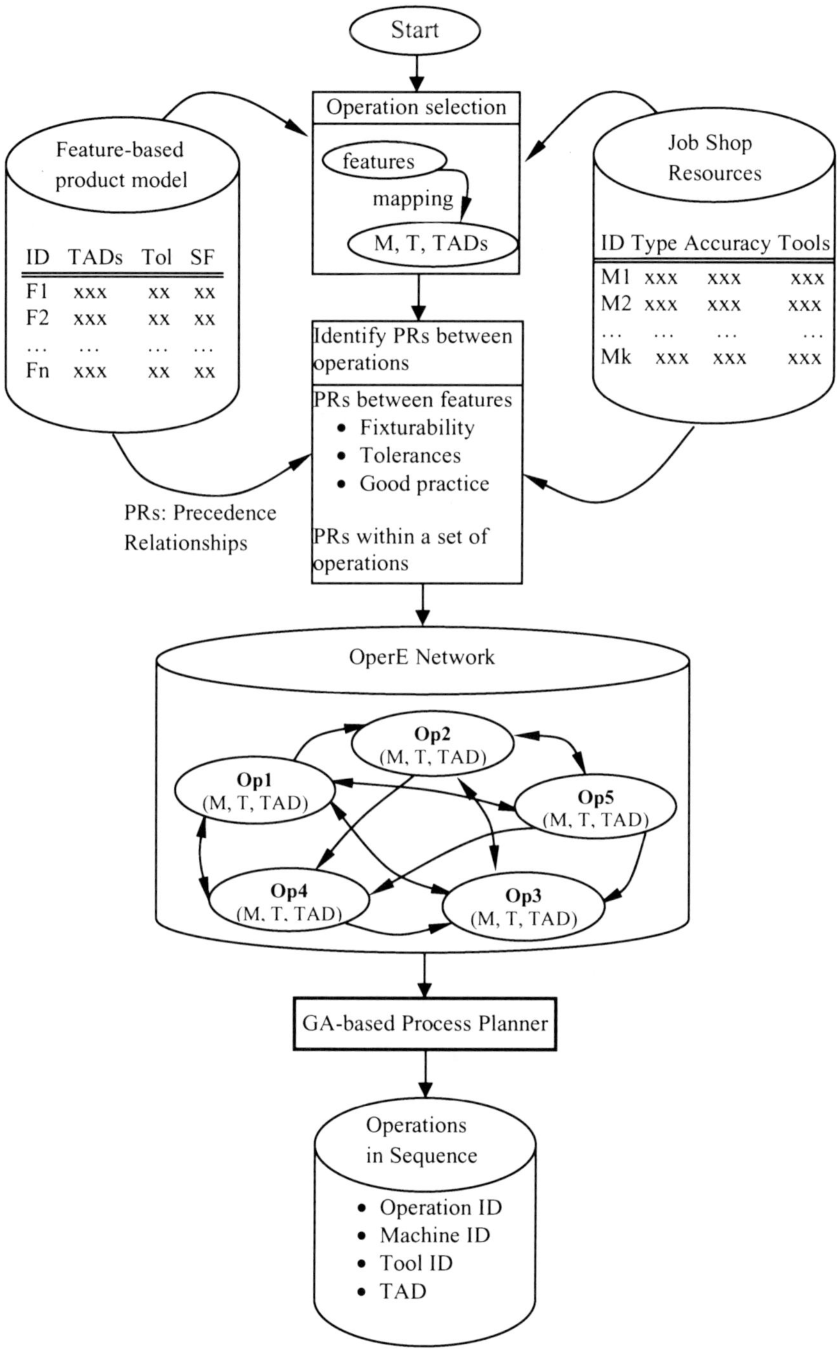

Fig. 9. Overview of the GA-based process planner.

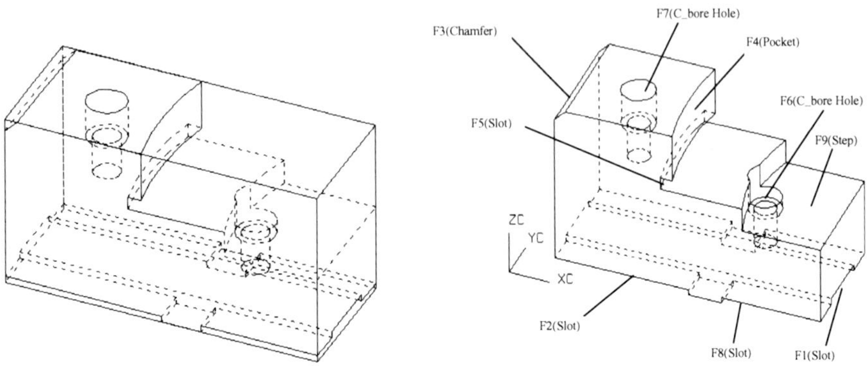

(a) Raw stock. **(b)** Machined part: chuck jaw.

Fig. 10. An example.

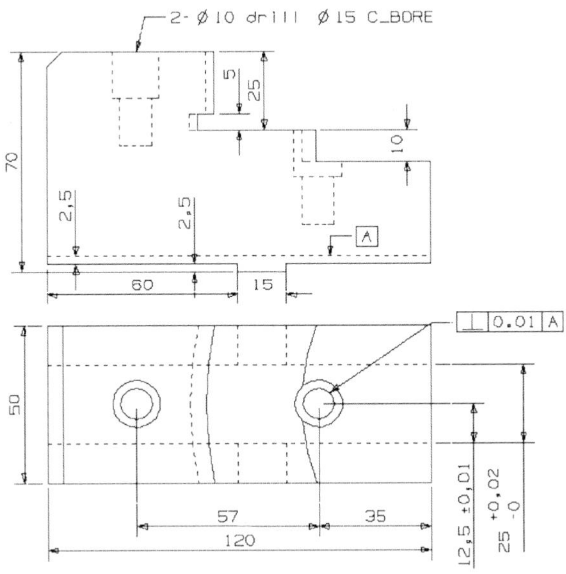

Fig. 11. Dimensions and tolerances of the chuck jaw.

2. Setting-2: M2 broken In this setting, it is assumed that M2 is down. The process planning information is listed in Table 7. The cost index is the same with those in setting-1 (see Table 5). The best solution found to date has a total cost of 1288. The obtained plan is shown in Table 8. Compared with the optimal plan of setting-1, the total cost increases from

Table 2. Precedence relationships between the features.

Feature	Predecessors	Feature	Predecessors
F1	F2, F9	F6	F1, F4
F2	Nil	F7	Nil
F3	Nil	F8	F6
F4	Nil	F9	
F5	F4		

Table 3. Machines and their cost indices used in the job shop.

Machines	Type	Table size	Travel Dim.	Accuracy	MCI
M1	CNC Vertical Mill	1400x650	1200x600x700	0.01	70
M2	Vertical Mill	1300x280	850x400x400	0.02	30
M3	Drill press	1000x280	850x400x400	0.1	10
M4	CNC Horizontal Mill	1300x550	930x750x1380	0.02	40
M5	Horizontal Mill	1800x1200	1400x1120x1000	0.01	85

Table 4. Cutting tools and their cost indices used in the job shop.

Tools	Type(diameter, flute length)	TCI	Tools	Type(diameter, flute length)	TCI
T1	End_mill(20,30)	10	T10	Centre_drill (20, 5)	2
T2	End_mill(30, 50)	10	T11	Angle_cutter (40, 45)	10
T3	End_mill(15, 20)	10	T12	Drill (70, 100)	5
T4	End_mill(40, 60)	12	T13	Drill (8, 30)	6
T5	Side_mill (50, 10)	8	T14	Drill (10, 35)	3
T6	T_slot_cutter (30, 15)	16	T15	T_slot_cutter (20, 5)	6
T7	Drill (20, 55)	3	T16	Drill (5, 30)	3
T8	Drill (30, 50)	3	T17	Drill (15, 50)	4
T9	Drill (50, 80)	4			

833 to 1288. All the operations are machined in M1 instead of M2. A near optimal plan obtained is shown in Table 9. This plan can be used when M1 is very busy; some of the drilling operations can be changed to the other machines such as drill press. The total cost increases; however, it will decrease the burden of M1 and improve the working efficiency. This demonstrates that the GA is able to handle changes in terms of machining resources and find a better plan with additional resources.

5.2 A GA Embedded in a Generative CAPP System

A generative CAPP system is built based on the GAs. The system is developed under Unigraphics (UG). The user interface is shown in Figure 12.

Table 5. The process planning information for Setting-1.

Operation	Feature	OpT	Ms	Ts	TADs	Cost index
Op1	F1	Milling	M1, M2	T1, T3	+z	MCI(M1)=70
			M4, M5	T5, T15	+y, -y	MCI(M2)=30
Op2	F2	Milling	M1, M2	T1, T2, T3, T4	+z, +x	MCI(M3)=10
			M4, M5	T5	+z	MCI(M4)=40
Op3	F3	Milling	M1, M2	T4	+y, -y	MCI(M5)=85
			M4, M5	T11	-z, +x	MCCI=150
Op4	F4	Milling	M1, M2	T1, T2, T4	-z	SCCI=90
Op5	F5	Milling	M1, M2	T15	-z	TCCI=20
Op6	F6	Center_drilling	M1, M2, M3, M4, M5	T10	-z	
Op7		Drilling	M1, M2, M4, M5	T14	-z	
Op8		Milling	M1, M2	T3	-z	
Op9	F7	Center_drilling	M1, M2, M3, M4, M5	T10	-z	
Op10		Drilling	M1, M2, M4, M5	T14	-z	
Op11		Milling	M1, M2	T3	-z	
Op12	F8	Milling	M1, M2,	T1, T2, T3, T4	-z	
Op13	F9	Milling	M1, M2,	T1, T2, T3, T4	-x, +y	
			M4, M5	T5	+z, -x	

Table 6. A plan found for Setting-1.

Operation	M	T	TAD	Summary
Op13	M2	T1	+z	Total Cost: 833
Op2	M2	T1	+z	
Op1	M2	T1	+z	No. of machine changes: 0
Op4	M2	T1	-z	
Op5	M2	T15	-z	No. of set-up changes: 2
Op9	M2	T10	-z	No. of tool changes: 5
Op6	M2	T10	-z	
Op10	M2	T14	-z	
Op7	M2	T14	-z	
Op8	M2	T3	-z	
Op11	M2	T3	-z	
Op12	M2	T3	-z	
Op3	M2	T4	+y	

Table 7. The process planning information for Setting-2.

Operation	Feature	OpT	Ms	Ts	TADs
Op1	F1	Milling	M1	T1, T3	+z
			M4, M5	T5, T15	+y, -y
Op2	F2	Milling	M1	T1, T2, T3, T4	+z, +x
			M4, M5	T5	+z
Op3	F3	Milling	M1	T4	+y, -y
			M4, M5	T11	-z, +x
Op4	F4	Milling	M1	T1, T2, T4	-z
Op5	F5	Milling	M1	T15	-z
Op6	F6	Centre_drilling	M1, M3, M4, M5	T10	-z
Op7		Drilling	M1, M4, M5	T14	-z
Op8		Milling	M1	T3	-z
Op9	F7	Centre_drilling	M1, M3, M4, M5	T10	-z
Op10		Drilling	M1, M4, M5	T14	-z
Op11		Milling	M1	T3	-z
Op12	F8	Milling	M1	T1, T2, T4	-z
Op13	F9	Milling	M1	T1, T2, T3, T4	-x, +y
			M4, M5	T5	+z, -x

Table 8. An optimal plan found for Setting-2.

Operation	M	T	TAD	Summary
Op13	M1	T1	+z	Total Cost: 1288
Op2	M1	T1	+z	
Op1	M1	T1	+z	No. of machine changes: 0
Op4	M1	T1	-z	
Op5	M1	T15	-z	No. of set-up changes: 2
Op9	M1	T10	-z	
Op6	M1	T10	-z	No. of tool changes: 5
Op10	M1	T14	-z	
Op7	M1	T14	-z	
Op8	M1	T3	-z	
Op11	M1	T3	-z	
Op12	M1	T3	-z	
Op3	M1	T4	+y	

Once the part is designed in the UG environment, the user can click "CAPP" button and the sub-menu will show up. The system contains several modules: feature analysis, process evaluation, plan generation and optimization, and database management. The architecture of the system is shown in Figure 13. In next section, detailed information about each module is introduced.

Table 9. A near-optimal plan found for Setting-2.

Operation	M	T	TAD	Summary
Op2	M1	T1	+z	Total Cost: 1308
Op13	M1	T1	+z	
Op1	M1	T1	+z	No. of machine changes: 2
Op4	M1	T1	-z	
Op5	M1	T15	-z	No. of set-up changes: 2
Op3	M1	T4	+y	
Op6	M4	T10	-z	No. of tool changes: 3
Op9	M4	T10	-z	
Op10	M4	T14	-z	
Op7	M4	T14	-z	
Op8	M1	T3	-z	
Op11	M1	T3	-z	
Op12	M1	T3	-z	

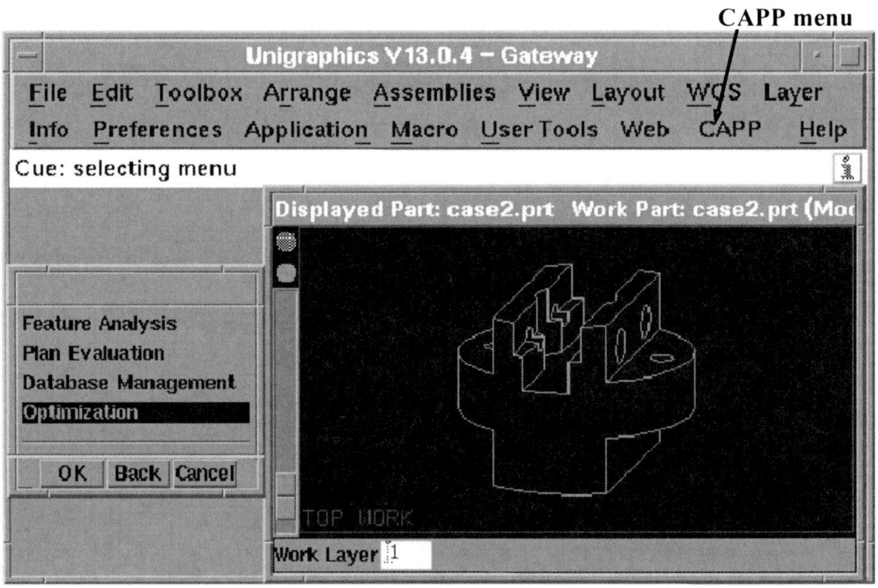

Fig. 12. System interface.

5.2.1 The Details of the CAPP System

There are four modules of the CAPP system. The functionalities of each module are list below.

1. **Feature Analysis** This module serves as a bridge from CAD to CAPP. It extracts the geometric information from the CAD model, interprets the

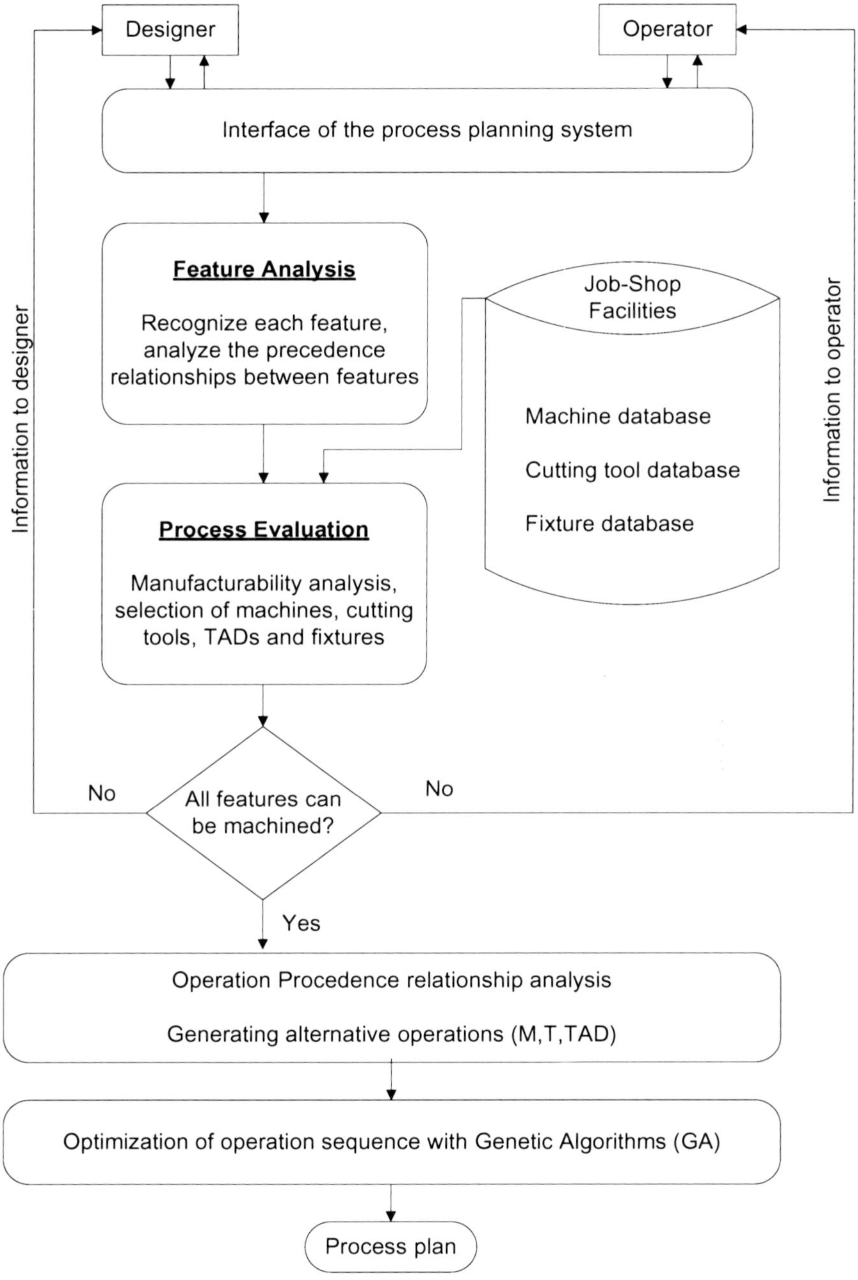

Fig. 13. System architecture.

manufacturing information, maps the design features to manufacturing features, generates the relationship between different features, and deals with the intersection features. Features that can be recognized by the system are shown in Figure 14.

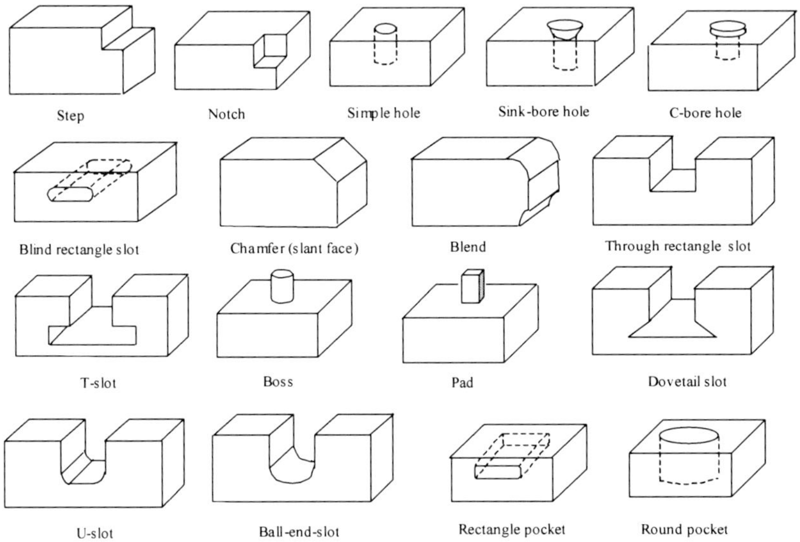

Fig. 14. Recognizable features.

2. **Process Evaluation** Due to the ability to generate several process plans simultaneously with this CAPP system, plan evaluation and manufacturability analysis can be integrated with CAPP. When a part comes into manufacturing, the process planner first checks the manufacturability by analyzing the geometry and tolerance to see whether the part can be made in the current job shop. This module determines the required machining operations, which include process selection, machine selection, cutting tool selection, operation selection, etc. These are tightly coupled among themselves and with the job shop database. If the part is machinable, the output of the module are alternative operations, which serve as the input information for the optimization module. If some of the features could not machined within the job shop environment, the output will be some feedback information. There are two kinds of information the process planner should provide as feedback: information for the designer and information for the operator.

   a. *Information to the designer* As mentioned before, some of the parts cannot be machined because the designer mainly considers the function, instead of the manufacturability of the part during design. For example, if a blind slot with sharp corners is designed, there is no way that the feature can be machined; therefore some suggestion should be made to the designer to round the sharp corners.

b. *Information to the process planner or job shop operator* Sometimes an operation cannot be performed due to lack of proper cutting tools or machines. This information should be given as feedback to the process planner or job shop operators. Some new tools are suggested to be added in the database.

3. **Plan Generation and Optimization** This module generates the optimal or near-optimal process plan for a part. The presented GAs are embedded in this module. The input of the module is the OpE network, and the output of the module is the sequence of the optimal or near optimal process plan.
4. **Job Shop Information and Database Management** This module contains all job shop resources. The job shop information, in terms of its machining capability and current status, is to be input by the user. An open architecture is provided to the user to input the currently available machining resources (machines, tools) along with their technological attributes, e.g. the maximum size, achievable accuracy and surface finish. The machines used include milling, drilling, grinding and CNC machining center.

   There are three databases: machine database, cutting tool database and fixture database. Users can construct their own database by selecting from the existing database, adding, deleting, and modifying the database. These databases are used for the plan evaluation module.

5.2.2 A Case Study

A part (socket) used in [8] (see Figure 15) is used here to test the developed system. After the part had been analyzed that all features can be machined, all available machining methods are generated by the plan evaluation module. A feature analysis module is applied to get the PRs between features. These then are converted to PRs between operations. The features and PRs among them are extracted and shown in Table 10. The available tools are assumed the same as in [8]. However, three machines (M1: vertical milling center, M2: 3-axis vertical milling, M3: drill machine) are assumed in the job shop, whereas only M2 is available in [8]. The cost indices are given in Table 11.

The CAPP system can generate a best plan based on the given job shop information shown in Table 11. The optimal process plan generated is shown in Table 12.With our GA embedded in the CAPP system, similar results, in terms of operation sequence, are obtained. Compared to Gupta's case that considers only one machine, the presented method can handle more complex job shop environments, i.e., more machines and cutting tools are available. In addition, in [8] each feature is machined by one single operation, which is not true for most practical cases. For example, holes are normally done by several steps based on the geometry and tolerance requirement.

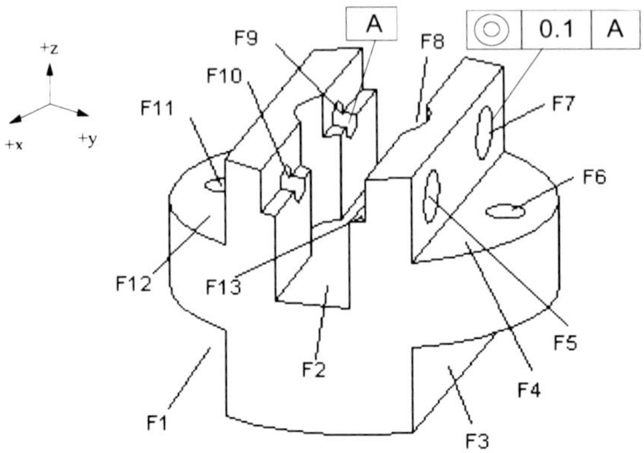

Fig. 15. A socket used in [8].

Table 10. PRs for the features shown in Figure 15.

Feature	Predecessors	Feature	Predecessors
F1	Nil	F8	Nil
F2	F8, F9, F10, F13	F9	F4, F12
F3	Nil	F10	F4, F12
F4	Nil	F11	F1, F12
F5	F4, F10, F12	F12	Nil
F6	F3, F4,	F13	F8, F9, F10
F7	F4, F9, F12		

Table 11. The process planning information.

Feature	TADs	Ms	Ts	Cost Index
F1	+z; +y	M2	Mil50	MCI(M1)=70
F2	-z	M2	Mil40	MCI(M2)=30
F3	+z; -y	M2	Mil50	MCI(M3)=15
F4	-z; -y	M2	Mil50	TCI(Mil50)=5
F5	-y	M2	Dril30	TCI(Mil40)=4
F6	+z; -z	M2	Dril20	TCI(Dril75)=5
F7	-y	M2	Dril30	TCI(Dril30)=3
F8	-z	M2	Dril75	TCI(Dril20)=2
F9	+y; -y	M2	Dril20	MCCI=150
F10	+y; -y	M2	Dril20	SCCI=90
F11	+z; -z	M2	Dril20	TCCI=20
F12	-z; +y	M2	Mil50	
F13	-z	M2	Mil50	

Table 12. A plan found for the socket shown in Figure 15.

Feature	M	T	TAD	Summary
F1	M2	Mil50	+y	Total Cost: 738 No. of machine changes: 0 No. of setup changes: 2 No. of tool changes: 6
F12	M2	Mil50	+y	
F4	M2	Mil50	-y	
F3	M2	Mil50	-y	
F10	M2	Dril20	-y	
F9	M2	Dril20	-y	
F7	M2	Dril30	-y	
F5	M2	Dril30	-y	
F8	M2	Dril75	-z	
F11	M2	Dril20	-z	
F6	M2	Dril20	-z	
F13	M2	Mil50	-z	
F2	M2	Mil40	-z	

In summary, by changing the job shop database dynamically, the CAPP system can react to dynamic events that occur in a job shop, such as machine breakdown. In addition, the ability to generate more than one process plan provides a good interface to integrate this CAPP system with scheduling and production control. For a more complex part case, please see [57, 58]. Detailed information about the presented work can be found in [59, 57, 60, 61, 62].

6 Conclusions

This chapter describes a process planning approach that integrates the tasks of routing and sequencing for obtaining a globally optimal process plan for a machined part. A GA-based algorithm has been developed to search for the globally optimal solution based on the proposed process-planning model. Results in the case studies suggest that the methods are effective. The approach employed has advantages over previous approaches in the following aspects:

1. The system is developed based on a customizable job shop environment so that users can modify the manufacturing database to suit their needs. This makes the system more realistic compared to the approaches in which a fixed machining environment is assumed.
2. By concurrently considering the selection of machines, tools, and TADs for each operation type and the sequence among the operations, together with the constraints of PR, the resulting process plan model successfully retains the entire solution space. This makes it possible to find a globally optimal process plan.
3. The system provides flexible optimization criteria that will satisfy the various needs from different job shops and/or job batches. The GA solves the problem effectively.

Our major contributions are:

- The process planning problem is modeled from an optimization perspective by considering all the possible combinations of M/T/TAD (operation selection) and operation sequencing concurrently.
- A GA has been successfully applied to solve this optimization problem by developing domain knowledge related GA operators.
- A CAPP system on top of a commercial CAD system was developed by embedding our GAs into the system.

7 Future Work

A limitation of this approach is that the dynamic aspects of the fixture constraints are not considered. First, only modular fixture elements are considered, and it is assumed that a fixture solution for an OpE can be found if the part under the TAD has sufficient base area, and clamping does not cause too much deformation. Secondly, it is assumed that two OpEs sharing the same machine and TAD also share the same fixture solution, therefore, no set-up change is required. These assumptions may affect the flexibility on fixture device choices and over-constrain the solution space. Currently, work is underway to consider other fixture devices and include the fixture solution as another attribute of OpEs, in addition to M, T, and TAD. The proposed methods to find the optimal plans for traditional manufacturing processes can be extended to the following areas:

- Integration of process planning with scheduling, design, and other activities,
- Green process planning,
- Process planning for layered manufacturing, and
- Process planning in the conceptual design phase.

The integration of process planning has been discussed in previous sections. Following are some details about the remaining areas.

7.1 Green Process Planning

Environmental factors are becoming an emerging dimension in manufacturing due to increasingly stringent regulations on health and safety of workers, the importance of manufacturing wastes on the product life cycle, emerging international standards on environmental performance, and a growing consumer preference for green products. Thus, cleaner manufacturing processes are becoming an important competitive advantage in many industries. Process planning provides a crucial framework to make robust decisions which decrease the environmental impact, while maintaining expected levels of production rate and quality. The criteria to evaluate a green process plan are quite different from the traditional process plan. However, using GAs, one can easily integrate "green evaluation" into an existing traditional CAPP system.

7.2 Process Planning for Layered Manufacturing

Layered manufacturing (LM) is a newly developed method to manufacture the part directly from geometry modeling. There are a number of LM processes currently available, such as stereo-lithography (SLA), fused deposition modeling (FDM), and 3D printing (3DP). The core of the process planning of layered manufacturing is the part orientation problem. The part orientation problem has multiple objectives such as minimum building time, better surface quality, etc. EAs can also be used to solve LM orientation problems.

7.3 Process Planning in the Conceptual Design Phase

Traditionally, a process plan will be generated after a design is finished and before the manufacturing starts. Sometimes it is too late to feed manufacturability information from CAPP back to the design engineers to modify the design. Thus there is a need to generate process plans in the early design phase or conceptual design phase.

Conceptual design is a key activity in early product development. It determines product functions, form, and the basic structure. Major manufacturing cost is committed in the early conceptual design process. The difficulty of generating a process plan in this phase is the lack of accurate design information. Because of this, a "co-evolution" concept will be introduced. The idea is to evolve the process plan along with the design development. In the conceptual design phase, the string to represent a process plan will be very different from the string to represent a traditional process plan, due to the lack of accurate design information, such as dimensions. Alternatively, a set of carefully designed string operators that represent the manufacturability of the part will be presented. Meanwhile, when the design evolves, the corresponding process plan will evolve as well.

References

1. Chang, T.C.: Expert Process Planning of Manufacturing. Addison-Wesley, Reading (1990)
2. ElMaraghy, H.A., Agerman, E., Davies, B.J.: Evolution and future perspective of capp. Annals of the CIRP 42(2), 739–755 (1993)
3. Alting, L., Zhang, H.C.: Computer aided process planning: The state-of-the-art survey. International Journal of Production Research 27(4), 553–585 (1989)
4. Bedworth, D., Henderson, M.R., Wolfe, P.M.: Computer-integrated Design and Manufacturing. McGraw-Hill, New York (1991)
5. Chen, C.L.P., LeClair, S.R.: Integration of design and manufacturing solving setup generation and feature sequencing using an unsupervised-learning approach. Computer Aided Design 26(1), 59–75 (1994)

6. Chu, C.C.P., Gadh, R.: Feature-based approach for set-up minimization of process design from product design. Computer Aided Design 28(5), 321–332 (1996)
7. Demey, S., van Brussel, H., Derache, H.: Determining set-ups for mechanical workpieces. Robotics & Computer-Integrated Manufacturing 12(2), 195–205 (1996)
8. Gupta, S.K.: Using manufacturing planning to generate manufacturability feedback. Journal of Mechanical Design 119, 73–80 (1997)
9. Khoshnevis, B., Park, J.Y., Sormaz, D.: A cost based system for concurrent part and process design. The Engineering Economist 40(1), 101–119 (1994)
10. Usher, J.M., Bowden, R.O.: The application of genetic algorithms to operation sequencing for use in computer-aided process planning. Computer and Industrial Engineering 30(4), 999–1013 (1996)
11. Hoi, D.Y., Dutta, D.: A genetic algorithm application for sequencing operations in process planning for parallel machining. IIE Transaction 28(1), 55–68 (1996)
12. Kiritsis, D., Porchet, M.: A generic petri net model for dynamic process planning and sequence optimization. Advances in Engineering Software 25, 61–71 (1996)
13. Motipalli, V.V.S.K., Krishnaswami, P.: Automation of process planning for boring of turned components with arbitrary internal geometry from a semi-finished stock. Journal of Computing and Information Science in Engineering 6(1), 49–59 (2006)
14. Waiyagan, K., Bohez, E.L.J.: Intelligent feature based process planning for five-axis mill-turn parts. Journal of Computers in Industry 60(5), 296–316 (2009)
15. Khoshnevis, B., Chen, Q.: Integration of process planning and scheduling functions. Journal of Intelligent Manufacturing 1, 165–176 (1990)
16. Krith, J.P., Detand, J.: A capp system for nonlinear process plans. Annals of the CIRP 41(1), 489–492 (1992)
17. Zhang, H.C.: Ippm–a prototype to integrate process planning and job shop scheduling functions. Annals of the CIRP 42(1), 513–518 (1993)
18. Huang, S.H., Zhang, H.C., Smith, M.L.: A progressive approach for the integration of process planning and scheduling. IIE Transactions 27, 456–464 (1995)
19. Kim, Y.K., Park, K., Ko, J.: A symbiotic evolutionary algorithm for the integration of process planning and job shop scheduling. Computers and Operations Research 30(8), 1151–1171 (2003)
20. Wang, Z.J., Tian, J., Chen, W.: Integration of process planning and production scheduling based on genetic algorithm. Journal of Communication and Computer 6(6), 12–16 (2009)
21. Hayes, C.C.: p^3 a process planner for manufacturability analysis. IEEE Transactions on Robotics and Automation 12(2), 220–234 (1996)
22. Tan, W.: Integrated Process Planning and Scheduling: a mathematical Programming Modeling Approach. PhD thesis, University of Southern California (1997)
23. Leung, H.C.: Annotated bibliography on computer-aided process planning. International Journal of Advanced Manufacturing Technology 12, 309–329 (1996)
24. Hayes, C.C., Wright, P.: Automating process planning: Using feature interactions to guide search. Journal of Manufacturing Systems 8(1), 1–14 (1990)
25. Gupta, S.K., Nau, D.S.: Systematic approach to analysing the manufacturability of machined parts. Computer–Aided Design 27(5), 323–342 (1995)
26. Rho, H.M., Geelink, R., et al.: An integrated cutting tool selection an doperation sequencing method. Annals of the CIRP 41(1), 517–520 (1992)

27. Palmer, G.J.: An Integrated Approach to Manufacturing Planning. PhD thesis, University of Huddersfield (1994)
28. Mettala, E.G., Hoshi, S.: A compact representation of alternative process plans/routings for fms control activities. Journal of Deisgn and Manufacturing 3, 91–104 (1993)
29. Irani, S.A., Koo, H.Y., Raman, S.: Feature based operation sequence generation in capp. International Journal of Production Research 33(1), 17–39 (1995)
30. Prabhu, P., Elhence, S., Wang, H., Wysk, R.: An operations network generator for computer aided process planning. Journal of Manufacturing Systems 9(4), 283–291 (1990)
31. Noto La Deiga, S., Perrone, G., Piacentini, M.: Multiobjectives approach for process planning selection in ims environment. Annuals of the CIRP 45(1), 471–474 (1996)
32. Zhang, H.C., Huang, S.H.: A fuzzy approach to process plan selection. International Journal of Production Research 32(6), 1265–1279 (1994)
33. Hayes, C.C.: Plan–based manufacturability analysis and generation of shape–changing redesign suggestion. Journal of Intelligent Manufacturing 7, 121–132 (1996)
34. Khoshnevis, B., Park, J.Y., Smoraz, D.: A cost based system for concurrent part and process design. The Engineering Economist 40(1), 101–119 (1994)
35. Wong, T.N., Siu, S.L.: A knowledge–based approach to automated manufacturing process selection and sequencing. International Journal of Production Research 33(12), 3465–3484 (1995)
36. Gu, P., Zhang, Y.: Operation sequencing in an automated process planning system. Journal of Intelligent Manufacturing 4, 219–232 (1993)
37. Váncza, J., Márkus, A.: Genetic algorithms in process planning. Computers in Industry 17(23), 181–184 (1991)
38. Husbands, P., Mill, F., Warrington, S.: Generating optimal process plans from first principle. In: Expert Systems for Management and Engineering, Ellis Horwood (1990)
39. Eiben, A.E., Smith, J.E.: Introduction to Evolutionary Computing. Springer, Heidelberg (2003)
40. Caponio, A., Cascella, G.L., Neri, F., Salvatore, N., Sumner, M.: A fast adaptive memetic algorithm for off-line and on-line control design of pmsm drives. IEEE Transactions on Systems, Man and Cybernetics – Part B, Special Issue on Memetic Algorithms 37(1), 28–41 (2007)
41. Goldberg, D.E.: Genetic Algorithms in Search, Optimization, and Machine Learning. Addison-Wesley, Reading (1989)
42. Bruns, R.: Direct chromosome representation and advanced genetic operators for production scheduling. In: The Fifth International Conference on Genetic Algorithms, pp. 352–359 (1993)
43. Moon, C., Seo, Y.: Evolutionary algorithm for advanced process planning and scheduling in a multi-plant. Computers and Industrial Engineering 48(2), 311–325 (2005)
44. Gonçalves, J.F., Resende, M.G.C.: An evolutionary algorithm for manufacturing cell formation. Computers and Industrial Engineering 47(1), 247–273 (2004)
45. Dereli, T., Filiz, H.I.: Optimization of process planning functions by genetic algorithm. Computers and Industrial Engineering 36(2), 281–308 (1999)
46. Gen, M., Lin, L., Zhang, H.: Evolutionary techniques for optimization problems in integrated manufacturing state-of-the-art-survey. Computers and Industrial Engineering 56(3), 779–808 (2009)

47. Brown, K., Cagan, J.: Optimized process planning by generative simulated annealing. Artificial Intelligent for Engineering Design, Analysis and Manufacturing 11, 219–235 (1997)
48. Rudolph, G.: Convergence properties of canonical genetic algorithms. IEEE Transactions on Neural Networks 5(1), 11–96 (1994)
49. Eiben, A.E., Aarts, E.H., Van Hee, K.M.: Global convergence of genetic algorithms: An infinite markov chain analysis. In: Proceeding of the First International Conference on Parallel Problme Solving from Nature, pp. 4–17. Springer, Heidelberg (1991)
50. Dagli, C., Sittisathanchai, S.: Genetic neuro-scheduler for job shop scheduling. Computers and Industrial Engineering 25(1/4), 267–270 (1993)
51. Alander, J.T.: On optimal population size of genetic algorithms. In: CompuEuro 1992, pp. 65–70 (1992)
52. Bäck, T., Schütz, M.: Intelligent mutation rate control in canonical genetic algorithms. In: The 9th International Symposium on Foundation of Intelligent Systems, pp. 155–167 (1996)
53. Reeves, C.: A genetic algorithm for flow shop sequencing. Computers and Operations Research 22, 5–13 (1996)
54. Fogel, D.B., Atmar, J.W.: Comparing genetic operators with gaussian mutations in simulated evolutionary processes using linear systems. Biological Cybernetics 63, 111–114 (1990)
55. Chu, P.C.H.: A Genetic Algorithm Approach for Combinatorial Optimization Problems. PhD thesis, University of London (1997)
56. Mattfeld, D.C.: Evolutionary Search and the Job Shop: Investigation on Genetic Algorithms for Production Scheduling. PhD thesis, University of Bremen (1995)
57. Zhang, F., Zhang, Y.F., Nee, A.Y.C.: Using genetic algorithms in process planning for job shop machining. IEEE Transactions on Evolutionary Computation 1(4), 278–289 (1997)
58. Zhang, F.: Genetic algorithm in computer-aided process planning. Master's thesis, National University of Singapore (1997)
59. Ma, G.H., Zhang, Y.F., Nee, A.Y.C.: A simulated annealing-based optimization algorithm for process planning. Internal Journal of Product Research 38(12), 2671–2687 (2000)
60. Zhang, Y.F., Ma, G.H., Nee, A.Y.C.: Modeling process planning problems in an optimization perspective. In: Proceedings of IEEE International Conference on Robotics and Automation, pp. 1764–1769 (1999)
61. Ma, G.H., Zhang, F., Zhang, Y.F., Nee, A.Y.C.: An automated process planning system based on genetic algorithm and simulated annealing. In: Proceedings of ASME Design Engineering Technical Conferences and Computer and Information in Engineering Conference, September 29-October 2 (2002)
62. Wang, L.: Database in process planning. Master's thesis, National University of Singapore (1999)

A Fitness Granulation Approach for Large-Scale Structural Design Optimization

Mohsen Davarynejad, Jos Vrancken, Jan van den Berg,
and Carlos A. Coello Coello

Abstract. The complexity of large-scale mechanical optimization problems is partially due to the presence of high-dimensional design variables, the nature of the design variables, and the high computational cost of the finite element simulations needed to evaluate the fitness of candidate solutions. Evolutionary algorithms are ruled by competitive games of survival and not merely by absolute measures of fitness. They can also exploit the robustness of evolution against uncertainties in the fitness function evaluations. This chapter takes up the complexity challenge of mechanical optimization problems by proposing a new fitness granulation approach that attempts to cope with several difficulties of fitness approximation methods that have been reported in the specialized literature. The approach is based on adaptive fuzzy fitness granulation having as its main aim to strike a balance between the accuracy and the utility of the computations. The adaptation algorithm adjusts the number and size of the granules according to the perceived performance and level of convergence attained. Experimental results show that the proposed method accelerates the convergence towards solutions when compared to the performance of other, more popular approaches. This suggests its applicability to other complex finite element-based engineering design problems.

Mohsen Davarynejad · Jos Vrancken · Jan van den Berg
Faculty of Technology, Policy and Management,
Delft University of Technology, NL-2600 GA, Delft, The Netherlands
e-mail: {m.davarynejad;j.l.m.vrancken;j.vandenberg}@tudelft.nl

Carlos A. Coello Coello
CINVESTAV-IPN, Departmento de Computación
(Evolutionary Computation Group), Av. IPN No. 2508,
Col. San Pedro Zacatenco,
México D.F. 07300, México
e-mail: ccoello@cs.cinvestav.mx

1 Introduction

Since the 1960s, and due to the significant development in numerical methods and computing, the finite element analysis (FEA) has become a frequent tool to solve engineering problems that arise in systems with several interacting components, complex geometries, and which are under the effect of different physical phenomena. These (complex) systems elude a thorough physical analysis with exact techniques which is made possible by means of a systematic discretization approach known as the finite element method (FEM) [1]. At the same time that the FEM was developed, efficient and fast optimization algorithms have arisen for solving various kinds of mathematical and optimization problems (OPs). Both trends contributed to the development of large-scale structural design and optimization problems (SDOPs) and to the discipline of structural optimization. The aim of structural optimization is to generate automated procedures for finding the best possible structure with respect to at least one criterion (the objective), and having to satisfy a set of constraints, by selecting from a set of geometrical dimensions, material properties and/or topological parameters [2].

Structural optimization problems are often challenging due to their high computational demands [1], multi-modality, non-convexity, high dimensionality, and multi-objectivity. Because of this, many structural optimization problems are weakly amenable to conventional mathematical programming approaches, which motivates the use of alternative solution methods.

Randomized search heuristics are among the simplest and most robust strategies that are applicable to a wide range of optimization problems including structural design (SD). While they can normally provide near optimal solutions, they cannot guarantee convergence to the optimum. Their computational requirements are typically high too. Among the randomized search heuristics currently available, evolutionary algorithms (EAs) have become very popular in the last few years, mainly because of their ease of use and efficacy. EAs are stochastic search techniques that operate on a set of solutions (the so-called population), which are modified based on the principles of natural evolution (i.e., the survival of the fittest) [3]. EAs have been commonly adopted for solving complex SD problems. For example, Walker and Smith [4] combined the FEM and EAs to minimize a weighted sum of the mass and deflection of fiber-reinforced structures. Similarly, Abe et al. [5] used the FEM and an EA for structural optimization of the belt construction of a tire. More recently, Giger and Ermanni [6] applied the FEM and an EA to minimize the mass of composite fiber-reinforced plastic (CFRP) rims subject to strength and stiffness constraints. However, EAs may suffer from a slow rate of convergence towards the global optimum, which implies that they may be too (computationally) expensive for certain SD problems. Consequently,

[1] FEA is computationally costly and may require several days to complete its calculations, even for a relatively simple problem.

it is challenging to develop computationally efficient evolution-based search methods.

To alleviate the problem of converging time of computationally expensive optimization problems, a variety of techniques have been proposed in the literature. Perhaps the most obvious choice is to use parallelization techniques [7]. However, another alternative is to rely on fitness approximation techniques, which avoid evaluating every individual in the population of an EA. In order to do this, these approaches estimate the quality of some individuals based on an approximate model of the fitness landscape. This is the sort of approach this chapter is focused on. Section 4 provides a review of fitness approximation techniques in evolutionary computation. When using fitness approximation techniques, it is necessary to strike a balance between exact fitness evaluation and approximate fitness evaluation. In this chapter, with a view to reducing computational cost, we employ the concept of fuzzy granulation to effectively approximate the fitness function. The advantage of this approach over others is the fact that no training samples are required, and the approximate model is dynamically updated with no or negligible overhead cost.

The remainder of this chapter is organized as follows. The following section elaborates upon four SD optimization problems before explaining the genetic algorithm (GA) approach proposed here for the SD optimization task (see Section 3). This is followed by a review of the variety of fitness approximation approaches that have been proposed for EAs in Section 4. In order to accelerate the convergence speed of the GA with a minimum number of fitness function evaluations, a novel method is presented in Section 5. The approach is based on generating fuzzy granules via an adaptive similarity analysis. To illustrate the efficiency of the proposed method in solving the four SD problems introduced in Section 2, the performance results of different optimization algorithms are presented in Section 6. A further statistical analysis confirms that the proposed approach reduces the computational complexity of the number of fitness function evaluations by over 50% while reaching similar or even better final fitness values. Finally, in Section 8 we provide our conclusions.

2 Structural Design Optimization Problems

Four SD optimization problems with increasing complexity are investigated here. They are the following: (1) the design of a *3-layer composite beam* with *two* optimization variables, (2) the design of an *airplane wing* with *six* decision variables, (3) the design of a *2D truss* frame with *36* decision variables, and (4) the *voltage/pattern design of piezoelectric actuators*. We discuss in more detail the last problem, because of its complexity. It consists of finding the best voltage and pattern arrangement for static shape control of a piezoelectric actuator with *200* design variables. Clearly, this is a more challenging and heavy optimization task from a fitness/computational perspective.

2.1 Easier/Smaller Problems

The first three SD problems are covered in this section. The ultimate goal in these optimization problems is to maximize the first natural frequency[2] of the given structure. To allow more space for the last problem (described in subsections 2.2 and 6.4), we limit ourselves here to a short description of the three problems.

2.1.1 3-Layer Composite Beam

A *multi-layered composite beam* is constructed from a combination of two or more layers of dissimilar materials that are joined together to act as a unit in which the resulting combination is lighter, stronger and safer than the sum of its parts. An FEA model has been developed to analyze the multi-layer composite beams and plates. The objective is to raise the first natural frequency of the beam.

2.1.2 Airplane Wing

An *airplane wing* is an elastic structure that, in the presence of aerodynamic loads, starts to vibrate. In this study, we treated the natural frequency as the design objective since it is quite intuitive and natural to raise the natural frequencies of the wing so that it is not easily excited by undesirable disturbances.

2.1.3 2D Truss Frame

Trusses are the most commonly used structure and in comparison to heavily-built structures, they have a relatively small *dead weight*. A truss consists of *bar-elements* (members) connected by *hinged joints* to each other and supported at the base. Truss design problems belong to the class of load-supporting structure design problems that are usually finite-dimension

[2] Resonance occurs when the excitation frequency is the same as the natural frequency. For the same excitation energy, the resulting vibration level at resonance frequency is higher than other exiting frequencies. The importance of maximizing the first natural frequency is to avoid the resonance phenomenon to occur.

optimization problems. The design of load-supporting structures plays a key role in engineering dynamics. The objective (fitness) here is to raise the structure's *first natural frequency* to reduce the vibration domain and to prevent the resonance phenomenon (in dynamic response) of the structure.

2.2 Voltage and Pattern Design of a Piezoelectric Actuator

Piezoelectric materials exhibit both direct (electric field generation as a response to mechanical strains) and converse (mechanical strain is produced as a result of an electric field) piezoelectric effects. The direct effect is used in piezoelectric sensors while the converse effect is used in piezoelectric actuators.

Apart from ultrasound applications, energy harvesting, sensor applications (e.g., strain gauges and pressure sensors), and vibration/noise control domains, piezoelectric materials are widely used as actuators in smart structures. Smart structures with integrated self-monitoring, self-diagnosis, and control capabilities have practical uses ranging from MEMS, biomedical engineering, control engineering, aerospace structures, ground transportation systems and marine applications. The smart structures technology is widely used in biomechanics, i.e., to expand obstructed blood vessels or to prevent further enlargement of blood vessels damaged by aneurysms [8] which most commonly occurs in arteries. Another apparent practical use of smart and adaptive structural systems is to properly control the undesirable motions of geometry-changing structures.

Piezoelectric actuators are also found in an enormous range of applications for distributed actuation and control of mechanical structures for shape correction and modification. One example for this is their use in flexible aircrafts where they improve the aerodynamic performance and deformation control of conformal antennas [9], through their incorporation within the structure. For instance, in [10], an optimization algorithm is used to deal with the shape control of functionally graded material (FGM) plates which are actively controlled by piezoelectric sensor and actuator patches. A computational intelligence-based algorithm is used to derive the optimal voltage distribution, by adopting the elements of the gain control matrix as the design variables.

The optimal shape control and correction of small displacements in composite structures using piezoelectric actuators concern complex engineering problems. To achieve a predefined shape of the structure of the metal plate, in this chapter we will present a fast converging global optimization algorithm to find the optimal actuation voltages that need to be applied to the piezoelectric actuators and to the pattern of piezoelectric patches.

3 GAs for Structural Optimization Problems

GAs are perhaps the most popular type of EAs nowadays and have been applied to a wide variety of problems [11]. The GA optimization procedure for solving SD problems begins with a set of randomly selected parents (design parameters). If any of these parents does not meet all the physical constraints, they are modified until they do. In subsequent generations, each offspring's phenotype is also checked for its feasibility. Furthermore, the fitness values of the parents and their offspring are compared and the worst individuals are rejected, preserving the remaining ones as parents of the new generation (known as steady-state population treatment). This procedure is repeated until a given termination criterion is satisfied.

Due to their robustness, GAs have been frequently used in a variety of real world optimization applications including optimizing the placement of actuators on large space structures [12], the design of a low-budget lightweight motorcycle frame with superior dynamic and mechanical properties [13], and the evolution of the structural configuration for weight minimization of a space truss structure [14]. The implementation of a GA can be summarized as follows:

1. **Initialization**: Initialize P design vectors $X = \{X_1, X_2, \ldots, X_i, \ldots, X_P\}$, where P is the population size.
2. **Constraints check**: If satisfied, continue, else modify X_i until the candidate solution becomes feasible.
3. **Evaluation (Analysis)**: Evaluate the fitness function $f(X_i)$, $i = \{1, 2, \ldots, P\}$.
4. **Convergence check**:
    a. **if satisfied** stop,
    b. **else** select the next generation parent design vectors, apply genetic operators (mutation, recombination) and generate the next offspring design vectors X. Go to step 2.

EAs in general are often expensive in the sense that they may require a high number of computationally costly objective function evaluations. As a result, it may be necessary to forgo an exact evaluation and use approximated fitness values that are computationally efficient. In the design of mechanical structures, for instance, each exact fitness evaluation requires the time-consuming stage of FEA which, depending on the size of the problem, may consume from several seconds up to several days. If we assume a conventional GA with a fixed and modest population size of 100, a maximum of 100 generations, and a very small-scale structural problem that requires 10 seconds for each fitness evaluation, the total execution of the GA would require 30 hours! This should make evident the inhibiting role of the computational complexity associated to GAs (and EAs, in general) for non-trivial and large-scale problems.

Since one of the crucial aspects for solving large-scale SD optimization problems using EAs is the computational time required, in the following section we outline a few existing strategies that have been proposed to deal with this issue.

4 Fitness Approximation in Evolutionary Computation

As indicated before, one possibility to deal with time-consuming problems using a GA is to avoid evaluating every individual and estimate instead the quality of some of them based on an approximate model of the search space. Approximation techniques may estimate an individuals' fitness on the basis of previously observed objective function values of *neighboring* individuals. There are many possible approximation models [15]. Next, we will briefly review some of the most commonly adopted fitness approximation methods reported in the specialized literature.

4.1 Fitness Inheritance

This is a very simple technique that was originally introduced by Smith et al. [16]. The mechanism works as follows: when assigning fitness to an individual, sometimes we evaluate the objective function as usual, but the rest of the time, we assign fitness as an average (or a weighted average) of the fitness of the parents. This fitness assignment scheme will save us one fitness function evaluation, and operates based on the assumption of similarity between an offspring and its parents. Clearly, fitness inheritance cannot be applied all the time, since we require some true fitness function values in order to obtain enough information to guide the search. This approach uses a parameter called *inheritance proportion*, which regulates how many times we do apply fitness inheritance (the rest of the time, we compute the true fitness function values). As will be seen next, several authors have reported the use of fitness inheritance.

Zheng et al. [17] used fitness inheritance for codebook design in data compression techniques. They concluded that the use of fitness inheritance did not degrade, in a significant way, the performance of their GA.

Salami et al. [18] proposed a Fast Evolutionary Strategy (FES) in which a fitness and associated reliability value were assigned to each new individual. Considering two decision vectors $p_1^i = (X_1^i, F_1^i, r_1^i)$ and $p_2^i = (X_2^i, F_2^i, r_2^i)$ where X_1^i and X_2^i are the chromosomes 1 and 2 at generation i with fitness values F_1^i and F_2^i and reliabilities r_1^i and r_2^i, respectively. In this scheme, the true fitness function is only evaluated if the reliability value is below a certain threshold. Otherwise, the fitness of the new individual and its reliability value is calculated from:

$$F^{i+1} = \frac{S_1 r_1^i F_1^i + S_2 r_2^i F_2^i}{S_1 r_1^i + S_2 r_2^i} \tag{1}$$

and

$$r^{i+1} = \frac{(S_1 r_1^i)^2 + (S_2 r_2^i)^2}{S_1 r_1^i + S_2 r_2^i} \tag{2}$$

where S_1 is the similarity between X_1^{i+1} and X_1^i and S_2 is the similarity between X_1^{i+1} and X_2^i. Also, they incorporated random evaluation and error compensation strategies. Clearly, this is another (more elaborate) form of fitness inheritance. Salami et al. reported an average reduction of 40% in the number of evaluations while obtaining similar solutions. In the same work, they presented an application of (traditional) fitness inheritance to image compression obtaining reductions ranging from 35% up to 42% of the total number of fitness function evaluations.

Pelikan et al. [19] used fitness inheritance to estimate the fitness for only part of the solutions in the Bayesian Optimization Algorithm (BOA). They concluded that fitness inheritance is a promising concept, because population-sizing requirements for building appropriate models of promising solutions lead to good fitness estimates, even if only a small proportion of candidate solutions is evaluated using the true fitness function.

Fitness inheritance has also been used for dealing with multi-objective optimization problems. Reyes-Sierra and Coello Coello [20, 21] incorporated the concept of fitness inheritance into a multi-objective particle swarm optimizer and validated it in several test problems of different degrees of difficulty. They generally reported lower computational costs, while the quality of their results improved in higher dimensional spaces. This was in contradiction with other studies (e.g., [22] as well as this chapter) that indicate that the performance of the parents may not be a good predictor for their children's composition in sufficiently complex problems, rendering fitness inheritance inappropriate under such circumstances.

4.2 Surrogates

A common approach to deal with expensive objective functions is to construct an approximation function which is much cheaper to evaluate (computationally speaking). In order to build such an approximation function which will be used to predict promising new solutions, several sample points are required. The meta-model built under this scheme aims to reduce the total number of (true objective function) evaluations performed, while producing results of a reasonably good quality.

Evidently, the accuracy of the surrogate model depends on the number of samples provided (and their appropriate distribution) and on the approximation model adopted. Since surrogate models will be used very frequently, it is very important that the construction of such models is computationally efficient [15]. The following are examples of the use of surrogates of different types.

Sano et al. [23] proposed a GA for optimization of continuous noisy fitness functions. In this approach, they utilized the history of the search to reduce the number of fitness function evaluations. The fitness of a novel individual is estimated using the fitness values of the other individuals as well as the sampled fitness values for it. So, as to increase the number of individuals adopted for evaluation, they not only used the current generation but also the whole history of the search. To utilize the history of the search, a stochastic model of the fitness function is introduced, and the maximum likelihood technique is used for estimation of the fitness function. They concluded that the proposed method outperforms a conventional GA in noisy environments.

Branke et al. [24] suggested the use of local regression for estimation, taking the fitness of neighboring individuals into account. Since in local regression it is very important to determine which solutions belong to the neighborhood of a given individual, they studied two different approaches for setting the value of the size of the local neighborhood (relative neighborhood and adaptive neighborhood). They concluded that local regression provides better estimations than previously proposed approaches. In their more recent work [25], a comparison between two estimation methods, interpolation and regression, is done. They concluded that regression seems to be slightly preferable, particularly if only a very small fraction of the individuals in the population is evaluated. Their experiments also show that using fitness estimation, it is possible to either reach a better fitness level in a given time, or to reach a desired fitness level much faster (using roughly half of the original number of fitness function evaluations).

Ong et al. [26] proposed a local surrogate modeling algorithm for parallel evolutionary optimization of computationally expensive problems. The proposed algorithm combines hybrid evolutionary optimization techniques, radial basis functions, and trust-region frameworks. The main idea of the proposed approach is to use an EA combined with a feasible sequential quadratic programming solver. Each individual within an EA generation is used as an initial solution for local search, based on Lamarckian learning. They employed a trust-region framework to manage the interaction between the original objective and constraint functions and the computationally cheap surrogate models (which consist of radial basis networks constructed by using data points in the neighborhood of the initial solution), during local search. Extensive numerical studies are presented for some benchmark test functions and an aerodynamic wing design problem. They show that the proposed framework provides good designs on a limited computational budget. In more recent work, Ong et al. [27] presented a study on the effects of uncertainty in the surrogate model, using what they call Surrogate-Assisted Evolutionary Algorithms (SAEA). In particular, the focus was on both the *curse of uncertainty* (impairments due to errors in the approximation) and *blessing of uncertainty* (benefits of approximation errors). Several algorithms are tested, namely the Surrogated-Assisted Memetic Algorithm (SAMA) proposed in [26], a standard GA, a memetic algorithm (considered as the standard hybridization of a GA and the feasible sequential quadratic programming solver used in [26]), and the SAMA-Perfect algorithm (which is the SAMA algorithm but using the exact fitness function as surrogate model), on three multi-modal benchmark problems (Ackley, Griewank and Rastrigin). The conclusion was that approximation errors lead to convergence at false global optima, but turns out to be beneficial in some cases, accelerating the evolutionary search.

Regis and Shoemaker [28] developed an approach for the optimization of continuous costly functions that uses a space-filling experimental design and local function approximation to reduce the number of function evaluations in an evolutionary algorithm. The proposed approach estimates the objective function value of an offspring by means of a function approximation model over the k-nearest previously evaluated points. The estimated values are used to identify the most promising offspring per function evaluation. A Symmetric Latin Hypercube Design (SLHD) is used to determine initial points for function evaluation, and for the construction of the function approximation models. They compared the performance of an Evolution Strategy (ES) with local quadratic approximation, an ES with local cubic radial basis function interpolation, an ES whose initial parent population is obtained from a SLHD, and a conventional ES (in all cases, the They used a (μ, λ)-ES with uncorrelated mutations). The algorithms were tested on a groundwater bioremediation problem and on some benchmark test functions for global optimization (including *Dixon-Szegö*, *Rastrigin* and *Ackley*). The obtained results (which include an analysis of variance to provide stronger and solid claims regarding the relative performance of the algorithms) suggest that the approach that uses SLHDs together with local function approximations has potential for success in enhancing EAs for computationally expensive real-world problems. Also, the cubic radial basis function approach appears to be more promising than the quadratic approximation approach on difficult higher-dimensional problems.

Lim et al. [29] presented a Trusted Evolutionary Algorithm (TEA) for solving optimization problems with computationally expensive fitness functions. TEA is designed to maintain good trustworthiness of the surrogate models in predicting fitness improvements or controlling approximation errors throughout the evolutionary search. In this case, the most interesting part was to predict search improvement as opposed to the quality of the approximation, which is regarded as a secondary objective. TEA begins its search using the canonical EA, with only exact function evaluations. During the canonical EA search, the exact fitness values obtained are archived in a central database together with the design vectors (to be used later for constructing surrogate models). After some initial search generations (specified by the user), the trust region approach takes place beginning from the best solution provided by the canonical EA. The trust region approach uses a surrogate model (radial basis neural networks) and contracts or expands the trust radius depending on the ability of the approximation model in predicting fitness improvements, until the termination conditions are reached. An empirical study was performed on two highly multi-modal benchmark functions commonly used in the global optimization literature (*Ackley* and *Griewank*). Numerical comparisons to the canonical EA and the original trust region line search framework are also reported. From the obtained results, the conclusion was that TEA converges to near-optimum solutions more efficiently than the canonical evolutionary algorithm.

4.2.1 Kriging

A more elaborate surrogate model that has been relatively popular in engineering is the so-called Gaussian Process Model, also known as Kriging [30]. This approach builds probability models through sample data and estimates the function values at every untested point with a Gaussian distribution.

Ratle [31] presented a new approach based on a classical real-encoded GA for accelerating the convergence of evolutionary optimization methods. A reduction in the number of fitness function calls was obtained by means of an approximation model of the fitness landscape using kriging interpolation. The author built a statistical model from a small number of data points obtained during one or more generations of the evolutionary method using the true fitness landscape. The model is updated each time a convergence criterion is reached.

4.3 Artificial Neural Networks

In the last few years, artificial neural networks (ANNs), including multi-layer perceptrons [32] and radial basis function networks [33] have also been employed to build approximate models for design optimization. Due to their universal approximation properties, ANNs can be good fitness function estimators if provided with sufficient structural complexity and richness in their training data set. Next, some representative applications of the use of ANNs for building approximate models will be briefly reviewed.

Khorsand et al. [34] investigated structural design by a hybrid of neural network and FEA that only selectively used the neuro-estimation when either interpolation was expected (interpolation is generally expected to be more accurate) or the individual was not deemed to be highly fit (error in estimation may not be important). The methodology used in [34] is presented

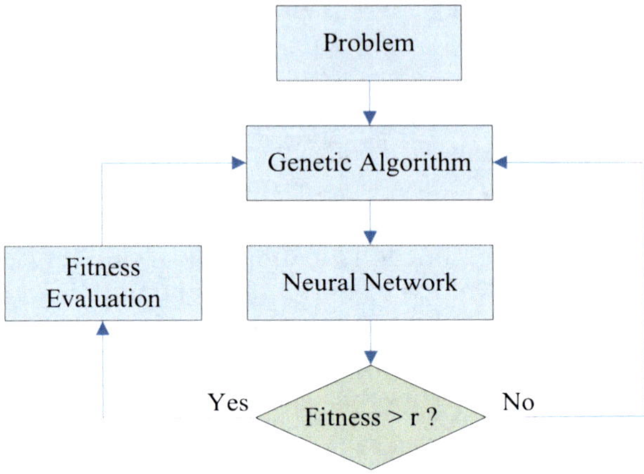

Fig. 1. The GA-ANN algorithm that is proposed in [34]. Only if the approximate fitness of an individual is better than the maximum fitness found in the last population, its fitness is re-evaluated in order to reflect its real fitness value.

in Figure 1 where r is considered as the maximum fitness of the individuals in the last generation. As with any other numerically driven approximation method, the performance of ANNs is closely related to the quality of the training data.

Jin et al. [35] investigated the convergence properties of an evolution strategy with neural network-based fitness evaluations. In this work, the concept of *controlled evolution* is introduced, in which the evolution proceeds using not only the approximate fitness function value, but also the true fitness function value. They also introduce two possibilities to combine the true with the approximate fitness function value: (1) the controlled individuals approach and (2) the controlled generation approach. Jin et al. defined "controlled" as evaluated with the true fitness function. Both approaches were studied and some interesting conclusions/recommendations for the correct use of such techniques are provided. A comprehensive survey of fitness approximation applied in evolutionary algorithms is presented in [36].

4.4 Final Remarks about Fitness Approximation

Lack of sufficient training data is the main problem in using most of the fitness approximation models currently available. Hence they may fail to build a model with sufficient approximation accuracy. Since evaluation of the original fitness function is very time-consuming and/or expensive, the approximate model may be of low fidelity and may even introduce false optima. Furthermore, if the training data does not cover all the domain range, large errors may occur due to extrapolation. Errors may also occur when the set of training points is not sufficiently dense and uniform. In such situations, a combination of methods may be more desirable. For example, Ong et al. [26] combined radial basis functions with transductive inference to generate local surrogate models.

Alternatively, if individuals in a population can be clustered into several groups as in [37], then only the individual that represents its cluster can be evaluated. The fitness value of other individuals in the same cluster will be estimated from the representative individual based on a distance measure. This is termed fitness imitation in contrast to fitness inheritance [15]. The idea of fitness imitation has been extended and more sophisticated estimation methods have been developed in [38]. A similarity based model is introduced in [39] and is applied to constrained and unconstrained optimization problems.

In multi-objective optimization problems (MOPs), the complexity of the problem is normally higher, compared to that of single-objective optimization problems (SOPs) [40]. In general, although the fitness approximation approaches used in SOPs can be simply extended and adapted for MOPs, such adaptation may require more elaborate mechanisms. One example of this is constraint-handling.[3] It is well-known that in real-world optimization problems there are normally constraints of different types (e.g., related to the geometry of structural elements to completion times, etc.) that must be satisfied for a solution to be acceptable. Traditionally, penalty functions have been used with EAs to handle constraints in SOPs [43]. However, because of the several problems associated to penalty functions (e.g., the definition of appropriate penalty values is normally a difficult task that has a serious impact on the performance of the EA), some researchers have proposed alternative constraint-handling approaches that require less critical parameters and perform well across a variety of problems (see for example [41, 44, 43]). However, when dealing with MOPs, many of these constraint-handling techniques cannot be used in a straightforward manner, since in this case, the best trade-offs among the objectives are always located in the boundary between the feasible and the infeasible region. This requires the development of different approaches specially tailored for MOPs (see for example [45, 46]). A similar problem occurs when attempting to migrate single-objective fitness approximation models to MOPs. For more details on this topic, see [47].

While the above methods aim to reduce the computational cost by approximating the fitness function, the prevalent problems with interpolation in rough surfaces remains. If the assumption of smooth continuity is not valid, interpolation may even yield values that are not physically realizable. Furthermore, we may be blinded to the optimal solutions using interpolation as

[3] Although constraint-handling techniques are very important in real-world optimization problems, their discussion is beyond the scope of this chapter, due to space limitations. Interested readers are referred to other references for more information on this topic (see for example [41, 42]).

interpolation assumes a pattern of behavior that may not be valid around optimal peaks. The next section addresses this problem by introducing the concept of information granulation.

5 Adaptive Fuzzy Fitness Granulation

Fuzzy granulation of information is a vehicle for handling information, as well as a lack of it (uncertainty), at a level of coarseness that can solve problems appropriately and efficiently [48]. In 1979, the concept of fuzzy information granulation was proposed by Zadeh [49] as a technique by which a class of points (objects) are partitioned into granules, with a granule being a clump of objects drawn together by indistinguishability, similarity, or functionality. The fuzziness of granules and their attributes is characteristic of the ways by which human concepts and reasoning are formed, organized and manipulated. The concept of a granule is more general than that of a cluster, potentially giving rise to several conceptual structures in various fields of science as well as mathematics.

In this chapter, with a aim to reducing computational costs, the concept of fitness granulation is applied to exploit the natural tolerance of EAs in fitness function computations. Nature's *survival of the fittest* is not about exact measures of fitness; rather it is about rankings among competing peers. By exploiting this natural tolerance for imprecision, optimization performance can be preserved by computing fitness only selectively and only to keep this ranking among individuals in a given population. Also, fitness is not interpolated or estimated; rather, the similarity and indistinguishability among real solutions is exploited.

In the proposed algorithm, an adaptive pool of solutions (fuzzy granules) with an exactly computed fitness function is maintained. If a new individual is sufficiently similar to a known fuzzy granule [49], then that granules' fitness is used instead as a crude estimate. Otherwise, that individual is added to the pool as a new fuzzy granule. In this fashion, regardless of the competitions' outcome, the fitness of the new individual is always a physically realizable one, even if it is a *crude* estimate and not an exact measurement. The pool size as well as each granules' radius of influence is adaptive and will grow/shrink depending on the utility of each granule and the overall population fitness. To encourage fewer function evaluations, each granule's radius of influence is initially large and gradually shrinks at later stages of the evolutionary process. This encourages more exact fitness evaluations when competition is fierce among more similar and converging solutions. Furthermore, to prevent the pool from growing too large, unused granules are gradually replaced by new granules, once the pool reaches a certain maturity.

5.1 Algorithm Structure

Given the general overview in the preceding section, the concrete steps of the algorithm are as follows:

Step 1: Create a random parent population $P_1 = \{X_1^1, X_2^1, \ldots, X_j^1, \ldots, X_t^1\}$ of design variable vector, where, more generally, $X_j^i = \{x_{j,1}^i, x_{j,2}^i, \ldots, x_{j,r}^i, \ldots, x_{j,m}^i\}$ is the jth parameter individual in the ith generation, $x_{j,r}^i$ the rth parameter of X_j^i, m is the number of design variables and t is the population size.

Step 2: Define a multi-set G of fuzzy granules (C_k, σ_k, L_k) according to $G = \{(C_k, \sigma_k, L_k) | C_k \in \Re^m, \sigma_k \in \Re, L_k \in \Re, k = 1, \ldots, l\}$. G is initially empty (i.e., $l = 0$). C_k is an m-dimensional vector of centers, σ_k is the width of membership functions (WMFs) of the kth fuzzy granule, and L_k is the granule's life index. A number of granules with different widths are shown in Figure 2.

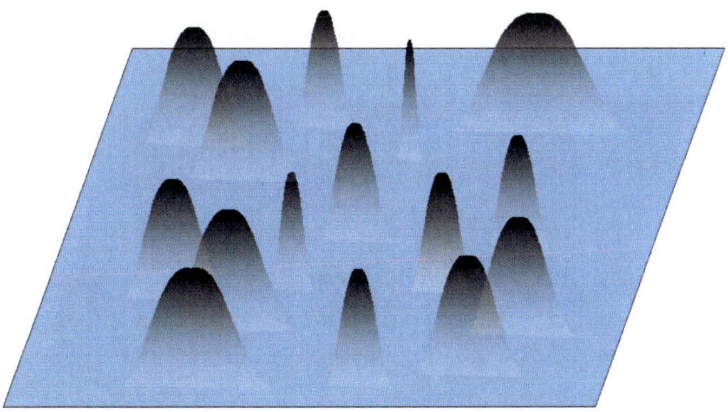

Fig. 2. A number of Gaussian granules with different widths in a 2-D solution space. Once a new individual is sufficiently similar to a granule in the granule pool, then that granules' fitness is used instead as a crude estimate. Otherwise, that individual is added to the pool as a new fuzzy granule. Each granules' radius of influence is determined based on equation (4).

Step 3: Choose the phenotype of the first chromosome $(X_1^1 = \{x_{1,1}^1, x_{1,2}^1, \ldots, x_{1,r}^1, \ldots, x_{1,m}^1\})$ as the center of the first granule $(C_1 = \{c_{1,1}, c_{1,2}, \ldots, c_{1,r}, \ldots, c_{1,m}\} = X_1^1)$.

Step 4: Define the membership $\mu_{k,r}$ of each $x_{j,r}^i$ to each granule member by a Gaussian similarity neighborhood function according to

$$\mu_{k,r}\left(x_{j,r}^i\right) = \exp\left(\frac{-\left(x_{j,r}^i - c_{k,r}\right)^2}{\left(\sigma_k\right)^2}\right), \quad k = 1, 2, \ldots, l \,, \tag{3}$$

where l is the number of fuzzy granules.

Remark: σ_k is the distance measurement parameter that controls the degree of similarity between two individuals. Like in [50], σ_k is defined based on equation (4). According to this definition, the granules shrink or enlarge in reverse proportion to their fitness:

$$\sigma_k = \gamma \frac{1}{\left(e^{F(C_k)}\right)^{\beta}}, \tag{4}$$

where $\beta > 0$ is an emphasis operator and γ is a proportionality constant. The problem arising here is how to determine the parameters β and γ as design parameters. The fact is that these two parameters are problem dependent and, in practice, a number of trials is needed to adjust them. This trial is based on a simple rule with respect to the acceleration of the parameter optimization procedure: high speed needs to have enlargement in the granule spread and, as a consequence of this, less accuracy is obtained in the fitness approximation, and vice versa. To deal with this rule, a fuzzy controller is presented in [50].

Step 5: Compute the average similarity of every new design parameter $X_j^i = \{x_{j,1}^i, x_{j,2}^i, \ldots, x_{j,r}^i, \ldots, x_{j,m}^i\}$ to each granule G_k using equation (5)

$$\overline{\mu}_{j,k} = \frac{\sum_{r=1}^{m} \mu_{k,r}\left(x_{j,r}^i\right)}{m} \tag{5}$$

Step 6: Either calculate the exact fitness function of X_j^i or estimate the fitness function value by associating it to one of the granules in the pool in case there is a granule in the pool with higher similarity to X_j^i than a predefined threshold, i.e.

$$f\left(X_j^i\right) = \begin{cases} f\left(C_k\right) & \text{if} \max\limits_{k \in \{1,2,\ldots,l\}} \{\overline{\mu}_{j,k}\} > \theta^i , \\ f\left(X_j^i\right) & \text{otherwise.} \end{cases} \tag{6}$$

where $f(C_x)$ is the fitness function value of the fuzzy granule, $f(X_j^i)$ is the real fitness calculation of the individual, $\theta^i = \alpha(\max\{f(X_1^{i-1}), f(X_2^{i-1}), \ldots, f(X_t^{i-1})\}/\overline{f}^{i-1})$, $K = \text{argmax}\{\overline{\mu}_{j,k}\}$ when $k \in \{1, 2, \ldots, l\}$, $\overline{f}^i = \sum_{j=1}^{i} f(X_j^i)/t$ and $\alpha > 0$ is a proportionality constant that is usually set at 0.9 unless otherwise indicated. The threshold θ^i increases as the best individual's fitness at generation i increases. As the population matures and reaches higher fitness values (i.e., while converging more), the algorithm becomes more selective and uses exact fitness calculations more often. Therefore, with this technique we can utilize the previous computational efforts during previous generations. Alternatively, if

$$\max_{k\in\{1,2,\ldots,l\}} \{\overline{\mu}_{j,k}\} < \theta^i \tag{7}$$

X_j^i is chosen as a newly created granule.

Step 7: If the population size is not completed, repeat **Steps 5 to 7**.

Step 8: Select parents using a suitable selection operator and apply the genetic operators of recombination and mutation to create a new generation.

Step 9: When termination/evolution control criteria are not met, then update σ_k using equation (4) and repeat **Steps 5 to 9**.

In [48] and [51], additional details on the convergence speed of the algorithm on a series of mathematical testbeds are provided along with a simple example to illustrate the competitive granule pool update.

5.2 How to Control the Length of the Granule Pool?

As the evolutionary procedures are applied, it is inevitable that new granules are generated and added to the pool. Depending on the complexity of the problem, the size of this pool can be excessive and become a computational burden itself. To prevent such unnecessary computational effort, a *forgetting factor* is introduced in order to appropriately decrease the size of the pool. In other words, it is better to remove granules that do not win new individuals, thereby producing a bias against individuals that have low fitness and were likely produced by a failed mutation attempt. Hence, L_k is initially set to N and subsequently updated as below,

$$L_k = \begin{cases} L_k + M & \text{if } k = K \ , \\ L_k & \textit{otherwise} \ , \end{cases} \tag{8}$$

where M is the life reward of the granule and K is the index of the winning granule for each individual at generation i. At each table update, only the N_G granules with the highest L_k index are kept, and the others are discarded. In [52], an example has been provided to illustrate the competitive granule pool update law.

Adding a new granule to the granule pool and assigning a life index to it, is a simple way of controlling the size of the granule pool, since the granules with the lowest life index will be removed from the pool. However, it may happen that the new granule is removed, even though it was just inserted into the pool. In order to prevent this, the pool is split into two parts with sizes εN_G and $(1-\varepsilon)N_G$. The first part is a FIFO (First In, First Out) queue and new granules are added to this part. If it grows above εN_G, then the top of the queue is moved to the other part. Removal from the pool takes place only in the $(1-\varepsilon)N_G$ part. In this way, new granules have a good chance to survive a number of steps. In all of the simulations that are conducted here, ε is set at 0.1.

The distance measurement parameter is completely influenced by the granule enlargement/shrinkage in the widths of the produced membership functions. As in [52], the combined effect of granule enlargement/shrinkage is in accordance with the granule fitness and it requires the fine-tuning of two parameters, namely β and γ. These parameters are problem dependent and it seems critical to have a procedure to deal with this difficulty. In [50] and [53], an auto-tuning strategy for determining the width of membership functions is presented which removes the need of exact parameter settings – without a negative influence on the convergence speed.

6 Numerical Results

To illustrate the efficacy of the proposed granulation algorithm, the result of applying it to the problems introduced in Section 2 are studied and analyzed in the two following sections. The commercial FEA software ANSYS [54] is used during the analysis and numerical simulation study.

The GA routines utilize initially random populations, binary-coded chromosomes, single-point crossover for the first three problems and 15-point crossover for the piezoelectric actuator design problem, mutation, fitness scaling, and an elitist stochastic universal sampling selection strategy. The crossover rate P_{XOVER} is set to 1, the mutation rate $P_{MUTATION} = 0.01$, and the population size chosen is 20. However, due to the number of parameters and complexity of the structural problems, the number of generations is set to 50 for the first three problems and 600 for the piezoelectric actuator design problem. These settings were determined during several trial runs to reflect the best performing set of parameters for the GA. Finally, the chromosome length varies depending on the number of variables in a given problem but each variable is still allocated 8 bits.

For performing the FES, a fitness and associated reliability value are assigned to each new individual that is truly evaluated if the reliability value is below a certain threshold T. The reliability value varies between 0 and 1 and depends on two factors: The first factor is the reliability of parents and the second one is how close parents and children are in the solution space, as explained in equation (2). Also, as mentioned in [18], $T = 0.7$ is used for the threshold as we empirically found that it generally produces the best results. The parameters of the GA-ANN are the same as in the GA alone. In the GA-ANN approach for solving optimization problems, a two-layer neural network is used, having as input the design variables and as outputs the fitness values.

Furthermore, due to the stochastic nature of EAs, each of the simulations was repeated ten times, and a paired Mann-Whitney U test was performed except for the last optimization problem. There, for each algorithm, it was performed only once, due to the running time needed. The significance level α represents the maximum tolerable risk of incorrectly rejecting the null hypothesis H_0, indicating that the mean of the 1st population is not significantly different from the mean of the 2nd population. The p-value or the observed significance level of a statistical test is the smallest value of α for which H_0 can be rejected. If the p-value is less than the pre-assigned significance level α, then the null hypothesis is rejected. Here, the significance level α was assigned, and the p-value was calculated for each of the following applications.

The results are presented in Tables 1, 2, 3 and 5, in which *FFE* stands for the number of fitness function evaluations needed to perform the optimization task and the *training data* column presents the number of initial input/output pairs needed in order to build up the approximation model. Since the most computationally expensive part of an evolutionary algorithm is usually, by far, its fitness evaluation, the convergence *time improvement* of different algorithms, compared to the standard GA, can be estimated in terms of the number of fitness evaluations. So, the *time improvement* percentage column is calculated as one minus the difference between the sum of *FFE* and *training data* divided by the number of FFE of the standard algorithm, i.e., a GA, multiplied by 100.

6.1 3-Layer Composite Beam

A 3-layer composite beam has been modeled numerically by using the ANSYS program. The composite layout is represented by the design variables that change in the region $[0, 180]$. The objective here is to raise the first natural frequency by appropriately choosing two composite layers' angles. In this example, the Young's modulus [55] is $EX = 210GPa$, $EY = 25GPa$, $EZ = 25GPa$, $GXY = GYZ = GXZ = 30GPa$, Poisson's ratio $\nu = 0.2$ and the density $\rho = 2100kg/m^3$. There are two design variables (two degrees of freedom) for this optimization problem each varying between 0 and 180. For this case, a 2-100-1 ANN architecture is consequently chosen and used for the optimization runs. The proposed algorithm (called AFFG, for adaptive fuzzy fitness granulation) is compared to other methods in Table 1. The results indicate that, while there is not a significant statistical difference between the three algorithms in terms of solution fitness, i.e., rigidity of the beam, the time savings provided by the proposed method are much higher than that of the GA-ANN. In particular, the proposed AFFG algorithm finds better solutions on the average with less computational time as compared with the GA-ANN. Also, while FES seems to have found better solutions, the proposed GA-AFFG used less than half as many evaluations.

6.2 Airplane Wing Design

Figure 3(a) shows the initial design of an airplane wing. The wing is of uniform configuration along its length, and its cross-sectional area is defined to

Table 1. Performance of the optimization methods (average of 10 runs) for the 3-layer composite beam, $\alpha = 0.9$, $\beta = 0.1$, $\gamma = 30$, $M = 5$, $N_G = 250$, $T = 0.7$.

	FFEs	Training data	Time improvement (%)	Optimum S^{-1}	p-value
GA	1000.0	Not Needed		19.3722	
FES	228.1	Not Needed	77.19	19.3690	0.0211
GA-ANN	155.9	100	74.41	19.3551	0.0026
GA-AFFG	97.5	Not Needed	90.25	19.3681	0.0355

be a straight line and a spline. It is held fixed to the body of the airplane at one end and hangs up freely at the other. The objective here is to maximize the wing's *first natural frequency* by appropriately choosing three key points of the spline. The material properties are: Young's modulus $= 261.820GPa$, density $\rho = 11031kg/m^3$, and Poisson's ratio $\nu = 0.3$.

The optimized shape found by a simple GA is shown in Figure 3(b) and that found by GA-AFFG is shown in Figure 3(c). A 6-100-1 architecture is chosen for the ANN used as fitness approximator. Table 2 illustrates that while the GA-ANN finds inferior solutions as compared with the GA, the use of the ANN significantly reduces computational time. The application of AFFG shows an improvement in the search quality while maintaining a low computational cost. Specifically, the average ten-run performance of the AFFG solutions is higher than that of any of the competing algorithms including the GA, FES and GA-ANN. Furthermore, while the Mann-Whitney U test confirms that the proposed algorithm solutions are at least as good as those produced by the GA, the proposed algorithm is over 82% faster.

Table 2. Performance of the optimization methods (average of 10 runs) for Airplane wing, $\alpha = 0.9$, $\beta = 0.5$, $\gamma = 1$, $M = 5$, $N_G = 250$, $T = 0.7$.

	FFEs	Training data	Time improvement (%)	Optimum S^{-1}	p-value
GA	1000.0			6.0006	
FES	481.6		51.84	5.9801	0.9698
GA-ANN	172.1	100	72.79	5.9386	0.4274
GA-AFFG	173.5		82.65	6.0527	0.3075

6.3 2D Truss Frame

A typical truss designed by an engineer is illustrated in Figure 4(a). The objective (fitness) here is to raise the structure's *first natural frequency* to reduce the vibration domain and to prevent the resonance phenomenon (in

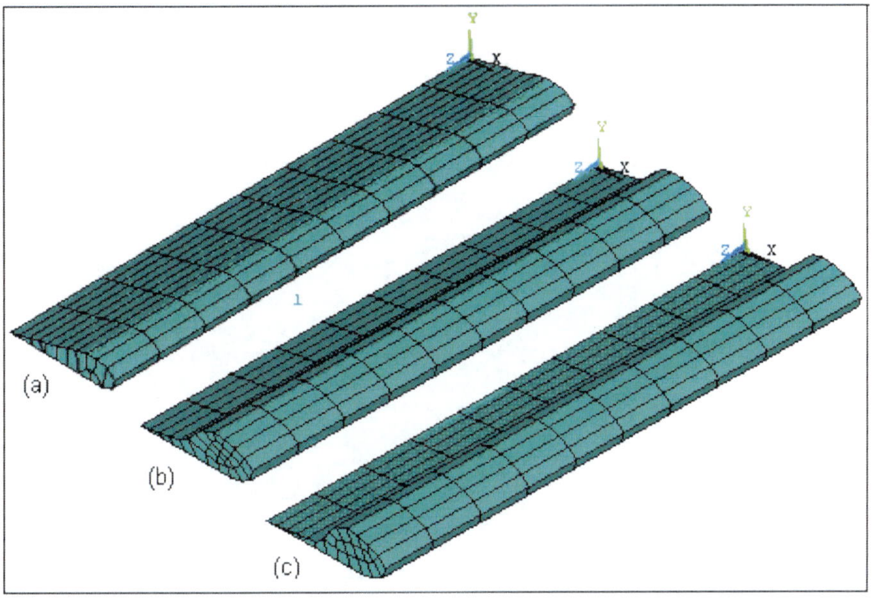

Fig. 3. Airplane wing: (a) initial shape, (b) GA optimized shape, and (c) GA-AFFG.

dynamic response) of the structure by appropriately choosing the 18 key point locations (our design variables) as illustrated in Figure 3(a).

In this benchmark, isotropic material properties are assumed (Young's modulus $E = 210GPa$, Poisson's ratio $\nu = 0.3$, and density $\rho = 7800kg/m^3$). The optimized shapes produced by the GA and the new proposed method AFFG are shown in Figures 4(b) and 4(c), respectively. The 36-100-1 ANN architecture is chosen and used for the optimization runs.

The search begins with an initial population. The maximum fitness in the initial population is nearly 9.32. Over several generations, the fitness gradually evolves to a higher value of 11.902. Figure 5 shows a plot of best, average and worst fitness vs. the generation number for one run of our GA-AFFG. This performance curve shows the learning activity or adaptation associated with the algorithm. The total number of generations is 50. For a population size of twenty, this requires 1000 (50×20) fitness evaluations for the GA while the proposed GA-AFFG required only 570.4 fitness evaluations. Figure 6 shows the plot of the number of FEA evaluations vs. generation number corresponding to one run [48].

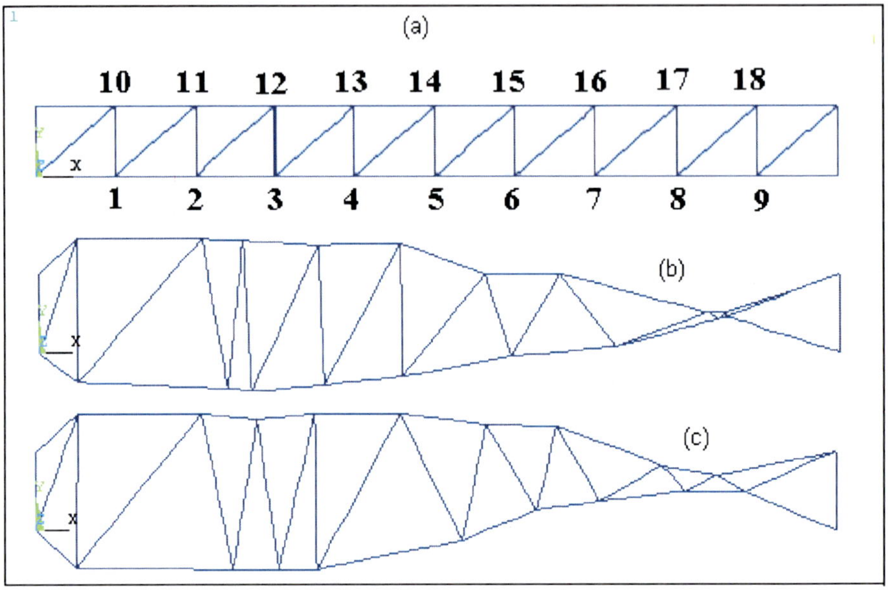

Fig. 4. 2D truss frame: (a) initial configuration, (b) GA optimized shape, and (c) GA-AFFG optimized shape.

Table 3. Performance of the optimization methods (average of 10 runs) for the 2D truss, $\alpha = 0.9$, $\beta = 0.11$, $\gamma = 3.05$, $M = 5$, $N_G = 550$, $T = 0.7$.

	FFEs	Training data	Time improvement (%)	Optimum S^{-1}	p-value
GA	1000.0			12.1052	
FES	1000.0		0.00	11.8726	0.0058
GA-ANN	293.0	100	60.66	11.8697	0.0257
GA-AFFG	570.4		42.96	12.1160	0.9097

6.4 Voltage and Pattern Design of Piezoelectric Actuator

Piezoelectric materials have coupled mechanical and electrical properties making them able to generate a voltage when subjected to a force or deformation (this is termed as the direct piezoelectric effect). Conversely, they exhibit mechanical deformation when subjected to an applied electric field (this is called the converse piezoelectric effect) [51]. Various applications of piezoelectric actuators/sensors have appeared in the literature. Lin et al. [56] investigated the modeling and vibration control of a smart beam by using piezoelectric damping-modal actuators/sensors. They presented theoretical

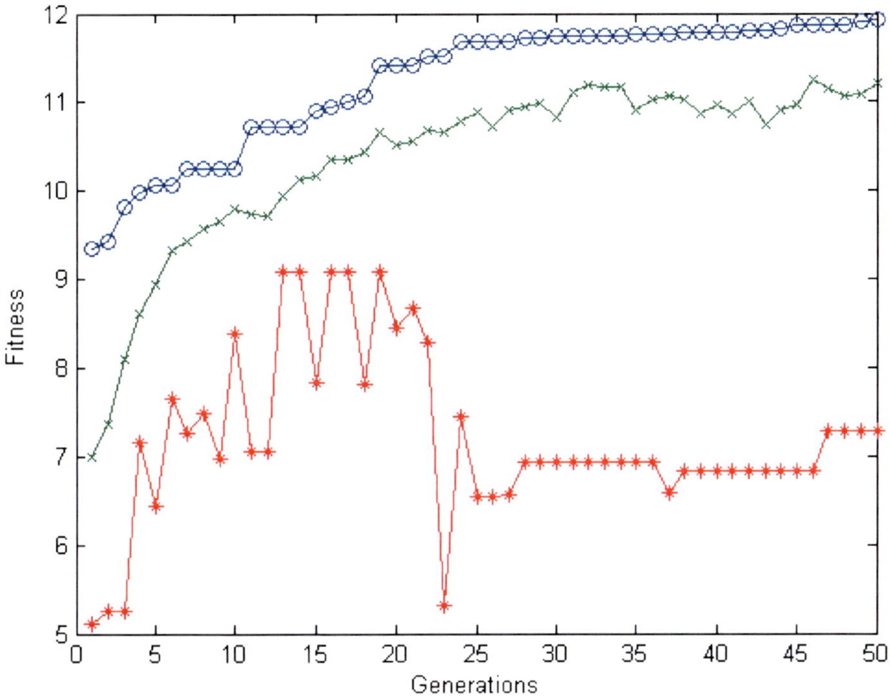

Fig. 5. Plot of generation number vs. fitness value for the 2D truss frame using GA-AFFG: best (circle), average (cross) and worst (asterisk) individuals at each generation.

formulations based on damping-modal actuators/sensors and numerical solutions for the analysis of a laminated composite beam with integrated sensors and actuators. A proof-of-concept design of an inchworm-type piezoelectric actuator of large displacement and force for shape and vibration control of adaptive truss structures is proposed by Li et al. in [57]. The applications of such actuators include smart or adaptive structural systems for the car and aerospace industries.

A fiber composite plate with initial imperfections and under in-plane compressive loads is studied by Adali et al. [58] with a view towards minimizing its deflection and optimizing its stacking sequence by means of the piezoelectric actuators and the fiber orientations. Krommer [59] studied a method to control the deformation of a sub-section of a beam. His intention was to apply a distributed control by means of self-stresses within the sub-section to keep the sub-section in its non-deformed state. In practical applications such as deformation control of conformal antennas, this strategy is highly valuable.

Global optimization algorithms [60] along with a finite element formulation are widely used in shape control. For instance in [10], a computational

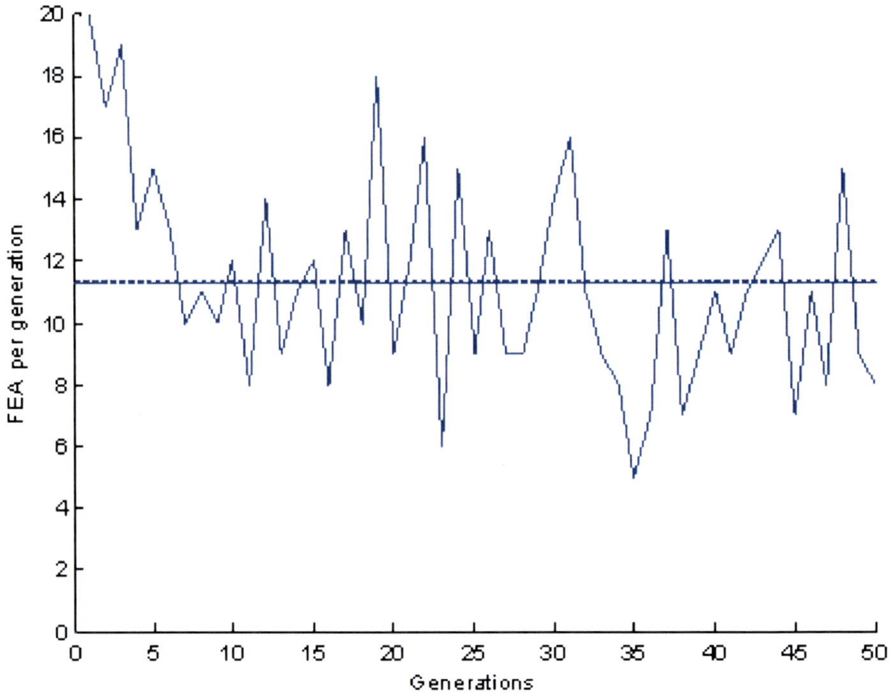

Fig. 6. Plot of the generation number vs. number of FEA evaluations for the 2D truss frame in a single run using GA-AFFG.

intelligence based optimization algorithm along with a modified finite element formulation is used to deal with the shape control of functionally graded material (FGM) plates that contain piezoelectric sensor and actuator patches. In this study, an optimal voltage distribution or a gain control matrix are used as design variables for the shape control of smart structures. Numerical simulations have been successfully carried out on the shape control of the FGM plates by optimizing the voltage distribution for the open loop shape control or gain values for the closed loop shape control. A finite element formulation with non-rectangular shaped actuators for laminated smart composite structure is studied in [61]. For smart cantilever plates, the actuated deflections are measured and are used to validate the presented formulation. They also investigated the effect of the actuator pattern on the optimum values of the applied voltages and the shape match factors. Numerical results showed that the actuator patterns may have an important influence on the values of the optimum voltages applied to each individual actuator and the final shape match factor.

6.4.1 Piezoelectric Equations (Constitutive Equations)

In this study, the assumption is that the thermal effect is negligible. The piezoelectric constitutive relationships describe how two piezoelectric mechanical and electrical quantities (stress, strain, electric displacement, and electric field) interact and it is expressed by the direct and the converse piezoelectric equations respectively [62]:

$$\{D\} = [e]\{\varepsilon\} + [\varepsilon]\{E\} \ , \tag{9}$$

$$\{\sigma\} = [Q]\{\varepsilon\} + [e]^T\{E\} \ , \tag{10}$$

where $\{\sigma\}$ is the stress vector, $[Q]$ is the elastic stiffness matrix, $\{\varepsilon\}$ is the strain vector, $[e]$ is the piezoelectric constant matrix, $\{E\} = -\nabla\varphi$ is the electric field vector. Also, φ is the electrical potential, $\{D\}$ is the electric displacement vector and $[\varepsilon]$ is the permittivity coefficient matrix. Equations (9) and (10) describe the electromechanical coupling. Assuming that a laminated beam consists of a number of layers and each layer possesses a plane of material symmetrically parallel to the x-y plane, the constitutive equations for the kth layer can be written according to [63] as:

$$\begin{Bmatrix} D_1 \\ D_3 \end{Bmatrix}_k = \begin{bmatrix} 0 & e_{15} \\ e_{31} & 0 \end{bmatrix}_k \times \begin{Bmatrix} \varepsilon_1 \\ \varepsilon_5 \end{Bmatrix}_k + \begin{bmatrix} \varepsilon_{11} & 0 \\ 0 & \varepsilon_{33} \end{bmatrix}_k \times \begin{Bmatrix} E_1 \\ E_3 \end{Bmatrix}_k \tag{11}$$

$$\begin{Bmatrix} \varepsilon_1 \\ \varepsilon_3 \end{Bmatrix}_k = \begin{bmatrix} Q_{11} & 0 \\ 0 & Q_{55} \end{bmatrix}_k \times \begin{Bmatrix} \varepsilon_1 \\ \varepsilon_5 \end{Bmatrix}_k + \begin{bmatrix} 0 & \varepsilon_{31} \\ \varepsilon_{15} & 0 \end{bmatrix}_k \times \begin{Bmatrix} E_1 \\ E_3 \end{Bmatrix}_k \tag{12}$$

where $Q_{11} = \frac{E_{11}}{1-v_{12}v_{21}}$ and $Q_{55} = G_{13}$ are the reduced elastic constants of the kth layer, E_{11} is the Young's modulus and G_{13} is the shear modulus. The piezoelectric constant matrix $[e]$ can be expressed in terms of the piezoelectric strain $[d]$ as:

$$[e] = [d][Q] \tag{13}$$

where

$$[d] = \begin{bmatrix} 0 & d_{15} \\ d_{31} & 0 \end{bmatrix} \tag{14}$$

Using the above piezoelectricity analysis and formulation, a finite element model (FEM) of piezoelectric patches and a metal plate [64] was built by ANSYS [54]. Also, a small deflection and the thin plate theory are assumed to hold for the FEM of the plate.

To validate the software, a clamped free aluminum plate with four piezoelectric patches is modeled and the results are compared with the experimental model of reference [65]. A close agreement between our model and our experimental results is observed. Also, in order to achieve an acceptable mesh density, a mesh sensitivity analysis[4] method is applied.

6.4.2 Piezoelectric Design for Static Shape Control

The shape control problem considered here is to find the optimal actuator pattern design vector P and exiting voltage vector V as design variables. This (quasi-) static shape control problem can be defined, in the context of an optimization formulation, as follows: Find $S = [P, V]^T$ to minimize:

$$f(S) = \sum_{j=1}^{N_x} \sum_{i=1}^{N_y} \frac{\left| d_{i,j}^d - d_{i,j}^f \right|}{\left| max\left(d_{i,j}^d\right) \right|} / \left(N_x \times N_y\right) \tag{15}$$

S is the design variable vector with two components: *i)* the pattern variable vector P, and *ii)* the applied voltage variable vector V. Here, $f(S)$ is the objective function. P is the distribution of the active piezoelectric actuator material (pattern variable) whereas the voltage variables in vector V are the electrical potentials applied across the thickness direction of each actuator. The objective function $f(S)$ in equation (15) is a weighted sum of all the absolute differences between the desired and designed shapes at all nodes. $d_{i,j}^d$ and $d_{i,j}^f$ are the desired and designed (calculated by the FE model) transversal displacements of the (i, j)-location, respectively and $max(d_{i,j}^d)$ is the maximum displacement in the desired structural shape. As the displacement is small here, there is no need to consider stress or strain constraint variables for the shape control problem.

6.4.3 Model Description

A cantilever plate clamped at its left edge and subjected to a non-applied mechanical load is assumed here. The plate has a length of 154 mm; width of 48 mm and consists of one layer of 0.5 mm in thickness. The piezoelectric actuators (thickness of 0.3 mm each) are attached to the top surfaces of the plate (Figure 7). The desired pre-defined surface [65] is defined as:

$$d_{i,j}^d = \left(1.91x^2 + 0.88xy + 0.19x\right) \times 10^{-4} \,. \tag{16}$$

The piezoelectric electro-mechanical actuators have the properties shown in Table 4 according to specification PX5-N from Philips Components. After a careful mesh sensitivity analysis, a FEM is built as illustrated in Figure 8.

[4] The mesh sensitivity analysis is used to reduce the number of elements and nodes in the mesh while ensuring the accuracy of the finite element solution [66].

Table 4. Material properties for the PX5-N piezoelectric material [65].

$C_{11}^{E}(N\,m^{-2})$	13.11×10^{10}	$d_{15}(m\,V^{-1})$	515×10^{-12}
$C_{12}^{E}(N\,m^{-2})$	7.984×10^{10}	$d_{31}(m\,V^{-1})$	-215×10^{-12}
$C_{13}^{E}(N\,m^{-2})$	8.439×10^{10}	$d_{33}(m\,V^{-1})$	500×10^{-12}
$C_{33}^{E}(N\,m^{-2})$	12.31×10^{10}	$\varepsilon_{11}^{t}/\varepsilon_0$	1800
$C_{44}^{E}(N\,m^{-2})$	2.564×10^{10}	$\varepsilon_{33}^{t}/\varepsilon_0$	2100
$C_{66}^{E}(N\,m^{-2})$	2.564×10^{10}	$\rho(kg\,m^{-3})$	7800

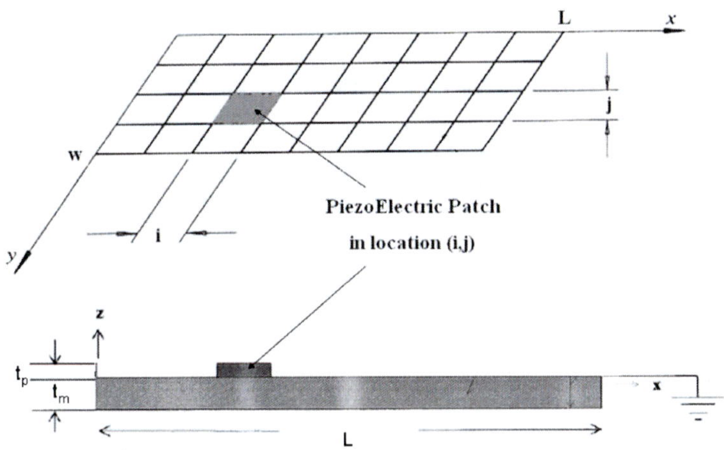

Fig. 7. Geometrical model of the piezoelectric patch adopted here.

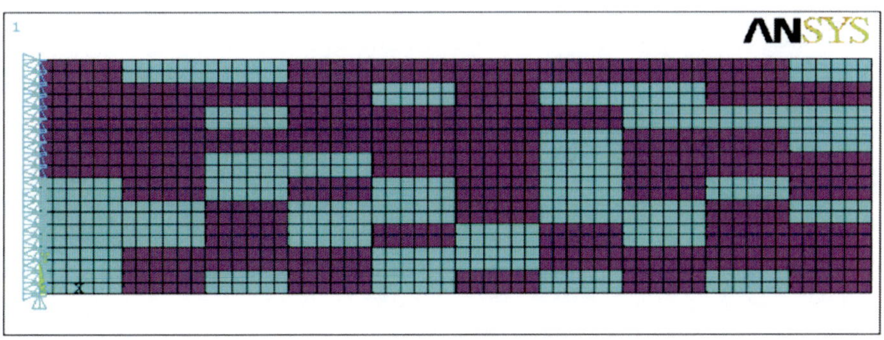

Fig. 8. Finite element model built by ANSYS.

For this SD and optimization problem, there are 200 design variables. Half of these design variables belong to actuation voltage of piezoelectric patches which vary between -10 and $20V$. The rest of the design variables are Boolean, indicating whether or not the voltage should be applied to the piezoelectric patches. When the ith $(i = 1, \ldots, 100)$ piezoelectric pattern variable is zero, the piezoelectric patch is not built so that there is no actuation voltage, and vice versa. Figure 9 shows the graph of the best, average and worst fitness vs. generation number and Figure 10 shows the number of FEA evaluations vs. the generation number for a single GA-AFFG run. Table 4 presents the results of the four optimization algorithms obtained from one run each.

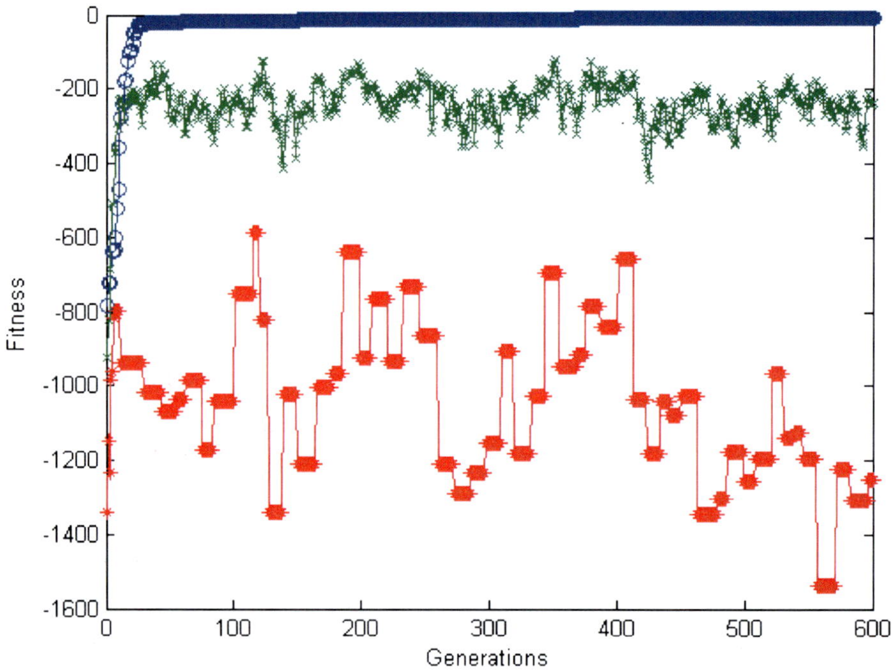

Fig. 9. Generation number vs. fitness for the piezoelectric actuator using our proposed GA-AFFG for a single run: best (circle), average (cross) and worst (asterisk) of individuals at each generation.

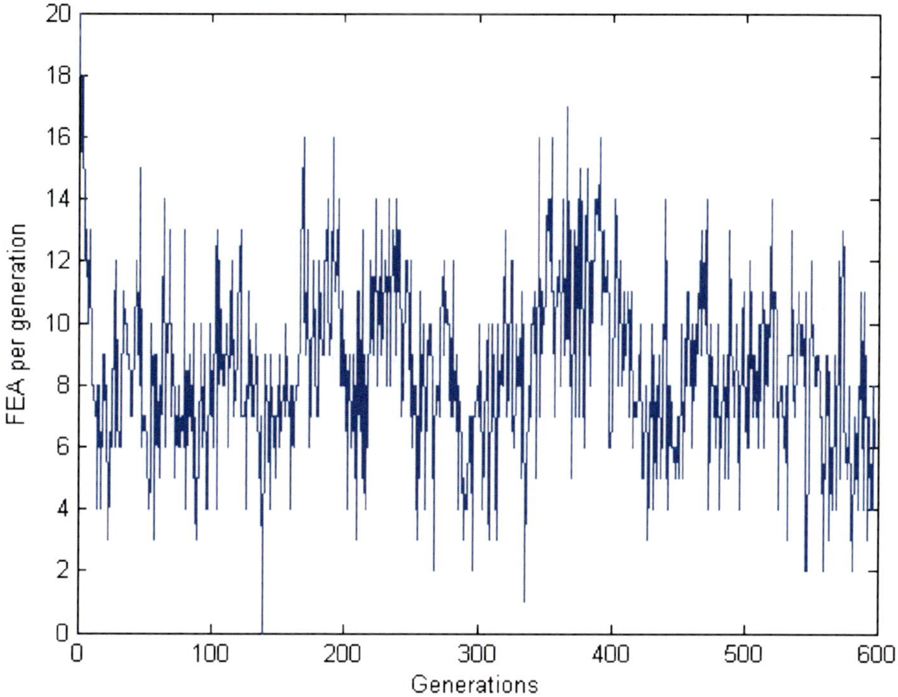

Fig. 10. Generation number vs. number of FEA evaluations, for the piezoelectric actuator, using our proposed GA-AFFG for a single run.

Table 5. Piezoelectric actuator performance of the optimization methods, $\alpha = 0.9$, $\beta = 0.11$, $\gamma = 3.05$, $M = 5$, $N_G = 550$, $T = 0.7$.

	FFEs	Training data	Time Improved (%)	Error (%)
GA	12000			7.313
FES	12000		0.00	12.820
GA-ANN	2617	5000	36.52	8.093
GA-AFFG	5066	Not needed	57.64	7.141

7 Analysis of Results

Tables 1, 2, 3 and 5, illustrate the performance of the proposed GA-AFFG method in comparison with a GA, FES, and GA-ANN [48] in terms of computational efficiency and performance for the 3-layer composite beam, the airplane wing, and the 2D truss design problems as well as for the piezoelectric actuator problem. Due to the stochastic nature of the GA, the first three design simulations are repeated ten times and a statistical analysis is performed. However, for the piezoelectic actuator we could not run the GA that many times, because of its high computational cost.

The second column in these tables makes a comparison of the three algorithms in terms of the number of FEA evaluations as compared to those of the GA, while the fourth column makes a comparison in terms of performance. Results indicate that our proposed GA-AFFG finds statistically equivalent solutions by using more than 90%, 82%, 42% and 57% fewer finite element evaluations. The GA-ANN also significantly reduces the number of FEA evaluations, but its average performance is inferior when compared with our proposed GA-AFFG due to the ANNs approximation error. It must be noted that the improvement in time by GA-ANN takes into account the time spent on constructing the training data set.

For the piezoelectric actuator design problem, Table 5 illustrates a comparison of the GA, FES, and GA-ANN [51] with respect to our proposed GA-AFFG in terms of computational efficiency and performance. The second column of this table makes a comparison of the four algorithms in terms of the number of FEA evaluations as compared with a GA, while the fifth column makes a comparison in terms of the quality of the optimal solutions. Results indicate that GA-AFFG finds at least equivalent solutions by using 57% fewer finite element evaluations as compared to the GA. Also, when compared with the GA-ANN, the proposed algorithm finds better solutions while being more computationally efficient. The main difference here is ANN's need for pre-training. Trying various sizes of initial training sets and considering the 200 design parameters, the ANN required at least 5000 training data pairs for comparable performance, see Table 5.

Overall, when compared with a GA, the two sets of applications indicate that FES, GA-ANN and GA-AFFG improve the computational efficiency by reducing the number of exact fitness function evaluations. However, the neuro-approximation as well as fitness inheritance fail with a growing size of the input-output space. Consequently, the utility of AFFG becomes more significant in larger and more complex design problems. Furthermore, our statistical analysis confirms that fitness inheritance has comparable performance when the size of the search space is small (Tables 1 and 2), but its efficiency deteriorates as the complexity of the problem increases (Tables 3 and 5).

A comparison of the number of exact fitness function evaluations in terms of mean and variance that shows the improvement in computational time is presented in Tables 6, 7 and 8 for the first three mechanical optimization problems described before. A Mann-Whitney U test is also performed to study the significance of lower computation cost. Since the fourth optimization problem (piezoelectic actuator design) could not be repeated due to its FEA time-consuming nature, a Mann-Whitney U test could not be performed in that case.

Table 6. A Mann-Whitney U test of the number of real fitness calculations for the 3-layer composite beam (10 runs).

3-layer composite beam	Simulation results		
	Mean	Var	p-Value
FES	228.1	4601.2	6.39×10^{-05}
GA-ANN	155.9	511.9	6.34×10^{-05}
GA-AFFG	97.5	406.7	6.39×10^{-05}

Table 7. A Mann-Whitney U test of the number of real fitness calculations for the airplane wing (10 runs).

Airplane wing	Simulation results		
	Mean	Var	p-Value
FES	481.6	38648.0	6.39×10^{-05}
GA-ANN	172.1	6392.1	6.39×10^{-05}
GA-AFFG	173.5	1600.3	6.39×10^{-05}

Table 8. A Mann-Whitney U test of the number of real fitness calculations for the 2D truss (10 runs).

2D truss	Simulation results		
	Mean	Var	p-Value
FES	100.0	0.0	Not available
GA-ANN	293.0	2394.2	6.39×10^{-05}
GA-AFFG	570.4	18477.0	6.39×10^{-05}

8 Conclusions

In this chapter, we have proposed a systematic and robust methodology for solving complex structural design and optimization problems. The proposed methodology relies on the use of the FEA and adaptive fuzzy fitness granulation. By considering the similarity/indistinguishability of an individual to a pool of fuzzy information granules, adaptive fuzzy fitness granulation provides a method to selectively reduce the number of actual fitness function evaluations performed. Since the proposed approach does not use approximation or online training, it is not caught in the pitfalls of these techniques such as false peaks, large approximation error due to extrapolation, and time-consuming online training.

The effectiveness and functionality of the proposed approach was verified through four structural design problems. In the first three of them, the objective was to increase the first natural frequency of the structure. In the last problem, a piezoelectric actuator was considered for the purposes of shape control and/or active control for correction of static deformations. The design variables were the voltage and the actuator locations and the performance index was considered as the square root of the error between the nodal predefined displacement and the observed displacement.

Acknowledgements. The work in this chapter is in the context of the *Con4Coord* project, supported by the European Community's Seventh Framework Programme (FP7/2007-2013) under grant agreement no. INFSO-ICT-223844, the Next Generation Infrastructures Research Program of Delft University of Technology and the Mexican CONACyT Project No. 103570.

References

1. Reddy, J.: Introduction to the Finite Element Method. McGraw-Hill, New York (1993)
2. Papadrakakis, M., Lagaros, N.D., Kokossalakis, G.: Evolutionary Algorithms Applied to Structural Optimization Problems. In: High Performance Computing for Computational Mechanics, pp. 207–233 (2000)
3. Michalewicz, Z.: Genetic algorithms + data structures = evolution programs. Springer-Verlag New York, Inc., New York (1994)
4. Walker, M., Smith, R.E.: A technique for the multiobjective optimisation of laminated composite structures using genetic algorithms and finite element analysis. Composite Structures 62(1), 123–128 (2003)
5. Abe, A., Kamegawa, T., Nakajima, Y.: Optimization of construction of tire reinforcement by genetic algorithm. Optimization and Engineering 5(1), 77–92 (2003)
6. Giger, M., Ermanni, P.: Development of CFRP racing motorcycle rims using a heuristic evolutionary algorithm approach. Structural and Multidisciplinary Optimization 30(1), 54–65 (2005)
7. Alba, E., Tomassini, M.: Parallelism and Evolutionary Algorithms. IEEE Transactions on Evolutionary Computation 6(5), 443–462 (2002)
8. Mackerle, J.: Smart materials and structures–a finite element approach–an addendum: a bibliography (1997- 2002). Modelling and Simulation in Materials Science and Engineering 11(5), 707–744 (2003)
9. Joseffsson, L., Persson, P.: Conformal Array Antenna Theory and Design. IEEE Press Series on Electromagnetic Wave Theory. Wiley-IEEE Press (2005)
10. Liew, K.M., He, X.Q., Ray, T.: On the use of computational intelligence in the optimal shape control of functionally graded smart plates. Computer Methods in Applied Mechanics and Engineering 193(42-44), 4475–4492 (2004)
11. Goldberg, D.E.: Genetic Algorithms in Search, Optimization and Machine Learning. Addison-Wesley Publishing Co., Reading (1989)
12. Furuya, H., Haftka, R.T.: Locating actuators for vibration suppression on space trusses by genetic algorithms, vol. 38. ASME-Publications-AD (1993)
13. Rodríguez, J.E., Medaglia, A.L., Coello Coello, C.A.: Design of a motorcycle frame using neuroacceleration strategies in MOEAs. Journal of Heuristics 15(2), 177–196 (2009)

14. Lemonge, A., Barbosa, H., Fonseca, L.: A genetic algorithm for the design of space framed structures. In: XXIV CILAMCE–Iberian Latin-American Congress on Computational Methods in Engineering, Ouro Preto, Brazil (2003)
15. Jin, Y.: A comprehensive survey of fitness approximation in evolutionary computation. Soft Computing 9(1), 3–12 (2005)
16. Smith, R., Dike, B., Stegmann, S.: Fitness inheritance in genetic algorithms. In: Proceedings of ACM Symposiums on Applied Computing, pp. 345–350. ACM, New York (1995)
17. Zhang, X., Julstrom, B., Cheng, W.: Design of vector quantization codebooks using a genetic algorithm. In: Proceedings of the IEEE Conference on Evolutionary Computation, pp. 525–529. IEEE, Los Alamitos (1997)
18. Salami, M., Hendtlass, T.: A fast evaluation strategy for evolutionary algorithms. Applied Soft Computing 2, 156–173 (2003)
19. Pelikan, M., Sastry, K.: Fitness inheritance in the Bayesian optimization algorithms. In: Genetic and Evolutionary Computation Conference, pp. 48–59. Springer, Heidelberg (2004)
20. Reyes Sierra, M., Coello Coello, C.A.: Fitness Inheritance in Multi-Objective Particle Swarm Optimization. In: 2005 IEEE Swarm Intelligence Symposium (SIS 2005), pp. 116–123. IEEE Press, USA (2005)
21. Reyes Sierra, M., Coello Coello, C.A.: A Study of Fitness Inheritance and Approximation Techniques for Multi-Objective Particle Swarm Optimization. In: Proceedings of the 2005 Congress on Evolutionary Computation (CEC 2005), pp. 65–72 (2005)
22. Ducheyne, E., De Baets, B., De Wulf, R.: Is fitness inheritance useful for real-world applications? In: Fonseca, C.M., Fleming, P.J., Zitzler, E., Deb, K., Thiele, L. (eds.) EMO 2003. LNCS, vol. 2632, pp. 31–42. Springer, Heidelberg (2003)
23. Sano, Y., Kita, H.: Optimization of noisy fitness functions by means of genetic algorithms using history. In: Deb, K., Rudolph, G., Lutton, E., Merelo, J.J., Schoenauer, M., Schwefel, H.-P., Yao, X. (eds.) PPSN 2000. LNCS, vol. 1917, pp. 571–580. Springer, Heidelberg (2000)
24. Branke, J., Schmidt, C., Schmeck, H.: Efficient fitness estimation in noisy environment. In: Spector, L. (ed.) Proceedings of Genetic and Evolutionary Computation Conference (GECCO), pp. 243–250. Morgan Kaufmann, San Francisco (2001)
25. Branke, J., Schmidt, C.: Fast convergence by means of fitness estimation. Soft Computing Journal 9(1), 13–20 (2005)
26. Ong, Y.S., Nair, P.B., Keane, A.J.: Evolutionary optimization of computationally expensive problems via surrogate modeling. American Institute of Aeronautics and Astronautics Journal 41(4), 687–696 (2003)
27. Ong, Y.S., Zhu, Z., Lim, D.: Curse and blessing of uncertainty in evolutionary algorithm using approximation. In: Proceedings of the 2006 Congress on Evolutionary Computation (CEC 2006), pp. 2928–2935 (2006)
28. Regis, R.G., Shoemaker, C.A.: Local function approximation in evolutionary algorithms for the optimization of costly functions. IEEE Transactions on Evolutionary Computation 8(5), 490–505 (2004)
29. Lim, D., Ong, Y.S., Jin, Y., Sendhoff, B.: Trusted evolutionary algorithm. In: Proceedings of the 2006 Congress on Evolutionary Computation (CEC 2006), pp. 149–156 (2006)

30. Sacks, J., Welch, W., Mitchell, T., Wynn, H.: Design and analysis of computer experiments (with discussion). Statistical Science 4, 409–435 (1989)
31. Ratle, A.: Accelerating the convergence of evolutionary algorithms by fitness landscape approximation. In: Eiben, A.E., Bäck, T., Schoenauer, M., Schwefel, H.-P. (eds.) PPSN 1998. LNCS, vol. V, pp. 87–96. Springer, Heidelberg (1998)
32. Hong, Y.-S., Lee, H., Tahk, M.-J.: Acceleration of the convergence speed of evolutionary algorithms using multi-layer neural networks. Engineering Optimization 35(1), 91–102 (2003)
33. Won, K.S., Ray, T., Tai, K.: A framework for optimization using approximate functions. In: Proceedings of IEEE Congress on Evolutionary Computation, pp. 1077–1084 (2003)
34. Khorsand, A.-R., Akbarzadeh, M.: Multi-objective meta level soft computing-based evolutionary structural design. Journal of the Franklin Institute, 595–612 (2007)
35. Jin, Y., Olhofer, M., Sendhoff, B.: On evolutionary optimization with approximate fitness functions. In: Proceedings of the Genetic and Evolutionary Computation Conference, pp. 786–792. Morgan Kaufmann, San Francisco (2000)
36. Shi, L., Rasheed, K.: A survey of fitness approximation methods applied in evolutionary algorithms. In: Hiot, L.M., Ong, Y.S., Tenne, Y., Goh, C.K. (eds.) Computational Intelligence in Expensive Optimization Problems. Adaptation Learning and Optimization, vol. 2, pp. 3–28. Springer, Heidelberg (2010)
37. Kim, H.-S., Cho, S.-B.: An efficient genetic algorithms with less fitness evaluation by clustering. In: Proceedings of IEEE Congress on Evolutionary Computation, pp. 887–894. IEEE, Los Alamitos (2001)
38. Bhattacharya, M., Lu, G.: A dynamic approximate fitness based hybrid ea for optimization problems. In: Proceedings of IEEE Congress on Evolutionary Computation, pp. 1879–1886 (2003)
39. Fonseca, L.G., Barbosa, H.J.C.: A similarity-based surrogate model for enhanced performance in genetic algorithms. Opsearch 46, 89–107 (2009)
40. Coello Coello, C.A., Lamont, G.B., Van Veldhuizen, D.A.: Evolutionary Algorithms for Solving Multi-Objective Problems, 2nd edn. Springer, New York (2007)
41. Mezura-Montes, E. (ed.): Constraint-Handling in Evolutionary Optimization. Springer, Berlin (2009); ISBN 978-3-642-00618-0
42. Runarsson, T.P.: Constrained evolutionary optimization by approximate ranking and surrogate models. In: Yao, X., Burke, E.K., Lozano, J.A., Smith, J., Merelo-Guervós, J.J., Bullinaria, J.A., Rowe, J.E., Tiňo, P., Kabán, A., Schwefel, H.-P. (eds.) PPSN 2004. LNCS, vol. 3242, pp. 401–410. Springer, Heidelberg (2004)
43. Coello Coello, C.A.: Theoretical and Numerical Constraint Handling Techniques used with Evolutionary Algorithms: A Survey of the State of the Art. Computer Methods in Applied Mechanics and Engineering 191(11-12), 1245–1287 (2002)
44. Runarsson, T.P., Yao, X.: Stochastic Ranking for Constrained Evolutionary Optimization. IEEE Transactions on Evolutionary Computation 4(3), 284–294 (2000)
45. Woldesenbet, Y.G., Yen, G.G., Tessema, B.G.: Constraint Handling in Multiobjective Evolutionary Optimization. IEEE Transactions on Evolutionary Computation 13(3), 514–525 (2009)

46. Kumar Singh, H., Ray, T., Smith, W.: C-PSA: Constrained Pareto simulated annealing for constrained multi-objective optimization. Information Sciences 180(13), 2499–2513 (2010)
47. Santana-Quintero, L.V., Arias Montaño, A., Coello Coello, C.A.: A Review of Techniques for Handling Expensive Functions in Evolutionary Multi-Objective Optimization. In: Tenne, Y., Goh, C.-K. (eds.) Computational Intelligence in Expensive Optimization Problems, pp. 29–59. Springer, Berlin (2010)
48. Davarynejad, M.: Fuzzy Fitness Granulation in Evolutionary Algorithms for Complex Optimization. Master's thesis, Ferdowsi University of Mashhad (June 2007)
49. Zadeh, L.A.: Fuzzy sets and information granularity. Advances in Fuzzy Set Theory and Applications, 3–18 (1979)
50. Davarynejad, M., Ahn, C.W., Vrancken, J.L.M., van den Berg, J., Coello Coello, C.A.: Evolutionary hidden information detection by granulation-based fitness approximation. Applied Soft Computing 10(3), 719–729 (2010)
51. Akbarzadeh-T, M.R., Davarynejad, M., Pariz, N.: Adaptive fuzzy fitness granulation for evolutionary optimization. International Journal of Approximate Reasoning 49(3), 523–538 (2008)
52. Davarynejad, M., Akbarzadeh-T, M.-R., Pariz, N.: A novel general framework for evolutionary optimization: Adaptive fuzzy fitness granulation. In: Proceedings of the 2007 Congress on Evolutionary Computation (CEC 2007), pp. 951–956 (2007)
53. Davarynejad, M., Akbarzadeh-T, M.R., Coello Coello, C.A.: Auto-tuning fuzzy granulation for evolutionary optimization. In: Proceedings of the 2008 Congress on Evolutionary Computation, pp. 3572–3579 (2008)
54. Ansys, I.: ANSYS users manual. ANSYS Inc., Southpointe, 275 (2004)
55. Freudenberger, J., Gllner, J., Heilmaier, M., Mook, G., Saage, H., Srivastava, V., Wendt, U.: Materials science and engineering. In: Grote, K.H., Antonsson, E.K. (eds.) Springer Handbook of Mechanical Engineering. Springer, Heidelberg (2009)
56. Lin, J., Nien, M.: Adaptive control of a composite cantilever beam with piezoelectric damping-modal actuators/sensors. Composite Structures Journal 70, 170–176 (2005)
57. Li, J., Sedaghati, R., Dargahi, J., Waechter, D.: Design and development of a new piezoelectric linear Inchworm actuator. Mechatronics Journal 15, 651–681 (2005)
58. Adali, S., Sadek, I., Bruch Jr., J., Sloss, J.: Optimization of composite plates with piezoelectric stiffener-actuators under in-plane compressive loads. Composite Structures Journal 71, 293–301 (2005)
59. Krommer, M.: Dynamic shape control of sub-sections of moderately thick beams. Computers & Structures 83(15-16), 1330–1339 (2005)
60. Weise, T.: Global Optimization Algorithms–Theory and Application. Abrufdatum, 1 (2008), `http://www.it-weise.de`
61. Nguyen, Q., Tong, L.: Shape control of smart composite plate with non-rectangular piezoelectric actuators. Composite Structures 66(1-4), 207–214 (2004)
62. Aryana, F., Bahai, H., Mirzaeifar, R., Yeilaghi, A.: Modification of dynamic characteristics of FGM plates with integrated piezoelectric layers using first- and second-order approximations. International Journal for Numerical Methods in Engineering 70(12), 1409–1429 (2007)

63. Khorsand, A.-R., Akbarzadeh-T, M.-R., Moin, H.: Genetic Quantum Algorithm for Voltage and Pattern Design of Piezoelectric Actuator. In: Proceedings of the 2006 Congress on Evolutionary Computation (CEC 2006), pp. 2593–2600 (2006)
64. Piefort, V.: Finite element modelling of piezoelectric active structures. PhD thesis, Université Libre de Bruxelles (2001)
65. da Mota Silva, S., Ribeiro, R., Rodrigues, J.D., Vaz, M.A.P., Monteiro, J.M.: The application of genetic algorithms for shape control with piezoelectric patches-an experimental comparison. Smart Materials and Structures 13, 220–226 (2004)
66. Kelly, D.W., De, J.P., Gago, S.R., Zienkiewicz, O.C., Babuska, I.: A posteriori error analysis and adaptive processes in the finite element method: Part i–error analysis. International Journal for Numerical Methods in Engineering 19, 1593–1619 (1983)

A Reinforcement Learning Based Hybrid Evolutionary Algorithm for Ship Stability Design

Osman Turan and Hao Cui

Abstract. Over the past decades, various search and optimisation methods have been used for ship design – a dynamic and complicated process. While several advantages of using these methods have been demonstrated, one of the main limiting factors of optimisation applications in ship design is the high runtime requirement of the involved simulations. This severely restricts the number of real applications in this area. This chapter presents a hybrid evolutionary algorithm that uses reinforcement learning to guide the search. Through giving and correcting the search direction, the runtime of optimisation can be effectively reduced. The NSGA-II, a well known multi-objective evolutionary algorithm, is utilised together with reinforcement learning to form the hybrid approach. As an important optimisation application field, the ship stability design problem has been selected for evaluating the performance of this new method. A Ropax (roll on/roll off passenger ship) damage stability problem is selected as a case study to demonstrate the effectiveness of the proposed approach.

1 Introduction

Ship design is an important optimisation application field in engineering. Especially in recent years, more and more optimisation methods have been introduced for ship design. However, due to its complex characteristics, most of the optimisation applications lack pertinence and are very time-consuming. As ship design is usually a large-scale and complex problem, naval architects often have to utilise professional naval architecture software suites to perform

Osman Turan · Hao Cui
Group of Marine Design, Operations and Human Factors,
Department of Naval Architecture and Marine Engineering,
University of Strathclyde, UK
e-mail: {o.turan,hao.cui}@strath.ac.uk

simulation and evaluation in order to solve practical problems. These design activities could take up a lot of time, which may make the optimisation runs consume several weeks or months.

Evolutionary Algorithms (EAs) have attracted much attention in ship design, but to obtain good results with them, a large population size and a high number generations are very important [1]. However, the fitness evaluations in this area usually depend on professional naval architecture software which takes plenty of time to run – this is a challenging issue that requires attention in practical optimisation.

Therefore, the questions of how to construct a practical framework for optimisation and how to select an efficient optimisation approach to reduce the time cost are extremely important. One possible answer is our combined reinforcement learning method, which can provide appropriate guidance during the search process.

Subdivision and damage stability are very important for the whole ship design. Over the last two decades, stability standards have increased significantly, which improves safety and prevents large scale maritime losses such as the Herald of Free Enterprise, Estonia, and Al Salem. The regulations and calculation methods on ship stability have been improved continuously by the International Maritime Organization (IMO), classification societies, and flag nations, resulting in stricter stability standards. The ship owners and designers have been seeking for the best compromise between ship safety and economic benefits, which is a typical multi-objective problem.

The IMO is a specialized agency of the United Nations devoted to maritime affairs. The Maritime Safety Committee (MSC) is the committee of the IMO concerned with maritime safety, and the MSC develops and implements marine safety standards in the form of a set of rules and regulations. Currently, a passenger ship has to satisfy the requirements of the related chapter (Chapter II-1) of SOLAS 2009 [2] on damage stability probabilistic regulations. The International Convention for the Safety of Life at Sea (SOLAS) is one of the most important international treaties protecting the safety of merchant ships. For the currently applied SOLAS 2009, the calculation of stability is complex and manual verification has become virtually impossible. Therefore the utilisation of optimisation methods to solve the subdivision and stability problem is important for practical ship design.

This chapter presents an efficient framework for design optimisation with regards to stability and subdivision via a new hybrid EA, which utilises a reinforcement learning approach. Q-learning, one of most effective reinforcement learning approaches, is selected to integrate with the NSGA-II, a popular Multi-Objective Evolutionary Algorithm (MOEA), to provide better searching ability and to reduce the runtime. A comparison between the original NSGA-II and the new improved method on real-world ROPAX subdivision design is presented to evaluate the performance of the proposed approach.

2 The Problem

2.1 Problem Definition

This chapter focuses on improving the performance of Ropax vessels in terms of not only maximizing the ship-related parameters but also reducing the time to perform the multi-objective design iteration. Ropax ships are vessels designed for freight transport, to carry vehicles, and passengers. As this study is based on the previous work of Olcer et al. [3], Turkmen et al. [4] and Cui et al. [5], the same problem is selected and used. The optimisation problem is an internal hull subdivision optimisation for a Ropax vessel and the main aim is to maximize the survivability as well as to improve the cargo capacity, which is an important indicator of economic performance. The optimisation problem has three objectives and 16 design variables.

The first objective is the attained subdivision index A. The index A is the main criterion of ship stability performance in SOLAS 2009 and is composed of the partial indices A_s, A_p, and A_l, which correspond to the subdivision, intermediate, and light draughts respectively. A_s is the partial index calculated according to the deepest subdivision draught (d_s) and d_s is the waterline which corresponds to the summer load line draught of the ship. A_l is the partial index and corresponds to the light service draught (d_l). A_p is calculated under the partial subdivision draught (d_p), which is the light service draught plus 60% of the difference between the light service draught and the deepest subdivision draught. The calculation formula of index A is defined in Equation 1.

$$A = 0.4A_s + 0.4A_p + 0.2A_l \tag{1}$$

Index R is the required index of subdivision defined in [2].

$$R = 1 - \frac{5000}{L_s + 2.5N + 15225} \tag{2}$$

where: $N = N_1 + 2N_2$. Here,
N_1: number of persons for whom lifeboats are provided.
N_2: number of persons (including officers and crew) the ship is permitted to carry in excess of N_1.

L_s is the subdivision length of the ship and it is defined in SOLAS 2009 as the greatest projected moulded length of that part of the ship, at or below deck or decks, limiting the vertical extent of flooding with the ship at the deepest subdivision draught. The numbers in Equation 2 are defined by SOLAS 2009.

The subdivision of a ship is considered sufficient if the attained subdivision index A is greater than the required subdivision index R [2]. In addition, for passenger ships, the partial indices A_s, A_p, and A_l should not be less than

0.9R. In this study, the simulation and calculation of index A are processed via the third party software NAPA [6].

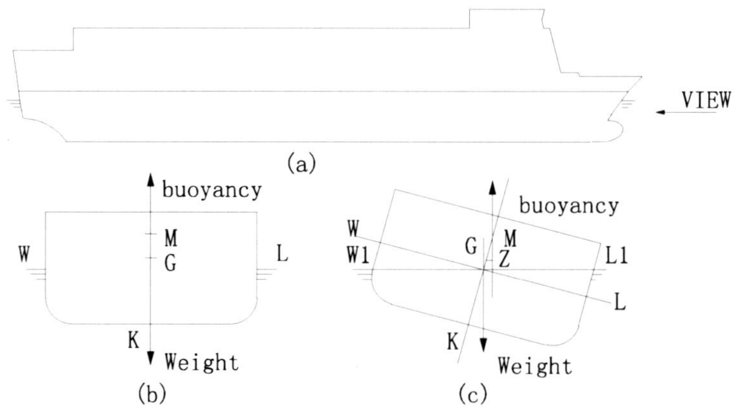

Fig. 1. Definition of KG, GM, and GZ in ship design.

The second objective is *limitingKG*. The parameter KG in ship design stands for the vertical centre of gravity for a ship. The *limitingKG* is the maximum value of KG which satisfies the requirement of the stability criteria. The *limitingKG* is used to check whether the ship is safe with certain loading conditions. In SOLAS 2009, it is also defined that the ship's master shall be provided with clear intact and damaged stability requirements which normally are achieved by determining limiting GM/KG curves, containing the admissible stability values for the draught and trim range to be covered. In ship stability design, the GM should equal or exceed the minimum required GM which ensures the ship has safe stability and the GZ is used to measure the stability at any particular angle of heel. The optimisation of *limitingKG* is seeking the maximum value. The calculation of *limitingKG*, GM, and GZ are again evaluated via the NAPA software. In Figure 1 (b), KG is the distance between K and G, in where, K usually means kneel and G is defined as the gravity centre of the ship. GM is the distance between G and M when M is the metacentre of the ship. GZ (shown in Figure 1 (c)) is known as the righting arm which is defined as the horizontal distance between the centre of buoyancy and the centre of gravity.

The last objective is the cargo capacity of the Ropax ships as the ship owners always want to maximise the cargo capacity while satisfying the stability requirements. When the main dimensions of the vessel are fixed, the designers always try to increase the number and length of the cargo lanes as much as possible to improve the cargo capacity. This objective is directly measured from the model in NAPA.

There are 16 design variables in this work. Three of them are shown in Figure 2: the depth of the ship to the car deck "Car Deck Height" which is the height from the baseline of ship to the car deck, the "Lower Hold Height", and the width of the side casing at the car deck, "Side Casing Width". The "Lower Hold Height" (shown in Figure 2) is usually used to extend the storing space in the ship. Here, the original design has a 9.7m depth to the car deck, a 2.6m lower deck height and no side casings at the car deck. The remaining 13 design parameters (as shown in Figure 3) are the locations of transverse bulkheads given in the format of frame numbers (starting from the stern of the ship).

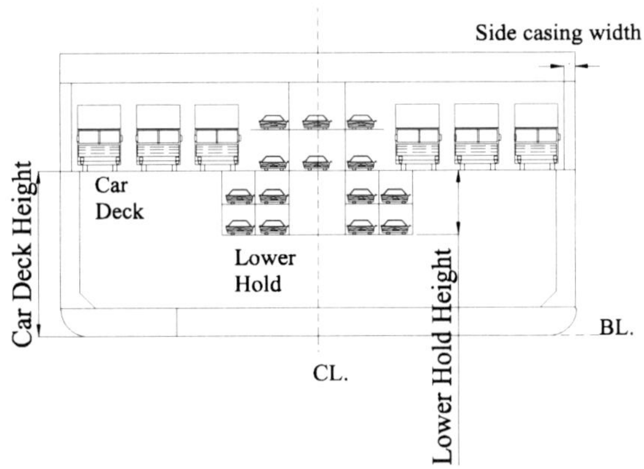

Fig. 2. Three design variables: Car Deck Height, Lower Hold Height and Side Casing Width.

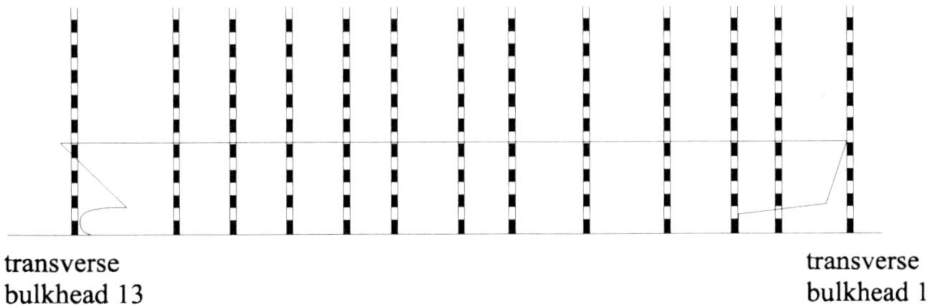

Fig. 3. The design variable of transverse bulkhead.

2.2 Motivation

As stability-based subdivision optimisation has become one of the critical design problems in ship design, naval architects have attempted to deploy different kind of approaches to deal with this problem. MOEAs have proven to be efficient in practical engineering applications and have gradually become a hot topic that attracts great attention. However, the runtime is still a restrictive factor for optimisation in ship stability. Following the introduction of SOLAS 2009, which is based on a probabilistic method, the stability calculations of passenger and cargo ships based on this probabilistic approach have become mandatory. This method makes a manual calculation impossible and so the computer-aided analysis becomes the only reasonable method. It does, however, require a lot of computing time. At the same time, MOEAs have provided limited improvement on reducing runtime. Combining MOEAs with a learning method from the area of artificial intelligence in order to speed up the optimisation is an effective way to improve their applicability in the ship design process.

3 Related Work

3.1 Background

History of Ship Design

Modern ship design, an era of complex concurrent engineering, begins with two significant events: the replacement of sail power with steam powered propulsion in the 1780s and William Froude's scientific approach to vessel performance prediction in the 1860s. Before that, ships were usually designed by simply developing a new ship based on an existing one. The new ship would be of the scale of the old designs with minimal alterations. The results were random because no scientific principles were applied.

In the middle of the twentieth century, computer technology brought a new leap in ship design. Computer-Aided Ship Design (CASD) has evolved since the 1960s with the synchronous development of Computer-Aided Design (CAD) [7]. To describe the ship design process, the design spiral was proposed by Professor J. Harvey Evans (shown in Figure 4) [8]. In this spiral, the ship design is viewed as a sequential iterative process (design spiral) and in every phase, ranging from concept design to detail design, every important aspect of the ship design is re-evaluated, starting from mission requirements to hydrostatics, powering, and cost estimates. In every cycle of the design spiral, the complexity increases and the number of possible designs decreases. In this design method, the computer became an effective tool to assist naval architects with their work on design. Buxton [9] proposed to embed cost estimates into the design spiral in 1972. Andrews [10, 11] proposed to include constraints into the ship design spiral concerning the three categories: design,

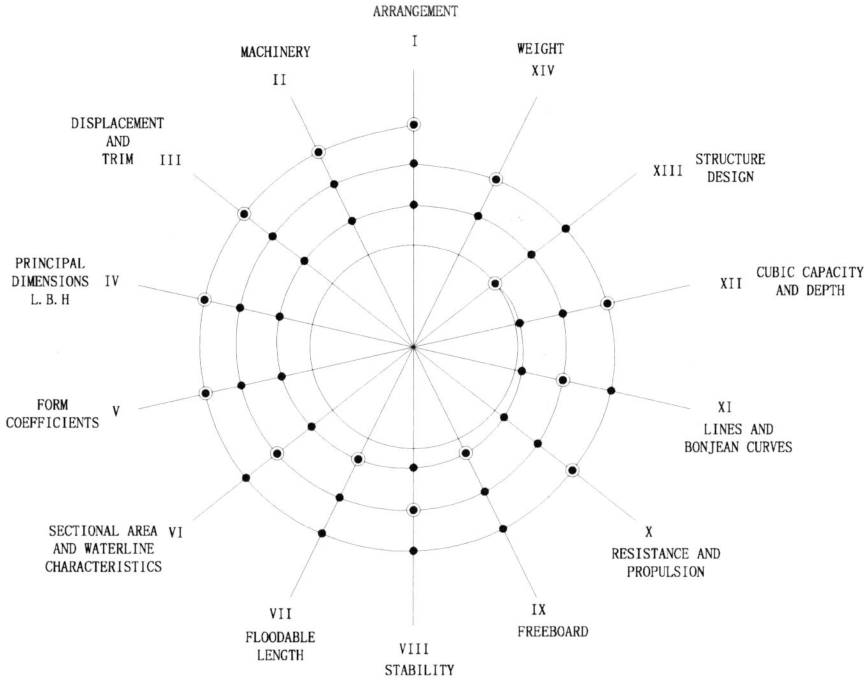

Fig. 4. The Ship Design Spiral.

design process, and design environment. This design spiral has been used until now.

In 1990, Mistree [12] proposed the decision-based design paradigm for the design of ships which encompasses systems thinking and embodies the concept of concurrent engineering design for the life cycle. In this new design method, the role of the computer would be changed from the tool to the partner of design work, and put the designer in the role of decision maker. Therefore the design procedure can be seen as an optimisation process and naval architects can decide the final designs from the optimisation solutions.

While the concept of probabilistic damage stability assessment of ships was introduced in the early 1960s, applications of risk-based approaches in the maritime industry were proposed only in 2009 [13]. This method supports a safety culture paradigm in the ship design process by treating safety as a design objective rather than a constraint. It is pointed out that the notion of "risk" is usually associated with undesirable events and shipping operations being undoubtedly "risky". So the ships should be designed with this as a basic principle and safety is the primary requirement of this method.

From the above introduction of the ship design history, it can be seen that computer-aided design and safety are the two most important trends of ship design.

There are various stages in modern ship design which may be referred to the different titles due to the varying design phases of different ships. At the same time the key points of design are different between navy ships and merchant ships, and this is also one of the reasons why there are different definitions of the design stages. The design stages adopted in this chapter are categorised as shown in Figure 5.

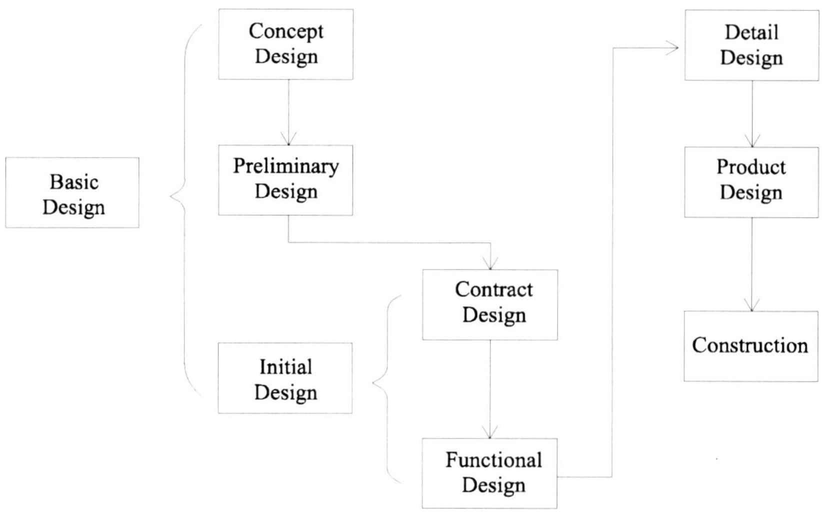

Fig. 5. Ship Design Phases.

Ship stability is one of the most important aspects of ship design and the stability design proposed in this chapter occurs at the stage of Basic Design in Figure 5. Especially for passenger ships including Ropax ships, the stability standards must be at a very high level. At the same time, the requirements of ship stability vary depending on the types of ships.

3.1.1 Development of Ship Damage Stability

Following the Titanic disaster in 1912, the SOLAS conference was organized and a series of safety regulations were developed. The first version of SOLAS was adopted in 1914, the second in 1929, the third in 1948, and the fourth in 1960. IMO, formerly known as the Inter-Governmental Maritime Consultative Organization (IMCO), was established in 1948, and then in 1982 the name of IMCO was changed to IMO.

The IMO performed a study on probabilistic damage stability regulations in 1960s and then published Resolution A.265 (VIII) which was adopted in 1973 [14]. The flooding and subsequent capsize of the roll on/roll off (RORO) passenger ferry The Herald of Free Enterprise in 1987 created a lot of attention for RORO ship design resulting in research and regulatory activities for the safety of RORO passenger ships. The SOLAS 90 and 92 amendments further improved the survivability characteristics of damaged passenger ships and cargo ships. SOLAS 2009 [2] was accepted as the probabilistic damage stability standard and introduced the index A (see Equation 1) as a measure of damage stability standards. As a result, the calculation of stability has become even more complex and naval architects have to perform the calculations using computers. Following the acceptance of the new stability assessment, the runtime to perform the calculations becomes a key factor that affects the optimisation of the ship design as the simulations and the calculations, not the optimisation, take a lot of time.

3.1.2 MOEAs

MOEAs are now used in Computer Aided ship Design (CAD) and Computer Aided Engineering (CAE). The recent trend in the ship industry is to build large-scale and high technology ships while the safety standards of ships are continually improved. At the same time, fast design and production become key issues to be addressed. Especially in ship stability-based hull subdivision design, with the increasing improvement of the standards of ship safety, the calculations become more and more complex, and the computational workload tends to become heavier when the ships get larger. Therefore, it is critical that a well developed methodology for this problem is found.

As an important sub-discipline of optimisation, the studying of MOEAs began in the 1960s, but the first complete algorithm, the Vector Evaluated Genetic Algorithm (VEGA) created by Schaffer, was not proposed until the mid-1980s [15]. Since then, MOEAs have remained an active research area, many MOEAs have been introduced, and more and more applications in engineering have been developed. The reasons for the fast development of MOEA are many, but one of the main reasons is that the MOEA can find the Pareto solutions [16] without being affected by the shape or continuity of the Pareto front. Hence a MOEA can find Pareto-optimal solutions even in complicated practical engineering situations.

Because of the specific requirements of different disciplines, there are several classification methods for MOEAs. Coello et al. [15] provided a very simple method for this purpose, based on the type of selection mechanism used by MOEAs. This chapter adopts this classification approach and divides existing MOEAs into three categories: Aggregating Function, Population-based Approaches, and Pareto-based Approaches.

Using an aggregating function means to directly combine all the objectives into a single function which is usually called an aggregating function.

This aggregating function can be linear or nonlinear. MOGA proposed by Fonseca and Fleming [17] can be seen as an example for this class of methods. Population-based approaches use the population of EAs to diversify the search.

Pareto-based approaches use Pareto optimality in the selection mechanism and are the most popular approaches in recent years. The Non-dominated Sorting Genetic Algorithm (NSGA) [18], NSGA-II [19], the Niched Pareto Genetic Algorithm (NPGA) [20], the Strength Pareto Evolutionary Algorithm (SPEA) [21], SPEA2 [22], the Pareto Archived Evolution Strategy (PAES) [23] all belong to this category.

3.2 Optimisation in the Subdivision of Ship Stability

In response to the new SOLAS 2009 regulations, the shipping industry is in search of new modern designs to match these high safety standards while maximizing the cargo capacity of vehicles in a cost effective approach. The changes in design focus on the damage stability and survivability, cargo, and passenger capacity. Therefore, the internal hull subdivision layout is an important problem especially for damage stability, survivability, internal cargo capacity, and the general arrangement of the vessel. Different solution methods have been proposed in the previous work. Sen and Gerick [24] suggested using a knowledge-based expert system for subdivision design using the probabilistic regulations for passenger ships. Zaraphonitis et al. [25] proposed an approach for the optimisation of RoRo ships in which centre-casing, side-casing, bulkhead deck height, and locations of transverse bulkheads are treated as optimisation variables. Olcer et al. [3] studied the subdivision arrangement problem of a Ropax vessel and evaluated conflicting designs in a totally crisp environment where all the parameters are deterministic. They also examined the same case study in a fuzzy multiple attributive group decision-making environment, where multiple experts were involved and available assessments were imprecise and deterministic. Turan et al. [26] studied the subdivision problem by the case-based reasoning approach. Turkmen et al. [4] proposed NSGA-II with TOPSIS (Technique for Order Preference by Similarity to Ideal Solution) to perform design optimisation for internal subdivision. Cui et al. [5] introduced a multi-PSO (particle swarm optimisation) for the same problem.

4 The Approach

4.1 Mathematical Description of Pareto-style Multi-objective Optimization

The goal of solving a general multi-objective optimisation problem (also called multiple criteria optimisation, multi-performance, or vector optimisation) is to find the design variable settings that optimize a vector objective function ($F(Y) = f_1, \ldots, f_t$) over the feasible design space [27]. The objective functions are the quantities that the designer wishes to minimize, maximize, or attain a certain value for. This problem can be formulated as follows:

$$\text{Minimize } F(Y) = \{f_1(Y), f_2(Y), \ldots, f_t(Y)\} \tag{3}$$

Subject to:
p inequality constraints $g_\delta(Y) \geq 0, \delta = 1, \ldots, p$
q equality constrains $h_\Phi(Y) = 0, \Phi = 1, \ldots, q$
where $Y = [y_1, y_2, \ldots, y_n]$ is the vector of decision variables.

In multi-objective optimisation, the objectives are usually in conflict with each other. The aim of multi-objective optimisation is to find a compromise solution which is acceptable for the decision makers. Design variables are the numerical quantities for which values are to be chosen in an optimisation problem. In most engineering applications, the design variables can be controlled by designers according to actual problems. Design variables usually have maximum and minimum boundaries which can be treated as separate restrictions. These restrictions are from the environment or resources (e.g., physical limitations, time restrictions, etc.), which must be satisfied by an acceptable solution in an optimisation problem. These restrictions are generally called constraints.

In multi-objective optimisation, the aim is not to only find a single globally optimal solution but to discover good compromises (or "tradeoffs"). Therefore, Pareto Optimality is introduced. For a multi-objective optimisation problem, any two solutions y_1 and y_2 can have one of two relations: one dominates the other or none dominates the other. In a minimization problem, without loss of generality, a solution y_1 dominates y_2 if the following two conditions are satisfied:

$$\forall \gamma \in \{1, 2, \ldots, t\} : f_\gamma(y_1) \leq f_\gamma(y_2) \tag{4}$$

$$\forall \lambda \in \{1, 2, \ldots, t\} : f_\lambda(y_1) < f_\lambda(y_2) \tag{5}$$

If any of the above conditions are violated, the solution y_1 does not dominate the solution y_2. If y_2 is not dominated by solution y_1, y_2 is called the non-dominated solution. The solutions that are non-dominated within the entire search space are denoted as Pareto-optimal and constitute the Pareto-optimal set or Pareto-optimal frontier [27].

4.2 Optimization Approach

To reduce the runtime in the application of stability optimisation, one of the most efficient methods is to find the hidden relationships among data via real-time learning during optimisation and to use these relationships to guide the search for optimisation. Reinforcement learning, as one of the most important machine learning approaches, is introduced here to assist the optimisation algorithm to find these relationships [28]. Reinforcement learning solves the problem via an agent reaction mechanism, which learns the behaviour through trial-and-error interactions and delayed rewards in a dynamic environment. In other words, the reinforcement learning tries to find a mapping from environment to action via reward. Sutton and Barto provide a detailed explanation of reinforcement learning in [29]. Here, a brief introduction is given to provide basic knowledge about reinforcement learning.

In a typical reinforcement learning model, an agent is connected to the environment via perception and action. In the model shown in Figure 6, B is an agent and T is the environment. In the first step, agent B receives an input i. In the second step, agent B chooses an action a to generate an output. This action a changes the environment T and in the third step, the value of this state transition is communicated to the agent B through a scalar reinforcement signal, r. The agent's behaviour, B, should choose actions that tend to increase the long-term sum of the values of reinforcement signal. It can learn to do this over time by systematic trial and error, guided by a wide variety of algorithms.

From the model in Figure 6, it can be seen that there are four basic modules in reinforcement learning: a policy, a reward function, a value function, and, optionally, a model of the environment [29]. The policy represents the mapping from environment to action. This module is the core of reinforcement learning and there are many ways to implement a policy, for example as a lookup table, as functions, or as an artificial neural network. The second module is the reward function which is the goal of reinforcement learning. The aim of an agent in a reinforcement learning scenario is to maximize the total reward in the lifetime run. The third factor, the function value, is a

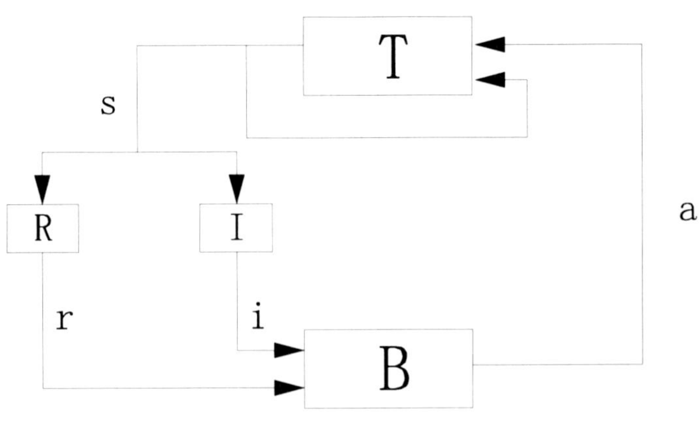

Fig. 6. The standard reinforcement learning model.

formula to synthesise all the rewards that an agent expects to get over the future.

Q-learning, as an efficient approach for reinforcement learning, is selected to assist the optimisation algorithm to find hidden relationships. Q-learning, developed by Watkins [30, 31], works via learning an action-value function that gives the expected utility of taking a given action in a given state and following a fixed policy thereafter. Q-learning is a form of model-free reinforcement learning, which means that it can compare the expected utility of the available actions without requiring a special environment. The requirements of Q-learning for the environment are flexible. However, this does not mean Q-learning is applicable for any situation. Compared to a continuous environment, a discrete environment is more suitable for the current version of Q-learning. Luckily, a discrete and finite engineering environment is one of the characteristics of ship stability design optimisation. The combination of Q-learning and ship stability design optimisation thus is reasonable.

In practice, it is computationally impossible to find the necessary integrals without additional knowledge or some modification. Q-learning solves the problem by taking the maximum value over a set of integrals. Rather than finding a mapping from states to state values (as in value iteration), Q-learning finds a mapping from state/action pairs to values (called Q-values). Instead of having an associated value function, Q-learning makes use of the Q-function. In each state, there is a Q-value associated with each action. The Q-value is defined as the sum of the (possibly discounted) reinforcements received when performing the associated action and then following the given policy thereafter. Likewise, the definition of an optimal Q-value is the sum of the reinforcements received when performing the associated action and then following the optimal policy thereafter. Equation 6 is a general expression for this situation.

$$Q(s_t, a_t) \leftarrow Q(s_t, a_t) + \alpha[r_{t+1} + \gamma \max_a Q(s_{t+1}, a_{t+1}) - Q(s_t, a_t)] \quad (6)$$

According to Equation 6, Q-learning differs from other value-iterating reinforcement learning methods in that it displays the relationship of given actions and expected values of the successor states. It does not require that each action is performed in a given state and the expected values of the successor states are calculated.

In the initial phase of to the Q-learning process (before the learning has started), Q returns a constant value chosen by the designer. In the following iterations, the agent is given a reward every time the state has changed and the Q value will be changed according to this reward. New values are calculated for each combination of a state 's' from the statement set 'S' and action 'a' from the action set 'A'. It assumes the old value and makes a correction based on the new information as shown in Equation 6. Equation 6 is equivalent to:

$$Q(s_t, a_t) \leftarrow Q(s_t, a_t)(1 - \alpha) + \alpha[r_{t+1} + \gamma \max_a Q(s_{t+1}, a)] \tag{7}$$

In this study, we use Equation 7.

The NSGA-II has been proposed by Deb et al. [19] as a fast and elitist multi-objective algorithm. This algorithm uses the crowding distance parameter to keep the population diverse and also defines the constraint-dominance principle in order to work better with constrained optimisation problems. It has been employed to solve the ship stability design problem in the past [4].

The NSGA-II can be divided into three main parts: non-dominated sorting, crowding distance assignment, and offspring selection with respect to fitness and crowding distance. In the sorting procedure, non-dominated individuals are found in the population and are assigned to the rank zero. Then the algorithm begins to seek the next front and so on. The crowding distance assignment is used to penalize the individuals that are close to each other.

The NSGA-II combines the parent and child populations for every generation. The child population selection in the NSGA-II can use roulette wheel selection, historically the default selection method used in GAs [16]. Here, a Q-learning operation is introduced to improve the child population selection.

According to the Q-learning theory, the child population can be generated by forecasting through learning from a known value. Since the changing steps are discrete, the hybrid algorithm will calculate the Q-value from every parent and utilize the previously calculated fitness to select the child population. As shown in Figure 7a, the evolutionary method adopted in the NSGA-II generates the offspring via the crossover and mutation operations. Two parents, *parent*1 and *parent*2 in Figure 7a, are randomly selected from the parent group. Then the crossover operation is applied to these two parents and two new individuals are obtained as the two children. In the next step, these new individuals will undergo mutation according to the mutation

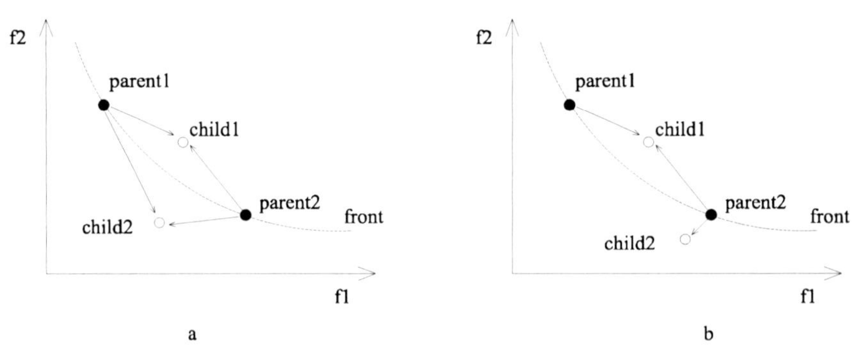

Fig. 7. The methods of selecting offspring.

rate [16]. In the last step, the two children, *child*1 and *child*2 form a new parent group with *parent*1 and *parent*2 in the NSGA-II. Figure 7b displays the new hybrid method which is generating offspring via Q-learning. The parents are randomly selected from the parent group. The crossover and mutation operation are applied as in the traditional EA, but this time, only one child is selected as offspring from *child*1 and *child*2. Here we assume that *child*1 is selected and *child*2 will be recalculated from Q-learning approach.

Figure 8 illustrates the method of direction selection on a parent point in the new hybrid algorithm. The hybrid algorithm will calculate the Q-value around *parent*1 and *parent*2. For sampling the process of calculation, the calculating range is usually defined by the user. Then the algorithm compares all calculated Q-values. The child which has the biggest Q-value will be selected as *child*2 in the next generation. The algorithm uses the Q-value calculation to replace random selection and a look-up table is used to calculate the Q-value.

If the Q-learning function in this new algorithm is studied, it can be seen that in the offspring selection period in the NSGA-II, Q-learning can forecast the moving direction of the parents. In other words, Q-learning can select the maximum expected utility of the moving direction of the parent point. When selecting offspring, Q-learning will first calculate the expected utility of all the possible moving directions of the parent point. Then the algorithm will compare these values and select the maximum expected utility. Finally, the algorithm will select the corresponding moving direction as a real action to determine the child point instead of random selection. It is noteworthy that this new algorithm is suitable for special optimisation problems which have two characteristic features.

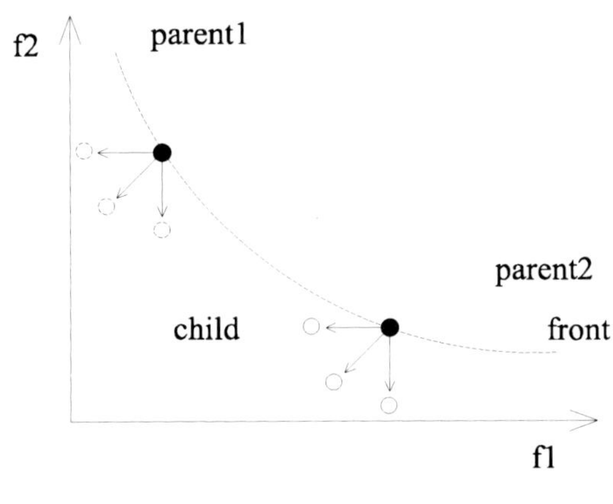

Fig. 8. The detailed method of selecting offspring in new hybrid algorithm.

One is that the ranges of design variables and searching step lengths are discrete and finite. For example, in the problem specified in Section 2, the locations of transverse bulkheads are limited in several integral states, which is the restriction of real engineering. At the same time, the changes that can be applied to the design variables are also limited to integer values.

The other feature is that the time cost of optimisation is smaller than the simulation necessary for the engineering process. For instance, in stability design, the time required for every simulation via the NAPA software needed for the fitness evaluation will be 4 to 8 minutes, which is far larger than the time requirement for the optimisation operations. Thus, the new hybrid algorithm will improve the time cost of optimisation by greatly reducing the simulation time which, in turn, will reduce the overall time.

4.3 Integration Optimization System

The language Java is used to implement the optimisation system based on a multi-agent structure and the calculation is processed between the optimisation system and the NAPA software. Visual Basic (VB) is used to form the interface between them. The optimisation system uses NAPA to calculate the Index A, *limitingKG*, and Cargo Capacity and after optimisation, the modified design variables are transferred to NAPA. The vessel is modelled in NAPA and can be modified for each design experiment (or design layout) with respect to each optimisation parameter via the NAPA macro language. For each design experiment, the relevant adjustments of draught and displacements are made during the optimisation process. The whole process is shown in Figure 9. *Notepad* is used as a media to transfer the information between the optimisation program and the NAPA software.

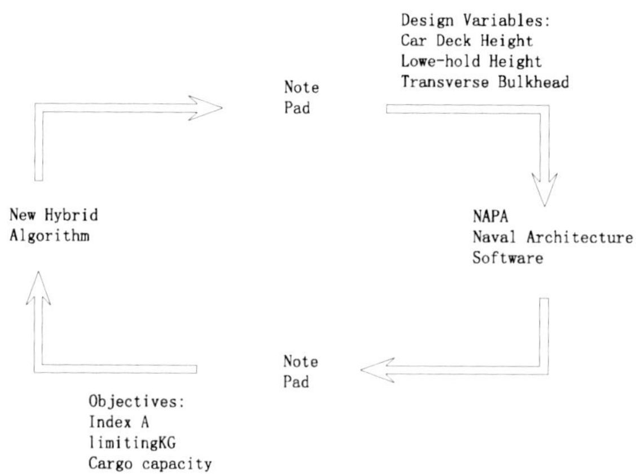

Fig. 9. The integration optimization system.

5 Results of the Case Study

5.1 Case Study Setting

The main dimensions of the Ropax ship used as basis for this case stury are listed in Table 1. Table 2 presents the design variables together with the lower and upper boundaries of the parameters with assigned increments to be used in the optimisation study.

Table 1. Main dimensions of the vessel in case study.

Length Overall(L_{oa})	194.4 m
Length between perpendiculars (L_{bp})	172.2 m
Breadth molded (B)	28.4 m
Depth to car deck	9.7 m
Lower Hold Height (m)	2.6 m
Depth to upper deck	15.0 m
Draught design (T)	6.6 m
Displacement	20200 ton
Max number of persons on board	2660
Number of car lanes	8

Table 2. Optimisation variables with their types, bounds, and objectives.

No	Variables	Original design	Type	Bounds		
			Discrete	Lower	Upper	Increment
1	Car deck height	9.7 m	√	9.6m	9.9m	0.025m
2	Lower-hold height	2.6m	√	2.6m	5.2m	2.6
3	Side casing width	No	√	1m	2m	0.5m
4	Transverse Bulkhead 02	27	√	25	29	1
5	Transverse Bulkhead 03	39	√	37	41	1
6	Transverse Bulkhead 04	51	√	49	53	1
7	Transverse Bulkhead 05	63	√	61	65	1
8	Transverse Bulkhead 06	81	√	79	83	1
9	Transverse Bulkhead 07	99	√	97	101	1
10	Transverse Bulkhead 08	117	√	115	119	1
11	Transverse Bulkhead 09	129	√	127	131	1
12	Transverse Bulkhead 10	141	√	139	143	1
13	Transverse Bulkhead 11	153	√	151	155	1
14	Transverse Bulkhead 12	165	√	163	167	1
15	Transverse Bulkhead 13	177	√	175	179	1
16	Transverse Bulkhead 14	189	√	187	191	1
Boundaries for transverse bulkheads are given in frame numbers						

In Table 2, there are sixteen design variables in this real-world case and all of them are discrete attributes which provide a perfect environment for Q-learning. The graphical illustrations of the first, second and third design variable are presented in Figure 2. The "car deck height" is limited between 9.6m and 9.9m with the increment of 0.025m as the original design is 9.7m. The performance of the ship (including sea keeping, stability, structural strength etc.) is very sensitive to the height of the car deck and therefore, these design variables have limited boundaries.

The second design variable "Lower-hold height" is an important parameter as it will determine the number of cars that can be placed in the lower hold. The lower-hold height, which is 2.6m in the existing design, limits the height of the cars that can be stored in this space. Due to the size of the lower hold, trailers cannot be stored in this space. In order to increase the car capacity, the height of the lower hold can be increased by 2.6m. In ship construction, an excessive size of lower hold is impractical due to the constructional and loading/unloading constraints. Therefore, the boundary for the lower hold height is set to 5.2m.

Increasing the third design variable, "side case", is beneficial to the stability in damaged conditions but reduces the cargo capacity. In other words, a bigger side case will decrease the total car lanes and therefore the boundaries are selected between 1m and 2m. The freedom of this design variable is very small because of the limitation imposed by the practical production requirements. The increment for side casing is set as 0.5m for the construction convenience and structural feasibility. All the design variables numbered between 4 and 16 are presented in Figure 3. The values of these design variables are frame numbers. As the frame numbers are integer numbers, the increments have to be integer numbers too, and in this situation the increment is set to 1. The NSGA-II uses the parameters setting of prior research in the same environment as shown in Table 3 [5].

Table 3. The parameters setting of NSGA-II.

Parameters Name	Parameters Value
SBX (Simulated binary crossover)	10
Polynomial mutation	20
Crossover probabilities	0.9
Mutation probabilities	0.1
Population	30
Generation	100

5.2 *Experimental Results*

All the experiments are processed in the same computer environment and run several times. Since we set a population size of 30 and a maximum generation number of 100 for our EA, it can at most explore 3000 different designs. With the NSGA-II, 2517 different designs were obtained in the final design space with 722 of them being infeasible designs resulting in 1795 (= 2517 − 722) feasible designs. The hybrid algorithm discovered 2326 different designs, including 603 infeasible ones, resulting in 1723 feasible designs. "different" in this context means that each design is unique. During the optimization process, designs usually are discovered more than once. The duplicate results are insignificant and should be deleted in the engineering application. The results of NSGA-II and of the new hybrid algorithm are presented in Table 4, while the convergence generation and runtime of both methods are listed in Table 5. In this study, in order to better judge the algorithm performance,

Table 4. Comparison between the new hybrid design and NSGA-II design.

No	Decision Variables	Original Design Design	NSGA-II-based Design	Hybrid-based Design
1	Car deck height	9.7 m	9.9m	9.9m
2	Lower-hold height (from car deck)	2.6m	5.2m	5.2m
3	Side Casing width(m)	No side-casing	1m	1m
	Watertight transverse bulkheads (In frame numbers)			
4	Transverse Bulkhead 02	27	27	28
5	Transverse Bulkhead 03	39	40	40
6	Transverse Bulkhead 04	51	52	52
7	Transverse Bulkhead 05	63	64	65
8	Transverse Bulkhead 06	81	82	82
9	Transverse Bulkhead 07	99	100	101
10	Transverse Bulkhead 08	117	118	118
11	Transverse Bulkhead 09	129	129	130
12	Transverse Bulkhead 10	141	142	143
13	Transverse Bulkhead 11	153	154	155
14	Transverse Bulkhead 12	165	165	166
15	Transverse Bulkhead 13	177	178	179
16	Transverse Bulkhead 14	189	189	190
Optimization Objectives				
1	Index A (m)	0.8028	0.8695	0.8664
2	limitingKG value (m)	14.012	14.8089	14.8762
3	Cargo capacity value (lines)	8	14	14
runtime (hours)			198	196

Table 5. The convergence generation and time cost.

Approach	Convergence Generation	Time (Hours)
NSGA-II	45-48	88-97
Hybrid Method	32-35	64-80

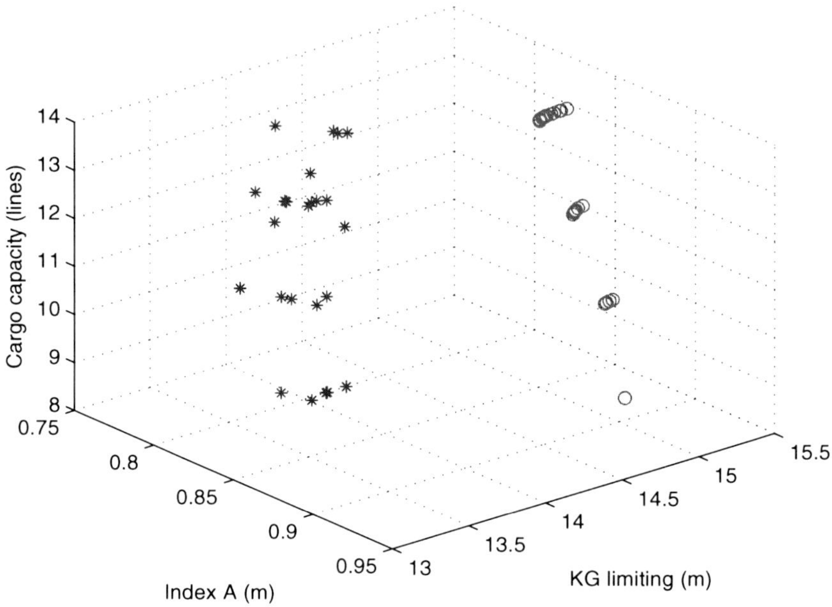

Fig. 10. Visualizations of the obtained Pareto frontiers.

the 'convergence generation' measure is used. During the data analysis after optimization, we check the Pareto line of every generation. If the Pareto line cannot move forward within five generations, we define the beginning generation of this phenomenon as convergence generation.

6 Comparison of the Results

In the solution presented in Table 4, compared to the original design, the index A using the hybrid algorithm is increased by almost 0.06 (7.5%) while the number of car lanes is increased from 8 to 14 (this is mainly due to the increase in the height of the lower hold). This means the final optimum design successfully improves the cargo capacity while the safety index A is increased. The bigger index A has many advantages in the concept design stage as well as during the ship's operational life. The index A may be decreased during further design stages and construction because of the possible deviations from the original estimations/plans (changes to compartment definitions, general arrangement and using material with different thicknesses than in the initial plan). In order to improve the robustness of the design and to provide a bigger stability margin, the designers rightly may seek a larger index A.

The *limitingKG* value is also increased significantly and provides flexibility for future modifications on the basis of changing passenger demands as well as machinery requirements without compromising the safety. When the NSGA-II and the new hybrid algorithm are compared, they provide very similar results for the final design. However, when the convergence generation and the runtime of both the NSGA-II and the hybrid algorithm are compared, it can be clearly seen that the convergence generation of the hybrid algorithm is reached 50% faster than the original NSGA-II, which means that the searching ability and speed of the hybrid algorithm is significantly better. In Figure 10, the "*" points are initial values and the "o" points are the obtained Pareto frontiers.

7 Conclusions and Future Work

In this chapter, we have presented a new hybrid algorithm that combines machine learning and the NSGA-II. The proposed hybrid approach was tested and validated successfully, utilizing the most widely used real-world task in ship design. In this study, internal subdivision optimisation of a Ropax vessel where there is a conflict between survivability, damage stability and cargo capacity was carried out. The proposed algorithm provides reasonably good results in terms of design objectives. In the study, the final solution is an improvement from the original design in every chosen objective with a significant margin, which clearly demonstrates the value of this system. The proposed algorithm is structured via a multi-agent system and every agent works remarkably well. It can be concluded that the proposed system has shown great potential and can be applied to similar and even more complex optimisation problems in ship design, as well as to related areas within the maritime industry.

Future work will focus on two aspects. The first being the selection principle and a sensitivity analysis of the parameters. In the current study, the parameters of Q-learning and NSGA-II have been selected according to the default setting for numerical calculations. The special character of the engineering application should be considered to improve the performance of this new hybrid algorithm. The comparisons of different parameters in terms of convergence rate are necessary. Moreover, a sensitivity analysis of parameters will be helpful in order to improve the performance of the algorithm.

The second aspect is to take into account a range of different applications in this area. The presented algorithm can be considered as very efficient for general ship design problems. However, we would also like to test its utility on problems that have features which are different from this kind of task. Applying this new algorithm to different ship design problems is useful to check the applicability of this algorithm.

References

1. Yang, X.: Engineering Optimization: An Introduction with Metaheuristic Applications. Wiley- Blackwell (2010)
2. International Maritime Orgnization (IMO). International convention for the safety of life at sea (SOLAS), 2009 amendment (June 2009)
3. Ölçer, A., Tuzcu, C., Turan, O.: Internal hull subdivision optimisation of Ro-Ro vessels in multiple criteria decision making environment. In: Proceedings of the 8th International Marine Design Conference, vol. 1, pp. 339–351 (2003)
4. Turkmen, B.S., Turan, O.: A new integrated multi-objective optimisation algorithm and its application to ship design. Ship and Offshore Structures 2, 21–37 (2007)
5. Cui, H., Turan, O.: An Improved PSO Approach in a Multi Criteria Decision Making Environment. Ship Technology Research 56(1), 14–21 (2009)
6. NAPA. NAPA Release (2009), `http://www.napa.fi/`
7. Nowacki, H.: Five decades of Computer-Aided Ship Design. Computer Aided Design (2009); doi:10.1016/j.cad.2009.07.006
8. Evans, J.H.: Basic design concepts. American Society of naval Engineers Journal, 671–678 (1959)
9. Buxton, I.L.: Engineering economics applied to ship design. Transactions of RINA 114, 409–428 (1972)
10. Andrews, D.: Creative ship design. Transactions of RINA 123, 447–471 (1981)
11. Andrews, D.: A comprehensive methodology for the design of ships (and other complex systems). Royal Society of London Proceedings 454, 447–471 (1988)
12. Mistree, F., Smith, W.F., Bras, B.A., Allen, J.K., Muster, D.: Decision-based design: a contemporary paradigm for ship design. Transactions of Society of Naval Architects and Marine Engineers 98 (1990)
13. Papanikolaou, A. (ed.): Risk-based ship design methods, tools and applications. Springer, Heidelberg (2009)
14. International Maritime Orgnization (IMO). Resolutions of A.265 (VIII) (November 1973)
15. Coello, C.A.C., Lamont, G.B., van Veldhuizen, D.A. (eds.): Evolutionary Algorithms for Solving Multi-Objective Problems, 2nd edn. Springer Science + Business Media (2007)
16. Weise, T.: Global optimisation algorithms: Theory and application, `http://www.it-weise.de/`
17. Fonseca, C.M., Fleming, P.J.: Genetic Algorithms for Multiobjective optimization: Formulation, Discussion and Generalization. In: Proceedings of the Fifth International Conference on Genetic Algorithms, pp. 416–423. Morgan Kauffman, St. Monteo (1993)
18. Srinivas, N., Deb, K.: Multiobjective Optimization Using Nondominated Sorting in Genetic Algorithms. Evolutionary Computation 2, 221–248 (1994)
19. Deb, K., Pratap, A., Agrawal, S., Meyarivian, T.: A fast and elitist multi-objective genetic algorithm: NSGA-II. IEEE Transactions on Evolutionary Computation 6(2), 182–197 (2002)
20. Horn, J., Nafpliotis, N., Goldberg, D.E.: A Niched Pareto Genetic Algorithm for Multiobjective Optimization. In: Proceedings of the First IEEE Conference on Evolutionary Computation, IEEE World Congress on Computational Intelligence, vol. 1, pp. 82–87 (1994)

21. Zitzler, E., Thiele, L.: Multiobjective Evolutionary Algorithms: A Comparative Case Study and the Strength Pareto Approach. IEEE Transactions on Evolutionary Computation 3(4), 257–271 (1999)
22. Zitzler, E., Laumanns, M., Thiele, L.: SPEA2: Improving the Strength Pareto Evolutionary Algorithm. In: EUROGEN 2001. Evolutionary Methods for Design, Optimization and Control with Applications to Industrial Problems, Athens, Greece, vol. 1, pp. 313–324 (2001)
23. Knowles, J.D., Corne, D.W.: M-PAES: A Memetic Algorithm for Multiobjective Optimization. In: 2000 Congress on Evolutionary Computation, Piscataway, New Jersey, vol. 1, pp. 325–332 (2000)
24. Sen, P., Gerigk, M.: Some aspects of a knowledge-based expert system for preliminary ship subdivision design for safety. In: The 6th International Symposium on Practical Design of Ships and Other Floating Sturctures PRADS 1992 , Newcastle upon Tyne, UK, vol. 2, pp. 1187–1197 (1992)
25. Zaraphonitis, G., Boulougouris, E., Papanikolaou, A.: An integrated optimization procedure for the design of Ro-Ro passenger ships of enhanced safety and efficiency. In: Proceedings of the 8th International Marine Design Conference , Athens, Greece, vol. 1, pp. 313–324 (2003)
26. Turan, O., Turkmen, B., Tuzcu, C.: Case-based reasoning approach to internal hull subdivision design. In: 4th International Conference on Advanced Engineering Design, Glasgow, UK (2004)
27. Ölçer, A.: A hybrid approach for multi-objective combinatorial optimisation problems in ship design and shipping. Computers & Operation Research 35(9), 2760–2775 (2008)
28. Kaelbling, L., Littman, M., Moore, A.: Reinforcement Learning: A Survey. Journal of Artificial Intelligence Research 4, 237–285 (1996)
29. Sutton, R.S., Barto, A.G. (eds.): Reinforcement Learning: An Introduction. MIT Press, Cambridge (1998)
30. Watkins, C.: Learning from delayed rewards. PhD thesis, University of Cambridge, UK (1989)
31. Watkins, C., Dayan, P.: Q-learning. Machine Learning 8(3-4), 279–292 (1992)

An Interactively Constrained Neuro-Evolution Approach for Behavior Control of Complex Robots

Christian Rempis and Frank Pasemann

Abstract. Behavior control of complex technical systems, such as robots, is a challenging problem. In this context, embodied neuro-control is a bio-inspired method for handling this type of problems, and evolutionary robotics has taken up some essential research topics in this field. However, for systems with many multi-modal sensor inputs and actuating outputs, new evolutionary methods have to be applied because the search spaces are high-dimensional and comprise many local optima. This becomes even harder when functional recurrent network *structures* cannot be given in advance and have to be evolved together with other parameters like synaptic weights and bias terms. This chapter describes a new evolutionary method, called *Interactively Constrained Neuro-Evolution* (ICONE), which restricts large search spaces by utilizing not only domain knowledge and user experience but also by applying constraints to the networks. The interactive use of this tool enables the experimenter to bias the solution space towards desired control approaches. The application of the ICONE method is demonstrated by evolving a walking behavior for a physical humanoid robot, for which a whole library of behaviors has been developed.

1 Introduction

Although a great deal of work has been done along this line of research, behavior control of complex technical systems remains a challenging problem. Apart from the classical artificial intelligence approaches, embodied neural control is one of the promising methods to further advance this field. On the

Christian Rempis · Frank Pasemann
Neurocybernetics Group, Institute of Cognitive Science,
University of Osnabrück
49069 Osnabrück, Germany
e-mail: {christian.rempis,frank.pasemann}@uni-osnabrueck.de

other hand, robots have served for a long time now as paradigmatic target systems for real-world applications and as versatile research platforms. In this context, evolutionary robotics [1, 2] takes up a manifold of different research problems associated with bio-inspired neural control techniques. To demonstrate the validity of this approach for the control of *complex* systems, one has to use evolutionary algorithms to generate neuro-controllers for robots with *many* sensors and actuators.

For relatively simple machines, evolutionary robotics has already proven to be able to generate interesting neural behavior control. Among the basic behaviors achieved for such robots were – depending on the provided sensors – all kinds of negative and positive tropisms. Often these behaviors relied on very simple neural controllers, for which specific relations concerning structure, dynamics and functions could be easily identified. Even more challenging problems could be addressed, for instance the goal-driven behavior and swarm behavior [3]. Although evolutionary robotics has already presented interesting capabilities demonstrating basic properties of embodied cognition, most of the targeted machines were quite simple when compared to the richness of sensors and actuators of animals.

Furthermore, truly autonomous systems will use internal needs for driving their behaviors; especially if they have to switch between different behaviors when interacting with their dynamically changing environments. Hence, they have to be equipped not only with exterior sensors, but also with proprioceptors which monitor their body states concerning body poses, energy consumption, temperature, and the like.

In using the artificial life approach to evolutionary robotics [4] to develop neural behavior control systems which come closer to the abilities of biological nervous systems, a new class of challenges emerges. This mainly originates from the fast growing search space when the robot complexity scales up by adding many new sensors and actuators. The chance of finding interesting, non-trivial neural controllers at a certain complexity level thereby often becomes unsatisfyingly low.

Therefore, at this point new evolutionary algorithms have to be investigated. It is suggested in this chapter that these new evolutionary algorithms have to make use of domain knowledge, such as insights in the organization of behaviors, local neural processing, symmetries, hierarchical feedback loops and other topological and hierarchical organizations of neural networks. Such algorithms may use neuro-modules that serve as hierarchical organization structures and functional building blocks. Using high-level neural building blocks as primitive elements instead of only single neurons allows larger, functionally interesting networks, while still ending up with a smaller search space. Also it is suggested that evolution is used rather as an assistant tool than as an autonomous problem solver, thus favoring user interactions and reactions to the progress of the evolution.

This chapter introduces a new evolutionary method called ***I**nteractively **Co**nstrained **N**euro-**E**volution* (ICONE) and demonstrates, as an example, how the method can be used to evolve neuro-controllers for a humanoid robot.

2 Evolving Large-Scale Recurrent Neural Networks for a Humanoid Robot

For the EU project ALEAR (Artificial Language Evolution on Autonomous Robots)[1], humanoid robots are used for so called *language games* [5]. In these experiments, humanoid robots interact with each other and develop a language, based on their actions and responses. Therefore the humanoid robots need a wide variety of behaviors and gestures, such as pointing, walking and hand waving.

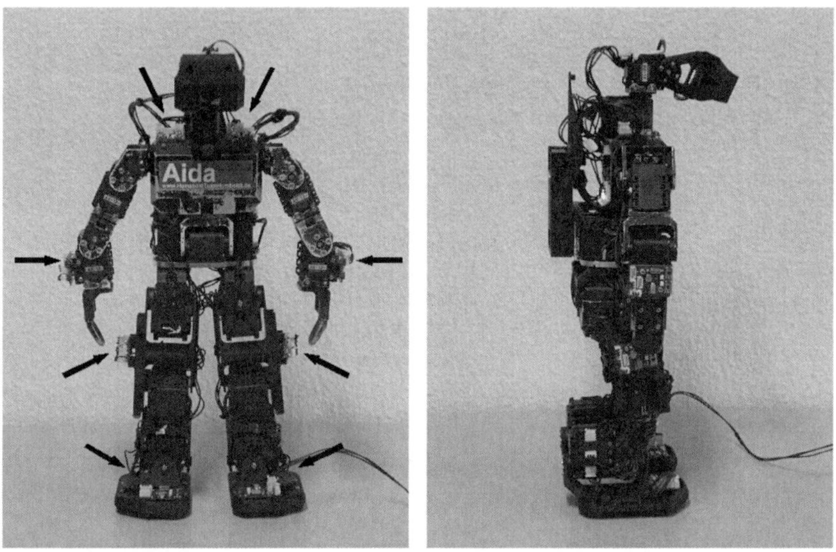

Fig. 1. The A-Series humanoid robot. The black arrows mark the locations of the *AccelBoards*.

The humanoid robots used during these language games are called A-Series robots [6] (Figure 1). They have been developed by the Neurorobotics Research Lab of Humboldt University Berlin based on the commercial robotics kit *Biolid*[2]. Each robot is about 45 cm high and equipped with 21 servo

[1] http://www.alear.eu
[2] http://www.robotis.com

motors, 21 angular sensors and 16 acceleration sensors. Eight so-called *AccelBoards*, small electronic boards responsible for the control of the motors, for reading the sensors and for calculating the activity of the artificial neural network controllers, are distributed over the body. The boards communicate over a synchronized bus with an update rate of 100 Hz. During each update step, each board knows the state of all motors and sensors of the robot. Because of the limited capacity of the *AccelBoards*, the neural controller networks have to be distributed over the available boards. Accordingly there is no central control, but rather a number of interacting autonomous networks. To allow the exchange of state information between the networks and thus a synaptic connection between the networks, each *AccelBoard* can communicate the state of up to four of its neurons to all other boards.

The control of such a robot is, because of its comparably large number of sensors and actuators, a challenging task. One common approach – as is used here – is to control the robots by artificial recurrent neural networks. This approach has the advantage, that controllers can be assembled from relatively simple, uniform processing elements. This facilitates the use of automatic network construction and optimization algorithms, including learning approaches and evolutionary algorithms [2, 7].

The A-Series robot requires already comparably large neural networks. Each motor of the A-Series robot is controlled by two motor neurons, one controlling the desired angular position, the other controlling the maximal motor torque during movements. Altogether the robot has 42 motor neurons, 37 sensor neurons and a number of hidden neurons for local processing, which make up a very large search space. Naive structure evolution approaches to construct networks by randomly inserting and altering synapses and bias values have no realistic chance to solve the targeted behavior problems in reasonable time.

Another problem is that evolution cannot practically take place on the physical robot. Accordingly a simulation of the robot has to be used. Because a perfect simulation of the physical robot is not feasible, neuro-controllers will adapt to the simulated system and hence may not work on the physical robot. Therefore it is important to evolve controllers to be robust with respect to the target platform, so that small differences between the platforms do not lead to an entire failure.

The remainder of this chapter describes the application of the evolutionary approach called the ICONE method, which allows the restriction of the search space of large networks based on modular structures, reuse of functional building blocks, domain knowledge and interactive control of the evolution process. This increases the chance of finding solutions and keeps evolved networks much more structured and comprehensible. With this method a number of behaviors for the A-Series have been realized and successfully transferred to the physical robot. Section 4 describes the method in detail. As an example, in Section 5 the application of the method is shown by evolving a walking behavior for the A-Series humanoid and by integrating it to the physical system. In the final sections 6 and 7, the results are discussed and an outlook for future work is given.

3 Related Work

The evolution of neuro-controllers for autonomous robots has a long tradition in the evolutionary robotics community [1, 8–10]. Embodied neural networks utilizing the sensors of a robot (sensori-motor loop [11]) can result in astonishing robust behaviors in noisy environments. Also, using neural networks for behavior control may give insights into the neural organization of biological organisms. However, the construction of such networks is – apart from simple behaviors like tropisms [12, 13] – difficult to design by analytical approaches. This requires algorithms and techniques to create and optimize neural networks in an efficient manner. Robot control tends to rely on *recurrent* neural networks, because behaviors are usually dependent on inner states and require the use of feedback loops in the network and through the environment. As a promising approach to create such networks, a number of different types of evolutionary algorithms have been developed in the last decades.

Evolutionary approaches for artificial neural networks can be roughly divided into *fixed topology* algorithms [14–18] and *structure construction* algorithms [19–30]. The former class of algorithms usually works with networks having a fixed number of fully interconnected neurons or with layers of neurons with a fixed interconnection pattern. One problem with fixed topology networks is that (1) the chosen number of neurons and their interconnection structure may not be appropriate and (2) that the search space is very large right from the beginning. Many structure construction algorithms on the other hand have the ability to start with a small network [31] and to add neurons and synapses to gradually increase the search space only when it seems necessary [32, 33]. These algorithms can also be used to shrink already well performing solutions to smaller, easier understandable network structures [32, 34].

The genome representation of neural networks differs widely between the different algorithms. As a main distinction two classes can be identified: *direct encoding* and *indirect encoding*. Direct encoding genomes directly specify the weights of all synapses of the network. In contrast, indirect encoding genomes specify parameters that are used to derive structure and synaptic weights. This may involve the application of grammars or rule sets to construct the network. Such algorithms are referred to as developmental algorithms [23, 28, 30, 35–37]. Developmental approaches have the advantage that the genome size is often much more compact compared to direct encoding if applied to large networks. This can speed up evolution, but usually also induces a strong bias to evolution, because often not all kinds of networks can be encoded with this type of algorithm. Some developmental algorithms, like simulated growing axons [38], may even slow down evolution, because the mapping from genotype to phenotype is computationally expensive.

As the experiments in evolutionary robotics become more and more difficult, neuro-evolution approaches try to increase the success rate of the evolutionary search with different strategies. Three important strategies should be mentioned here: *structure reuse*, *search space restriction* and *diversity conservation*.

Structure reuse means, that a single neural structure encoded directly or indirectly in the genome, can be reused in different parts of the network. This often relates to modular structures, or neuro-modules [39]. The main advantage of reusing sub-networks is the possibility to develop functional building blocks, that can be reused in many parts of the network, without being reinvented by evolution multiple times. An example in biological systems are the cortical columns in the brain of mammals [40]. Modular sub-networks are explicitly addressed in algorithms like Modular NEAT [41], ModNet [29], CoSyNE [15], Cellular Encoding [22, 37], ENSO [42], ESP [43] and to some extent in SGOCE [30]. Another approach is the usage of CPPNs [17] and the related evolution method HyperNEAT [18, 44, 45]. HyperNEAT does not evolve neural networks, but evolves function networks (CPPNs) that generate the weights of a fixed network topology, resulting in repetitive, symmetric structures similar to neuro-modules.

In fact, structure reuse is one efficient method to reduce the search space. However the evolutionary search can be enhanced significantly by further restricting the search space. One common strategy is the incremental complexification of networks, as done in NEAT [33] and its derivates, ESP [43], ModNet [29], ENS^3 [32, 46, 47] and most developmental approaches. Evolution starts with minimal networks – for instance only including the sensor and motor neurons – and explores this search space until no further improvement is observed. Then, repeatedly, the search space is extended by adding additional neurons and synapses and the new search space is explored. The assumption is, that if solutions with simple network structures exist, then it would be beneficial, if these are found first, because they are expected to be evolved faster than complex networks doing the same task. This strategy has been shown to accelerate the evolution process in a number of experiments [33].

Another way to reduce the structural search space is the consideration of symmetries and other topological peculiarities, like the positions of neurons. Developmental approaches with growing synapses for instance have a higher probability to connect neurons close by [24, 30, 38]. Also HyperNEAT depends strongly on the positions of the neurons and can exploit topological neighborhoods. Symmetries can be used in many experimental scenarios, because the bodies of most robots have one or more symmetry axes. This can be reflected in the network topology and shrinks the search space by the number of neurons affected by the symmetry. ENSO [42] is an example of an algorithm that systematically breaks down symmetry step by step, starting with a highly symmetric network, hereby exploring the more symmetric and thus smaller search spaces first. Other search space restrictions may include the temporary protection of certain neurons or synapses from being changed during evolution, as with ENS^3.

The conservation of a certain level of diversity seems to be very beneficial for neuro-evolution. One reason is that networks with very different topologies and weight distributions can show a very similar behavior. Furthermore even small enhancements of the behavior may require large changes to the network structure. Therefore a premature convergence to a single class of topologies easily leads to an invincible local optimum. To protect different lineages during evolution, NEAT uses a niching method based on historical markings [33], which allow a dynamic grouping of the networks into similarity classes. Then, only similar networks have to compete against each other. ESP [16] and SANE [48] increase the diversity by not evolving complete neural networks, but instead by evolving single neurons, that are assembled to a full network for each evaluation (cooperative co-evolution). Because the neurons are always combined in other ways, the resulting networks remain diverse longer. In addition ESP makes use of fixed niches that further enhance diversity. Using a similar approach on the synaptic level, CoSyNE [15] also uses niches to delay convergence.

4 The Approach: Interactively Constrained Neuro-Evolution

The ICONE method (**I**nteractively **CO**nstrained **N**euro-**E**volution) tackles the described difficult problem domain by applying a number of strategies to structure the networks and to restrict the search space [49]. Restrictions are based on domain knowledge, user experience and on-line operations on the evolving individuals during evolution.

4.1 Genome

Genomes in ICONE use direct encoding, not making distinctions between the genome representation and the actual phenotype of an individual. Therefore no separate mapping operator from genotype to phenotype is necessary. The genomes are fully configured neural networks, building up an object structure from *primitive elements*, like neurons, synapses, neuron-groups and neuro-modules. Each of these elements provides a number of evolvable attributes, such as the bias and the transfer function of a neuron, and the weight of a synapse. All mutation operators work directly on these networks. In addition to the evolvable attributes, each element provides a list of properties, which can be used to control details of the neural network or the evolution process. These properties comprise string pairs, which combine a property key with a property value. With this, all kinds of tags and markings can be attached to any element by the user or the evolution operators. Table 1 shows an overview of the main attributes and properties of each element.

During evolution, both the network weights and the network structure can be evolved, so neurons, synapses and neuro-modules can be added and removed by the mutation operators. Exceptions are the so-called *interface neurons*, which are marked with a specific tag (*Input/Output*). These neurons are the interface to the motors and sensors of the robot and thus cannot be removed during evolution.

Table 1. An overview of the network elements, their attributes and their supported main properties. The meanings of specific properties are explained in the text.

Element	Attributes	Common Properties
Neuron	Bias TransferFunction ActivationFunction	ProtectBias ValidBiasRange ProtectTransferFunction ProtectActivationFunction NoSynapseSource/Target Input/Output ModuleInput/ModuleOutput ExtendedModuleInterface FlipActivity
Synapse	Weight	ProtectWeight ValidWeightRange
NeuronGroup	Member Neurons Constraints	
Neuro-Module	Member Neurons Constraints	ModuleType CompatibleTypes PreventModuleReplacements MaxNumberOfSubModules MaxNumberOfMemberNeurons
All Elements	Position	Protected ProtectExistence

4.2 Neuro-Modules

Neuro-modules are groups of neurons with a fixed interface to the surrounding network parts. A neuron can be a member of only one neuro-module at the same time. That way neuro-modules partition the network into well-defined sub-networks. As neuro-modules can be composed of other sub-modules, neuro-modules provide a way to organize a neural network in a specific topological and hierarchical manner.

Neuro-modules are usually treated as functional building blocks, e.g. an oscillator, a classifier, a reflex loop or a memory unit. Because of this, the neural implementation of the module – thus its inner structure – is not of relevance for other parts of the neural network, which utilize such a module. Important for using a module is solely the neural interface it provides. Neurons in a module can be tagged as *module interface* neurons (*ModuleInput/ModuleOutput*) which makes them visible from outside of the module. By default a module interface neuron is visible only for all neurons of the direct parent module. To increase the visibility of an interface neuron, it may be tagged as *ExtendedModuleInterface*, associated with a number, specifying the depth of its visibility. Figure 2 shows some examples.

Module interfaces reduce the search space, because they reduce the number of valid synaptic connections to a well defined subset, preventing arbitrary connections, which would make the development of functionally distinct sub-networks more difficult.

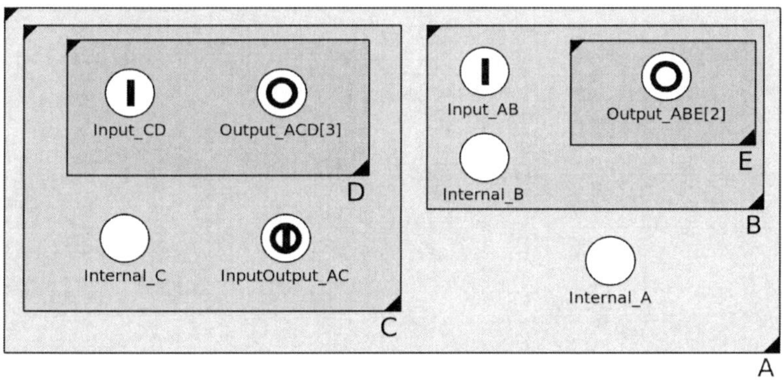

Fig. 2. Visibility of interface neurons in neuro-modules. Neurons marked as input (output) may be the target (source) of synapses from neurons outside of a module. The visibility is indicated by the module name after the underscore. The extended interface depth is given in square brackets. Neuron Output_ACD is also visible outside of A.

4.3 Constraints

To restrict the search space further, the network can be constrained by the user. One method to do this is the usage of the already mentioned property lists of each element. With this mechanism local properties of the evolution can be influenced, e.g. to prevent certain connections or to forbid the manipulation of specific network attributes.

Apart from these *low level restrictions* ICONE also supports *high level constraints* on neuron groups and neuro-modules. Table 2 shows some useful examples.

In addition to such standard constraints, the user may define *custom constraints*, that can be very specific. This allows the definition of functional

Table 2. Some useful neuron group constraints to restrict the search space.

Name	Constraint
Cloning	Clones all elements of a group or module to the target group.
Symmetry	Makes two modules or groups symmetric/antisymmetric horizontally or vertically. The symmetry may include incoming, outgoing and mutual synapses. Symmetry may be on structure only or including weights and biases.
PreventConnections	Prevents synaptic connections between members of the group.
ForceConnections	Forces synaptic pathways between all members of two groups.
AttributeCalculation	Evaluates equations to calculate the weight or bias of neurons and synapses depending on other synapses and neurons.

modules, that may be mutated during evolution, but whose function is preserved by a specialized constraint. An example may be an oscillator: changes of the synaptic weights during evolution may influence frequency and amplitude of the oscillation, while destructive changes are prevented by a constraint through corrections at the synapses.

Allowing fully configured functional neuro-modules as building blocks during the evolution speeds up the search process. Thus, already known neural structures do not have to be reinvented over and over again. Therefore, experiences from already completed experiments can be transferred to a *library of modules*, which may be exploited during future experiments.

Constraints are the main contributors to search space restriction. Applying for instance a *symmetry constraint* on parts of the control network reduces the search space by half of the constrained network section. Using the *clone constraint* to reuse a module in many places of the network without losing its evolvability reduces the search space to the original module only. All cloned modules are automatically adapted and can benefit from enhancements immediately. And even the simple constraints to prevent or force connections between neurons avoid the evaluation of individuals whose structure does not fulfill the requirements to be a potential solution.

4.4 ICONE Evolution Algorithm

The main algorithm of ICONE is – similar to most evolutionary algorithms – a cyclic application of evaluation, selection, reproduction and variation to a population of individuals (Algorithm 1). The initial networks hereby are networks prepared by the user with the search space restriction methods described above. The initial generation contains all starting networks and a number of randomly modified versions of these networks.

Algorithm 1. Main Evolution Loop

```
Data: I_init a set of domain specific, constrained start networks
begin
    P ← I_init
    if sizeOf(P) < popSize then
        P ← P ∪ completeInitialPopulation(popSize − sizeOf(P))
    repeat
        foreach individual p_i ∈ P do
            Evaluate(p_i)
        P_best = keepBestIndividuals(P)
        P_new = CreateOffspring(P, popSize − sizeOf(P_best))
        foreach individual pn_i ∈ P_new do
            VariationChain(pn_i)
        P ← P_best ∪ P_new
    until sufficient solution found
end
```

4.4.1 Evaluation

During the `Evaluation` phase, each individual is evaluated in one or more slightly randomized environments, so that the start conditions for each experiment vary from trial to trial. The fitness of each individual is calculated from all trials during its evaluation. Usually this is the minimum or mean fitness of all trials, because this supports the evolution of controllers, being robust to environmental noise. Fitness functions can have parameters that can be adjusted during the evolution to adapt the fitness to the progress of the evolution. Because of the many necessary evaluations and the related personnel and material overhead, networks cannot be practically evaluated directly on the physical hardware. Therefore neuro-controllers are evaluated in a physical simulator. In our implementation, the evaluation of the individuals is additionally distributed on a computer cluster, so that the interactive control of the algorithm is possible in real-time.

4.4.2 Selection

Each generation is composed of the best individuals of the parent generation (elitist selection) and new individuals, which are created as variations of selected parents. The selection method in ICONE can be exchanged and supports approaches like tournament selection, rank based selection and selection with ecological niches [50,51]. The chosen selection method creates the required number of offspring (`CreateOffspring` phase), which only contain pointers to their parents, but not already a network. The actual network is later created during the reproduction and variation phase (see Section 4.4.3).

4.4.3 Reproduction and Variation

The genomes of new individuals are created by the so-called `VariationChain` (Algorithm 2). This algorithm first prepares new individuals without genomes by cloning their first parent. Then modular crossover is applied (see Section 4.4.4) to exchange genetic material with the second parent. The resulting network then gets modified through the actual variation chain.

Algorithm 2. VariationChain(P)

```
Data: P a set of individuals
Data: M an ordered list of mutation operators
begin
    foreach individual p_i ∈ P do
        if hasNoGenome(p_i) then
            setGenome(p_i, CloneFirstParent(p_i))
        if hasTwoParents(p_i) then
            ModularCrossover(p_i)
        n ← maxIterations
        repeat
            valid ← true
            foreach mutation operator m_j ∈ M do
                executeOperator(m_j, p_i)
                if isNotValid(m_j, p_i) then
                    valid ← false
            execute ConstraintResolver(p_i)
            if ConstraintResolver failed then
                valid ← false
            n ← n − 1
        until valid = true or n = 0
        if valid = false then
            remove p_i from P
end
```

This chain contains an extensible number of variation and filter operators. These operators can directly modify the neural network, e.g. to add neurons, remove synapses or change properties. Common operators are listed with a brief description in Table 3.

The variation chain is executed at least once on each genome. However, variation operators may reject a genome, if it does not meet operator specific requirements. If at least one variation operator rejects a genome, then the entire variation chain is applied again to give all operators the chance to further modify the genome until all requirements of all operators are met. If this cannot be achieved during a given maximal number of chain iterations, then

Table 3. Variation operators of the variation chain.

Name	Function
Add/Remove Neuron	Adds or removes neurons to/from the network.
Add/Remove Synapse	Adds or removes synapses to/from the network.
Add Neuro-Modules	Inserts predefined functional neuro-modules from a library.
ChangeBias/Weight	Changes neuron bias or synapse weights.
Custom Filters	Rejects networks not fulfilling user defined requirements.

the individual is entirely removed from the generation. This avoids the evaluation of networks that do not fulfill user defined requirements and therefore cannot be a solution to the problem.

The final operator in this chain is the `ConstraintResolver` (Algorithm 3). This operator tries to resolve all constraints in the target network. Because constraints may influence each other by modifying the network, all constraints are applied repeatedly until all constraints have been successfully resolved, or if the maximum number of iterations has been exceeded. Individuals with unresolved constraints are removed from the generation, hence networks with conflicting constraints cannot evolve.

4.4.4 Modular Crossover

The exchange of genetic material between individuals has the potential to speed up evolution, because useful sub-networks do not have to be reinvented in each direct lineage. To prevent destructive effects during crossover, the modular crossover (Algorithm 4) operator exchanges only compatible sub-modules while keeping the in- and outgoing synapses.

The compatibility of two modules is defined by the *module type*, a module's *type compatibility list* and the module's *interface* structure. If a module m_1 has a type listed in the compatibility list of module m_2 and if the number of input and output neurons in the modules are equal, then m_1 is compatible with m_2 and therefore can replace m_2. That way, modules can be separated into *compatible module classes*, which enhances the probability that networks with exchanged modules still perform well.

4.4.5 Incremental Complexification and User Interaction

To keep the search space size low, evolution is started with as few neurons and synapses in the networks as possible. The insertion probabilities for neurons and synapses should be low and thoroughly controlled by the user during evolution. Smaller networks do not only lead to a smaller search space, but also to networks, for which structure and dynamical properties can be understood more easily.

Algorithm 3. ConstraintResolver(p)

```
Data: p a single individual
Data: C the set of all constraints used in the network of individual p
begin
    n ⟵ maxIterations
    repeat
        modified ⟵ false
        foreach constraint c_i ∈ C do
            applyConstraint(c_i)
            if c_i made changes to the network then
                modified ⟵ true
        n ⟵ n − 1
    until modified = false or n = 0
    if modified then
        report resolver failed
    else
        report success
end
```

Algorithm 4. ModularCrossover(p)

```
Data: p a single individual
Data: N_1 the neural network of individual p
Data: N_2 the neural network of the second parent of p
begin
    M1 ⟵ all unprotected modules of N_1
    M2 ⟵ all unprotected modules of N_2
    foreach module m_i ∈ M1 do
        if rand() < crossoverProbability then
            M_match ⟵ all modules of M2 matching type and interface of m_i
            m_flip ⟵ randomly chosen module from M_match
            replace m_i in M1 with a copy of m_flip
end
```

During evolution, the user can observe and examine each individual and react on the development process. This is assisted by using a computer cluster to evaluate the individuals. Interactions may involve adaptations of parameters of the mutation operators or the fitness function, implantations of promising network structures into individuals, removal of undesired structures and changes of the simulation environment. Thus, evolution is not used (only) as a batch process, that autonomously finds solutions to the problem. Rather it serves as a search assistant to support the user to validate the suitability of certain neural approaches. The focus in searching with ICONE is not only to get a solution to the problem, whatever it may be. In contrast, the problem should be solved in very specific, previously specified ways. This allows the systematic investigation of specific neural paradigms and organizations.

5 Evolving a Walking Behavior for the A-Series Robot

The evolution of behaviors for the A-Series robot is shown exemplarily with the evolution of a humanoid walking behavior. Walking, as the main locomotion behavior for a humanoid robot, has been realized with evolutionary approaches in different ways for a number of robots. Most of the approaches evolve walking controllers in simulation with a simplified robot morphology and a small neural interface. A popular approach is the evolutionary optimization of weights in predefined fixed-topology neural networks [52–55] or the optimization of network model parameters [56, 57]. The evolution of such networks is often successful when the network topology is known or the number of fully interconnected neurons is low. When the morphology of the robot is not fixed, then co-evolving the robot body and the controller becomes an option [58, 59]. This can lead to simpler controllers due to the evolutionary development of bodies that are optimized for walking. However evolving a neural controller for a given physical humanoid robot using structure evolution is difficult to achieve.

The evolution of a walking behavior for the A-Series humanoid is a challenging task because of several reasons. First, almost all joints are involved in the behavior and hence also their sensors. Because of the many input (38) and output neurons (42), the search space is quite large. Second, the behavior is very sensitive to the implementation of the motor and sensor models and the modeling of the physical body properties. Small differences between the simulated model and the physical robot can easily prevent a successful transfer of the evolved controllers to the physical machine. Third, trying to evolve walking from scratch without restrictions in a single step often leads to many undesired local optima, which do not represent behaviors close to the desired type of walking.

To cope with the given problems – large search space, difficult fitness function, transfer to hardware – the experiment has been divided into three consecutive evolution experiments: learning to march in place (Section 5.1), learning to walk forwards (Section 5.2), and transferring the walking behavior to the physical robot (Section 5.3). This incremental evolution approach [12, 31, 60] allows a start with a simple, relatively easy solvable task, which is then iteratively made more complex. This leads to simpler fitness functions and an improved convergence behavior. Furthermore, partial solutions can be used as starting points for many different consecutive evolution experiments. Therefore they serve as a kind of fall-back point for different routes towards the desired target behavior.

The evolution experiment takes place in a physical simulator for the A-Series humanoid. The simulator uses the *Open Dynamics Engine* (ODE)[3] to simulate the rigid body dynamics. The simulation models of motors and sensors have been modeled according to data recorded on the physical robot. The interface to the neural networks is similar to the one on the physical robot, so that neuro-controllers from simulation can be transferred to the robot without modifications. To obtain a similar behavior on both target platforms, the limited accuracy of weights, bias values and neural activities on the AccelBoards have also been modeled in the simulator. All recurrent neural network controllers use the additive discrete-time neuron model with *tanh* transfer function [61]. Both, the simulation and the ICONE evolution algorithm have been implemented within the *Neurodynamics and Evolutionary Robotics Development kit* (NERD) [62].

5.1 Learning to March in Place

The A-Series robot cannot – due to its physical constraints – stand stable on a single leg. The motors of the A-Series, especially in the ankles, are too weak to simultaneously counteract the weight of the robot and to do fine control of a desired angular position. Therefore walking is not assumed to be a static movement from a stable one-legged standing position to another, as often seen in humanoid robots. Instead a dynamic, pendulum-like walking is implemented. The stability for the A-Series robot has to originate primarily from the dynamics of the pendulum-like lateral swinging of the robot while staggering from one leg to the other.

As starting point for walking it is assumed that the robot should be able to lift its legs in a regular, periodic way. To achieve this the first task for the robot is to step from one leg to the other on the spot without falling over.

5.1.1 Simulated Environment and Fitness Function

The environment of the robot consists of four balks at 10 cm height that restrict the operational range of the robot (Figure 3). These balks support the development of stepping behaviors that keep the robot near its starting position. All collisions between the robot and the balks or the ground (except with its feet) stop the evaluation immediately. The approach to stop evaluation at the violation of hard constraints has shown its benefit in many other experiments. It speeds up evolution by preventing wasteful evaluation time on controllers that do not fulfill all constraints on the behavior. Combined with a suitable fitness function, that prefers individuals with longer evaluation time, the evolution of controllers outside the desired specifications can be avoided right from the beginning.

[3] http://www.ode.org/

Fig. 3. The simulated A-Series robot in its evaluation environment for marching in place with constrained operational range.

The fitness function combines several aspects of the stepping motion, that can be weighted separately with parameters δ and γ. The first aspect is the maximal stepping height h during one step. This favors controllers that lift the leg as high as possible. The second aspect is the step duration d. The longer a footstep takes, the better, because large steps are desired for walking, which require some time.

We use footsteps j as the measuring unit in the fitness function. Due to missing foot contact sensors, footsteps are detected monitoring the height of the feet. A new footstep is assumed to take place when the difference between the minimal heights of both feet changes its sign. To avoid the false detection of footsteps caused by noise or *vibrating* behaviors, the feet have to reach a minimal distance $|d_{min}|$ after the sign change. d_{min} hereby can be adjusted as a parameter of the fitness function. The current footstep count at time step i is s. The fitness at time step i is given by

$$f_i = \sum_{j=0}^{s-1} (\delta h_j + \gamma d_j) \tag{1}$$

that is, f_i is the sum of the maximal heights h_j at footsteps j and their duration d_j up to the previous footstep $(s - 1)$. Only summing up to the previous footstep avoids controllers that try to maximize duration or height in an uncorrectable one-time attempt, such as lifting the leg very high by falling over to the side. Fitness therefore is only gained for motions that lead to another footstep.

5.1.2 Neural Network Templates

The starting network for this behavior is constrained with the techniques described in section 4. As can be seen in Figure 4 the basic neural network

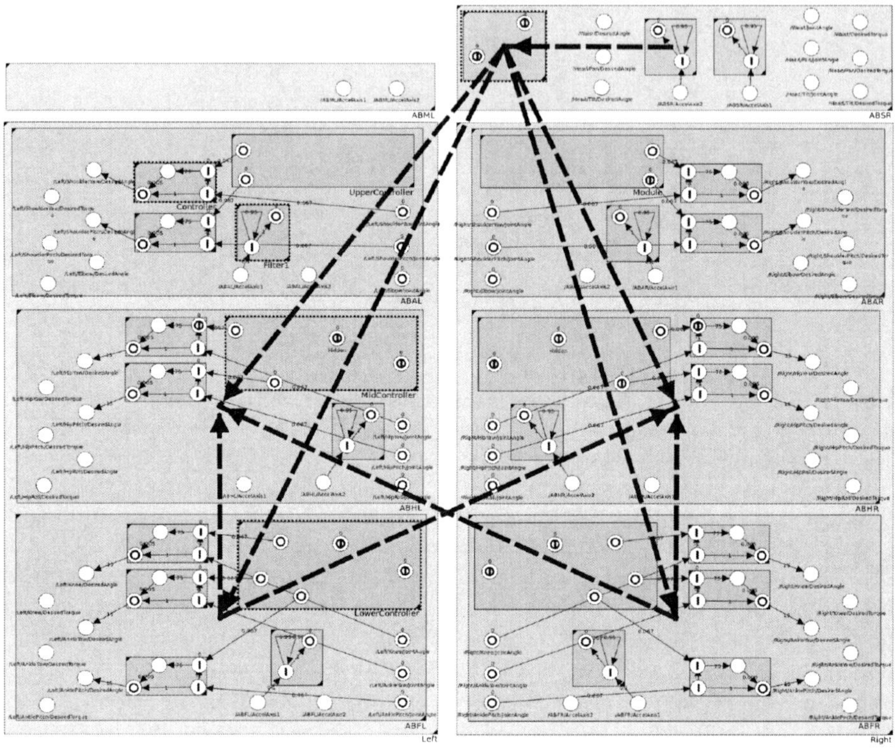

Fig. 4. The neural network template for marching in place. Neurons with names are interface neurons of the robot (motors and sensors). The actual names are not important here. Interface neurons of modules are marked with an I (input) or O (output). Allowed synaptic pathways are indicated by black dotted lines.

with its 38 sensor and 42 motor neurons would be difficult to understand without structuring the network.

All modules with names starting with AB (the abbreviation for *Accel-Board*) represent the hardware controller boards on the robot. Using modules to structure the network according to the hardware boards allows an intuitive understanding of the location of the sensors and motors. Furthermore, constraints on these modules enforce evolved networks to satisfy all constraints originating from the hardware, like the maximum number of neurons and synapses per board (limited by memory and execution time), and the maximum number of neurons visible to other boards (limited by the communication bus). Obviously, neurons outside of the AB** modules are prevented, because otherwise the network could not be transferred to the hardware.

To restrict the search space for the evolutionary algorithm, all motors and sensors not needed for the transversal movements, have been *protected* and hence cannot be targeted by mutation operators. This affects all motors and sensors, except the transversal hip and ankle motors, including their corresponding angular sensors, and the transversal acceleration sensors at one shoulder and the feet. Arms and knees have been fixed at suitable angles to support the task statically. The motor torque neurons have been fixed with a bias of 1.5, so the networks in this experiment are forced to control the motor angles with maximum torque. As stated in section 4.4.5 this induces a certain kind of desired solution, namely to control the motion with the angular motors, based on acceleration sensors. Different constraints would lead to very distinct solutions, e.g. when the motors would be forced to be controlled by torque or by including other sensors.

As an additional constraint the lower six AB** modules have been organized into two larger modules to realize symmetry between the left and the right side. All elements on the left side are horizontally mirrored to the right side. All incoming synapses of these two modules have been chosen to be anti-symmetric, i.e. they get synapses coming from the same external neurons, but with reverse signs. All mutual synapses between both sides have been chosen to be symmetric.

The A-Series motor neurons represent the desired joint angle of a motor. Thus no additional controller is required to hold a given angle. Nonetheless, it makes sense to connect each motor neuron with a controller that limits the rate of change of an angle setting to smoothen the motions, in order to protect the motors and to simplify the transfer of the controllers to the hardware later on. The latter is the case because the motors behave less predictably near their operational limits and hence are difficult to simulate adequately for this case.

The structure of these controller modules is given in advance, but the synapse weights are open for mutations to manipulate their reactivity and characteristics. Because each motor neuron should be equipped with such a controller, it makes sense to use only a single mutable controller prototype in the network, and a clone of this prototype in each place where a controller is needed. That way only a single controller module is part of the search space.

The same holds true for the filter modules used at each acceleration sensor. The signals of the acceleration sensors are not smooth and hence difficult to use. Filtering the signal reduces the effect of spikes, but induces a delay. Therefore, the filter properties of one prototypic filter module should be open for evolution to find the best suitable filter behavior, while every other acceleration sensor is filtered by a clone of this mutable module.

Finally synaptic pathways (black dotted arrows) have been introduced to restrict the possible connections between modules. These pathways force all new synaptic connections to be added only between the specified modules, including visible interface neurons of their sub-modules. Here, only connections from the shoulder sensors to the hip and feet modules, and from the feet modules to the hip modules, are allowed.

The search space is illustrated in Figure 4: all modules that can be modified during evolution are highlighted with black dotted boundaries. All other modules are either fixed, or depend on one of the other modules due to a constraint. However some of the fixed modules contain mutable elements, which are also part of the search space.

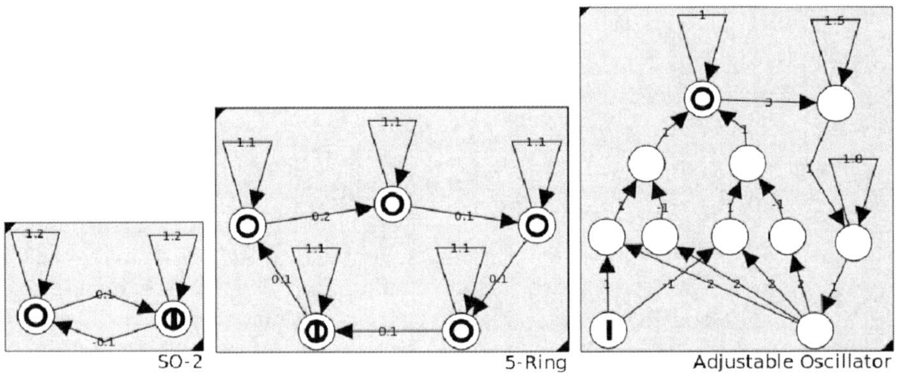

Fig. 5. Some examples of neural oscillator modules [63, 64] that can be used as building blocks during evolution.

Other network templates used for the experiments were variations of the above described network. In some networks, additional functional modules have been allowed, such as oscillatory pattern generators like those shown in Figure 5. These networks can exploit and adapt the periodic pattern signals for the behaviors, without having to (re-)invent the oscillators during evolution.

5.1.3 Evolution Settings

The experiments have been conducted with a variety of parameter settings for the evolutionary algorithm (see Table 4). These parameters were interactively adapted during the evolution according to the convergence behavior of the algorithm. Therefore instead of exact parameter settings for each experimental run only general ranges are given, in which the parameters have been varied during each evolution run.

As a rule of thumb, all operators mentioned in Table 4 have been active, starting with very low probabilities for search space extensions (Add Neuron, Add Synapse) and with a relatively high probability for changes of synapse weights and bias values (Change Bias, Change Weight). During the evolution, the probabilities for structural changes were increased when the results

Table 4. Settings of the main evolution operators. The settings are given as ranges in which the parameters have been varied during interactive evolution. The functions of the operators are listed in Table 3

Operator	Parameter	Setting
General	Population Size	[100, 200]
	Max Simulation Steps per Trial	[500, 3000]
	Number of Trials per Evaluation	[2, 5]
Tournament Selection	Tournament Size	[3, 5]
	Keep Best Parents	1
Modular Crossover	Crossover Probability	0.5
	Crossover Probability per Module	0.5
Add Synapse	Probability	[0, 0.02]
	Number of Insertion Trials	[0, 5]
Add Neuron	Probability	[0, 0.005]
	Number of Insertion Trials	[0, 2]
Change Bias	Change Probability	[0.005, 0.015]
	Deviation	[0.005, 0.02]
Change Weight	Change Probability	[0.01, 0.2]
	Deviation	[0.005, 0.2]

converged to an undesired local optimum, so that more complex structures could evolve to overcome the local optimum. The probabilities of weight and bias changes and their average amount of change were decreased when a promising area of the solution space was reached, so that the behavior of the network could be fine-tuned.

The probability for modular crossover usually was about 0.5 to support the transfer of genetic material between lines of ancestries. The number of trials indicates how often each individual is evaluated with slightly randomized environments, e.g. alterations of the starting angles. As selection method an implementation of the standard *Tournament* selection [65] was used.

5.1.4 Results

The evolution was performed 33 times for about 100 to 300 generations per evolution, depending on the observed progress. In 21 cases, networks have been found that solve the task and provide a valid starting condition for the next evolution scenario. The fitness progress during the evolution (maximal and mean fitness, variance) for the best solution is shown in Figure 6. The progress of the maximal fitness for the next best 10 evolution runs are shown in Figure 7.

Because of the constrained network, the implemented strategies are not too surprising. In the networks driven by the acceleration sensors, the main strategy was to destabilize the robot with the transversal hip or ankle motors

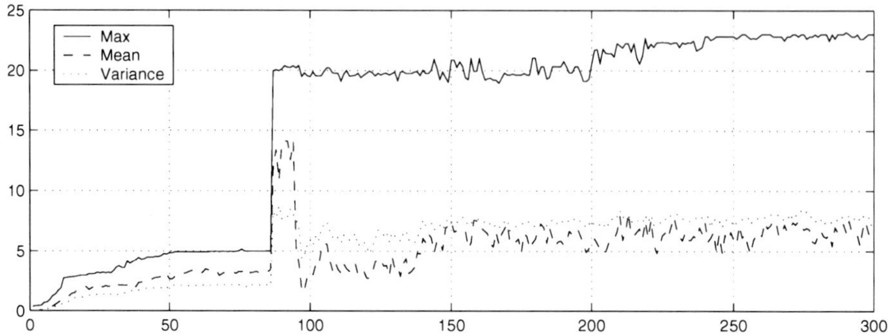

Fig. 6. Maximal and mean fitness of the best evolution run for the march-in-place task.

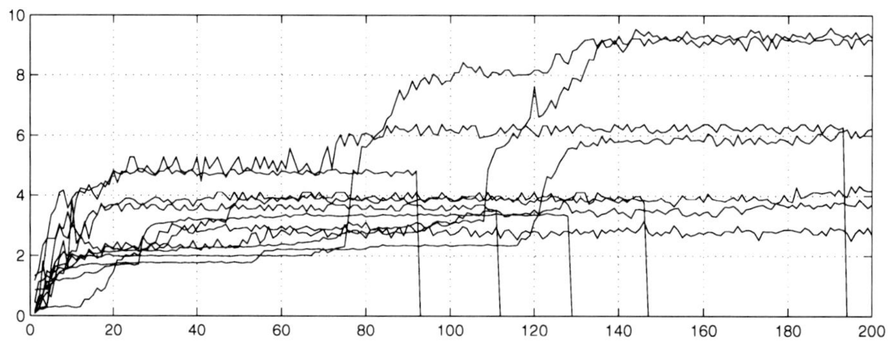

Fig. 7. Maximal fitness of the 10 best evolution runs not including the very best run (Figure 6) for scaling reasons. Fitness curves dropping to zero indicate evolution runs that were interactively stopped due to stagnation.

according to the swing phase. Once swinging, the transversal acceleration sensors provide an oscillatory signal, that is used to control the hip or ankle motors. An example of such an acceleration sensor driven behavior is shown in the time series in Figure 8.

However, 12 evolutions did not come up with satisfactory solutions. Even with the symmetry constraints between the left and the right side, many behaviors resulted in unstable, irregular motions. This was especially the case when the acceleration sensors of the feet were connected to the ankle motors. As the movement of the ankle motors directly influences the acceleration sensors on the feet, this leads to an isolated, unsynchronized swinging behavior local to each leg, which could not generate a stable global behavior. This suggests that the feet sensors may only be useful if the feet remain at fixed angles while their sensor signal is used to control the hips. Such solutions can be

Fig. 8. Time series of the robot controlled by a resulting network for the march-in-place task.

avoided interactively during evolution by preventing this kind of connections and, therefore, by excluding this type of frequent local optimum.

Network solutions based on neural oscillators (Figure 5) usually worked out fine for a while. But because these networks did not respond to the sensors, these pattern generator-based networks tended to be unstable on the long run due to their inability to adapt to the (noisy) state of the robot's motion.

Figure 7 gives a rough overview on the performance of the algorithm for this experiment. As comparisons with other algorithms are difficult due to the interactive approach and therefore due to the involved user experience, these graphs should give a general idea about the usability of the method. Nevertheless, one observation can be made: Applying the ENS^3 algorithm or NEAT to an unconstrained, empty starting generation with networks of this size, solutions for this problem have not been found at all.

5.2 Learning to Walk Forwards

Based on the results of Section 5.1 the next step towards walking was conducted: modifying a stepper network to move forwards. Adding forward movement is only one next possible step. A possibility could have been first to stabilize the stepping behavior to make it more robust to small bumps on the ground or a shaking floor. Another next step could have been an optimization of the leg lifting by involving the motors and sensors of the sagittal plane of the knees, ankles and the hip. However in this document we continue directly with walking forward due to space limitations.

5.2.1 Simulated Environment and Fitness Function

The environment for the walking experiment (Figure 9) gives the robot space to walk forwards, but still restricts its operational range to the sides and backwards. In these directions, balks obstruct the path of the robot. As in the experiment before, collisions with these balks immediately stops evaluation. Thus, undesired behaviors, like moving backwards or in circles, can be avoided efficiently. To avoid a common local optimum during evolution, namely moving by *vibrations* instead of steps, obstacles have been introduced in regular intervals. To overcome these obstacles, the robot has to lift its legs high enough to get the feet over the obstacle. To avoid the robot from tilting when the obstacle gets below the feet, the obstacles are implemented as sliders, that are lifted to their target height with a soft spring. Stepping on such an obstacle just makes it slide back below the ground without resulting in a bump. Therefore the obstacles hinder the feet only when colliding horizontally, not vertically.

Fig. 9. The simulated A-Series robot in its evaluation environment for walking with constrained operational range and obstacles.

In the fitness function a new term is introduced: x_{max} is the maximally reached distance along the x-axis up to time step i. The fitness function extends the one from Section 5.1.1 by a weighted factor that rewards moving along the x-axis:

$$f_i = \sigma x_{max} \sum_{j=0}^{s-1} (\delta h_j + \gamma d_j) \tag{2}$$

that is, f_i is the sum of feet height and footstep duration multiplied by the weighted distance x_{max}. The distance from the origin at time step i, x_i, is the minimum of the distances of the head, waist and both feet from the origin at that time step. Taking the minimum distance of several parts of the body prevents the robot from becoming easily trapped in a common local optimum, where the robot catapults the single relevant body part, e.g. the head or one of the feet, as far as possible. Such optima have to be avoided right from the start, because they are often dominant in the beginning of the evolution and in general do not provide a suitable path towards desired solutions.

5.2.2 Neural Network Templates

Based upon solution networks for the previous task, the network templates for the walking experiments have been derived (Figure 10). It was chosen to use one of the accelerator sensor driven solutions. The base network was pruned to get the smallest working version of the network, so that evolution may start with a minimal network again.

The major modification of the network constraints is the activation of additional motors and sensors by removing their *protected* property. The networks now can also connect synapses to the sagittal motors of the hip, the feet, the knees and the arms, and to their corresponding angular sensors. This enables the robot to bend its knees to make higher steps, to move the legs forwards and backwards, to use the feet tilt to push the robot forwards, or to use the arms to keep balance during walking.

To prevent evolution from destroying the already achieved basic motion, the existence of all involved synapses and neurons has been protected.

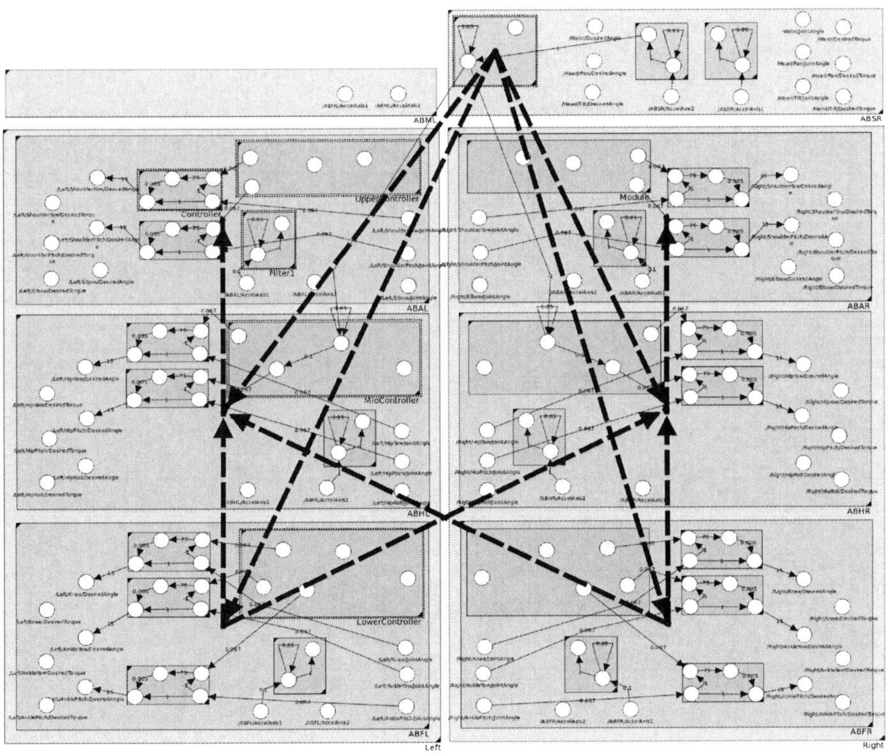

Fig. 10. The neural network template for walking, based on a cleaned, slightly adapted solution of the march-in-place task.

However the weights of these synapses and the bias values of the neurons have not been fixed and remain mutable.

In the upper modules, additional module interface neurons have been introduced. This new interface facilitates new neural structures in the upper part of the body to make use of the arm motors and sensors. To include these arm control modules, the synaptic pathways have been adapted, so that a new pathway leads to these modules. It is assumed that the arm movement may depend on the state of the hip joints, so the synaptic pathway runs from the middle module to the arm module.

5.2.3 Evolution Settings

The evolution settings were similar to the previous experiments (see Table 4). Again, for the evolution no fixed parameter sets have been used, because the parameters are targets of an online observation and modification.

5.2.4 Results

The evolution was able to generate walking control networks in 26 of 100 performed evolution runs. The fitness progress of the best 10 evolution runs is shown in Figure 11. The higher number of unsuccessful evolution runs may be partly a result of the interactive evolution. Undesired approaches, that do not seem to lead to a solution, can be stopped in early generations. Therefore, evolution runs have been stopped prematurely to focus the search on more promising areas of the search space. In fact, all unsuccessful evolutions together had a average runtime of 44 generations, which is low compared to the successful evolution runs, that had an average runtime of 156 generations.

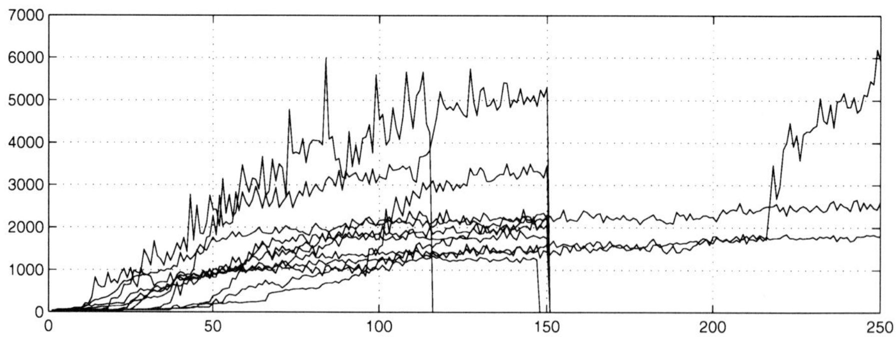

Fig. 11. Maximal fitness of the 10 best evolution runs to solve the walking problem. Fitness curves dropping to zero indicate evolution runs that were interactively stopped due to stagnation or runtime limitations (at 150 generations).

As expected, the sagittal hip or feet motors usually were used to realize the forward motion. In some networks (like in Figure 13) also the knees and arms were utilized to get a better walking performance. The time series of such a network is shown in Figure 12.

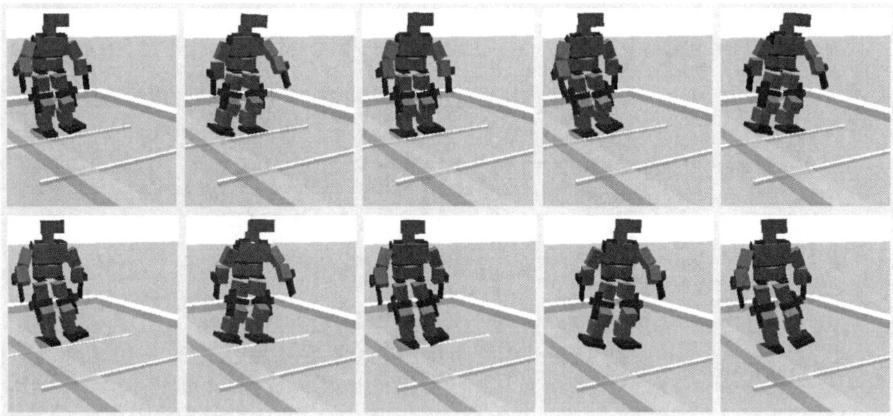

Fig. 12. Time series of the robot controlled by a resulting network for the walking task.

The walking behavior still is not very human-like or elegant. Also, depending on noise and the obstacles on the ground, the robot turns from time to time by a few degrees and continues its movement. This behavior is difficult to overcome, because the robot does not have feedback sensors for the direction and therefore is not able to correct such deviations. On the other hand, taking the robot's limited motor and sensor equipment into account, these behaviors seem quite satisfactory.

5.3 Controller Transfer to the Physical A-Series Robot

Controllers evolved in the simulator usually cannot be transferred without modifications, because the differences between the simulated robot model and the physical machine are significant. Although the main behavior in simulation and on the physical robot is very similar, even small differences between the two target platforms may cause a dynamic behavior – like walking – to fail. Some differences may be reduced by implementing more and more detailed models of the motors, the sensors, and the body structure. In practice, however, this approach is limited by the performance of the utilized computers and the time spent for the implementation of such a model. Some aspects may not be taken into account at all, like small variations during the assembly of the robot, fabrication related differences in the motor or sensor behavior, or just behavioral differences caused by aging or heat. Other aspects are restricted by the utilized physics engine, which may not support elastic material, such as the plastic material of the robot's body parts. Therefore adaptations of the behaviors to the target robot are usually unavoidable.

5.3.1 Evolving Robustness through Robot Model Rotation

Using simulated robots during evolution will provide neuro-controllers that adapt to the simulated robot, but not necessarily to the physical one. Controller networks will optimize for all aspects of the simulated system, taking advantage of any implementation detail. This includes modeling errors and simplifications of the model. Because a convenient, error free robot model is not feasible, any model will have implementation details that can – and will – be exploited by evolution.

To reduce this effect, a number of approaches have been proposed, such as adding sensor noise [66], (post-)evolving or adapting individuals directly on the hardware [67], or co-evolving the simulation properties along with the behavior controllers [68]. We use an approach called *Model Rotation*. The idea is not to evolve controllers for a single simulated robot, but for a variety of similar, but slightly differing robots. The fitness of a controller is then the minimum achieved on all of the target platforms. Thus, the fitness corresponds to the robot model with the weakest performance. To get a high fitness, a neuro-controller has to perform well on all given robot models.

Because of this, the behaviors cannot exploit flaws of the models, as long as each model has different weaknesses and strengths. Resulting controllers are expected to work on a variety of similar robots, which means, that they are robust to small variations between the robots. Such robust networks have a higher chance to work also on other similar robots not used during evolution, including the physical robot.

The robot models used during model rotation should not be seen just as random variations (noise). Simple randomization works out only for some parameters, like the angles at which body parts are assembled to each other or the absolute size of the body parts. Randomly changing parameters of more complex parts of the robot, like the motor and sensor models, often does not lead to behaviors similar to the physical robot. This is due to the high dependencies between parameters. Therefore entire sets of parameters, or even different model implementations, have to be found that produce a similar behavior, but have significant modeling differences. During evolution, each controller then can be evaluated with each of the given model parameter sets.

Model rotation can be used during the entire evolution. This avoids partial solutions that are too dependent on a specific motor model. On the other hand, model rotation results in a much higher number of evaluations and slows down the evolution process. Therefore it is often useful to start without or with a limited model rotation and do a separate final evolution with full model rotation to optimize the results for robustness.

5.3.2 Manual Adaption

The final step to transfer a controller to the physical robot is manual adaption. Even robust controllers often do not work out-of-the-box and require some fine-tuning. During this step, the hierarchical, modular structure of the networks is advantageous. The structured networks are easier to understand, they assist in identifying the sub-networks responsible for specific functions, and help to isolate the synapses that have to be modified to optimize the performance of the controllers for the physical robot.

5.3.3 Results

For the transfer to the physical A-Series humanoid, the network in Figure 13 was chosen, because its pruned structure is easy to understand and allows some adaptations if necessary. Figure 15 shows the time series of the transferred walking controller shown in Figure 13.

This network was optimized for robustness with model rotation. Five distinct parameter sets for the motors have been used during the optimization, each behaving closely to the physical machine for some test behaviors, but each differing in friction and control parameters. Figure 14 shows the fitness progression of the model rotation optimization. As can be seen the fitness started low compared to the fitness achieved without model rotation. This is due to the fact that the minimal performance determines the fitness of a controller. Therefore a single failing attempt reduces the fitness significantly. During evolution, the fitness increased to a fitness level close to the one without model rotation. This indicates that the controllers became more robust with respect to small differences in the motor model.

For some solutions, adaptations were necessary due to the flexible body parts and the stronger friction on the ground. The legs had to be straddled wider and the upper part of the body had to be bent a bit forwards. Apart from these small bias changes, the behavior control performed well on the physical robot. Some solutions, as the one used in Figure 15, did not require any changes to the network at all and worked directly on the physical robot. Nevertheless, due to friction, elastic body parts and a missing vertical stabilization behavior, walking on the physical robot is by far not as stable as in simulation, where the simulated robot could walk hundreds of footsteps without falling over.

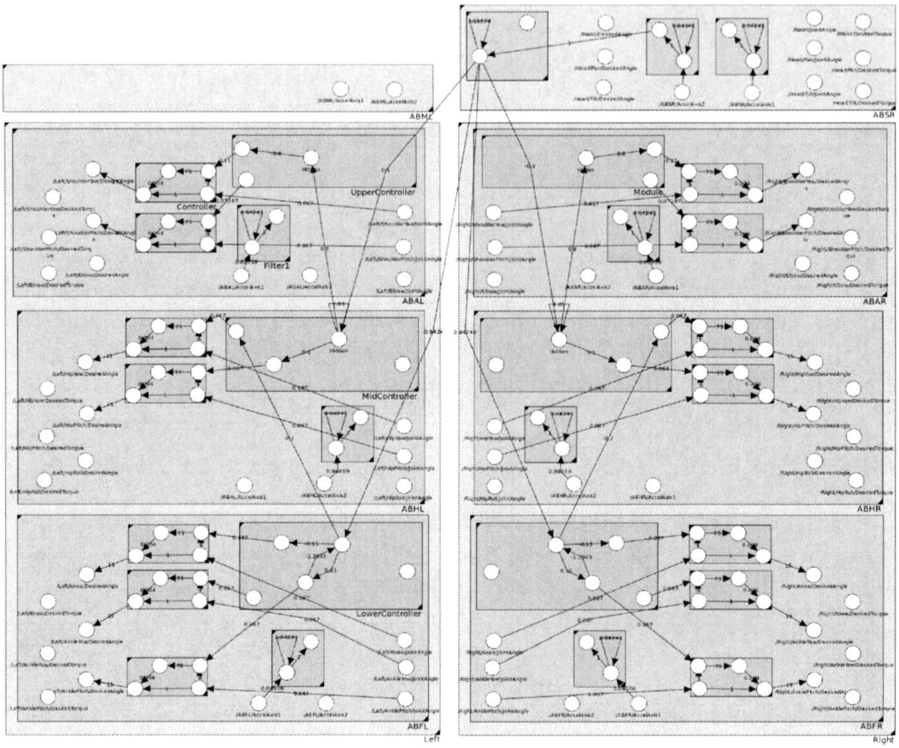

Fig. 13. Successful acceleration sensor based walking network for the A-Series humanoid.

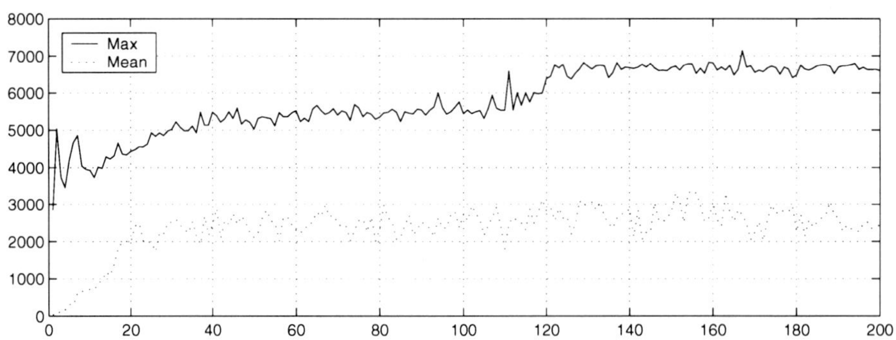

Fig. 14. Fitness progress of a walking network during robustness evolution using model rotation.

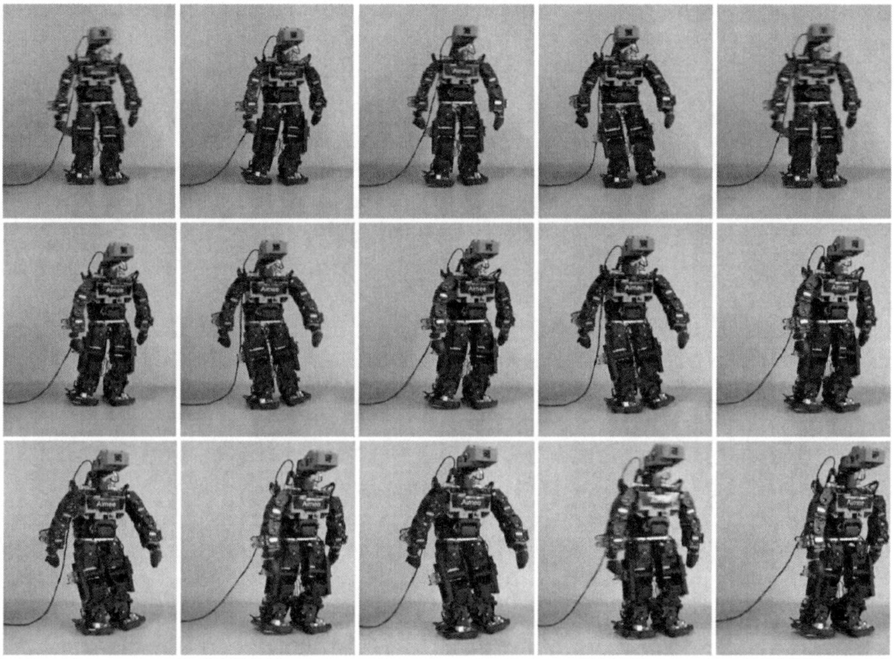

Fig. 15. Time series of the physical robot controlled by the walking network of Figure 13.

6 Discussion and Conclusion

This chapter introduced the interactive neuro-evolution method ICONE designed for the development of large-scale networks serving as control systems of autonomous robots. For demonstration, a humanoid robot with 37 sensors (proprioceptors) and 21 motors was chosen as a target system. The goal then was to generate recurrent neural networks able to control specific behaviors of this robot. Because an effective structure of such neuro-controllers cannot be given right from the beginning, the connectivity also has to be evolved. The challenge for evolutionary algorithms is to cope with the corresponding high-dimensional search spaces.

Our previous neuro-evolution technique, the ENS^3 algorithm [32], works well for small neural networks with the network size having less than about 30 neurons. In this domain many interesting experiments have been conducted [32, 69–71]. But with more complex robots, such as humanoid robots, having a much larger number of neurons, the probability to produce a successful behavior drops significantly. In this domain the algorithm cannot practically be used to evolve the structure of the network. The optimization of the weights of predefined network topologies often still leads to successful solutions, but with the drawback that there is no evolutionary discovery of interesting neural structures.

The ICONE approach tries to tackle these problems by a number of search space restriction strategies. The method does, of course, not solve the scaling problem of evolutionary robotics, but may shift the frontier of solvable behavioral problems a bit further.

These strategies however also restrict the way a behavioral problem can be solved. This aspect may be an advantage, because it allows the experimenter to specify own solution strategies in advance and guide the evolution through a predefined solution space. A difficult behavior can then be tackled with a number of neural approaches using, for instance, networks driven by acceleration sensors, central pattern generators (CPG), feedback-loops or sensor-modulated CPGs. Such an approach can be specified by a suitable restriction of the search space and by selecting a suitable set of building blocks (neuro-modules) for evolution. The usage of functional neuro-modules also allows the reuse of neural structures from previous experiments, even from very different target systems. To do so, general functional neuro-modules can be isolated and worked up to neural building blocks, which can be utilized by evolutionary operators during upcoming experiments. That way, even complex structures, like oscillators, memory structures, filters, controllers or behavior coordination modules can be exploited during evolution.

As the evolution process is interactive and highly dependent on the experience of the experimenter, this method is difficult to compare with other (batched) algorithms. Also, in the proposed high-dimensional search space, the goal of the evolution is not so much to find solutions with focus on evolution performance, but on finding solutions at all. Therefore, the given empirical information about run-time and number of evaluations of the experiments are rather rough indications of the performance, than detailed performance ratings. They should give the reader a general idea about the usability of the method.

In the discussed example of a humanoid robot it was demonstrated how a difficult behavior – like walking – is divided into easier to evolve successive subtasks, and how the resulting behavior can be integrated on the physical machine. The walking behavior is only one example of a number of behaviors that have been realized for the A-Series robot. These behaviors include:

- different kinds of walking,
- stabilized standing on a shaking and tilting platform,
- stabilized squatting to arbitrary knee angles in rough terrain,
- dynamic pointing to objects and appropriate switching of the pointing arm,
- gestures and fun motions, like hand waving, playing air guitar, saluting, taking a bow, and others.

All of these behaviors were successfully transferred to the physical robot. The corresponding library of neural behaviors now is part of the A-Series humanoid platform and can be used to conduct experiments with communicating robots.

7 Future Work

The presented ICONE method is now applied also to other types of advanced machines like the modular multi-legged walking machine *Octavio*[4] and also to a new class of mid-scale humanoid robots, the so called Myon humanoids [72]. These 1.25 meter tall Myon robots are developed, as the A-Series robots, by the Humboldt University of Berlin, and have been engineered from scratch, integrating a number of interesting new features.

Equipped, for instance, with force sensors in the feet, current consumption sensors in the motors, the use of multiple separate motors to control single joints, the use of spring couplings at motors and the possibility to let joints swing with minimal friction, this humanoid robot allows the exploration of new types of controllers which can be very different from those developed for the A-Series robot. Certainly, the corresponding neural networks again become much larger, using now 178 sensor neurons and 80 motor neurons. Therefore additional constraint techniques will be examined and integrated into the ICONE method, so that a library of new functional units and whole neural networks for behavior control can be realized.

Acknowledgements. This work was partly funded by EU-Project Number ICT – 214856 (ALEAR Artificial Language Evolution on Autonomous Robots. http://www.alear.eu). Thanks go to Verena Thomas, Ferry Bachmann, Robert Muil and Chris Reinke for contributions to the simulation environment, and to Tanya Beck for proofreading the chapter. We also thank the entire team of the *Neurorobotics Workgroup* at the Humboldt University of Berlin for providing and maintaining the A-Series robot hardware.

References

1. Nolfi, S., Floreano, D.: Evolutionary Robotics. MIT Press, Cambridge (2004); ISBN-13: 978-0-262-14070-6
2. Floreano, D., Dürr, P., Mattiussi, C.: Neuroevolution: from architectures to learning. Evolutionary Intelligence 1(1), 47–62 (2008)
3. Trianni, V.: Evolutionary Swarm Robotics: Evolving Self-Organising Behaviours in Groups of Autonomous Robots. SCI. Springer Publishing Company, Heidelberg (2008)
4. From animals to animats 10, Proceedings of 10th International Conference on Simulation of Adaptive Behavior, SAB 2008, Osaka, Japan, July 7-12 (2008)

[4] http://www.ikw.uni-osnabrueck.de/~neurokybernetik/projects/octavio.html

5. Steels, L.: Language games for autonomous robots. IEEE Intelligent Systems 16(5), 16–22 (2001)
6. Hild, M., Meissner, M., Spranger, M.: Humanoid Team Humboldt Team Description 2007 for RoboCup 2007, Atlanta, USA (2007)
7. Yao, X.: Evolving artificial neural networks. Proceedings of the IEEE 87(9), 1423–1447 (1999)
8. Harvey, I., Husbands, P., Cliff, D., Thompson, A., Jakobi, N.: Evolutionary robotics: the sussex approach. Robotics and Autonomous Systems (1997)
9. Harvey, I., Di Paolo, E., Wood, R., Quinn, M., Tuci, E.: Evolutionary robotics: A new scientific tool for studying cognition. Artificial Life 11(1-2), 79–98 (2005)
10. Lungarella, M., Mettay, G., Pfeifer, R., Sandiniy, G.: Developmental robotics: a survey. Connection Science 15(4), 151–190 (2003)
11. Pfeifer, R.: On the role of embodiment in the emergence of cognition: Grey walter's turtles and beyond. In: Proc. of the Workshop The Legacy of Grey Walter (2002)
12. Hülse, M., Wischmann, S., Pasemann, F.: The Role of Non-linearity for Evolved Multifunctional Robot Behavior. In: Moreno, J.M., Madrenas, J., Cosp, J. (eds.) ICES 2005. LNCS, vol. 3637, pp. 108–118. Springer, Heidelberg (2005)
13. Braitenberg, V.: Vehicles: Experiments in Synthetic Psychology. MIT Press, Cambridge (1984)
14. Yamauchi, B.M., Beer, R.D.: Sequential behavior and learning in evolved dynamical neural networks. Adaptive Behaviour 2(3), 219–246 (1994)
15. Gomez, F., Schmidhuber, J., Miikkulainen, R.: Accelerated neural evolution through cooperatively coevolved synapses. Journal of Machine Learning Research 9, 937–965 (2008)
16. Gomez, F. J.: Robust Non-Linear Control through Neuroevolution. PhD thesis, August 1, Tue, 6 Jan 104 19:10:41 GMT (2003)
17. Stanley, K.O.: Compositional pattern producing networks: A novel abstraction of development. Genetic Programming and Evolvable Machines 8(2), 131–162 (2007)
18. Stanley, K.O., D'Ambrosio, D.B., Gauci, J.: A hypercube-based encoding for evolving large-scale neural networks. Artificial Life 15(2), 185–212 (2009)
19. Koza, J.R., Rice, J.P.: Genetic generation of both the weights and architecture for a neural network. In: International Joint Conference on Neural Networks (1991)
20. Pasemann, F., Steinmetz, U., Hülse, M., Lara, B.: Robot control and the evolution of modular neurodynamics. Theory in Biosciences 120(3-4), 311–326 (2001)
21. Angeline, P.J., Saunders, G.M., Pollack, J.P.: An evolutionary algorithm that constructs recurrent neural networks. IEEE Transactions on Neural Networks 5(1), 54–65 (1994)
22. Gruau, F.: Neural Network Synthesis using Cellular Encoding and the Genetic Algorithm. Laboratoire de l'Informatique du Parallilisme, Ecole Normale Supirieure de Lyon (1994)
23. Hornby, G.S., Pollack, J.B.: Body-brain co-evolution using L-systems as a generative encoding. In: Spector, L., Goodman, E.D., Wu, A., Langdon, W.B., Voigt, H.-M., Gen, M., Sen, S., Dorigo, M., Pezeshk, S., Garzon, M.H., Burke, E. (eds.) Proceedings of the Genetic and Evolutionary Computation Conference (GECCO 2001), July 7-11, pp. 868–875. Morgan Kaufmann, San Francisco (2001)

24. Cangelosi, A., Parisi, D., Nolfi, S.: Cell division and migration in a 'genotype' for neural networks. Network: Computation in Neural Systems 5(4), 497–515 (1994)
25. Belew, R.K.: Interposing an ontogenetic model between genetic algorithms and neural networks. In: Advances in Neural Information Processing Systems 5, NIPS Conference, p. 106. Morgan Kaufmann, San Francisco (1992)
26. Moriaty, D.E.: Symbiotic Evolution of Neural Networks in Sequential Decision Tasks. PhD thesis, The University of Texas at Austin, 1 (1997)
27. Clune, J., Beckmann, B.E., Pennock, R.T., Ofria, C.: HybrID: A hybridization of indirect and direct encodings for evolutionary computation. In: Kampis, G., Karsai, I., Szathmáry, E. (eds.) ECAL 2009, Part II. LNCS, vol. 5778, pp. 134–141. Springer, Heidelberg (2011)
28. Inden, B.: Stepwise Transition from Direct Encoding to Artificial Ontogeny in Neuroevolution. In: Almeida e Costa, F., Rocha, L.M., Costa, E., Harvey, I., Coutinho, A. (eds.) ECAL 2007. LNCS (LNAI), vol. 4648, pp. 1182–1191. Springer, Heidelberg (2007)
29. Doncieux, S., Meyer, J.-A.: Evolving modular neural networks to solve challenging control problems. In: Proceedings of The Fourth International ICSC Symposium on Engineering of Intelligent Systems (EIS 2004), Acta Press (2004)
30. Meyer, J.-A., Doncieux, S., David, Guillot, A.: Evolutionary approaches to neural control of rolling, walking, swimming and flying animats or robots. Biologically Inspired Robot Behavior Engineering, 1–43 (2003)
31. Jeffrey, L., Elman, J.L.: Learning and development in neural networks: The importance of starting small. Cognition 48, 71–99 (1993)
32. Hülse, M., Wischmann, S., Pasemann, F.: Structure and function of evolved neuro-controllers for autonomous robots. Connection Science 16(4), 249–266 (2004)
33. Stanley, K.O., Miikkulainen, R.P.: Efficient evolution of neural networks through complexification. PhD thesis, The University of Texas at Austin (2004)
34. Lee Giles, C., Omlin, C.W.: Pruning recurrent neural networks for improved generalization performance. IEEE Transactions on Neural Networks/A Publication of The IEEE Neural Networks Council 5(5), 848 (1994)
35. Bongard, J.C., Pfeifer, R.: Evolving complete agents using artificial ontogeny. Morpho-Functional Machines: The New Species Designing Embodied Intelligence, 237–258 (2003)
36. Nolfi, S., Parisi, D.: Evolving Artificial Neural Networks that Develop in Time. In: Proceedings of the Third European Conference on Advances in Artificial Life, p. 367. Springer, Heidelberg (1995)
37. Gruau, F.: Automatic definition of modular neural networks. Adaptive Behaviour 3(2), 151–183 (1995)
38. Nolfi, N., Parisi, D.: Growing neural networks. Technical Report PCIA-91-15, Institute of Psychology (December 1991)
39. Pasemann, F.: Neuromodules: A dynamical systems approach to brain modelling. In: Herrmann, H.J., Wolf, D.E., Poppel, E. (eds.) Workshop on Supercomputing in Brain Research: From Tomography to Neural Networks, November 21-23. World Scientific Publishing Co., Germany (1995)
40. Horton, J.C., Adams, D.L.: The cortical column: a structure without a function. Philosophical Transactions of the Royal Society B: Biological Sciences 360(1456), 837 (2005)

41. Reisinger, J., Stanley, K.O., Miikkulainen, R.: Evolving Reusable Neural Modules. In: Deb, K., Poli, R., Banzhaf, W., Beyer, H.-G., Burke, E., Darwen, P., Dasgupta, D., Floreano, D., Foster, J., Harman, M., Holland, O., Lanzi, P.L., Spector, L., Tettamanzi, A., Thierens, D., Tyrrell, A., et al. (eds.) GECCO 2004. LNCS, vol. 3103, pp. 69–81. Springer, Heidelberg (2004)
42. Valsalam, V.K., Miikkulainen, R.: Evolving symmetric and modular neural networks for distributed control. In: Proceedings of the 11th Annual Conference on Genetic and Evolutionary Computation, pp. 731–738. ACM, New York (2009)
43. Gomez, F., Miikkulainen, R.: Incremental evolution of complex general behavior. Technical Report AI96-248, The University of Texas at Austin, Department of Computer Sciences, June 1, November 7, 106 21:26:08 GMT (1997)
44. Gauci, J., Stanley, K.O.: Generating large-scale neural networks through discovering geometric regularities. In: Proceedings of the 9th Annual Conference on Genetic and Evolutionary Computation, p. 1004. ACM, New York (2007)
45. David, B., D'Ambrosio, D.B., Stanley, K.O.: A novel generative encoding for exploiting neural network sensor and output geometry. In: Genetic and Evolutionary Computation Conference: Proceedings of the 9th Annual Conference on Genetic and Evolutionary Computation. Association for Computing Machinery, Inc., New York (2007)
46. Dieckmann, U.: Coevolution as an autonomous learning strategy for neuromodules. In: Herrmann, H.J., Wolf, D.E., Poppel, E. (eds.) Workshop On Supercomputing In Brain Research: From Tomography To Neural Networks, November 21-23. World Scientific Publishing Co., Germany (1995)
47. Pasemann, F., Steinmetz, U., Dieckman, U.: Evolving structure and function of neurocontrollers. In: Angeline, P.J., Michalewicz, Z., Schoenauer, M., Yao, X., Zalzala, A. (eds.) Proceedings of the Congress on Evolutionary Computation, July 6-9, vol. 3, pp. 1973–1978. IEEE Press, USA (1999)
48. Moriarty, D.E., Miikkulainen, R.: Efficient reinforcement learning through symbiotic evolution. Technical Report AI94-224, The University of Texas at Austin, Department of Computer Sciences (September 1, 1994)
49. Christian, W., Rempis, C.W., Pasemann, F.: Search space restriction of neuro-evolution through constrained modularization of neural networks. In: Mandai, K. (ed.) Proceedings of the 6th International Workshop on Artificial Neural Networks and Intelligent Information Processing (ANNIIP), in Conjunction with ICINCO 2010, pp. 13–22. SciTePress, Portugal (2010)
50. Mahfoud, S.W.: Niching methods for genetic algorithms. Department of Computer Science, University of Illinois at Urbana-Champaign (1995)
51. Hancock, P.J.B.: An empirical comparison of selection methods in evolutionary algorithms. To appear in the Proceedings of the AISB Workshop on Evolutionary Computation, vol. 1 (1994)
52. Reil, T., Husbands, P.: Evolution of central pattern generators for bipedal walking in a real-time physics environment. IEEE Transactions on Evolutionary Computation 6(2), 159–168 (2002)
53. Hein, D., Hild, M., Berger, R.: Evolution of biped walking using neural oscillators and physical simulation. In: Visser, U., Ribeiro, F., Ohashi, T., Dellaert, F. (eds.) RoboCup 2007: Robot Soccer World Cup XI. LNCS (LNAI), vol. 5001, pp. 433–440. Springer, Heidelberg (2008)
54. Geng, T., Porr, B., Wörgötter, F.: A reflexive neural network for dynamic biped walking control. Neural computation 18(5), 1156–1196 (2006)

55. Manoonpong, P., Geng, T., Kulvicius, T., Porr, B., Wörgötter, F.: Adaptive, Fast Walking in a Biped Robot under Neuronal Control and Learning. PLoS Computational Biology 3(7) (2007)
56. McHale, G., Husbands, P.: GasNets and other evolvable neural networks applied to bipedal locomotion. In: From Animals to Animats 8: Proceedings of the Seventh [ie Eighth] International Conference on Simulation of Adaptive Behavior, p. 163. The MIT Press, Cambridge (1994)
57. Ishiguro, A., Fujii, A., Hotz, P.E.: Neuromodulated control of bipedal locomotion using a polymorphic cpg circuit. Adaptive Behavior 11(1), 7 (2003)
58. Hase, K., Yamazaki, N.: Computational evolution of human bipedal walking by a neuro-musculo-skeletal model. Artificial Life and Robotics 3(3), 133–138 (1999)
59. Josh, C.: Making evolution an offer it can't refuse: Morphology and the extradimensional bypass. Advances in Artificial Life, 401–412 (2001)
60. Cliff, D., Harvey, I., Husbands, P.: Incremental evolution of neural network architectures for adaptive behaviour. In: Proceedings of the First European Symposium on Artificial Neural Networks, ESANN 039, vol. 93, pp. 39–44. D facto Publishing (1992)
61. Pasemann, F.: Complex dynamics and the structure of small neural networks. Network: Computation in Neural Systems 13(2), 195–216 (2002)
62. Rempis, C., Thomas, V., Bachmann, F., Pasemann, F.: NERD Neurodynamics and Evolutionary Robotics Development Kit. In: Ando, N., Balakirsky, S., Hemker, T., Reggiani, M., von Stryk, O. (eds.) SIMPAR 2010. LNCS, vol. 6472, pp. 121–132. Springer, Heidelberg (2010)
63. Pasemann, F.: Characterization of periodic attractors in neural ring networks. Neural Networks 8(3), 421–429 (1995)
64. Pasemann, F., Hild, M., Zahedi, K.: So(2)-networks as neural oscillators. In: Computational Methods in Neural Modeling, vol. 2686, pp. 144–151 (2003)
65. Miller, B.L., Goldberg, D.E.: Genetic algorithms, tournament selection, and the effects of noise. Urbana 51, 61801 (1995)
66. Jakobi, N.: Evolutionary robotics and the radical envelope-of-noise hypothesis. Adaptive behavior 6(2), 325 (1997)
67. Pollack, J.B., Lipson, H., Ficici, S., Funes, P., Hornby, G., Watson, R.A.: Evolutionary techniques in physical robotics. Evolvable Systems: from biology to hardware, 175–186 (2000)
68. Bongard, J.C., Lipson, H.: Automating Genetic Network Inference with Minimal Physical Experimentation Using Coevolution. In: Deb, K., et al. (eds.) GECCO 2004. LNCS, vol. 3102, pp. 333–345. Springer, Heidelberg (2004)
69. von Twickel, A., Pasemann, F.: Reflex-oscillations in evolved single leg neurocontrollers for walking machines. Natural Computing 6(3), 311–337 (2007)
70. Rempis, C.W.: Short-term memory structures in additive recurrent neural networks. Master's thesis, University of Applied Sciences Bonn-Rhein-Sieg, Germany (2007)
71. Wischmann, S., Pasemann, F.: The emergence of communication by evolving dynamical systems. From Animals to Animats 9, 777–788 (2006)
72. Sidel, T., Hild, M., Weidner, M.: Concept and Design of the Modular Actuator System for the Humanoid Robot MYON. In: International Conference on Intelligent and Applications, ICIRA 2011 (2011)

A Genetic Programming-Based Approach for the Performance Characteristics Assessment of Stabilized Soil

Amir Hossein Alavi*, Amir Hossein Gandomi, and Ali Mollahasani

Abstract. This chapter presents a variant of genetic programming, namely linear genetic programming (LGP), and a hybrid search algorithm coupling LGP and simulated annealing (SA), called LGP/SA, to predict the performance characteristics of stabilized soil. LGP and LGP/SA relate the unconfined compressive strength (UCS), maximum dry density (MDD), and optimum moisture content (OMC) metrics of stabilized soil to the properties of the natural soil as well as the types and quantities of stabilizing additives. Different sets of LGP and LGP/SA-based prediction models have been separately developed. The contributions of the parameters affecting UCS, MDD, and OMC are evaluated through a sensitivity analysis. A subsequent parametric analysis is carried out and the trends of the results are compared with previous studies. A comprehensive set of data obtained from the literature has been used for developing the models. Experimental results confirm that the accuracy of the proposed models is satisfactory. In particular, the LGP-based models are found to be more accurate than the LGP/SA-based models.

Amir Hossein Alavi
School of Civil Engineering, Iran University of Science and Technology, Tehran, Iran
e-mail: ah_alavi@hotmail.com, am_alavi@civileng.iust.ac.ir

Amir Hossein Gandomi
Intelligent Structural Engineering and Health Monitoring Laboratory, Department of Civil Engineering, University of Akron, Akron, OH, USA
e-mail: a.h.gandomi@gmail.com

Ali Mollahasani
Department of Civil, Environmental and Material Engineering, University of Bologna, Bologna, Italy
e-mail: ali.mollahasani2@unibo.it

* Corresponding author.

1 Introduction

Chemical stabilization involves the addition of chemicals such as lime, cement, and asphalt, or a combination of these to soil. When mixed into a particular type of soil, the additives react with natural soil constituents. This results in strength increases, changes in porosity, volume, permeability and density, waterproofing, and reduction of surface abrasion including cementation of the soil particles [1, 2, 3]. The chemical stabilization of soils often leads to savings in construction costs of various civil engineering applications such as road construction, earth wall construction, foundation, and other earthworks purposes. A variety of properties including measures of strength, density, durability, and shrinkage can be measured as outcomes of the soil stabilization.

Besides the emphasis on other characteristics of the stabilized soil, the experts generally agree on the unconfined compressive strength (UCS) and maximum dry density (MDD) as the most important measures for the outcome of the stabilization [1, 2, 4]. The UCS and MDD of the stabilized soils are important factors in determining the soil suitability or unsuitability for a stabilization process. The optimum moisture content (OMC) of soil-stabilizer mixes is also an explanatory variable in stabilizing soils. The moisture content influences the properties of a compacted soil mixture stronger than any other factor. Bryan [5] and Osula [6] employed the OMC as a discriminator variable to assess the suitability of some soil samples for the stabilization purposes. In order to obtain effective compaction in the chemical stabilization, the OMC of the compacted soil must be measured for any of the available soils. The UCS, MDD, and OMC of soils are generally determined from extensive and cumbersome laboratory tests on every new construction site. The appropriate selection of a chemical stabilizer is also a major concern for successful soil stabilization. Therefore, it is desirable to develop precise mathematical models relating UCS, MDD, and OMC to the properties of natural soil as well as quantities and types of stabilizers.

There were several studies in the literature dealing with the estimation of the compaction parameters (OMC and MDD) of natural soil using empirical regression models [7, 8, 9, 10, 11] or neural networks (NNs) [12, 13]. Most of the models in the stabilization literature were particularly developed to relate the strength to the important variables involved [14, 15, 16, 17]. Nearly all of the available models are essentially statistical correlations between the natural and stabilized soil compaction properties and quality control parameters (i.e., UCS, MDD, and OMC), classification data (or index properties) of soil, and stabilizer type or quantity [5, 7, 8, 9, 10, 11, 16, 17]. The classification data used generally include plasticity characteristics (such as the liquid limit, the plastic limit, the shrinkage limit, and the plasticity index), the specific gravity, and the grain size distribution. It can be observed that the specific index properties adopted in various correlation equations differ significantly.

Although the conventional statistical models can provide reasonable predictions, they have significant limitations. Different statistical models need to be developed for different conditions of the same soil. Moreover, several assumptions are incorporated into these models. On the other hand, NNs are black-box models. That is, they usually do not give a deep insight into the way in which they use the available information to obtain a solution. Hence, they do not provide a better understanding of the nature of the derived relationship between different interrelated input and output data.

Genetic programming (GP) [18, 19] is another alternative approach for behavior modeling of civil engineering problems. GP belongs to the family of evolutionary algorithms (EA) [20], which are inspired by Darwin's evolution theory. It may generally be defined as a supervised machine learning technique that searches a program space instead of a data space [19]. Some example application areas of GP include symbolic regression, grammar induction, data mining and data analysis, circuit design and layout, and evolving game players [21]. GP has a great ability to learn from data examples presented to it in order to capture the subtle functional relationships among the data. Linear genetic programming (LGP) [22] is a GP approach where programs have a linear structure similar to string genomes employed in the earliest EAs. LGP is a machine learning approach that uses sequences of imperative instructions as genetic material. More specifically, LGP operates on programs that are represented as linear sequences of instructions of an imperative programming language [22, 23].

Simulated annealing (SA) is a general stochastic search algorithm used for solving optimization problems. The Metropolis algorithm, the foundation of SA, was proposed by Metropolis et al. [24] to simulate the annealing process in metalworking. The SA algorithm was first applied to optimization problems by Kirkpatrick et al. [25] and Cerny [26]. SA is very useful for solving different types of optimization tasks with nonlinear functions and multiple local optima [27, 28]. The ability and shortcomings of SA are well summarized by Ingber in [29]. Folino et al. [30] combined GP and SA to produce a hybrid algorithm with better efficiency. They used the SA acceptance strategy to select new individuals and showed that introducing this strategy into the GP process improves its resulting outcome.

This study investigates the potential of LGP and a hybrid search algorithm that couples LGP and SA, called LGP/SA, in simulating the nonlinear behavior of the UCS, MDD and OMC of the soil-stabilizer mixes. These techniques are useful in deriving empirical models for UCS, MDD, and OMC by directly extracting the knowledge contained in the experimental data. The rest of the chapter is organized as follows: Section 2 presents brief descriptions of the traditional GP, LGP, and LGP/SA algorithms. Section 3 outlines the model development using LGP and LGP/SA and reviews the existing results. The detailed performance analyses of the proposed models are then discussed in Section 4. The results of sensitivity and parametric analyses are given in

Sections 5 and 6, respectively. Finally, concluding remarks and future research directions are outlined in Section 7.

2 Genetic Programming

GP is a symbolic optimization technique that creates computer programs to solve a problem using principles copied from the Darwinian theory of evolution in nature. The breakthrough in GP came in the late 1980s with the experiments of Koza [18] on symbolic regression [21]. Like genetic algorithms (GAs), GP belongs to the family of evolutionary algorithms. This classical tree-based GP technique is also referred to as standard GP [18]. In GP, an initial, highly diverse population of individuals (computer programs) is created randomly. A population member in standard GP is a hierarchically structured tree comprising functions and terminals. These functions and terminals are selected from a set of functions and a set of terminals. For example, the function set F can contain the basic arithmetic operations (+, −, ×, /, etc.), Boolean logic functions (∧, ∨, ¬, and so on), or any other mathematical functions. The terminal set T contains the arguments for the functions and can consist of, for instance, numerical constants, logical constants, and variables. The functions and terminals are chosen at random and constructed together to form a tree-like structure with a root point with branches extending from each function and ending in a terminal. An example instance of such a tree representation of a program is illustrated in Figure 1.

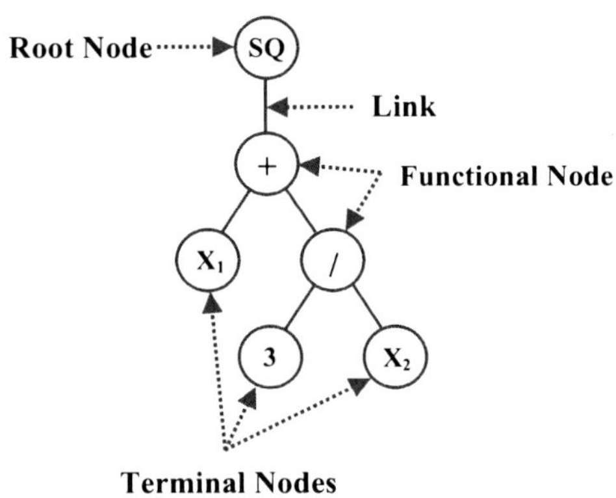

Fig. 1. The tree representation of a GP model $(X_1 + 3/X_2)^2$.

In addition to traditional tree-based GP, there are other types of GP where programs are represented in different ways (see Figure 2). These are linear and graph-based GP [31]. The emphasis of the present study is placed on the linear GP techniques. Several linear variants of GP have been proposed such as linear genetic programming (LGP) [22] and multi-expression programming (MEP) [32]. The linear variants of GP make a clear distinction between the genotype and phenotype of an individual. In these variants, individuals are represented as linear strings [33]. Such linear programs can have a complex control flow similar to the trees of standard GP when executed. There are some main reasons for using linear GP. Basic computer architectures are fundamentally the same now as they were twenty years ago, when GP began. Almost all architectures represent computer programs in a linear fashion. In other words, computers do not naturally run tree-shaped programs. Hence, slow interpreters have to be used as part of tree-based GP. Hence, by evolving the machine code patterns actually obeyed by the computer, the use of an expensive interpreter (or compiler) is avoided and GP can run several orders of magnitude faster [31].

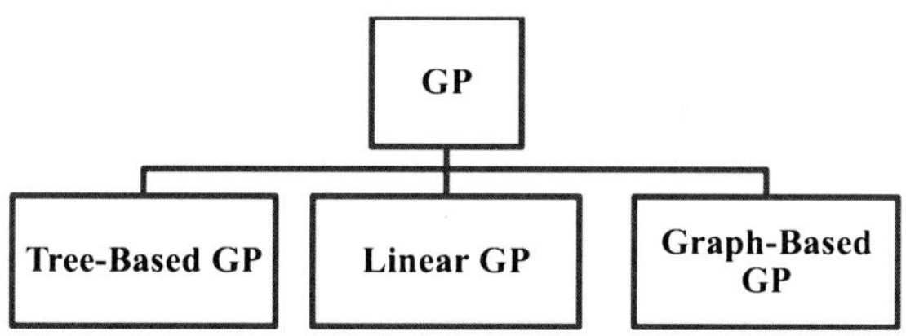

Fig. 2. Different types of genetic programming.

2.1 Linear Genetic Programming

LGP is a subset of GP with a linear representation of individuals. The main characteristic of LGP in comparison with traditional tree-based GP is that expressions of a functional programming language (like LISP) are substituted by programs of an imperative language (like C/C++) [22, 23]. Figure 3 presents a comparison of the program structures in LGP and tree-based GP. A linear genetic program can be seen as a data flow graph generated by multiple usage of register content. That is, on the functional level, the evolved imperative structures denote directed graphs. In tree-based GP, the data flow is more rigidly determined by the tree structure of the program [23].

In the LGP system described here, an individual program is interpreted as a variable-length sequence of simple C instructions. The instruction set or function set of LGP consists of arithmetic operations, conditional branches, and function calls. The terminal set of the system is composed of variables and constants. The instructions are restricted to operations that accept a minimum number of constants or memory variables, called registers (r), and assign the result to a destination register, e.g., $r[0] = r[1] + 1$. Part of a program in linear representation in C code is represented in Figure 4. In this figure, register $r[0]$ holds the final program output.

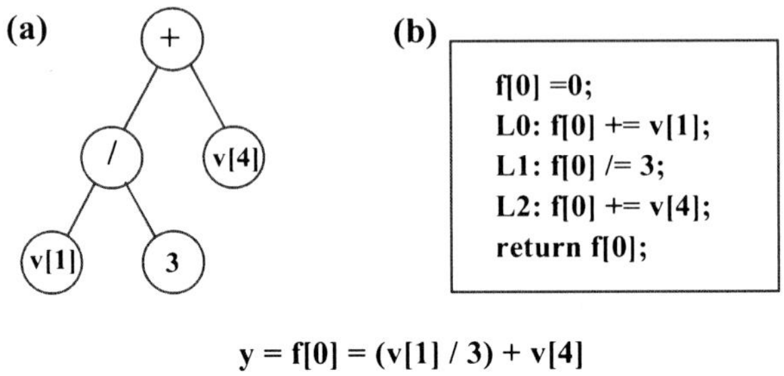

Fig. 3. Comparison of GP program structures. (a) Tree-based GP (b) LGP.

```
void LGP (double r[5])
{ ...
r[0] = r[5] + 70;
r[5] = r[0] - 50;
if (r[1] > 0)
if (r[5] > 2)
r[4] = r[2] × r[1];
r[2] = r[5] + r[4];
r[0] = sin(r[2]);
}
```

Fig. 4. An excerpt of a linear genetic program.

The Automatic Induction of Machine code by Genetic Programming (AIMGP) system is a well-known implementation of LGP. In AIMGP, evolved programs are stored as linear strings of native binary machine code and are directly executed by the processor during fitness calculation. The absence of an interpreter and complex memory handling results in a significant speedup in the AIMGP execution compared to tree-based GP [23]. This machine-code-based LGP approach searches for the computer program structure and the constants at the same time. The machine-code based LGP performs the following steps for a single run [23, 34]:

I. A population of randomly generated programs is initialized and the fitness values of these individuals are calculated.
II. Select four programs randomly from the population for a tournament. Based on their fitness, two programs are picked as the winners and two as the losers.
III. After that, the two winner programs are copied and transformed in a randomized fashion as follows:

- Parts of the winner programs are exchanged with each other to create two new programs (crossover), and/or

- Each of the tournament winners are modified randomly to create two new programs (mutation).

IV. The loser programs of the tournament are replaced with the transformed winner programs. The winners of the tournament remain without change.

V. Repeat steps two through four until the termination or convergence conditions are satisfied.

Crossover occurs between instruction blocks. Figure 5 demonstrates a two-point linear crossover used in LGP for recombining two tournament winners. As it is seen, a segment of random position and arbitrary length is selected in each of the two parents and exchanged. If one of the two children would exceed the maximum length, crossover is aborted and restarted with exchanging equally sized segments [23]. The mutation operation occurs on a single instruction. Two types of standard LGP mutations are commonly used: micro and macro mutation. Micro mutation changes an operand or an operator of an instruction [23]. The macro mutation operation inserts or deletes a random instruction [23]. Comprehensive descriptions of the basic parameters used to direct the search in linear genetic programming can be found in [22]. According to Francone and Deschaine [35], the LGP system can be regarded as an efficient modeling tool for complex problems for several reasons including:

- Its speed permits conducting many runs in realistic timeframes. This leads to deriving consistent, high-precision models with little customization;
- it is well-designed to prevent overfitting and to evolve robust solutions; and
- the solutions evolved by the LGP system execute very quickly when called by an optimizer.

2.2 Hybrid Linear Genetic Programming/Simulated Annealing Algorithm

In this chapter, LGP with a SA-based selection strategy is employed for developing the prediction models. In this coupled algorithm, the SA strategy is used to select new individuals [30, 36]. SA makes use of the Metropolis algorithm [24] for computer simulation of annealing. Annealing is a process in which a metal is heated to a high temperature and then is gradually cooled to relieve thermal stresses. During the cooling process, each atom takes a specific position in the crystalline structure of the metal. By changing the temperature this crystalline structure changes to a different configuration. An internal energy, E, can be measured and assigned to each state of crystalline structure of the metal which is achieved during the annealing process.

At each step of the cooling process, if the temperature does not decrease too quickly, the atoms are allowed to adjust to a stable equilibrium state of least energy. It is evident that changing of the crystalline structure of

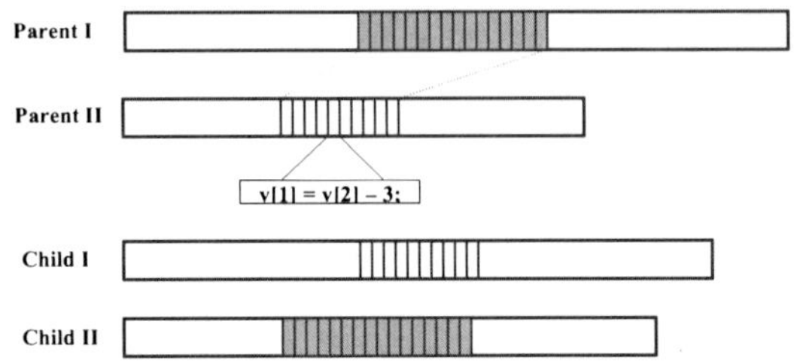

Fig. 5. Crossover in LGP [23].

a metal, through the annealing, is associated with changing of the internal energy by an amount ΔE. As the metal temperature drops down gradually, the overall trend of changing internal energy follows a decreasing process. However, sometimes the energy may also increase by chance. The probability of acceptance of an increase in internal energy by ΔE is given by Boltzmann's probability distribution function as follows:

$$P(\Delta E) = e^{(\frac{-\Delta E}{KT})} \qquad (1)$$

where T is the temperature of the metal measured in Kelvin and K is the Boltzmann's constant. The crystalline structure of a metal achieves near global minimum energy states during the process of annealing. This process is simulated by SA to find the minimum of a function in a certain design space. The objective function corresponds to the energy state and moving to any new set of design variables corresponds to a change of the crystalline structural state.

2.2.1 A Coupled Algorithm of LGP and SA

Considering the above explanations for LGP and SA, the coupled LGP/SA algorithm uses the following main steps to evolve a computer program [30, 36]:

I. A single program is initially created at random. This is the "parent" program for the first repetition of the learning cycle.
II. The parent program is copied.
III. A search operator – crossover or mutation – transforms the copy of the parent program. The transformed copy is called "child" program or "offspring" program. The crossover operator produces two children programs. But only one of these programs is compared with the parent as a candidate to replace the parent program. Which of the two children is used depends on the value of the offspring choice parameter.
IV. The fitness values of both, the parent and the child program, are calculated.

V. Based on these values, the SA algorithm decides whether to replace the parent program with the child program. If the child has better fitness than the parent, the child always replaces its parent. If the child has worse fitness than the parent, the child replaces the parent probabilistically. The probability of replacement depends on how much worse the fitness of the child is than the parent and also on the temperature parameter T in SA. As the annealing process continues, T is gradually reduced at each nth iteration. This means that, for the program, the probability of replacing a better parent with a worse child gets lower and lower as the run continues. If the child program replaces the parent program, then the child program becomes the new parent for the next cycle. Alternatively, if the parent program is not replaced by the child, it remains as the parent program for the next cycle.

VI. If the termination or convergence conditions are satisfied, the process is terminated. Otherwise, the process is continued at step III.

2.3 Related Works on Applications of GP, LGP, and LGP/SA

For the last ten years, tree-based GP has been pronounced as an alternative and robust method for solving civil engineering problems. Some of these applications include structural optimization [37], modeling of a municipal wastewater treatment plant [38], the prediction of the soil-water characteristic curve [39], and the evaluation of liquefaction induced lateral displacements [40]. Narendra et al. [17] have recently conducted research with the specific objective of applying tree-based GP to the soil stabilization problem.

Unlike tree-based GP and other soft computing tools like NNs, applications of LGP are restricted to relatively fewer areas. They include the prediction of compressive and tensile strength of limestone [41], behavior appraisal of steel semi-rigid joints [42], the formulation for compressive strength of CFRP confined concrete cylinders [43], time-series modeling of daily flow rates [44], and the prediction of circular pile scour [45]. Applications of LGP/SA to solve problems in civil engineering are conspicuous by their near absence. Recently, Gandomi et al. [46] utilized this hybrid method to simulate the behavior of the beam-column steel joints.

3 Modeling of Performance Characteristics of Stabilized Soil

The complexity of analysis of the chemical stabilization is due to the participation of a considerable number of factors in this process. UCS, MDD, and OMC, as the performance characteristics of the stabilized soil, are functions of variables such as properties of the soil used, and the type and quantity of the stabilizer used. It is difficult to isolate the effects of these parameters on the stabilized material. This chapter considers the feasibility of using the LGP and LGP/SA algorithms to predict the UCS, MDD, and OMC metrics of the stabilized soil. The most important factors representing the behavior of UCS, MDD, and OMC were selected based on an extensive trial study and literature review [5, 13, 16, 17, 47]. Consequently, the general models for UCS, MDD, and OMC were considered to be as follows:

$$\begin{pmatrix} UCS \\ MDD \\ OMC \end{pmatrix} = f(LL, PI, LS, C, S, LC, CC, AC) \tag{2}$$

where,

LL (%): Liquid limit
PI (%): Plasticity index
LS (%): Linear shrinkage
C (%): Clay/silt content
S (%): Sand content
LC (%): Lime content
CC (%): Cement content
AC (%): Asphalt content

The above variables represent the natural soil properties such as textural properties, plasticity and linear shrinkage, and different types of stabilizers. After developing and controlling several models with different combinations of the input parameters, the best models were selected and presented.

3.1 Experimental Database and Data Preprocessing

The models were evolved with LGP and LGP/SA based on a comprehensive database obtained from the literature. Burroughs [47] provided a total of 230 determinations of the UCS of 28-day cured stabilized earth samples. These data have been accumulated over a period of 8 years from 29 rammed earth construction sites in New South Wales, Australia. The database includes measurements of percentages of liquid limit (%LL), plastic limit (%PL), plasticity index (%PI), linear shrinkage (%LS), gravel (%G), clay/silt (%C), sand (%S), lime (%LC), cement (%CC), and asphalt (%AC). UCS (MPa), MDD (ton/m^3) and OMC (%) were also measured as the quality control parameters of the stabilized soil. Data samples that missed the required data were eliminated. Out of 230 data samples, 219 samples were considered for developing the UCS prediction models and 192 samples were taken for developing the prediction models for MDD and OMC.

It is noteworthy that some of the soil property variables are fundamentally interdependent. This interdependency can cause problems in analysis as it will tend to exaggerate the strength of the relationships between the variables. Gravel is calculated by subtracting the sum of sand and clay/silt from 100, and the plastic limit is the difference between the plasticity index and the liquid limit. Hence, both gravel and plastic limit were excluded from the models. Sand, clay/silt, liquid limit, or plasticity index could potentially have been excluded rather than the gravel and plastic limit. They were retained since traditionally, they have been more frequently used as indicators of soil condition. Also, they have more favorable distributions of values. The ranges of different input and output parameters involved in the model development are given in Table 1.

The minimum ratio of the number of objects over the number of selected variables for model acceptability is 3, but often a suffer value of 5 is suggested [48]. In the present study, this ratio is much higher and is at least equal to $192/8 = 24$. Although normalization is not strictly necessary in the GP-based analysis, better results are usually reached after normalizing the variables. This is mainly due to influence of unification of the variables, no matter their range of variation. Thus, the input and output variables were normalized in this study. After controlling several normalization methods [49, ?], the following method was used to normalize the variables to a range of $[L, U]$:

$$X_n = ax + b \tag{3}$$

Table 1. The ranges of different input and output parameters.

Parameters	Minimum	Maximum	Standard deviation	Mean
LL (%)	18.00	95.00	36.54	14.67
PI (%)	0.00	70.00	17.30	13.73
LS (%)	1.00	19.80	7.54	4.43
C (%)	5.00	53.00	24.98	10.54
S (%)	30.00	94.00	64.98	15.59
UCS (MPa)	1.00	5.40	2.65	0.96
MDD (ton/m^3)	1.52	2.20	0.16	1.88
OMC (%)	6.00	22.70	3.20	10.63
LC (0%, 2%, 3%, 4%, 5%, and 6%)				
CC (0%, 2%, 3%, 4%, 5%, and 6%)				
AC (0%, 3%)				

where $a = (U - L)/(X_{\max} - X_{\min})$ and $b = U - aX_{\max}$, in which $X_{\max}$ and $X_{\min}$ are the maximum and minimum values of the variable and X_n is the normalized value. In the present study, L and U were respectively set to 0.05 and 0.95. 0 and 1 were not selected as the lower and upper limits since some of the functions used by the LGP and LGP/SA algorithms are sensitive to 0 (e.g., division).

3.2 Performance Measures

The correlation coefficient (R), mean squared error (MSE) and the mean absolute error (MAE) were used to evaluate the capabilities of the LGP and LGP/SA models. R, MSE and MAE are calculated using the following equations:

$$R = \frac{\sum_{i=1}^{n}(h_i - \overline{h}_i)(t_i - \overline{t}_i)}{\sqrt{\sum_{i=1}^{n}(h_i - \overline{h}_i)^2 \sum_{i=1}^{n}(t_i - \overline{t}_i)^2}} \tag{4}$$

$$MSE = \frac{\sum_{i=1}^{n}(h_i - t_i)^2}{n} \tag{5}$$

$$MAE = \frac{1}{n}\sum_{i=1}^{n}|h_i - t_i| \tag{6}$$

where h_i and t_i respectively are the experimental and predicted output values for the ith output; $\overline{h}_i$ is the arithmetic mean of the experimental outputs, and n is the number of sample.

3.3 Model Development Using LGP

The available database was used for developing comprehensive LGP-based models relating the UCS, MDD, and OMC of the chemically stabilized soil to LL (%), PI(%), LS(%), S(%), C(%), LC(%), CC(%), and AC(%). Three LGP prediction models were separately developed, one each for UCS, MDD, and OMC. The parameters of the applied LGP algorithm are shown in Table 2. The parameter selection will affect the model generalization capability of LGP. In this study, four basic arithmetic operators (+, −, ×, /), basic mathematical functions ($\sqrt{}$, ||, power, sin, cos), a comparison instruction (<), and conditional branches (if <=) were utilized to get the optimum LGP models. Several runs were conducted to come up with a parameterization of LGP that provided enough robustness and generalization to solve the problem. The LGP parameters were changed for different runs. Three levels were set for the population size and two levels were considered for the maximum program size and crossover rate. One level was considered for the other parameters based on some previously suggested values [41, 42, 43, 50] and also after performing several preliminary runs and observing the performance behavior. The number of demes presented in Table 2 defines how the population of programs is divided. Demes are semi-isolated subpopulations. Evolution proceeds faster in such smaller populations in comparison to a single big population of equal total size [22]. There are $3 \times 2 \times 2 = 12$ different combinations of the parameters. All of these combinations were tested and for each combination, 10 replications were carried out. This makes 120 runs for each of UCS, MDD, and OMC. Therefore, the overall number of runs is equal to $120 \times 3 = 360$ (the considered outputs is 3). A fairly large number of generations were tested on each run to find models with minimum error. For each case, the program was run until there no longer was any significant improvement in the performance of the models. Each run was observed for overfitting while in progress. In order to do this, situations were searched in which the fitness on the samples for the training of LGP was negatively correlated with the fitness on the validation data sets. For the runs showing signs of overfitting, the LGP parameters were progressively changed so as to reduce the computational power available to the LGP algorithm until the observed overfitting was minimized. The resulting run was then accepted as the production run. The programs with the best performance on both of the training and validation data sets were finally selected as the outcomes of the experiment. For the LGP-based analysis, the Discipulus Pro^{TM} [51] software was used which works on the basis of the AIMGP platform.Two types of LGP models were used in this chapter:

1. a program model/evolved program, which is a single solution, and
2. a team model, which is a combination of single program models.

Table 2. Parameter settings for the LGP algorithm.

Parameters	Settings
Population size	2500, 3500, 6000
Generations since start of run	1000
Crossover rate (%)	90
Homologous crossover (%)	30
Mutation rate %)	30
Block mutation rate (%)	30
Instruction mutation rate (%)	30
Data mutation rate (%)	40
Maximum program size	256, 512
Initial program size	80
Function set	$+, -, \times, /, \sqrt{}, \|\|$, power, sin, cos, $<$, if $<=$
Numerical Constants	Randomize $(\min - : 10, \max + : 10)$
Number of demes	20
Fitness function	Squared error

The team solution employs a combination of an odd number of single solutions (minimum 1; maximum 9). After completing an experiment, the evolved single and team solutions are transformed in Java, C++, or Intel assembler source code [52]. Discipuls' interactive evaluator mode was employed to run the resulting C codes evolved by LGP.

Overfitting is one of the principal problems in machine learning generalization. It is a case in which the error on the training set is driven to a very small value, but when new data is presented to the model, the error is large. An efficient approach to prevent overfitting is to test other individuals from the run on a validation set to find a better generalizer [19]. This technique was used in this study for improving the generalization of the models. For this purpose, the available data sets were randomly divided into training, validation and testing subsets. The training data were used for learning (genetic evolution). The validation data were used to specify the generalization capability of the evolved programs on data they did not train on (model selection). In other words, the training and validation data sets were used to select the best evolved programs. This technique provides decent results as long as the models perform well on the training data sets [19]. The testing data were finally used to measure the performance of the models obtained by LGP on data that played no role in building the models. Out of 219 data samples for the prediction of UCS, 111 vectors were used for the training, 54 samples for the validation, and 54 for the testing of the models. Out of 192 data samples for the prediction of MDD and OMC, 100 data vectors were used for the training, 46 for the validation, and 46 for the testing of the models. In order to obtain a consistent data division, several combinations of the training, validation and testing sets were considered. The selection was such that the maximum, minimum, mean, and the standard deviation of parameters were consistent in the datasets.

3.3.1 LGP Single Solutions

The model architectures that gave the best results for predicting UCS, MDD, and OMC were built using all the eight parameters. An excerpt of the best LGP single solution for the prediction of UCS is given in Appendix as C++ source. Comparisons of the experimental and predicted UCS (MPa), MDD (ton/m^3), and OMC (%) values using the best single solutions found by LGP are illustrated in Figures 6a-f.

3.3.2 LGP Team Solutions

In the best LGP team solutions, too, eight variables were used as the inputs. The prediction results obtained by the best LGP team solutions are illustrated in Figures 7a-f.

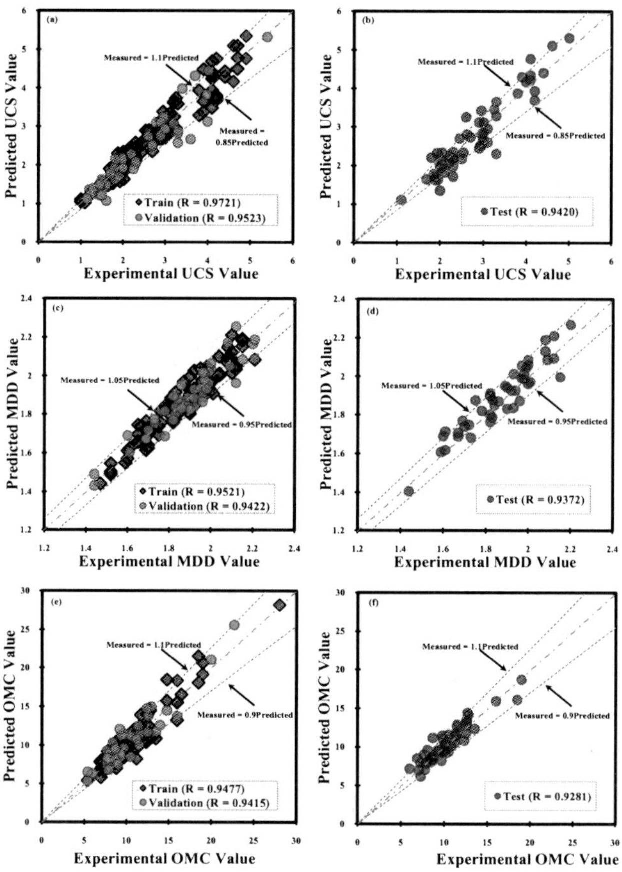

Fig. 6. Experimental versus predicted UCS, MDD, and OMC using the LGP single solutions.

3.4 Model Development Using LGP/SA

The LGP/SA-based models were trained, validated and tested using the available stabilization experimental results. The eight predictor variables used in the LGP/SA modeling process were LL (%), PI(%), LS(%), S(%), C(%), LC(%), CC(%), and AC(%). Three LGP/SA prediction models were separately developed for UCS, MDD, and OMC. Various parameters involved in the LGP/SA prediction algorithm are shown in Table 3. Again, basic arithmetic operators and mathematical functions were used to get the optimum LGP/SA models. Similar to LGP, several runs were conducted considering different values for the control parameters of the LGP/SA algorithm. The proper number of temperature levels until the algorithm should terminate

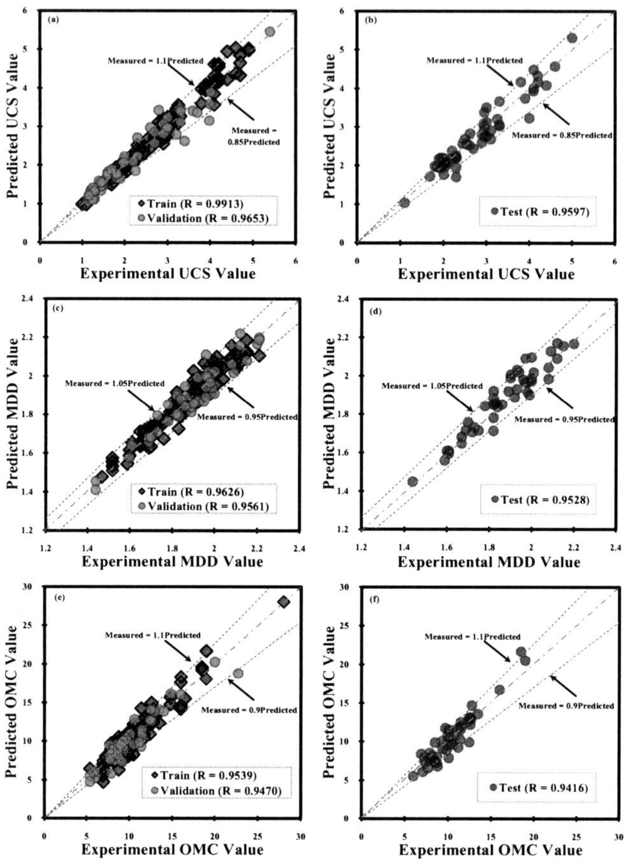

Fig. 7. Experimental versus predicted UCS, MDD, and OMC using the LGP team solutions.

depends on the number of possible solutions and complexity of the problem. The number of iterations per temperature level defines the number of times a new child program is created from the parent program at each temperature level. In order to find models with minimum error, each run was performed with large numbers of temperature levels and iterations. The program was run until no significant error decrease could be observed anymore.

Three different values were tested as number of temperature levels. 1500 iterations were considered for each temperature level. The crossover rate parameter in the LGP/SA algorithm sets the balance between the uses of the search operators (crossover and mutation). The crossover rate was set to values between 50% and 95%. A value of 50% means that 50% of time the used search operator will be the crossover operator. The mutation operator will therefore be employed in the other 50% of reproduction steps by the LGP/SA

Table 3. Parameter settings for the LGP/SA algorithm.

Parameters	Settings
Number of temperature levels	9000, 1100, 12000
Number of iterations per temperature level	1500
Start temperature	5
Stop temperature	0.01
Crossover rate (%)	50, 95
Homologous crossover (%)	95
Probability of randomly generated parent in crossover (%)	99
Mutation rate (%)	90
Block mutation rate (%)	30
Instruction mutation rate (%)	30
Data mutation rate (%)	40
Offspring choice rate (%)	50
Replacement scaling factor	1
Maximum program size	256, 512
Initial program size	80
Function set	$+, -, \times, /, \sqrt{}, \|\|$, power, sin, cos, tan
Numerical Constants	Randomize (min : $-10, Max : +10$)
Fitness function	Squared error

algorithm [36]. Two levels were set for the maximum program size. The values of the other involved parameters were selected based on some previously suggested values [45] and also after performing many preliminary runs and checking the performance. Hence, in total, there are $3 \times 2 \times 2 = 12$ different parameter settings. All of these combinations were tested and 10 replications were performed for each combination. Thus, 120 runs were carried out with the LGP/SA algorithm for each quantity subject to prediction. Therefore, the total number of runs was equal to $120 \times 3 = 360$ since there are three metrics (UCS, MDD, OMC). The best programs during the training and validation processes were chosen as the outcomes of the LGP/SA algorithm. A similar strategy to that considered for deriving the LGP models was followed during the LGP/SA runs to prevent overfitting. The LGP/SA algorithm was implemented using the Discipulus Lite^{TM} [53] software. The programs evolved by LGP/SA are written in C++ or inline assembler code. The TurboC environment was employed to run the C codes evolved by LGP/SA. Similar to the LGP models, the data sets were divided into the training, validation and testing subsets.

3.4.1 LGP/SA Solutions

The model architectures that gave the best results for predicting UCS, MDD, and OMC were built using eight predictor variables. An excerpt of the best LGP/SA solution for the prediction of UCS is given in the Appendix as C++ source. Comparisons of the predicted and experimental UCS (MPa), MDD (ton/m^3) and OMC (%) of the soil-stabilizer mixes are illustrated in Figures 8 a-f.

4 Performance Analysis of the Models

The LGP and LGP/SA prediction models were developed to relate UCS, MDD, and OMC to a number of influencing variables. Detailed performance statistics of the models are summarized in Tables 4, 5, and 6. Based on a logical hypothesis [54], if a model gives $R > 0.8$, and the error values (e.g. MSE and MAE) are at minimum, there is a strong correlation between the predicted and measured values. The model can therefore be judged as very good. It can be observed in the Figures 6, 7, and 8 and Tables 4, 5, and 6 that the LGP and LGP/SA models with high R and low MSE and MAE values predict the target values with high accuracy. The results indicate that the models have memorized the training examples and have also learned to generalize to new situations. As it is seen, the best team solutions obtained by LGP have produced the best results on the training, validation and testing data for the UCS, MDD, and OMC prediction. Also, the best single solutions evolved by LGP perform superior than the LGP/SA solutions in nearly all cases. The exception is the prediction results for the UCS and MDD testing data sets. Considering the corresponding MSE and MAE values for these cases, the performance of the LGP/SA solutions is slightly better than or similar to the LGP single solutions.

Furthermore, new criteria recommended by Golbraikh and Tropsha [55] were checked for external verification of the LGP and LGP/SA solutions on the testing data sets. It is suggested that at least one slope of regression lines (k or k$'$) through the origin should be close to 1. Also, the performance indexes of m and n should be lower than 0.1. Recently, Roy and Roy [56] introduced a confirmation indicator of the external predictability of models (R_m). For $R_m > 0.5$, the condition is satisfied. Either the squared correlation coefficient (through the origin) between predicted and experimental values (Ro^2), or the coefficient between experimental and predicted values (Ro'^2) should be close to 1. The considered validation criteria and the relevant results obtained by the models are presented in Table 7. As can be observed in this table, the derived solutions satisfy the required conditions. The validation phase ensures the LGP and LGP/SA-based solutions are strongly valid, have prediction power, and are not chance correlations.

Multilayer perceptrons (MLP) [57] are one of the most widely used classes of neural networks (NNs). An MLP is essentially capable of approximating

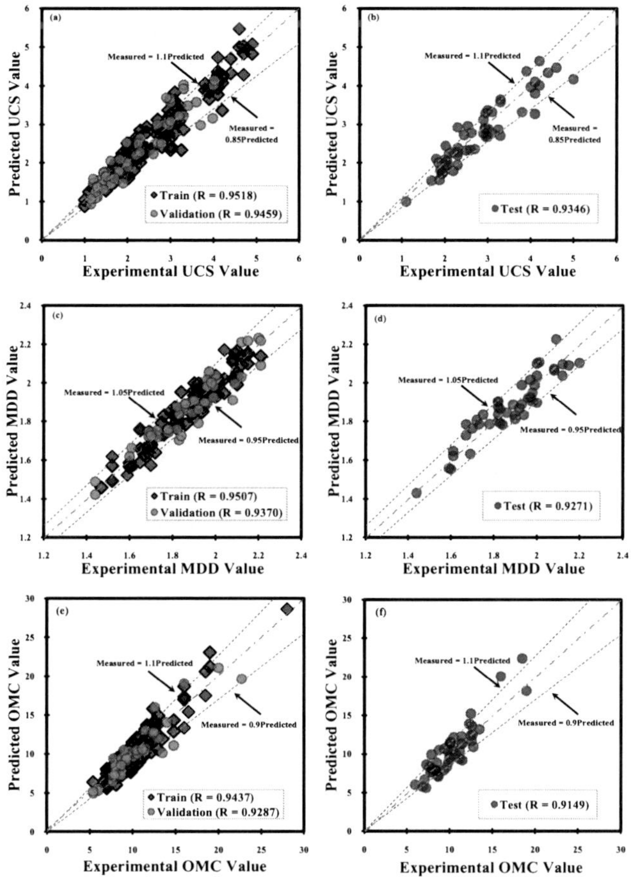

Fig. 8. Experimental versus predicted UCS, MDD, and OMC using the LGP/SA evolved solutions.

any continuous function to an arbitrary degree of accuracy [57]. The LGP and LGP/SA models were benchmarked against MLP models derived by Alavi et al. [58] for predicting the MDD and OMC of the stabilized soil. A new MLP-based prediction model was also developed for UCS in this study. The MLP models were trained using the total of the training and validation data sets used for developing the LGP and LGP/SA models. The generalization capabilities of the MLP models were tested on the same testing data considered for the GP-based solutions. The results made by the MLP models are presented in Tables 4, 5, and 6. The prediction results for UCS, MDD, and OMC indicate that, in all cases, the results obtained by the LGP team solutions are better than or comparable with those of the MLP models. The MLP models with higher R and lower MSE and MAE values provide slightly better results than the single solutions obtained by LGP and LGP/SA. In case of OMC, the MAE values of the LGP and LGP/SA models are lower than those of the MLP model.

Although NNs are successful in prediction, they usually do not give a definite solution to calculate the outcome using the input values. They also require the structure of the neural network (e.g. number of inputs, transfer functions, number of hidden layers, etc.) to be identified a priori. On the other hand, LGP and LGP/SA introduce completely new characteristics and traits. The best solutions evolved by these techniques are determined after controlling numerous preliminary models, even millions of linear and non-linear models. LGP and LGP/SA provide transparent human-interpretable solutions which can be represented in Java, C++, or assembler codes. The evolved solutions are a knowledge representation of the underlying data information. They can be used for further analysis of the UCS, MDD, and OMC of the soil-stabilizer mixes. The LGP and LGP/SA methods are especially practical for cases where the behavior is too complex and the conventional models are unable to effectively describe various aspects of the behavior. However, it is notable that the LGP and LGP/SA techniques are extremely parameter sensitive, especially when difficult experimental training data sets like the one used in this chapter are employed. Using any form of optimally controlling the parameters of the run (e.g., GAs), can significantly improve the performance of these algorithms.

Table 4. Performance statistics of the models for the prediction of UCS.

Models	Training				Validation			Testing			
	LGP (S[a])	LGP (T[b])	LGP /SA	MLP	LGP (S)	LGP (T)	LGP /SA	LGP (S)	LGP (T)	LGP /SA	MLP
R	0.972	0.991	0.952	0.979	0.952	0.965	0.946	0.942	0.960	0.935	0.944
MSE	0.065	0.031	0.091	0.043	0.092	0.063	0.139	0.114	0.072	0.099	0.086
MAE	0.203	0.132	0.234	0.181	0.218	0.184	0.256	0.267	0.205	0.252	0.228
SD[c]	0.087	0.058	0.113	0.084	0.129	0.104	0.135	0.140	0.094	0.117	0.111

[a] Single
[b] Team
[c] Standard deviation.

5 Sensitivity Analysis

A sensitivity analysis is of utmost concern for selecting the important input variables. The contribution of each input parameter in the LGP and LGP/SA models was evaluated through such an analysis. In order to evaluate the importance of the input parameters, their frequency values [35, 43] were obtained. A frequency value equal to 1.00 for an input indicates that

Table 5. Performance statistics of the models for the prediction of MDD.

Models	Training				Validation			Testing			
	LGP (S[a])	LGP (T[b])	LGP /SA	MLP	LGP (S)	LGP (T)	LGP /SA	LGP (S)	LGP (T)	LGP /SA	MLP
R	0.952	0.963	0.951	0.958	0.942	0.956	0.937	0.937	0.953	0.927	0.954
MSE	0.003	0.002	0.003	0.003	0.003	0.003	0.004	0.004	0.003	0.004	0.003
MAE	0.042	0.037	0.044	0.044	0.047	0.044	0.05	0.054	0.041	0.054	0.042
SD[c]	0.028	0.024	0.029	0.028	0.031	0.027	0.034	0.032	0.028	0.034	0.027

[a] Single
[b] Team
[c] Standard deviation.

Table 6. Performance statistics of the models for the prediction of OMC.

Models	Training				Validation			Testing			
	LGP (S[a])	LGP (T[b])	LGP /SA	MLP	LGP (S)	LGP (T)	LGP /SA	LGP (S)	LGP (T)	LGP /SA	MLP
R	0.948	0.954	0.944	0.950	0.941	0.947	0.929	0.928	0.942	0.915	0.944
MSE	1.337	1.243	1.652	0.882	1.574	1.881	1.712	1.043	1.386	2.052	0.896
MAE	0.856	0.907	1.054	1.184	1.007	1.056	0.903	0.813	0.936	1.122	1.130
SD[c]	0.122	0.106	0.122	0.099	0.114	0.108	0.114	0.099	0.121	0.131	0.090

[a] Single
[b] Team
[c] Standard deviation.

this variable has been appeared in 100% of the best thirty programs evolved by LGP and LGP/SA.

The frequency values of the input parameters of the UCS, MDD, and OMC prediction models are presented in Figures 9 a-c. According to these results, it can be found that the UCS of the soil-stabilizer mix is more sensitive to LS and S in comparison with the other inputs. For MDD and OMC, the results are somewhat different and indicate that C and LL are the most significant variables. Since LL and PI represent the Atterberg limits of soils, they were categorized into one group referred to as Plasticity. LS was individually considered as a category named as Linear Shrinkage. Similarly, S and C are the soil particle size distribution parameters and were categorized into a separate group as Textural Properties. In addition, AC, CC, and LC illustrate the type and quantity of stabilizing additives and were categorized into another separate group as Stabilizer Treatments.

The essential observation from the results of the sensitivity analysis is that different stabilizer types or quantities (LC, CC and AC), for the ranges investigated, are much less important to explain variations in UCS, MDD,

Table 7. Statistical parameters of the LGP and LGP/SA solutions for external validation.

Item	Formula	Condition	UCS			MDD			OMC		
			LGP (S[a])	LGP (T[b])	LGP /SA	LGP (S)	LGP (T)	LGP /SA	LGP (S)	LGP (T)	LGP /SA
1	$k = \frac{\sum_{i=1}^{n}(h_i \times t_i)}{h_i^2}$	$0.85 < K < 1.15$	0.990	1.001	1.012	0.988	0.994	0.999	0.985	0.978	0.965
2	$k' = \frac{\sum_{i=1}^{n}(h_i \times t_i)}{t_i^2}$	$0.85 < K' < 1.15$	0.997	0.992	0.977	1.012	1.005	1.000	1.006	1.010	1.020
3	$m = \frac{R^2 - Ro^2}{R^2}$	$m < 0.1$	- 0.126	-0.086	-0.143	-0.116	-0.097	-0.163	-0.16	-0.12	-0.17
4	$n = \frac{R^2 - Ro'^2}{R^2}$	$n < 0.1$	- 0.127	-0.085	-0.138	-0.120	-0.098	-0.163	-0.16	-0.13	-0.19
5	$R_m = R^2 \times (1 - \sqrt{\lvert R^2 - R0^2 \rvert})$	$R_m > 0.5$	0.591	0.662	0.564	0.598	0.638	0.537	0.545	0.598	0.520
where	$Ro^2 = 1 - \frac{\sum_{i=1}^{n}(t_i - h_i^0)^2)}{\sum_{i=1}^{n}(t_i - \overline{t}_i)^2}$, $h_i^0 = k \times t_i$		0.999	1.000	0.998	1.012	1.005	1.000	0.997	0.993	0.980
	$Ro'^2 = 1 - \frac{\sum_{i=1}^{n}(h_i - t_i^0)^2)}{\sum_{i=1}^{n}(h_i - \overline{h}_i)^2}$, $t_i^0 = k' \times h_i$		1.000	0.999	0.994	0.981	0.996	1.000	0.999	0.999	0.996

[a] Single
[b] Team.

and OMC than the soil type. There are earlier findings for UCS and MDD that are in close agreement with this observation. Burroughs [47] assessed the contribution of the predictor variables to explain the variations in the UCS and MDD of the chemically stabilized soil using p-values. The p-values measure the degree of significance of each variable by means of analysis of covariance (ANCOVA). Table 8 presents the p-values obtained from ANCOVA. This table decomposes the variability of UCS and MDD into contributions due to three stabilizers and five soil property variables. The contribution of each variable is measured having removed the effects of all other variables, and the p-values reported test the statistical significance of each of the variables. As can be observed from this table, overall, there are no statistically significant differences in UCS and MDD between different percentages of stabilizer treatment for any of the stabilizers, as the relevant p-values all exceed 0.05. LS and C respectively have the most significant effects on the variations of the UCS and MDD values regarding their significant p-values. The data analysis results presented in Table 8 clearly indicate that the soil type, not the stabilizer type or quantity, exerts dominant influence on the variations of UCS and MDD. Similar results were obtained by Osula [6], Bryan [59], and Walker [60] for lime and cement stabilization.

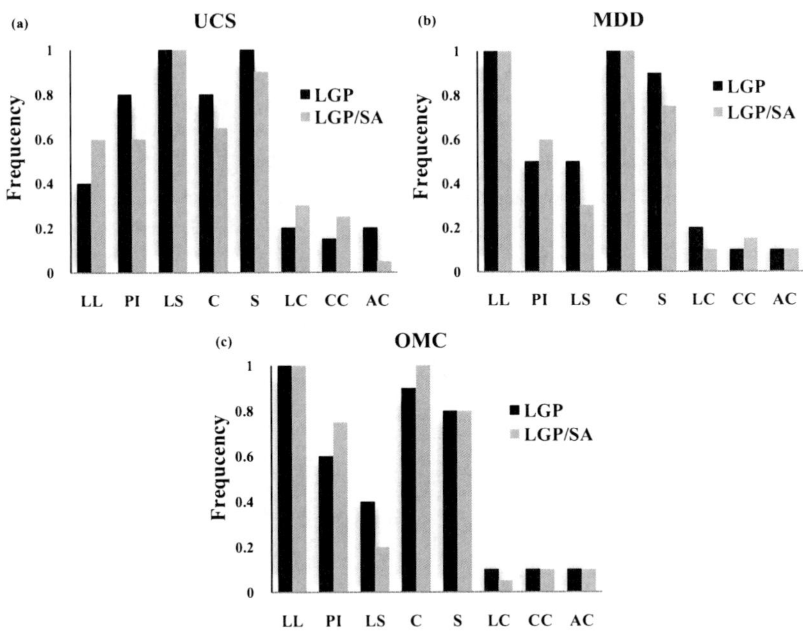

Fig. 9. Contributions of the predictor variables in the LGP and LGP/SA models.

Table 8. Pearson correlations and p-values from ANCOVA (after [47]).

Item	UCS		MDD		OMC	
	P-value	PCC[a]	P-value	PCC	P-value	PCC
LL	0.536	-0.27	0.323	-0.37	NA[b]	0.38
PI	0.473	-0.35	0.680	-0.4	NA	0.46
LS	0.000	-0.46	0.689	0.4	NA	0.46
C	0.012	-0.1	0.046	-0.14	NA	0.24
S	0.001	-0.22	0.422	-0.05	NA	0.03
LC	0.726	NA	0.866	NA	NA	NA
CC	0.556	NA	0.71	NA	NA	NA
AC	0.982	NA	0.755	NA	NA	NA

[a] Pearson correlation coefficient
[b] Not available.

6 Parametric Analysis

For further verification of the proposed models, a parametric study was performed using the solutions evolved by LGP and LGP/SA. The main goal was to find the effect of each input parameter on the values of UCS, MDD, and OMC. The methodology was based on the change of only one input variable at a time while other input variables were kept constant at the average values of their entire data sets. Figures 10 to 12 present the predicted values of UCS, MDD, and OMC as functions of each parameter, respectively. The sensitivity of prediction to each of the input parameters, LL(%), PI(%), LS(%), S(%), C(%), LC(%), and CC(%) can be determined according to these figures. Since there were not significant variations in the percentages of asphalt (AC) in the database, no sensitivity analysis was done on it.

The results of the parametric analysis indicate that UCS continuously decreases due to increasing LL, PI, LS, C, and S. It can be observed that, within the ranges of LC and CC used for the training of the models evolved with LGP and LGP/SA, UCS is not sensitive to the changes in the percentages of lime and cement.

The results of the parametric study for the MDD prediction models indicate that MDD continuously decreases due to increasing LL, PI, C, and S. The results of the LGP models demonstrate that MDD continuously increases due to increasing LS. Those of the LGP/SA model indicate that MDD increases when LS increases up to about 10% and afterwards its increment rate declines for larger LS. It can be seen that MDD slightly increases when LC and CC increase up to about 3% and 2%, respectively, and then it starts decreasing. It is obvious that MDD is not sensitive to the changes in LC and CC. Considering the results of the parametric analysis for OMC, the moisture content increases due to increasing LL, PI, LS, C, S, LC, and CC. Similar to

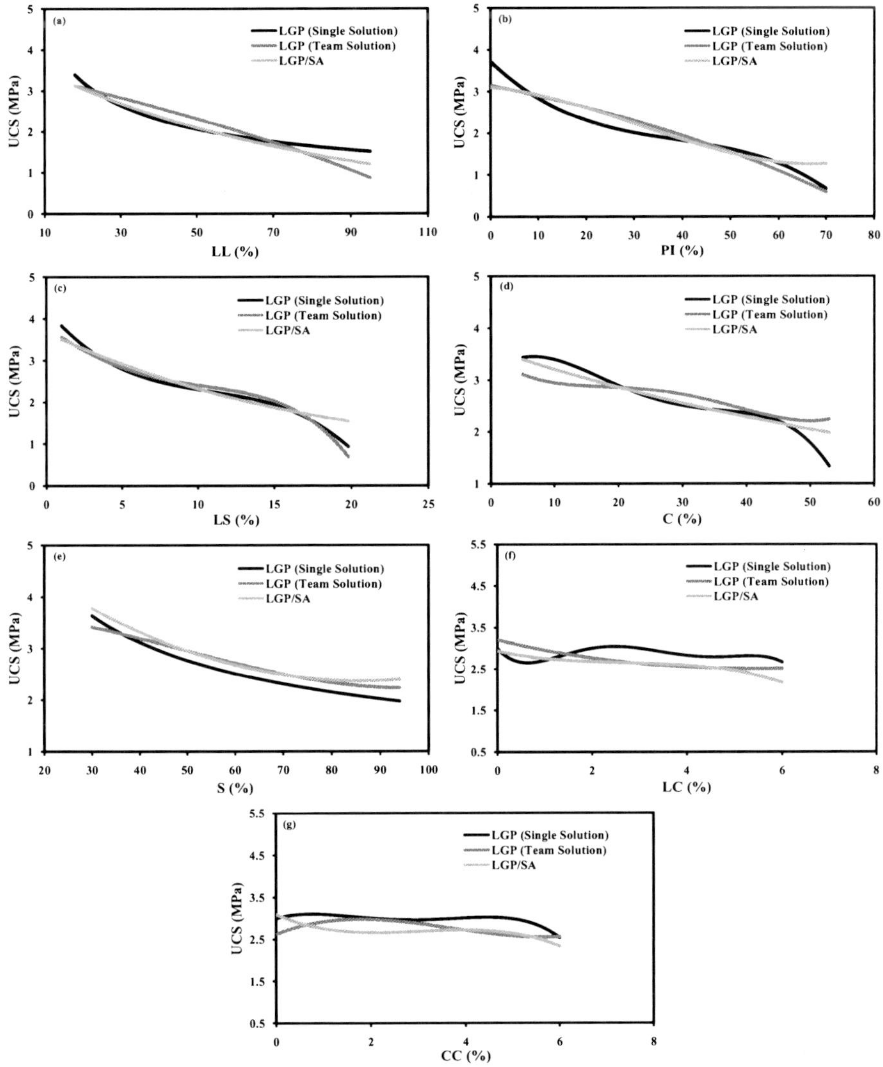

Fig. 10. Parametric analysis of UCS in the LGP and LGP/SA models.

UCS and MDD, it can be seen that OMC is less sensitive to LC and CC compared to other inputs. The results of our parametric study for the stabilizer types are also in agreement with those of the sensitivity analysis.

Table 8 shows the Pearson correlation coefficients between the soil property variables (LL, PI, LS, C, and S) and the UCS, MDD, and OMC of the chemically stabilized soil. These correlations give a simple insight into the degree of (linear) relationship between two variables. As can be seen in this

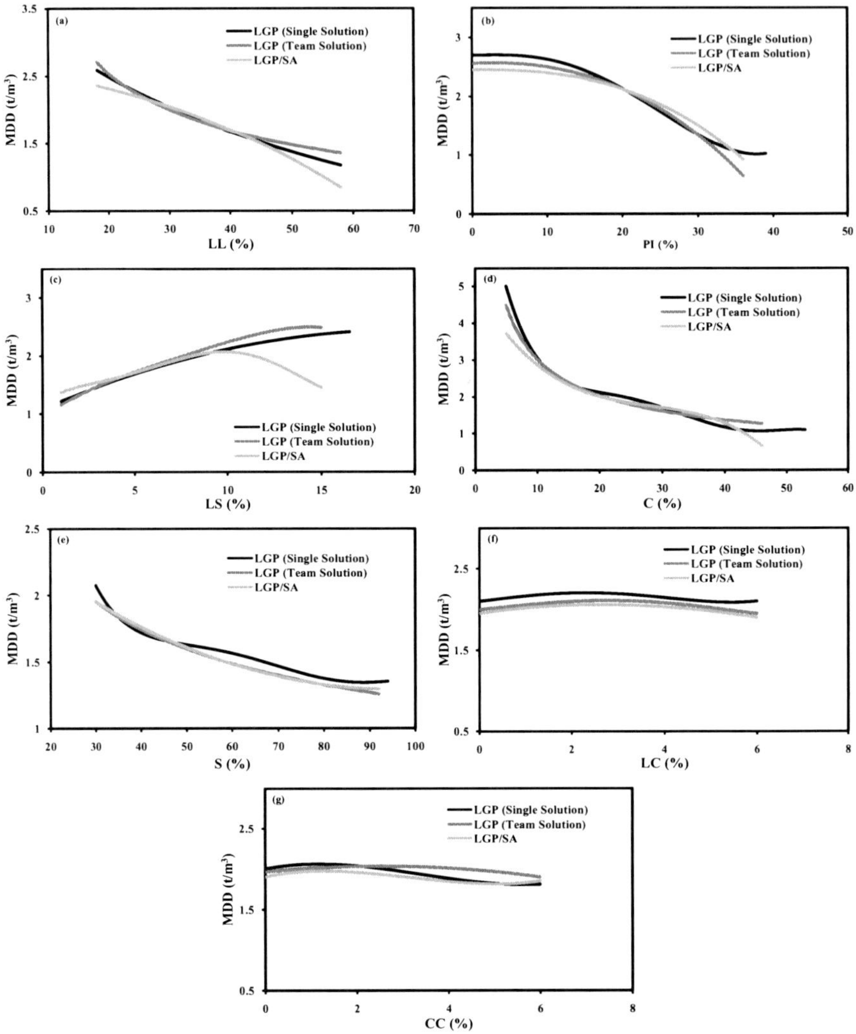

Fig. 11. Parametric analysis of MDD in the LGP and LGP/SA models.

table, all of the soil properties are negatively correlated with UCS, with the highest negative correlations being between UCS, LS, and PI. The correlation of MDD with LS is highly positive while the other soil properties are negatively correlated with MDD. The highest negative correlations for MDD are with PI and LL. Table 8 also reveals that the correlation of OMC with the soil properties is positive. Contrary to those of UCS, the highest positive correlations for OMC are with LS and PI. It is obvious that the results of the parametric analysis of UCS, MDD, and OMC are in close agreement with those presented by Burroughs [47]. The only difference is that the results

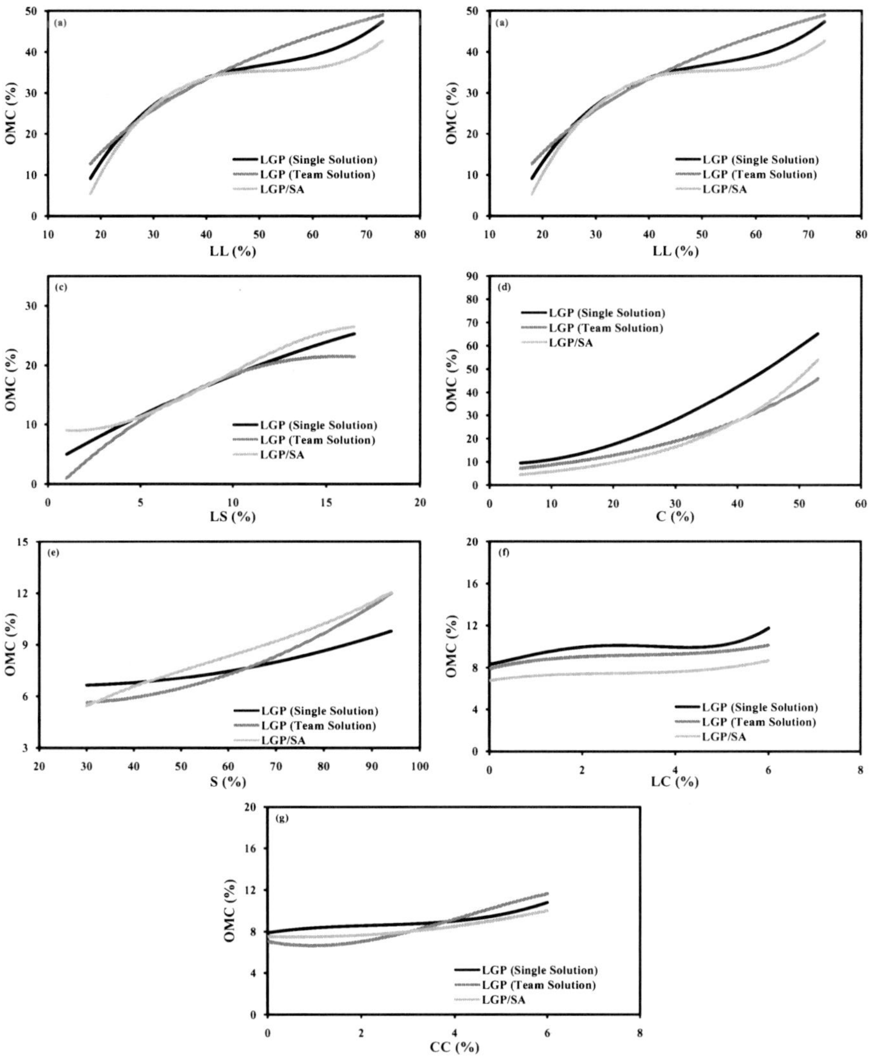

Fig. 12. Parametric analysis of OMC in the LGP and LGP/SA models.

obtained by the LGP/SA model (Figure 11c) indicate that LS is positively correlated with MDD just up to about 10% and thereafter the correlation becomes negative. Another observation from the parametric study results is that increases in LL, PI, C, and S respectively decrease and increase the MDD and OMC. These trends indicate to the high negative correlation between MDD and OMC as presented by other researchers [2, 47].

7 Conclusions

In this chapter, a linear variant of genetic programming, namely LGP, and a hybrid search algorithm combining LGP and SA, called LGP/SA, were employed for the prediction of UCS, MDD, and OMC of soil-stabilizer mixes. A large number of soil types and stabilizers and considerable variations in their characteristics were considered to develop the models. The database was obtained from previously published test results. Results of the LGP and LGP/SA analyses provided six sets of prediction models. The following conclusions may be drawn based on the results presented:

i. Despite high nonlinearity in the geotechnical behavior of the stabilized soil, the proposed models give precise estimates of the target UCS, MDD, and OMC values. The team solutions evolved by LGP have produced the best results followed by the LGP single solutions and the LGP/SA-based solutions. The performance of the LGP team solutions is better than or similar to that of the models developed using MLP.
ii. In general, LGP has better behavior compared to LGP/SA. This indicates that applying the SA strategy to the LGP process (LGP/SA) does not improve the efficiency of the LGP algorithm for the investigated problem. A possible reason is that LGP/SA learns by randomly creating a single parent program rather than using a population of programs like LGP.
iii. Unlike the majority of the constitutive modeling approaches, LGP and LGP/SA simultaneously take into account the role of several important factors representing the engineering behavior of the chemically stabilized soil.
iv. LGP and LGP/SA simultaneously take into account the role of several important factors representing the engineering behavior of the chemically stabilized soil.
v. An important finding from the results of the sensitivity analysis is that the types and quantities of stabilizers are less important variables to explain variations in UCS, MDD, and OMC as compared to the other properties.
vi. The sensitivity of the proposed models to the variation of the influencing parameters was evaluated through a parametric analysis. The results were confirmed with the results obtained by other researchers.
vii. In addition to the acceptable accuracy, LGP and LGP/SA provide transparent programs of an imperative language or machine language. These programs can easily be inspected and evaluated. The interested readers may consult the corresponding author for free C++ codes of the LGP and LGP/SA evolved solutions.
viii. A major distinction of LGP and LGP/SA for determining the stabilized soil performance characteristics lies in their powerful ability to model the mechanical behavior without any need to establish a pre-defined function.

ix. Using the LGP and LGP/SA algorithms, UCS, MDD, and OMC can be estimated without carrying out sophisticated and time-consuming laboratory or field tests.

Further research can focus on both the problem domain and the computing one. As more data become available, including those for other test conditions, the LGP and LGP/SA-based models can be improved to make more accurate predictions for a wider range of the data. LGP and LGP/SA are quite robust in the modeling of nonlinear relationships. However, the underlying assumption that the input parameters are reliable is not always the case. Since fuzzy logic can provide a systematic method to deal with imprecise and incomplete information, the process of developing hybrid fuzzy and linear GP models can be a suitable topic for further studies. Hybridizing LGP with other optimization algorithms such as Tabu Search can also be investigated in order to improve the proficiency of LGP.

References

1. Akpokodje, E.G.: The stabilization of some arid zone soils with cements and lime. Quarterly Journal of Engineering Geology London 18, 173–180 (1985)
2. Bell, F.G.: Lime stabilization of clay minerals and soils. Engineering Geology 42(40), 223–237 (1996)
3. Soil engineering and stabilization. Technical report, USACE (2000)
4. Ngowi, A.B.: Improving the traditional earth construction: a case study of Botswana. Construction and Building Materials 11(1), 1–7 (1997)
5. Bryan, A.J.: Criteria for the suitability of soil for cement stabilization. Building and Environment 23(4), 309–319 (1988)
6. Osula, D.O.A.: A comparative evaluation of cement and lime modification of laterite. Engineering Geology 42, 71–81 (1996)
7. Jumikis, A.R.: Geology and soils of the Newark (NJ) metropolitan area. Journal of the Soil Mechanics and Foundations Division ASCE 93 (SM2), 71–95 (1946)
8. Linveh, M., Ishai, I.: Using indicative properties to predict the density-moisture relationship of soil. Transportation Research Record 60P, 22–28 (1978)
9. Wang, M.C., Huang, C.C.: Soil compaction and permeability prediction models. Journal of Environmental Engineering ASCE 110, 1063–1083 (1984)
10. Blotz, L.R., Benson, C.H., Boutwell, G.P.: Estimating optimum water content and maximum dry unit weight for compacted clays. Journal Geotechnical and Geo-Environmental Engineering ASCE 124(9), 907–912 (1998)
11. Sridharan, A., Nagaraj, H.B.: Plastic limit and compaction characteristics of fine-grained soils. Ground Improvement 9(1), 17–22 (2005)
12. Sinha, S.K., Wang, M.C.: Artificial neural network prediction models for soil compaction and permeability. Geotechnical and Geological Engineering 26, 47–64 (2008)
13. Gunaydin, O.: Estimation of soil compaction parameters by using statistical analyses and artificial neural networks. Environmental Geology 10, 1300–1306 (2008)
14. Horpibulsuk, S.: Analysis and assessment of engineering behavior of cement stabilised clays. PhD thesis, Saga University, Japan (2001)

15. Tan, T.S., Goh, T.L., Yong, K.Y.: Properties of singapore marine clays improved by cement mixing. Geotechnical Testing Journal 25, 422–433 (2002)
16. Horpibulsuk, S., Bergado, D.T., Lorenzo, G.A.: Compressibility of cement-admixed clays at high water content. Geotechnique 54(2), 151–154 (2004)
17. Narendra, B.S., Sivapullaiah, P.V., Suresh, S., Omkar, S.N.: Prediction of unconfined compressive strength of soft grounds using computational intelligence techniques: A comparative study. Computers and Geotechnics 33, 196–208 (2006)
18. Koza, J.R.: Genetic programming: On the programming of computers by means of natural selection. MIT Press, Cambridge (1992)
19. Banzhaf, W., Nordin, P., Keller, R.E., Francone, F.D.: Genetic Programming – An introduction on the automatic evolution of computer programs and its application. Morgan Kaufmann Publishers, Heidelberg, San Francisco (1998)
20. Bäck, T.: Evolutionary algorithms in theory and practice: Evolution strategies, evolutionary programming, genetic algorithms. Oxford University Press, USA (1996)
21. Weise, T.: Global optimization algorithms-Theory and application (2007), `http://www.it-weise.de`
22. Brameier, M., Banzhaf, W.: Linear genetic programming. Springer Science + Business Media, New York (2007)
23. Brameier, M., Banzhaf, W.: A comparison of linear genetic programming and neural networks in medical data mining. IEEE Transactions on Evolutionary Computation 5(1), 17–26 (2001)
24. Metropolis, N., Rosenbluth, A.W., Rosenbluth, M.N., Teller, A.H., Teller, E.: Equation of state calculations by fast computing mechanics. Journal of Chemical Physics 21(6), 1087–1092 (1953)
25. Kirkpatrick, S., Gelatt, C.D., Vecchi, M.P.: Optimization by simulated annealing. Science 220(4598), 671–680 (1983)
26. Cerny, V.: Thermo-dynamical approach to the traveling salesman problem: An efficient simulation algorithm. Journal of Optimization Theory and Applications 45, 41–52 (1985)
27. Aarts, E.: Simulated annealing and boltzmann machines: A stochastic approach to combinatorial optimization and neural Computing. Wiley, New York (1989)
28. Kita, H.: Simulated annealing. Japan Society for Fuzzy Theory and Intelligent Informatics 9(6), 870–875 (1997)
29. Ingber, L.: Simulated annealing: Practice versus theory. Mathematical and Computer Modeling 18(11), 29–57 (1993)
30. Folino, G., Pizzuti, C., Spezzano, G.: Genetic programming and simulated annealing: A hybrid method to evolve decision trees. In: Poli, R., Banzhaf, W., Langdon, W.B., Miller, J., Nordin, P., Fogarty, T.C. (eds.) EuroGP 2000, vol. 1802, pp. 294–303 (2000)
31. Poli, R., Langdon, W.B., McPhee, N.F., Koza, J.R.: Genetic programming: An introductory tutorial and a survey of techniques and applications. Technical Report ces-475, UK: University of Essex (2007)
32. Oltean, M., Dumitrescu, D.: Multi expression programming. Technical Report UBB-01-2002, Babeş-Bolyai University, Cluj-Napoca, Romania (2002)
33. Oltean, M., Grosan, C.: A comparison of several linear genetic programming techniques. Complex Systems 14(4), 1–29 (2003)
34. Francone, F.D.: Discipulus(TM) owner's manual, version 4.0. Register Machine Learning Technologies (2001)

35. Francone, F.D., Deschaine, L.M.: Extending the boundaries of design optimization by integrating fast optimization techniques with machine-code-based, linear genetic programming. Information Sciences 161, 99–120 (2004)
36. Francone, F.D.: Discipulus Lite(TM) owner's manual, version 4.0. Register Machine Learning Technologies (2004)
37. Soh, C.K., Yang, Y.: Genetic programming-based approach for structural optimization. Journal of Computing in Civil Engineering 14(1), 31–37 (2000)
38. Hong, Y.S., Rao, B.: Evolutionary self-organising modeling of a municipal wastewater treatment plant. Water Research 37(6), 1199–1212 (2003)
39. Johari, A., Habibagahi, G., Ghahramani, A.: Prediction of soil-water characteristic curve using genetic programming. Journal of Geotechnical and Geoenvironmental Engineering ASCE 132(5), 661–665 (2006)
40. Javadi, A.A., Rezania, M., Mousavinezhad, M.: Evaluation of liquefaction induced lateral displacements using genetic programming. Computers and Geotechnics 33(4-5), 222–233 (2006)
41. Baykasoglu, A., Gullub, H., Canakci, H., Ozbakir, L.: Prediction of compressive and tensile strength of limestone via genetic programming. Expert Systems with Applications 35(1-2), 111–123 (2008)
42. Gandomi, A.H., Alavi, A.H., Kazemi, S., Alinia, M.M.: Behavior appraisal of steel semi-rigid joints using linear genetic programming. Journal of Constructional Steel Research 65(1-2), 1738–1750 (2009)
43. Gandomi, A.H., Alavi, A.H., Sahab, M.G.: New formulation for compressive strength of CFRP confined concrete cylinders using linear genetic programming. Materials and Structures 43(7), 963–983 (2010)
44. Guven, A., Azamathulla, H.M., Zakaria, N.A.: Linear genetic programming for prediction of circular pile scour. Ocean Engineering 36(12-13), 985–991 (2009)
45. Gandomi, A.H., Sahab, M.G., Alavi, A.H., Heshmati, A.A.R., Gandomi, M., Arjmandi, P.: Application of a coupled simulated annealing-genetic programming algorithm to the prediction of bolted joints behavior. American-Eurosian Journal of Scientific Research 3(2), 153–162 (2008)
46. Burroughs, V.S.: Quantitative criteria for the selection and stabilization of soils for rammed earth wall construction. PhD thesis, University of New South Wales, Australia (2001)
47. Frank, I.E., Todeschini, R.: The data analysis handbook. Elsevier, Amsterdam (1994)
48. Swingler, K.: Applying neural networks a practical guide. Academic Press, New York (1996)
49. Mesbahi, E.: Application of artificial neural networks in modelling and control of diesel engines. PhD thesis, University of Newcastle upon Tyne (2000)
50. Guven, A.: Linear genetic programming for time-series modelling of daily flow rate. Journal of Earth System Science 118(2), 137–146 (2009)
51. Conrads, M., Dolezal, O., Francone, F.D., Nordin, P.: Discipulus-fast genetic programming based on aim learning technology. Technical report, Register Machine Learning Technologies Inc., Littleton, CO (2000)
52. Deschaine, L.M.: Using genetic programming to develop a c/c++ simulation model of a waste incinerator. Technical report, Science Applications International Corp. (2000)
53. Conrads, M., Dolezal, O., Francone, F.D., Nordin, P.: Discipulus-fast genetic programming based on aim learning technology. Technical report, Register Machine Learning Technologies Inc., Littleton, CO. (2004)

54. Smith, G.N.: Probability and statistics in civil engineering. Collins, London (1986)
55. Golbraikh, A., Tropsha, A.: Beware of q2. Journal of Molecular Graphics and Modelling 20(4), 269–276 (2002)
56. Roy, P.P., Roy, K.: On some aspects of variable selection for partial least squares regression models. QSAR & Combinatorial Science 27, 302–313 (2008)
57. Cybenko, J.: Approximations by superpositions of a sigmoidal function. Mathematics of Control, Signals, and Systems 2, 303–314 (1989)
58. Alavi, A.H., Gandomi, A.H., Mollahasani, A., Heshmati, A.A.R., Rashed, A.: Modeling of maximum dry density and optimum moisture content of stabilized soil using artificial neural networks. Journal of Plant Nutrition and Soil Science 173(3), 368–379 (2010)
59. Bryan, A.J.: Soil/cement as a walling material. i: Stress/strain properties. Building and Environment 23(4), 321–330 (1988)
60. Walker, P.J.: Strength durability and shrinkage characteristics of cement stabilised soil blocks. Cement and Concrete Composites 17, 301–310 (1995)

Appendix

The optimum LGP and LGP/SA programs can be run in the Discipulus interactive evaluator mode or may be compiled in the C++ environment. (Note: $v[0], \ldots, v[7]$ respectively represent LL, PI, LS, S, C, LC, CC, and AC in their normalized forms. f[0] is the normalized UCS.)

```
float DiscipulusCFunction(float v[])
{
 long double f[8];
 long double tmp = 0;
 int cflag = 0;
f[0]=f[1]=f[2]=f[3]=f[4]=f[5]=f[6]=f[7]=0;
 L0: if (!cflag) f[0] = f[3];
 L1: f[0]=sin(f[0]);
 L2: f[0]+=f[1];
 L3: f[0]=sqrt(f[0]);
 L4: f[0]+=v[1];
 L5: f[0]=sqrt(f[0]);
 L6: f[0]*=pow(2,TRUNC(f[1]));
 L7: f[0]=-f[0];
 L8: f[0]=sin(f[0]);
 L9: f[0]-=-1.026116609573364f;
 L10: f[0]=fabs(f[0]);
 L11: f[0]*=pow(2,TRUNC(f[1]));
 L12: f[0]*=pow(2,TRUNC(f[1]));
 L13: f[2]/=f[0];
 L14: f[0]=-f[0];
 L15: f[2]-=f[0];
 L16: cflag=(f[0] < f[3]);
 L17: f[0]*=f[0];
 L18: cflag=(f[0] < f[3]);
 L19: f[0]-=-1.026116609573364f;
 L20: tmp=f[1]; f[1]=f[0]; f[0]=tmp;
 L21: f[0]*=pow(2,TRUNC(f[1]));
 L22: f[0]*=pow(2,TRUNC(f[1]));
 L23: f[0]=fabs(f[0]);
 L24: f[0]=-f[0];
 L25: f[0]=sin(v[4]);
 L26: f[0]-=-1.026116609573364f;
```

Fig. 13. Part of the best LGP single solution for the prediction of UCS.

```
float DiscipulusCFunction(float v[])
{
 long double f[8];
 long double tmp = 0;
 f[0]=f[1]=f[2]=f[3]=f[4]=f[5]=f[6]=f[7]=0;
 l0:  f[0]-=v[0];
 l1:  f[0]*=f[0];
 l2:  f[0]*=v[0];
 l3:  f[0]+=f[1];
 l4:  f[3]-=f[0];
 l5:  f[0]+= v[7];
 l6:  f[0]+=f[1];
 l7:  tmp=f[0]; f[0]=f[0]; f[0]=tmp;
 l8:  tmp=f[0]; f[0]=f[0]; f[0]=tmp;
 l9:  f[0]-=-1.115510215114008f;
 l10: f[0]=fabs(f[0]);
 l11: tmp=f[1]; f[1]=f[0]; f[0]=tmp;
 l12: f[0]*=f[1];
 l13: f[0]-=v[6];
 l14: f[0]*=v[3];
 l15: f[0]=sin(f[0]);
 l16: f[0]-=v[2];
 l17: f[0]-=f[1];
 l18: tmp=f[0]; f[0]=f[0]; f[0]=tmp;
 l19: f[0]+=f[3];
 l20: f[0]-=v[7];
 l21: f[0]/=f[1];
 l22: f[0]-=v[0];
 l23: f[0]-=0.005032832622528f;
 l24: f[0]-=v[0];
 l25: f[0]-=v[4];
 l26: f[0]+=f[1];
 l27: f[0]/=f[1];
 l28: f[0]=-f[0];
```

Fig. 14. Part of the best LGP/SA solution for the prediction of UCS.

Evolving Cellular Neural Networks for the Automated Segmentation of Multiple Sclerosis Lesions

Eleonora Bilotta, Antonio Cerasa, Pietro Pantano, Aldo Quattrone, Andrea Staino, and Francesca Stramandinoli

Abstract. This chapter presents an innovative approach for the segmentation of brain images that contain multiple sclerosis (MS) white matter lesions. Quantitative research of Magnetic Resonance Images (MRI), aimed at detecting and studying lesion load and tissue volumes, has turned out to be very useful for the re-evaluation of patients and clinical assessment of therapy. Until now, the standard procedure for this purpose has been the manual delineation of MS lesions, which makes the analysis a time-consuming process. The application presented in this work is a genetic algorithm (GA) that evolves a Cellular Neural Network (CNN) for pattern recognition. This network is capable to automatically segment the brain areas affected by lesions in MRI and also to immediately eliminate the parts of the brain that are not directly connected to the disease (like the skull, the optic nerve, etc.) in the segmentation process. In comparison to manual segmentations, the proposed method shows a very high level of reliability. It must also be reported that the relative algorithm is more accurate and it adapts to different conditions of the stimulus. Furthermore, it can create 3D images of the brain regions affected by MS, providing new perspectives of the diagnostic analysis of this disease. The work has practical applications in the medical field. Future industrial development of this work could lead to the embodiment of the algorithm directly into the MRI equipment, because CNNs can be implemented in hardware (via discrete off-the-shelf components) or fabricated as a Very Large Scale Integrated (VLSI) chip.

Eleonora Bilotta · Pietro Pantano · Andrea Staino · Francesca Stramandinoli
Evolutionary Systems Group, University of Calabria,
87036, Arcavacata di Rende, Italy
e-mail: {bilotta,piepa}@unical.it,
{andreastaino,francescastramandinoli}@gmail.com

Antonio Cerasa · Aldo Quattrone
Neuroimaging Research Unit, Institute of Neurological Sciences,
National Research Council, 88100 Catanzaro, Italy
e-mail: {a.cerasa,a.quattrone}@isn.cnr.it

1 Introduction

The introduction of functional brain imaging technology into the neuroscience field has opened the door to new frontiers in the diagnosis of brain diseases. However, it has also created a major problem affecting the medical diagnosis, which consists of distinguishing, identifying, and recognizing the salient parts of the information produced by the technology, within a very large amount of available data. Indeed, one of the most important problems in medicine is to carry out an accurate diagnosis, based on this huge amount of data. Human beings make mistakes because of their limited cognitive capacity in managing all this information. Furthermore, subjectivity is a significant drawback in medical diagnosis. Often, the physicians do not use objective criteria in extracting salient clues from Magnetic Resonance Images (MRI), but simply their experience closely related to other, previously made, diagnoses. This means that the diagnostic results do not depend on quantitative data in a clear systematized diagnostic picture, but rely merely on the interpretation of signals from the patient, correlated with data from the MRI [1].

Usually, Artificial Neural Networks (ANNs) are a very effective tool in medical diagnosis and in predicting of clinical outcomes. As well as in other important fields of research such as business, finance, artificial intelligence, mathematics, cognitive modelling, and evolutionary robotics, the most important strong points of ANNs are:

1. they can be trained on examples instead of rules;
2. they do not have the human limitations;
3. they can identify the problem quickly;
4. they can achieve a real-time analysis and make the specific performance related to the application field.

Until recently, ANNs were considered to be a suitable system for the recognition of brain damage in neurodegenerative diseases [2, 3] and for the management of huge amount of data coming from MRI [4, 5, 6]. Currently, this paradigm has been proved to be ineffective in managing and processing of information on a wide scale [7]. This is why other paradigms are taking the place of ANNs. In particular, for their exceptional performance as well as for their ability to process data in parallel and in order to ease the use and implementation in both hardware and software, Cellular Neural Networks (CNNs) [8, 9, 10] start to become a dominant model in the field of neuroscience [7, 11, 12, 13].

A CNN "is a large-scale non linear analog circuit which processes signals in real time. Like Cellular Automata (CA), it is made of a massive aggregate of regularly spaced circuit clones, called cells, which communicate directly only with their nearest neighbors. Each cell is made of a linear capacitor, a nonlinear voltage-controlled current source, and a few resistive linear circuit elements" [8]. In comparison with other kinds of neural networks, these systems have the advantage of being easily implemented in a silicon chip. The dynamics of a CNN are always governed by a set of Ordinary Differential Equations (ODEs), which is as large as the number of cells in the system. The important characteristic of a CNN is to be a meta-model for other systems, both discrete as CAs [14] and continuous [15, 16]. This makes them particularly suitable for image processing tasks.

The latest evolution of this system, which is the CNN Universal Machine, is a powerful computer built on a chip. The system has a compiler, an operating system, and a programming system, based on the C language, that enables us to implement any kind of algorithm for image processing. The sequence of algorithms for image processing is incorporated into the chip, which has the same functions as a digital computer. The CNN paradigm is achieving more and more importance in image processing applications; in fact, a great number of algorithms based on it can be found in the literature and are used for the treatment of images in the many different disciplines.

In this chapter, our focus is to use CNNs in the field of medical diagnosis, where very encouraging and satisfactory results have been obtained. Medical images segmentation and signal processing form one of the most important areas of medical diagnosis, which has been successfully supported by CNNs [17]. This diagnosis involves magnetic resonance imaging [7, 13], computed tomography (CT) [18, 19] and Electro Encephalo Gramme (EEG) [11]. A CNN-based approach to classify MRI with respect to the presence of mesial temporal sclerosis has been presented by Döhler et al. in 2008 [7]. Multiple sclerosis (MS) is a demyelinizing disease of the central nervous system that leads to inflammatory pathology. MS pathology is primarily expressed as focal lesions in the white matter of the brain. Because of its superior contrast, MRI serves as the modality of choice for clinical evaluation of MS. Generally, manual delineation of MS lesions is time-consuming, because three-dimensional information from several magnetic resonance contrasts must be integrated.

In recent years, there has been increasing interest in developing novel techniques for automated MS lesions segmentation. This chapter presents an automated approach to segment MS lesions in MRI. Through the evolution of CNNs, based on genetic algorithms (GAs), we have developed a system that precisely identifies demyelinized brain areas due to the neurodegenerative disease. This approach has been tested on a dataset of 11 patients, who underwent structural MRI and who have been diagnosed with MS. For each patient, 24 slices which cover the whole brain were analyzed. The results show a very high percentage of agreement (almost 80%) with manual segmentation. The exposed method has given satisfactory results, showing that, after the learning process, the CNN is capable to detect MS lesions with different shapes and intensities. The technique we propose is a fully automatic method and does not require manually segmented data.

The rest of this chapter is organized as follows. After the introduction, the second section clearly identifies the problem we intend to address and solve. It also presents the most relevant aspects of this research field and their implications for the future development of this sector. Related works, itemizing the most important issues concerning both the medical diagnosis and the use of CNNs in the neuroscience discipline are presented in the third section. The fourth section discusses the approach we have used. This section presents the mathematical aspects of both CNN and GA models, the optimization approach we have implemented and the integration in the real life context of medical diagnosis. The fifth section presents the experimental results on the performance measurements, the statistical evaluation of the results, and the comparison with manual segmentation. Conclusions and future work, contained in sections six and seven respectively, finalize the chapter.

2 Automatic Segmentation of MS by Using a CNN

MS is a debilitating and progressive autoimmune disease that causes inflammation, demyelination, and axonal damage of the Central Nervous System (CNS) [20, 21, 22]. Used in most clinical trials, the Expanded Disability Status Scale (EDSS) is a method to analyze the functional damage to motor, cognitive and sensory systems [23] in people with MS damages. This method not only provides a quantification of disability, but also serves to monitor changes in the level of disability over time. Devised by John Kurtzke in 1983 [24] and following his previous assessment system (Disability Status Scale, or DSS), the EDSS ranges from 0 to 10, with increments of 0.5 units. The score is based on an examination done by a neurologist and based mainly on the impairment of motor activity. EDSS steps 1.0 – 4.5 identify people who are able to walk without help, even if they have MS. EDSS steps 5.0 – 9.5 categorize the complete impairment of motor activity. The method also recognizes eight functional systems (FS) that may be affected, which can produce a huge amount of symptoms, from the speech to the sensory modality impairments, without any patterns. Each functional system is scored on a scale from 0 (no disability) to 5 or 6 (severe disability).

Regarding the cognitive disability, the Paced Auditory Serial Addition Test (PASAT) [25] is a frequently used test for capturing deficits of attention and working memory in MS subjects. The assessment of onset and progression of MS diseases is critically dependent on the identification of lesions or changes in white matter regions of the brain, through many techniques of brain imaging, and especially through MRI. The most important problems that researchers encounter in this domain are the following:

1. the inhomogeneous, dramatically complex structure of the brain [26];
2. the unpredictability of the disease that affects different areas of the brain and other nervous structures, without a defined pattern;
3. the lack of clear correlation between the amount of damaged brain areas and the related clinical symptoms [27];

4. the lack of highly defined technical standards in brain imaging processing, that introduces in MS lesions segmentation methods a wide range of technological bias [28].

Many mathematical models have been used to develop methods for brain MRI segmentation [29, 30]. They are concerned to discriminate lesions for different diseases [31, 32, 33], to identify changes in anatomical structures [34], and also to create hardware and software applications for 3D brain visualization. The process of image segmentation precisely allows to achieve this analysis both for training the neurologists and for research purposes (The Whole Brain Atlas, `http://www.med.harvard.edu/AANLIB/home.html`).

As Filippi et al. [35] pointed out, MS diagnosis can be done by analyzing T2-weighted images, while for monitoring the evolution of the disease, it is necessary to obtain efficient measures of the Total Lesion Load (TLL) by using lesion segmentation techniques based on signal intensity thresholds. To analyze these problems, Souplet et al. [36] present a comprehensive *state of the art* classification of the most representative systems for MS lesions segmentation.

Shiee et al. [37] use a segmentation method based on topological and statistical atlases, while all the segmented structures are topologically constrained, allowing a further segmentation of shapes. In the literature on these topics, it is noted that the processes of segmentation usually require a trade-off between accuracy and computation time. Due to these problems, many segmentation algorithms, based on artificial neural networks, have failed to give satisfactory results [7].

ANNs, developed in the past decades, are connectionist systems that simulate the biological mechanisms of information processing. They have been widely used in medical science and biomedical research, especially as pattern recognition systems, as medical expert systems to make predictions, as systems to monitor the status of some diseases such as cancer and as MRI segmentation systems. Recently, ANNs with many layers have been established [38], and usable algorithms have been implemented. Although the ANNs, one of the leading paradigms of machine learning during the 1980's [39], have been dropped in favor of simpler methods of classification to define and manage the tasks of recognition / segmentation / forecast, methods more efficient from the computational point of view, achieving superior performance compared to ANN approaches, have been implemented. The main reason lies in the fact that the algorithm of back propagation is not practical for training a network that has more than two or three layers, and the computational load of multi-layers ANN architectures is very high.

CNNs allow to solve, or at least, to mitigate, those problems; they can be regarded as one of the paradigms of the Science of Chaos and Complexity. CNNs are massively parallel systems, easily integrated in silicon, and mainly used for advanced image recognition, especially in medical and graphic fields.

In this chapter, we describe a new application based on GAs that evolves a CNN capable to automatically determine the TLL in MS patients. Our study aims to analyze MRI and to obtain reliable and reproducible measures of the total lesion load in patients with MS. The process of segmentation, that we have developed, is based on adaptive approaches. These approaches can be applied to different clinical settings, having as output different types of images. It must be noted that CNNs are chaotic nonlinear systems. The introduction of nonlinear dynamics in the field of brain imaging will lead to the acquisition of already well established methods and concepts [40].

Many important results can be achieved by using the methods, concepts and visualization systems already developed in the field of nonlinear dynamic systems [41, 42, 43, 44, 45, 46]. The segmentation of MRI, the study of brain shapes and their 3D visualization, can be revisited in the light of the chaos theory, which may represent a powerful tool to study the dynamics and the patterns with which the neurodegenerative diseases occur. The future implications of those new methods and concepts in the field of brain imaging are promising because, by improving the traditional approach, they could lead to the development of fully automated tools, highly efficient from the computational point of view and very reliable in diagnosing the evolution of the disease.

The scientific potentials in the intersection of Chaos and Complexity Theories and the brain imaging technologies are to be fully explored and exploited for the future development of this sector.

3 Related Works

3.1 Related Work in the Problem Domain

In the last decades, the rapid collection of brain images of healthy and unhealthy people has inspired the development of mathematical algorithms, that compare sets of brain data in a huge number of subjects all over the world. New methods of machine vision [47, 48, 49, 50, 51], new interfaces [52], the construction of the anatomical model [53], and the use of differential geometry [54, 55] were developed to represent the great variations in the brain, and to identify disease specific models [56]. These models can also identify qualitative and quantitative patterns of their anatomy in relation to their function [57], highlighting the surprising relationships between genotype and phenotype.

As we have already said in the previous sections, MS pathology presents focal lesions in the white matter of the brain. For its particular properties of providing much greater contrast for the different soft tissues of the body, magnetic resonance is the technology used for clinical evaluation of MS. The delineation of the disease is performed manually by experienced neurologists. It varies from person to person and it is a very labor-intensive and time-consuming task. Many mathematical models have been used to implement automated MS lesion segmentation algorithms and systems. There are many segmentation methods: the automated methods and the Atlas-based segmentation methods. Among the automated methods, there are also several classes. The first one classifies the lesions as outliers of the normal brain tissues, together with a system for the delineation of the lesion borders [58, 59, 60, 61, 62]. Contrariwise, the second class of methods instead shapes the lesions as a separate set [63]. All these methods try to establish an increasingly precise delineation to make the process of identification of MS lesions accurate and to define the total MS lesion load. These techniques have many problems [37]. Several of these methods are only based on the lesions segmentation process, ignoring the quantification of the volume of the brain injured area [64]. These methods also do not pay attention to the sub-cortical structures of the brain, to the cortical surface analysis and they do not consider the degree and the location of brain atrophy, which is a crucial variable for evaluating the evolution of MS. None of these methods consider the special topology of patients with MS, altered by the lesions. Furthermore, in order to define the outliers, many algorithms are heavily dependent on the threshold choice, and often the process is not completely automated.

3.2 Related Work in the Optimization Domain

The domain of optimization of this chapter relates to processes for automatic segmentation of MRI. As we said earlier, the mapping of the brain covers often hundreds or even thousands of images [65, 66, 67]. To address these problems in terms of processing speed, computational effectiveness, automated image registration and warping methods, many algorithms must filter the information from these images automatically [68]. For fast image segmentation and labeling, other techniques were also developed and even the modern technology of 'IC was used in this field.

Databases with topics related to active brain are active and they increase worldwide at a almost exponentially rate (`http://www.sfn.org/index.aspx?pagename=NDG_main`; `http://brancusi.usc.edu/bkms/`). A huge community of researchers analyzes brain images through client-server software technologies. The intensive analysis can be performed on a remote server, with the help of supercomputing resources, aimed to discover many general trends for many degenerative diseases [69]. These advances in technology and infrastructure have enabled the creation of a population-based atlas of the brain [70]. Such atlases combine imaging data from healthy and diseased populations and are leading to the creation of communities of users in the field of neuroscience. The atlases are an important source of information for scientists as they describe how the brain varies with age, sex and demographics. They also provide a comprehensive approach to the study of a particular subgroup of the population, with a specific disease or a specific disorder, such as that discussed in this chapter on MS. Generally in the MRI segmentation domain, supervised segmentation algorithms normally work according to one of two paradigms for guiding the automatic or semiautomatic process. The steps are as follows:

1. Specification of sections of the borders of the chosen object or of a closed complete border that develops to the chosen border;
2. Specification of a small set of pixels, belonging to the chosen object and to a set of pixels, belonging to the background.

Furthermore, any of the automatic segmentation algorithms might be considered supervised by following expert clinician's selection of the desired segment. However, if the desired object is not an entire segment, another clustering/segmentation algorithm must be used to divide or combine the automatic segments. In fact, while semiautomatic methods are highly dependent on the choice of an appropriate threshold (to effectively detect lesions) and on the experts selection of the desired target brain areas, our algorithm, improving the method already developed by Döhler [7], allows for obtaining the desired output by programming a fully automated strategy on the entire dataset, without the need of external calibration. The images, acquired by the magnetic resonance scanning, have, from patient to patient, sometimes significant differences in the intensity of grey and in the area of the injuries. Differences can be observed also from a slice to another one. For this reason, it is necessary to train the CNN to become adapted to different input conditions, not encountered before. By evolving the CNN templates, we have been able to determine most of the lesions in all the patients, optimizing the present problem. The system could provide a useful support tool for the evaluation of lesions in MS, and particularly to assess the evolution of the lesions. It is worth noting that our analysis was carried out on two-dimensional slices.

Another improvement, about what was already reported in the work of Döhler et al. [7], is the possibility to extend our results by working on volumes of data, rather than processing planar images. 3D CNNs [71] may represent a new and powerful tool for the development of applications for supporting of medical diagnosis. By exploiting information provided by a three dimensional representation of the brain, the evolution of the 3D network can lead to a significant improvement of the performances. For this reason, by modifying the architecture of the standard CNN, we propose a 3D CNN model that is able to handle and perform different functions on objects in a three dimensional space. An innovative library of 3D templates has been implemented, that allows the execution of interesting and efficient operations. We used GAs for 3D template learning. This means that even a 3D CNN can be trained by a learning algorithm in such a way that the network learns the configuration for performing the desired operation during the training process.

4 The Approach

The approach, adopted in this work, is based on GAs and CNNs, that represent the mathematical tools that we have used to tackle the problem of automatically detecting MS lesions in MRI, which presence is revealed by regions in the brain that are brighter than their surroundings (Figure 1).

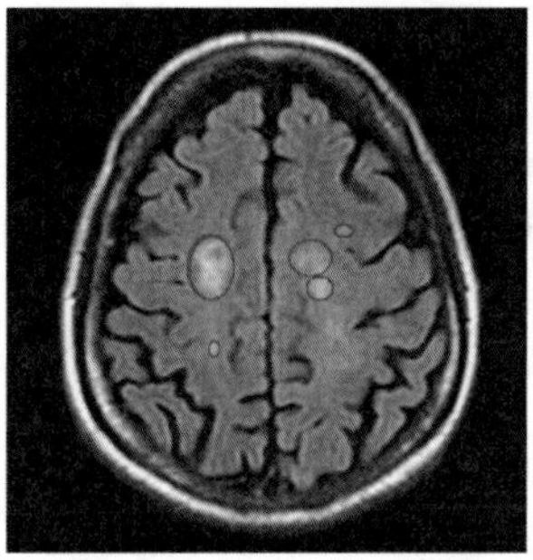

Fig. 1. MS lesions in the white matter of the brain. Lesions are brighter than other tissues in MRI.

In what follows, we give a short description of the CNN paradigm and we present the key features of GAs. CNNs [8] are an array of nonlinear programmable analog processors, called cells, that perform parallel computation. The key idea behind the CNN paradigm was that to combine the main advantages of cellular automata and artificial neural networks [8]. Like neural networks, a CNN is a nonlinear analog circuit which is capable of processing a large amount of data in real time and in an asynchronous way. At the same time, the structure of CNNs is similar to that found in cellular automata, because interactions between cells are only local, that is each cell is physically connected only with its nearest neighbors. From a mathematical point of view, each cell is a dynamical system whose state evolves in time, according to a specific mathematical model, and whose output is a nonlinear function of the state. For the image processing purpose, the most usual architecture is a regular two dimensional grid, in which each processing unit directly interacts only with the neighboring cells, located within a prescribed sphere of influence; given a CNN of $M \times N$ cells, the neighborhood $S_{ij}(r)$ of radius $r \geq 0$ for the cell C_{ij} is the set of cells, satisfying the following property:

$$S_{ij}(r) = \{C_{kl} : \max(|k-i|, |l-j|) \leq r\}, \quad 1 \leq k \leq M\,,\ 1 \leq l \leq N \qquad (1)$$

The grey box in Figure 2 highlights the sphere of influence $S_{ij}(1)$ of the cell at position (i,j), whose neighborhood comprises nine elements (the eight adjacent units and the cell itself).

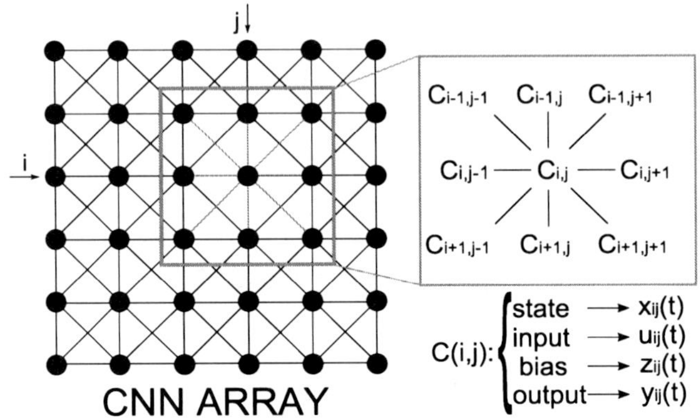

Fig. 2. A cellular array whose elements represent nonlinear dynamic systems.

A sphere of influence of radius $r = 1$ corresponds to a 3×3 neighborhood, one of radius $r = 2$ to a 5×5 neighborhood and so on. Referring to the central grey unit, neighborhood of different sizes are shown (Figure 3).

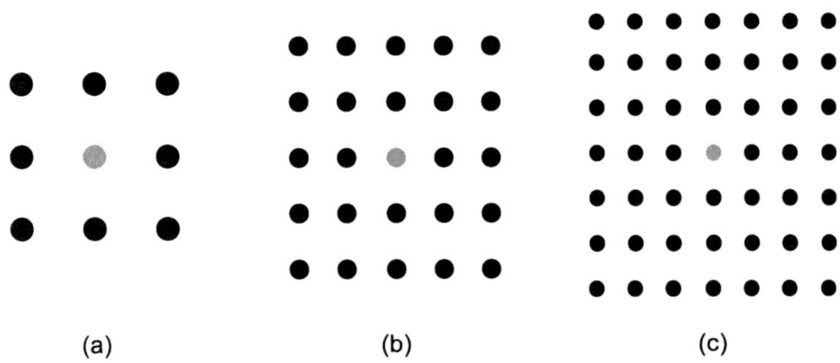

Fig. 3. Neighbourhood of radius r=1 (a), r=2 (b) r=3 (c) for the considered grey unit.

Each cell C_{ij} is characterized by four time variables, called the *state* $x_{ij}(t) \in \mathbb{R}^n$, that generally is not observable from the outside, the *input* $u_{ij}(t) \in \mathbb{R}^u$, corresponding to external stimuli provided to the cell, the *output* $y_{ij}(t) \in \mathbb{R}^p$, that represents the value that can be observed and measured, and an additional input $z_{ij}(t) \in \mathbb{R}^z$ called *bias*. The dynamics of a $M \times N$ CNN are described by $M \times N$ coupled differential equations (2), modeling the evolution of the state of each cell and its interaction with its neighbors ("coupling laws").

$$\dot{x}_{ij}(t) = g(x_{ij}(t),\ z_{ij}(t), \underline{u}_{kl}(t),\ f(\underline{x}_{kl}(t))), \quad (k,l) \in S_{ij}(r), \qquad (2)$$
$$i = 1 \ldots M\ , j = 1 \ldots N$$

where $\underline{u}_{kl}(t)$ and $f(\underline{x}_{kl}(t))$ denote vectors, whose components are the input and the output of the neighbors of C_{ij}. The output of each cell is obtained by applying the nonlinear algebraic function $f(x_{ij}(t))$, such that:

$$y_{ij}(t) = f(x_{ij}(t)) \qquad (3)$$

Standard nonlinearity for the output equation is given by the following expression:

$$y_{ij}(t) = f(x_{ij}(t)) = \frac{1}{2}\left[|x_{ij}(t) + 1|\right] - \left[|x_{ij}(t) - 1|\right] \qquad (4)$$

which characteristic is shown in Figure 4.

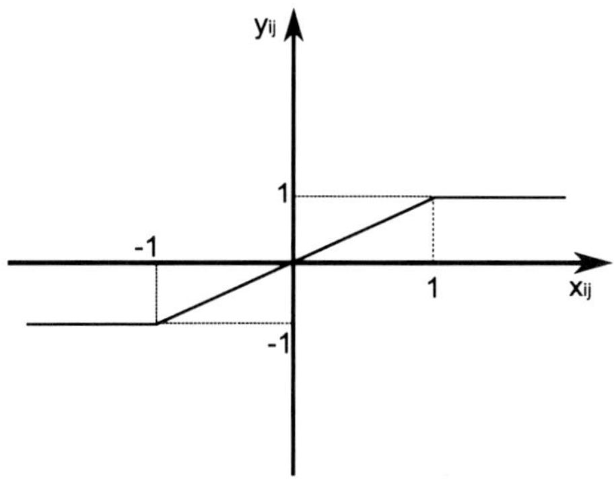

Fig. 4. Standard nonlinearity for the output equation.

It is worth noting that (1) is not completely defined for cells lying on the boundaries of the array, whose sphere of influence extends beyond the boundary of the given CNN. Therefore, it is necessary to introduce some additional elements to the array, called virtual cells, whose state and input are given in accordance with the boundary conditions of the network. The following types of boundary conditions are proposed:

1. Fixed (or Dirichlet): the values of the *virtual cells* are given as a prescribed constants;
2. Zero-flux (or Neumann): the values of the *virtual cells* are the same as the corresponding *boundary cells*;
3. Periodic (or toroidal): values of the virtual cells are the same as the boundary cells on the opposite side (e.g., top *virtual cells* have the value of bottom *boundary cells*).

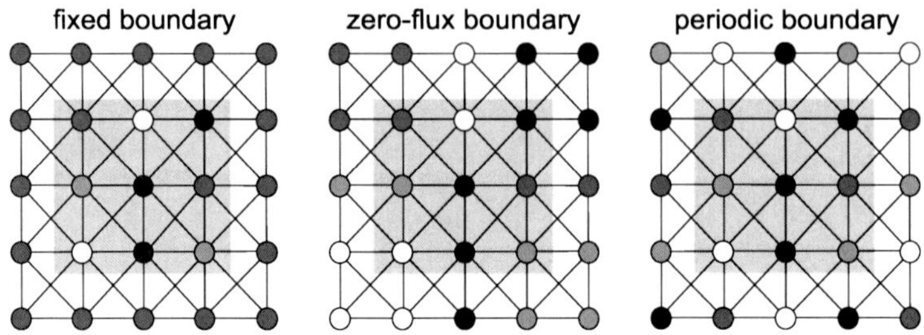

Fig. 5. Boundary conditions commonly used for CNNs.

Different kinds of boundary conditions for a 3×3 CNN (whose cells belong to the shaded grey area) are shown in Figure 5; the values for the virtual cells are set according to the corresponding chosen boundary conditions. By choosing an appropriate g function in (2), different models of CNNs can be defined. In this work, we will deal with the **standard CNN model**, in which each cell C_{ij} is a first order dynamical system, i.e., its state is a scalar quantity ($x_{ij}(t) \in \mathbb{R}$) that can be associated with the intensity of a physical variable at the corresponding point in space at time t. For ease of notation, we will omit the time argument of the variables associated to C_{ij}. The set of ODEs that describes the dynamics of a standard CNN is:

$$\frac{\mathrm{d}x_{ij}}{\mathrm{d}t} = -x_{ij} + \sum_{C(k,l)\in S_r(i,j)} A(i,j;k,l;t)y_{kl} + \sum_{C(k,l)\in S_r(i,j)} B(i,j;k,l;t)u_{kl} + z_{ij} \tag{5}$$

where $A(i;j;k;l;t)$ and $B(i;j;k;l;t)$ represent respectively the weights by which outputs and inputs of the cells in the neighborhood contribute to changes in the state of C_{ij}. The feedback coupling parameters $A(i,j;k,l,t)$ and input coupling parameters $B(i,j;k,l,t)$ can be used to change and control the strength of interactions between cells and, in general, can vary both in space and time. Being given input, initial state, and boundary conditions for each cell C_{ij} such that $1 \leq i \leq M$, $1 \leq j \leq N$, the dynamics of a two-dimensional standard CNN are uniquely specified by the synaptic weights between a cell and its neighbors. These parameters, together with the bias $z_{ij}(t)$, define a CNN template that can be expressed in the form $\{A(i,j;k,l,t), B(i,j;k,l,t), z_{ij}(t)\}$. The operation performed by a CNN on the input data is fully defined by the set of coefficients in the CNN template. If the pattern of interconnection is the same for each cell and does not vary in time, the weighting coefficients can be arranged in a feedback matrix $A \in \mathbb{R}^{(2r+1)\times(2r+1)}$, and a feed-forward or control matrix $B \in \mathbb{R}^{(2r+1)\times(2r+1)}$, while the bias $z_{ij}(t) = z \in \mathbb{R}$ for each (i,j) and t; in this case, the template reduces to the triple $\{A, B, z\}$. As we will see, an evolutionary approach can be used in order to find a template that allows obtaining a desired operation.

As it was stated in the introduction, the key feature of CNNs is their ease in implementation in VLSI chips; in fact, the main difference between CNN and other neural network paradigms is that in the former information is directly exchanged just between neighboring units so, when it comes to physical realization, CNNs show more flexibility than ANNs, allowing easier integration in silicon devices and greater efficiency in computation. In the original model [8], each CNN cell is a simple nonlinear analog circuit (Figure 6), composed of a linear capacitor, an independent current source, an independent voltage source, two linear resistors, and at most $2m$ linear voltage-controlled current sources, m being the number of neighbor cells of the considered unit. The voltage $v_{x_{ij}}(t)$ across the capacitor is the state of the cell C_{ij}, while $v_{u_{ij}}$ and $v_{y_{ij}}(t)$ represent the input and the output respectively.

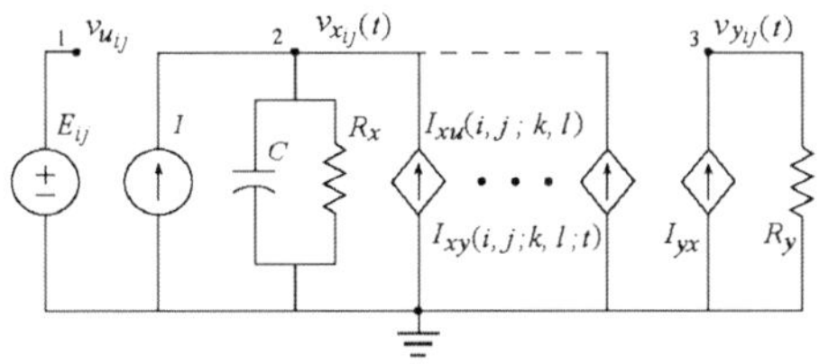

Fig. 6. Original CNN cell model.

The characteristics of the generators $I_{xy}(i, j; k, l; t)$ and $I_{xu}(i, j; k, l)$ are defined as:

$$\begin{aligned} I_{xy}(i, j; k, l; t) &= A(i, j; k, l)\, v_{y_{kl}}(t) \\ I_{xu}(i, j; k, l) &= B(i, j; k, l)\, v_{u_{kl}} \end{aligned} \tag{6}$$

The equations (6) state that the current sources $I_{xu}(i, j; k, l)$ are controlled by the input voltage of the neighbors C_{kl}, while the others $I_{xy}(i, j; k, l; t)$ get a feedback from the output voltages of the neighbor cells. In such a way, it is possible to control the strength of interactions between cells by setting the coupling parameters $A(i, j; k, l)$ and $B(i, j; k, l)$. The output $v_{y_{ij}}(t)$ is determined by the nonlinear voltage controlled current source I_{yx} that is the only nonlinear element of the cell and it is characterized by the following equation:

$$I_{yx} = \frac{1}{R_y} f(v_{x_{ij}}(t)) \tag{7}$$

where f is the characteristic function of the nonlinear controlled current source defined as:

$$f(v_{x_{ij}}(t)) = \frac{1}{2}\left(|v_{x_{ij}}(t) + 1| - |v_{x_{ij}}(t) - 1|\right) \tag{8}$$

Using the Kirchhoff laws, the state of a CNN cell can be described by the following nonlinear differential equation:

$$C\,\dot{v}_{x_{ij}}(t) = -\frac{1}{R_x} v_{x_{ij}}(t) + I + \sum_{C_{kl} \in S_{ij}(r)} \left(A(i,j;k,l) f\left(v_{x_{kl}}(t)\right) + B(i,j;k,l) v_{u_{kl}} \right) \tag{9}$$

that, by considering $1 \leq i \leq M$, $1 \leq j \leq N$, corresponds to the system of ODEs (5). Chua and Roska [10] propose CNNs as a parallel computing paradigm, especially suited for processing analog array signals, with important applications in image processing, pattern recognition, numerical solution of PDEs and investigation of nonlinear phenomena. CNNs have been successfully applied in various image processing applications, especially because of the high pay-off offered by the CNN based architectures [40].

CNNs can be trained by learning algorithms in such a way that, during the training process, the network learns which is the best configuration for performing a given task. In this work, GAs have been applied in order to evolve a CNN capable of detecting MS lesions from MRI. As is shown in the work of J.H. Holland in 1975 [72], GAs are computational bio-inspired methods for solving problems; inspired by Darwinian evolution, GAs are based on the principles of *genetic variation* and *natural selection*. These algorithms simulate the evolution of a population of individuals (Figure 7), which represent possible solutions to a given problem.

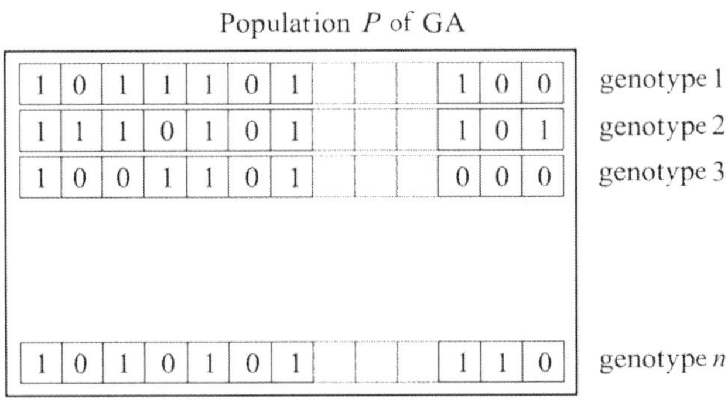

Fig. 7. Population of a GA.

To evaluate the performance of each individual in relation to the problem, it is possible to define an appropriate fitness function, which quantitatively measures the performance of each individual, in a given generation and for all the generations. The standard method for developing a GA is to choose a genetic representation, a fitness function and then proceeding with the following steps:

1. Generating a random number of strings (initial population), that encode possible solutions to the problem;
2. Decoding of the genotypes of the population and assessment of each individual (phenotype), according to the fitness function;

3. If the current population contains a satisfactory solution, the algorithm stops.
4. If the system does not find a *good* solution, a new evolution starts, generating a new population of individuals, by applying the operators of selection, crossover and mutation.

The process continues with the evaluation of new individuals through the fitness function cyclically in this manner until a satisfactory solution to a given problem is obtained (Figure 8).

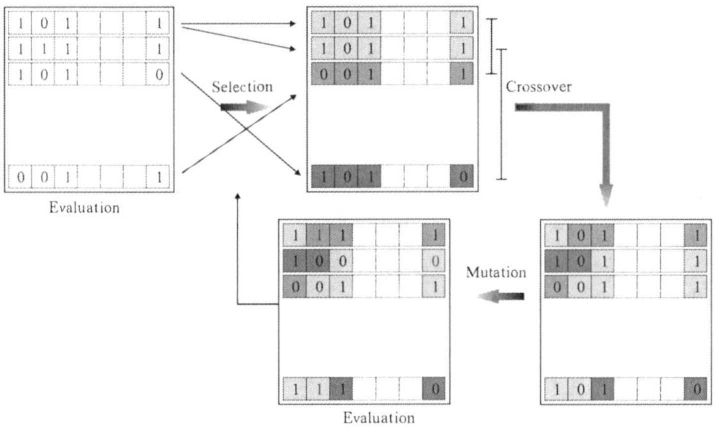

Fig. 8. Steps of a GA.

The **Selection**, **Crossover**, and **Mutation** operators proposed by Holland are inspired by natural selection and genetics; in particular:

1. **Selection**: takes place according to a probability that is proportional to the fitness value, in order to give preference to best performing individuals, allowing them to transmit their genes to the next generation.
2. **Crossover**: two individuals are randomly chosen from the population, using the selection operator and new "child" genotypes are created.
3. **Mutation**: once new individuals are generated by the crossover operator, some of their genes may undergo a mutation process, which means that, according to a given probability, newly created genotypes will have some of their bits flipped.

In image processing applications, a neighborhood of radius $r = 1$ is commonly used and, in most cases, space-invariant templates are chosen, that is the operators $A(i, j; k, l)$ and $B(i, j; k, l)$ depend only on the relative position of a cell with respect to its neighbors. With this assumption, the whole system is characterized by a 3×3 feedback matrix A, a 3×3 control matrix B and a scalar z. Therefore, 19 parameters are needed to "program" a CNN;

this means that, once the initial state and boundary conditions have been assigned, the operation performed by the CNN on a given input image is determined only by 19 real values that completely define the properties of the network.

For our aim, which is to design a GA to search for the weights of a standard two-dimensional space invariant CNN, in which each cell has a radius of influence $r = 1$, it is convenient to adopt a representation of templates in vector form. To this purpose, the 19 parameters that define the triple $\{A, B, z\}$ are arranged in an array consisting of 9 feedback synaptic weights defining the A matrix, 9 control synaptic weights defining the B matrix and the threshold z (Figure 9). These 19 coefficients represent a gene for the CNN, associated with a particular function performed by the network.

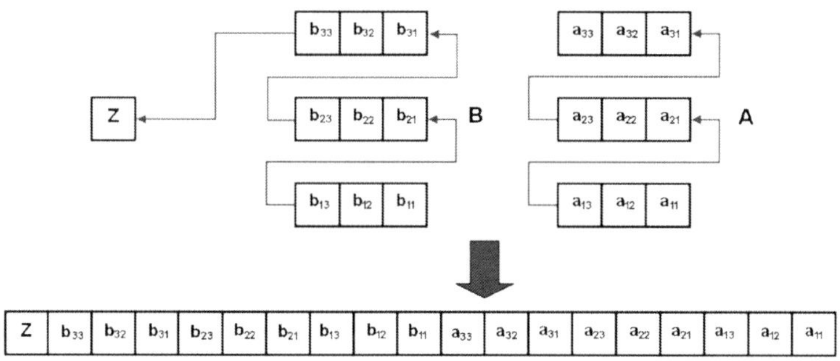

Fig. 9. Representation of a CNN template in vector form.

The GA has been designed to get a template to be used for image processing applications. For this reason, we have chosen to impose that the matrices A and B are symmetric with respect to their central element. In this way, we set the conditions for the stability of the CNN, provided in the complete stability theorem [8], which ensures the convergence of the network. It also reduces the computational load of the algorithmic search, since it is necessary to determine only 11 coefficients, 5 belonging to the matrix A, 5 to the matrix B and 1 corresponding to the threshold z. Each genotype is therefore represented by a vector G of 11 elements:

$$G = [a_{11}\, a_{12}\, a_{13}\, a_{21}\, a_{22}\, b_{11}\, b_{12}\, b_{13}\, b_{21}\, b_{22}\, z]. \tag{10}$$

To assess the fitness of a CNN gene compared to an assigned problem, we introduce a target image T of $M \times N$ pixels to be used for training the network. Applying the template corresponding to G and providing an input image to the CNN, it generates as an output an image I^G which can be compared with T, through the cost function:

$$\mathrm{diff}(G) = \sum_{i=1}^{M}\sum_{j=1}^{N} I_{ij}^{G} \oplus T_{ij}\,. \tag{11}$$

where the operator $\oplus$ denotes the logical exclusive or (xor) between the element in position (i, j) of the target image and the corresponding pixel in the CNN output. The fitness function for each phenotype CNN^G, then, is evaluated by calculating the number of equal pixels between T and the CNN output:

$$\mathrm{fitness}(\mathrm{CNN}^G) = M \,\mathrm{x}\, N - \mathrm{diff}(G)\,. \tag{12}$$

Hence, the fitness measures the number of equal pixels between the target image and that obtained from the CNN simulation. In this way, higher values of fitness are associated with phenotypes corresponding to templates that produce outputs with a high number of pixels, that coincide with the image target.

4.1 Optimization Approach

A set of evolutionary runs has been performed in order to reach the best level of performance of the developed systems. In our implementation, we have run an initial random population of $35 - 60$ individuals, making them evolve for a maximum of 500 generations; "*weighted roulette wheel selector*" and "*best chromosome selector*" were used as selection methods; mutations and elitism strategies were applied. In order to reduce the computational effort due to the large search space, we have chosen to constrain the elements of each genotype to be in the range $[-8, 8]$.

The GA was conducted as follows: after evaluating the fitness of each phenotype, the elite individual, i.e., the best performing one, has been directly copied in the next generation; a number of single-point crossover operations, depending on the population size, has been performed. In our experiments, we have used a crossover rate in the range from 30% to 70%. Mutations have been randomly applied in order to prevent trapping into local minima. The elements of the genotypes in the population have been randomly mutated according to a given mutation rate, each coefficient had a given probability of being changed by a randomly selected real number that falls in the chosen interval $[-8, 8]$. Using a mutation rate of 0.05, each component had 5% probability of being changed, resulting in one every twenty of the parameters being mutated on average.

Once genetic operators have been applied, a fixed number of genotypes has been selected and moved on the next generation. Obviously, the selection has been guided by the fitness, i.e. higher probabilities of survival have been associated to phenotypes providing higher fitness values. Figures 10-11 show two of the evolutionary runs that we have performed.

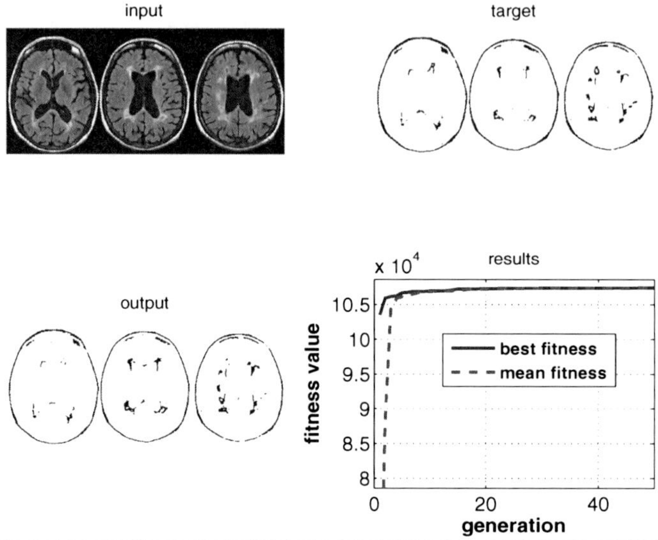

Population	Generations	% Crossover	Mutation rate	Selection method	Max fitness
35	50	35	0.05	weighted roulette wheel	107464

Fig. 10. Results of an evolutionary run that we have performed. The initial population is composed of 35 individuals; the run has evolved for 50 generations with a mutation rate of 0.05 and a 35% as crossover rate. The *weighted roulette wheel selector* has been the selection method we have used. The highest fitness reached is 107464; by referring that value to the size of the target image, we have obtained an overlapping of 0.9851 with respect to the target (1 being a total agreement).

A number of trials have been performed according to the GA scheme described above, some of which are reported in Table 1. In our experiments, the GA achieved a maximum fitness value of 0.9851, resulting in 98.51% of overlapping between the CNN output and the corresponding target image. The average fitness value obtained over the evolutionary runs performed is 0.9820.

Table 1. Evolutionary runs performed according to the proposed GA based learning process.

Population	Generations	% Crossover	Mutation rate	Selection method	Max fitness	Normalized fitness
35	50	0.35	20	weighted roulette wheel	107464	0.9851
35	100	0.35	20	weighted roulette wheel	289617	0.9833
30	300	0.3	20	best chromosomes	107392	0.9845
30	500	0.5	7	best chromosomes	142902	0.9830
60	50	0.5	5	best chromosomes	142456	0.9799
40	40	0.7	5	best chromosomes	142697	0.9816

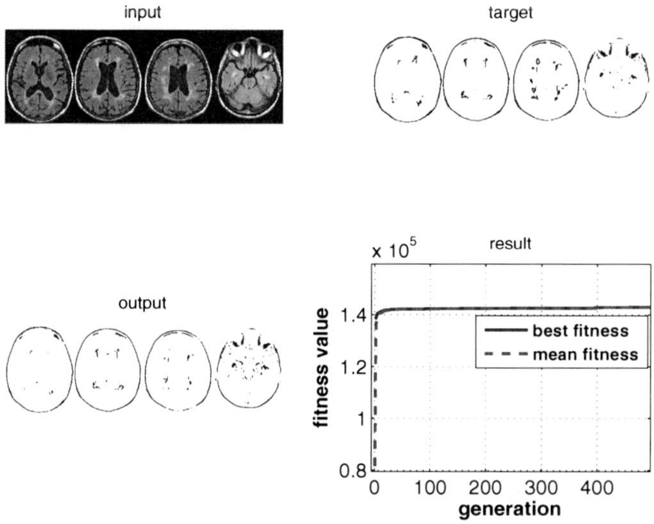

Population	Generations	% Crossover	Mutation rate	Selection method	Max fitness
30	500	50	0.14	best chromosomes	142902

Fig. 11. Evolutionary run we have performed. The initial population is composed of 30 individuals; the run has been evolved for 500 generations with a mutation rate of 1/7 and a 50% crossover rate. The *best chromosomes selector* has been the selection method we have used. The highest fitness reached is 142902, resulting in 0.9830 of overlapping with the target.

At the end of the training process, the following template has been selected:

$$A = \begin{bmatrix} -3.51879 & 3.42019 & -3.48386 \\ 6.47032 & 7.75293 & 6.47032 \\ -3.48386 & 3.42019 & -3.51879 \end{bmatrix} \quad B = \begin{bmatrix} 1.33076 & -3.86887 & 1.53728 \\ -2.30849 & -7.76398 & -2.30849 \\ 1.53728 & -3.86887 & 1.33076 \end{bmatrix} \quad z = -4.81797 \tag{13}$$

Given a MRI slice, by tuning the CNN standard model (5) according to the parameters (13), the system is able to generate images in which MS lesions are isolated from the brain matter, as learned during training; this is the key-step on which the segmentation algorithm is based.

4.2 Application of the Optimization Approach

One practical application based on the optimization performed by GAs consists of evolving a CNN capable of supporting neurological diagnosis in order to determine the lesion load in patients affected by MS. To evaluate the evolution of the disease, images acquired by magnetic resonance can be used by neurologists in order to detect the presence of lesions in the white matter of the brain. The main issue is to develop a CNN algorithm for supporting image analysis and identifying brain areas affected by lesions in MRI. Magnetic resonance scanners acquire gray scale images; each image represents a "slice" corresponding to a volumetric portion of the brain. The main purpose of the algorithm, for each slice, is to generate a binary image as output, in which the lesions are isolated from healthy tissue. In this way, knowing the size of the voxels of the slice, it is possible to calculate an estimate of the lesion load due to the pathology. Although the pixels corresponding to the lesions are always brighter than the others in healthy tissue, they are not the brightest in the image, therefore it is not enough to simply apply a single template that performs a binarization according to a given threshold. The CNN, therefore, must be trained to distinguish and classify only the areas actually affected by injuries.

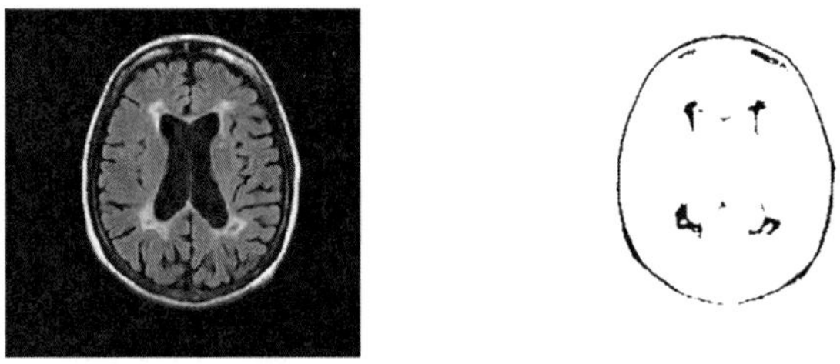

(a) input slice **(b)** lesion detection and binarization

(c) lesions extraction and isolation **(d)** final result of processing

Fig. 12. CNN algorithm for lesions detection in patients affected by MS.

As shown in Figure 12, the algorithm proposed for the segmentation of the MS lesions consists of three principal steps:

1. Lesions detection and their binarization;
2. Segmentation of the white matter of the brain;
3. Lesions extraction and isolation.

The first step of the algorithm has been faced by using GAs, to determine the weights that allow CNN to generate as output a binary image in which the healthy brain matter is removed. Then, templates in the CNN library [19] were applied for extracting features of interest, eliminating the pixels corresponding to the skull. Lesions have different shapes and intensities that vary from slice to slice, even for the same patient. It is thus necessary to iterate the training process of the network more times, for each iteration providing target images showing different characteristics. Then, the CNN can be able to "learn" how to perform the task in a more accurate way (Figure 13).

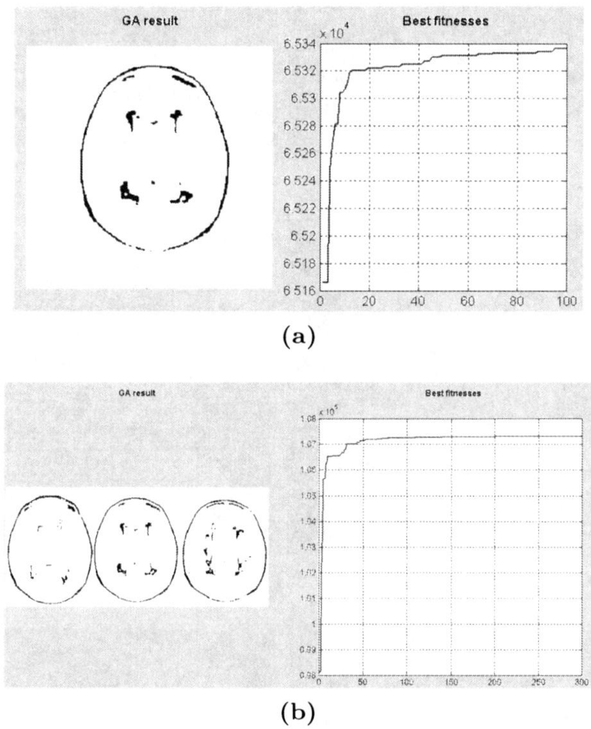

Fig. 13. Training of CNN by GA for the detection of lesions in MRI.

To implement the second step of the algorithm, we have performed a masking operation based on the templates proposed in the library [19] for the extraction of objects. The idea is to extract, from the image obtained in the previous step, only objects lying in the slice corresponding to the brain matter; thus at the end of the simulation only lesions will emerge, while areas related to the skull will be deleted.

Simulations (Figure 14) have allowed to verify the validity of the proposed algorithm; the output generated by CNN can be viewed in the MRICro medical image viewer (`http://www.mricro.com`). Once the number of pixels corresponding to the injury has been determined and knowing the size of the voxel in the performed scan, it is possible to estimate the total lesion load

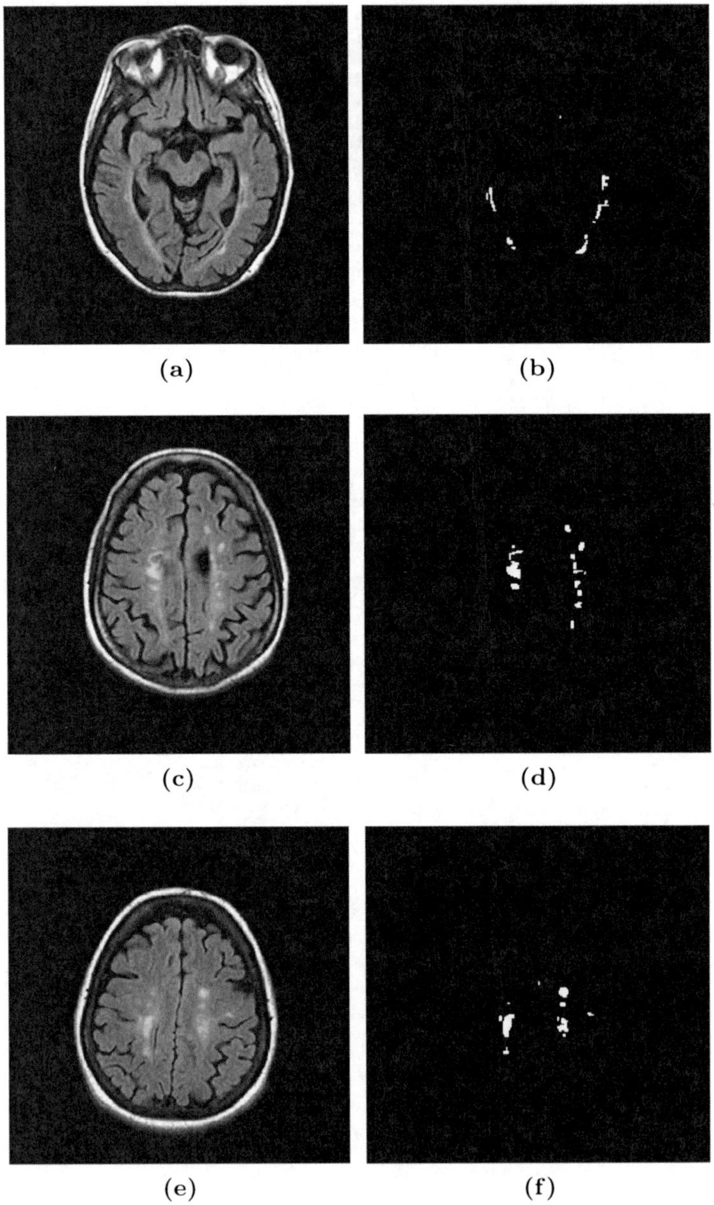

Fig. 14. Segmentation of MRI for MS lesions detection by using CNN.

for a given patient. This provides an important parameter to monitor the progression of the disease. It should be emphasized that the results in Figure 14 were obtained without operating any manual thresholding operation; in other words, the proposed algorithm allows to fully automate the process of estimating the lesion load for a given patient. Obviously, operating a manual tuning of the network, the algorithm is able to produce more accurate results. Overlapping the slices and the output of CNN, is possible to obtain a 3D reconstruction of the brain of the patient, which displays an estimate of brain tissue sclerosis (Figure 15).

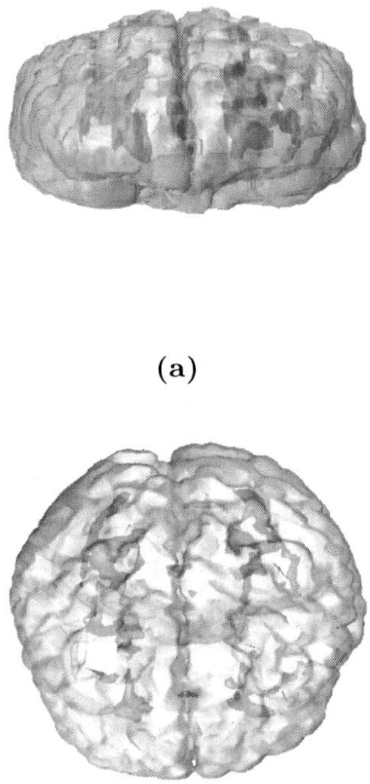

(a)

(b)

Fig. 15. Three-dimensional recontruction of the lesions detected by using CNN.

Sometimes, the images acquired by MRI systems have, from patient to patient, significant differences in the intensity of grey and injuries, as indeed it has been observed also from a slice to another. For this reason, it is necessary to train the network for adapting it to conditions not encountered before. However, the results obtained by applying the proposed algorithm have been very convincing, since CNN can determine most of the lesions and thus it could provide a useful support tool for the diagnosis of pathology in particular to assess the presence and the evolution of the lesions. Efforts are under way for improving templates obtained by GA. Some snapshots of the software we have implemented are shown in Figure 16.

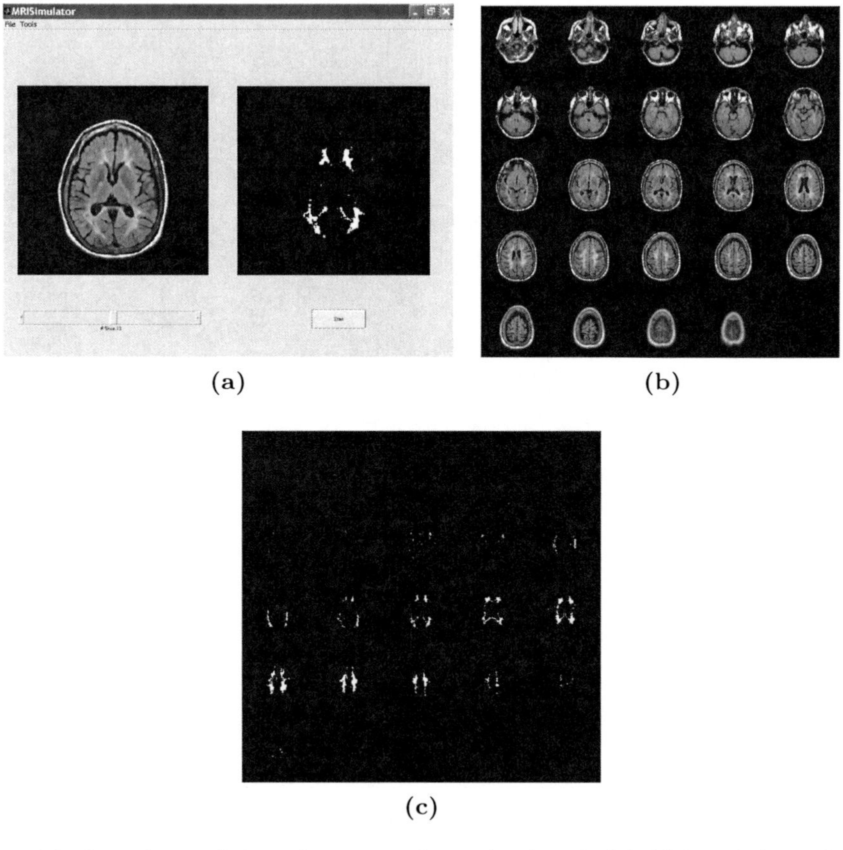

(a) **(b)**

(c)

Fig. 16. Snapshots of the software we have developed. (a) Main window of the software (b) Input of the simulation (c) Output of the simulation.

The experiments presented above have been carried out by using an algorithm based on two-dimensional data. The new idea is to extend our results by working on volumes of data, rather than processing planar images. 3D CNNs may represent a new and powerful tool for the development of applications for supporting medical diagnosis; in fact, by exploiting information provided by a three dimensional representation of the brain, the evolution of a 3D network can lead to a significant improvement of the performances. For this reason, by modifying the architecture of the standard CNN, we propose a 3D-CNN model that is able to handle and perform different functions on objects in a three-dimensional space. We consider an array of $M \times N \times P$ cells representing dynamical systems. In this case, each cell is referenced by the triple (i, j, k), and the evolution of the state is described by the following set of ODEs:

$$\begin{aligned}\frac{dx_{ijk}(t)}{dt} = & -x_{ijk}(t) + \\ & \sum_{C(l,m,n)\in S_r(i,j,k)} A(i,j,k;l,m,n)\, y_{lmn}(t) + \\ & \sum_{C(l,m,n)\in S_r(i,j,k)} B(i,j,k;l,m,n)\, u_{lmn} + z_{ijk}\end{aligned}$$

$$y_{ijk}(t) = f(x_{ijk}(t)) = \frac{1}{2}[|x_{ijk}(t)+1|] - [|x_{ijk}(t)-1|] \tag{14}$$

where A and B contain the synaptic weights between C_{ijk} and its neighbors, and z_{ijk} denotes the bias of the cell. We define the sphere of influence of radius r of C_{ijk}, denoted by $S_{ijk}(r)$, as the set of cells satisfying the following property:

$$S_{ijk}(r) = \{C(l,m,n) \mid \max_{1\le l\le M, 1\le m\le N, 1\le n\le P}\{|l-i|, |m-j|, |n-k|\} \le r\}, \quad r \ge 0 \tag{15}$$

This means that, assuming $r = 1$, a $3 \times 3 \times 3$ neighborhood is associated to each cell, i.e., each processing unit has 27 neighbors (including the cell itself). Figure 17 shows the $3 \times 3 \times 3$ neighborhood for the cell at the position (i, j, k) (reported as grey). To avoid cluttering, only the connections between C_{ijk} and its neighbors are shown.

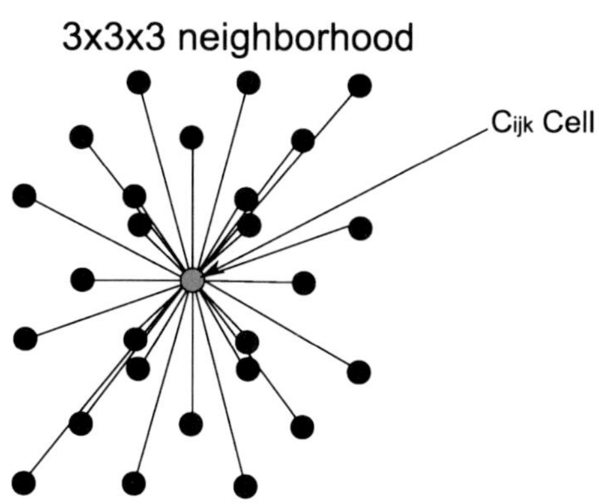

Fig. 17. Neighbourhood for the cell C_{ijk} of a 3D-CNN.

Therefore, for a space-invariant 3D-CNN, the feedback operator is given by the array $A \in \mathbb{R}^{3\times3\times3}$, the control operator is $B \in \mathbb{R}^{3\times3\times3}$ and so 55 parameters are needed in order to define a template $\{A, B, z\}$ for a 3D-CNN.

An innovative library of 3D templates has been implemented, that allows to execute interesting and efficient operations, as shown in Figures 18 and 19.

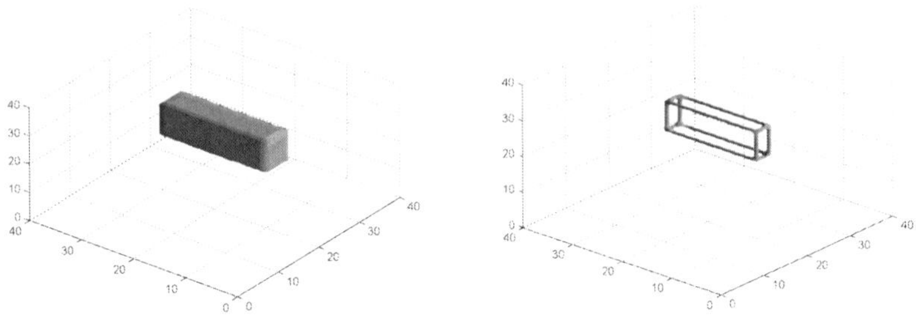

(a) Input (b) Output

Fig. 18. Edge corners detection.

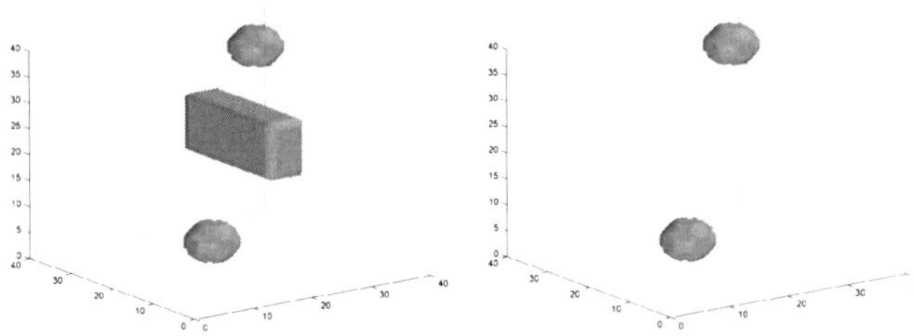

(a) Input (b) Output

Fig. 19. Spherical objects detection.

We used GAs for 3D template learning. This means that even a 3D-CNN can be trained by a learning algorithm in such a way that, during the training process, the network learns which the configuration for performing the desired operation is. The slices acquired from a MRI scanning can be recasted into a three dimensional array and provided as input to the 3D-CNN; a 3D visualization of the input, obtained by superimposing 24 slices for a given patient, is shown in Figure 20. In this way, the analysis of MRI images can be performed according to the CNN model (14). The template used for the segmentation of MS lesions on planar images (13) has been adapted in order to construct a $3 \times 3 \times 3$ template to be used for 3D-CNNs.

Again, after detecting lesions inside the brain area (Figure 21(a)), by applying 3D templates we have found for operations on volumetric data, "noise removal" is performed and voxels corresponding to MS lesions are segmented, as shown in Figure 21(b).

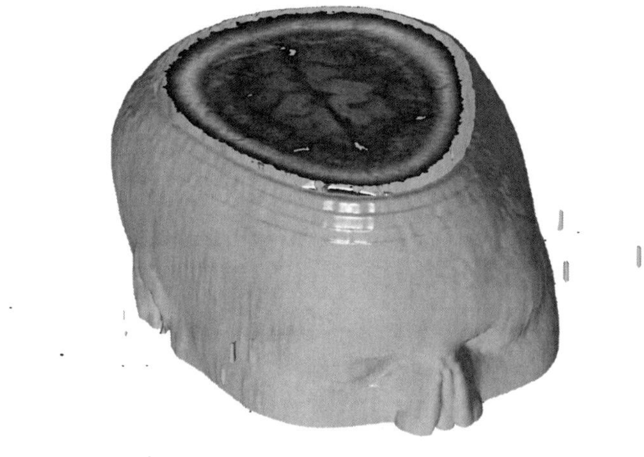

Fig. 20. 3D visualization of MRI slices re-casted into an array 256 × 256 × 24.

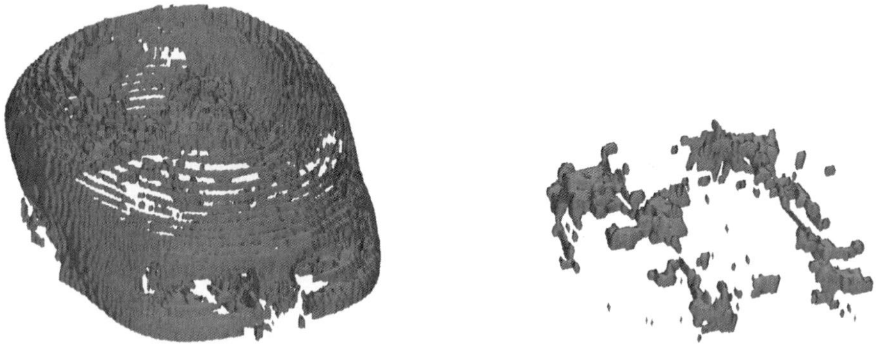

(a) **(b)**

Fig. 21. CNN processing of MRI volumetric data. A 3D surface representation of sub-cortical structure and lesions, generated from the results of our method. Note how the 3D rendering outlines the global structure of the lesions.

4.3 Integration into the Productive System and with other Software or Hardware Components

We are integrating this system within the Institute of Neurological Sciences, National Research Council, Piano Lago di Mangone, Cosenza, in collaboration with the Faculty of Medicine, University "Magna Græcia" of Catanzaro. Meanwhile we are expanding the number of patients to least 10 for each step of the EDSS scale, for a total of 100 patients. We are trying to correlate TLLs, found via the automated approach that we have developed, with clinical signs of MS, in correlation with the EDSS. We are also conducting a study on the evolution of the disease, in correlation with the increase of lesion load. We are trying to find "the clinical impossible", i.e., intermediate lesion load observed in the scale of functional systems, that cannot be assessed by neurologists. Instead, by using the CNN automated segmentation method, we could research interstitial values of lesion load related to different disabilities in both cognitive and motor skills.

These results could be very useful in the management of MS patients' therapies, thus allowing this method to work in a real life situation. Moreover, given the possibility of implementing in hardware, the CNNs-based algorithm could even be integrated on chip as a system embedded in the MRI machine. This method can also be generalized to other diseases and other forms of brain imaging technologies. Furthermore, the method has practical application as it can be used both for diagnostic purposes and for training medical professionals.

5 Experimental Results

The experiment described in this section compares the accuracy of the CNN-based segmentation method, we have presented in the previous sections, with manual delineations of MS lesions. The second approach is similar to the CNN based approach but it consists of different non-automatic steps. Both methods utilize the same set of data, which captures the spatial distributions of the TLL in brain white matter. For both methods, the training data of the experiment consists of sequences of magnetic resonance images, as the examples in Figures 13. The data are generated through the approach suggested by Warfield et al. [73], which aligns the training subjects to a pre-selected training case and then measures the overlap between the corresponding segmentations. Each method segments the same number of cases. Magnetic resonance imaging scanning was performed on a 1.5 T Unit (Signa NV/i, General Electric, Milwaukee, Wisconsin) using a standard quadrature head coil. 2D fast fluid-attenuated inversion-recovery (FLAIR) axial images (TR $8000ms$, TE $120ms$; 256×224 image matrix, FOV: $24cm$; 24 slices, $4mm$ slices, $1-mm$ gap) oriented along the AC-PC line were used to calculate the hyper-intense lesion load. We have evaluated the accuracy of the approaches by measuring the agreement of the automatic segmentations of the TLL to the manual ones. Note that the other method greatly depends on the precise segmentation of white matter as the MS disease is characterized of weakly visible boundary. The performances of the process have been quantitatively evaluated by comparing the CNN output and the expert's manual delineation of MS lesions, using the Dice coefficient [74] as a metric. The Dice coefficient D is a statistic measure used for comparing the extent of spatial overlap between two binary images. It is commonly used in reporting performance of segmentation and its values range between 0 (no overlap) and 1 (perfect agreement). In this work, the Dice values, expressed as percentages, are computed as follows:

$$D = \frac{2|L^{CNN} \cap L^G|}{|L^{CNN}| + |L^G|} \times 100 \quad (16)$$

where L^{CNN} is the automated segmentation result and L^G the manual one. The graph in Figure 22 shows the Dice measure for the three compared templates in the 11 cases.

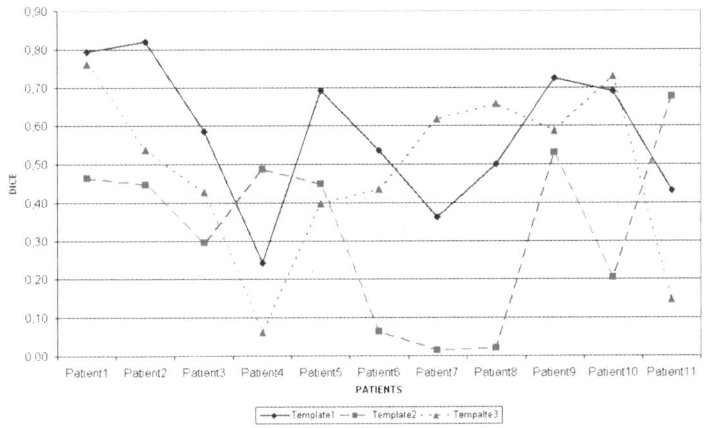

Fig. 22. Dice measures for Templatel, Template2, Template3.

We have compared quantitatively the performance of the three templates with the gold standard represented by the manual measures. The results can be summarized as follows:

1. The template which appears to be the best is Templatel, with an average overlap ratio 0.58 compared to the gold standard handbook (1 perfect overlap, 0 no overlap); Templatel is absolutely the best in $6/11 = 55\%$ of the subjects, respectively, followed by Template3 in $3/11 = 27\%$ of the subjects, and Template2 in $2/11 = 18\%$.
2. Template2 is more prone to under-segmentation.
3. Template3 is more prone to over-segmentation.

Proportionally, the average error of under-segmentation of Template2 (1.25) is greater than that of over-segmentation of Template3 (0.89). From Figure 22, it is possible to observe that, by applying Templatel, the highest Dice measure is achieved for Patient2 (~ 0.82), while the algorithm produces the lowest value (~ 0.24) on Patient4. The method we have implemented for automatically segmenting MS lesions gives satisfactory results, showing that after the learning process the CNN is capable of detecting MS lesions with different shapes and intensities, even in MRI slices with different contrasts between white and grey matter, with respect to the images used during the genetic training process. Thus, the proposed algorithm provides a good robustness.

It is worth noting that training images, as well as the ones used for testing, come from a dataset of real MS patients and the MRI are not pre-processed. The algorithm is run directly on the slices as they are acquired by the magnetic resonance. The aim is to provide a tool that does not require any manual operation for the physicians, and the training process is still underway in order to further improve the robustness. The vast majority of the lesion load, detected by the CNN for the described sample, ranges from $D = 0.6$ to $D = 0.8$. The technique we propose for segmenting white matter lesions in MS is a fully automatic method and does not require manually segmented data. Semiautomatic methods are highly dependent on the choice of an appropriate threshold to effectively detect lesions. Thresholds usually may vary between different slices even for the same patient, thus leading to a time consuming task. Our algorithm, however, allows for obtaining the desired output by programming a fully automated strategy on the entire dataset, without the need of external calibration.

Simulations have allowed verifying the validity of the algorithm described above. The output generated by the CNN can also be viewed in the MRIcro medical image viewer. Calculating the number of pixels corresponding to the injury and knowing the size of the voxel used in the scanning, it is possible to estimate the TLL for any patient. This method provides an important parameter to monitor the progress of the pathological disease. It should be emphasized that the results were obtained without changing the template from one slice to another.

6 Conclusions

In this chapter, we have presented a new, completely automatic segmentation technique for detecting MS lesions in white matter regions of the brain. The performance of the method has been evaluated and compared to manual segmentation. The results, obtained by applying the proposed algorithm, are very convincing, since the CNN system can determine most of the lesions in all the patients. The method we have implemented could be a useful MRI support tool for the evaluation of lesions in MS, particularly to assess the evolution of the lesions. From a comparison with other existing methods in the literature on this topic, we can say that the results are valuable and the threshold of recognition is currently at 70%. Furthermore, it should be emphasized the real improvements of the proposed method, with respect to our previous work, are the greater accuracy of the system, its adaptation to different conditions of the stimula, and its ability to create 3D images of the injured areas of the brain. It thus effectively supports medical diagnosis. With this method, slighter changes in the total lesion load can be detected, improving both our knowledge of the disease and our ability to monitor it.

7 Future Work

MS lesions have a complex evolution that begins with an initial disruption of blood brain barrier, accompanied by demyelination, inflammation, and axonal damages. Later on the disease appears to regress partially. So far, clinical studies and the characteristics of the MRI technology have allowed the identification of the evolution of this disease and three main groups of lesions have been identified. This classification includes:

1. Acute lesions demonstrated by losses in the blood-brain barrier, detected in contrast-MR Advanced imaging (enhancing lesions);
2. Chronic wounds severely damaged, hypointense so-called “black holes”, visible on T1-weighted MR images (T1WI);
3. Hyperintense lesions, detectable in T2 (T2 lesions), visible on T2-weighted MRI (T2WI).

This classification of MS lesions in these subtypes is widely accepted and has demonstrated good clinical utility in the related studies [75, 76]. For future work we plan to use the CNNs based method to automatically segment MS lesions into the three subtypes mentioned above, with respect to different brain tissue compartments. To do this, we shall use different sequences of magnetic resonance images that identify different stages of evolution of the disease. We also want to divide MRI by using a grid, like a chessboard, in order to locate in which quadrant of the grid the lesions settled. This would enable us to correlate the location of the lesions in the brain with a more accurate identification of the specific disabilities of each patient. We also want to see if it is possible to study the patterns of this disease through the application of methods of chaos and nonlinear dynamical systems, intercepting sensitivity to initial data, variables for each subject, the possibility of developing the disease within a parameter space, the evolution of the patterns of MS injuries.

References

1. Lanzarini, L., De Giusti, A.: Pattern recognition in medical images using neural networks, `http://journal.info.unlp.edu.ar/journal/journal4/papers/pap4.pdf`
2. Wismuller, A., Vietze, F., Dersch, D.R.: Segmentation with Neural Networks. In: Handbook of Medical Image Processing and Analysis. ch. 7, pp. 113–143. Elsevier, Johns Hopkins University, USA, Baltimore (2008)

3. Suri, J.S., Wilson, D.L., Laxminarayan, S.: Segmentation Models, Part B. In: Handbook of Biomedical Image Analysis. ch. 7, vol. 2, pp. 315–368. Kluwer Academic, Plenum Publishers, New York (2005)
4. Wismuller, A., Meyer-Bease, A., Lange, O., Auer, D., Reiser, M.F., Sumners, D.: Model-free functional MRI analysis based on unsupervised clustering. Journal of Biomedical Informatics 37, 10–18 (2004)
5. Leinsinger, G.L., Wismuller, A., Joechel, P., Lange, O., Heiss, D.T., Hahn, K.: Evaluation of the motor cortex using fMRI and image processing with self-organized cluster analysis by deterministic annealing. Radiology, 221–487 (2001)
6. Wismüller, A., Dersch, D.R., Lipinski, B., Hahn, K., Auer, D.: Hierarchical Clustering of Functional MRI Time-Series by Deterministic Annealing. In: Brause, R., Hanisch, E. (eds.) ISMDA 2000. LNCS, vol. 1933, pp. 49–54. Springer, Heidelberg (2000)
7. Döhler, F., Mormann, F., Weber, B., Elger, C.E., Lehnertz, K.: A cellular neural network based method for classification of magnetic resonance images: Towards an automated detection of hippocampal sclerosis. Journal of Neuroscience Methods 170(2), 324–331 (2008)
8. Chua, L.O., Yang, L.: Cellular Neural Networks: Theory. IEEE Transactions on Circuits and Systems 35(10), 1257–1272 (1988)
9. Chua, L.O.: CNN: A paradigm for Complexity. World Scientific Series on Nonlinear Science (1996)
10. Chua, L.O., Roska, T.: Cellular Neural Networks and Visual Computing: Foundations and Applications. Cambridge University Press, Cambridge (2004)
11. Niederhoefer, C., Gollas, F., Tetzlaff, R.: EEG analysis by multi layer Cellular Nonlinear Networks (CNN). In: Biomedical Circuits and Systems Conference, November 29–December 1. IEEE, Los Alamitos (2006)
12. Schwarz, T., Heimann, T., Tetzlaf, R., Rau, A.M., Wolf, I., Meinzer, H.P.: Interactive Surface Correction for 3D Shape Based Segmentation. Medical Imaging (2008)
13. Bilotta, E., Cerasa, A., Pantano, P., Quattrone, A., Staino, A., Stramandinoli, F.: A CNN Based Algorithm for the Automated Segmentation of Multiple Sclerosis Lesions. In: EvoStar 2010 Conference (2010)
14. Chua, L.O.: A Nonlinear Dynamics Perspective of Wolfram's New Kind of Science. World Scientific Publishing Co., Singapore (2007)
15. Roska, T., Chua, L.O., Wolf, D., Kozek, T., Tetzlaff, R.: Simulating Nonlinear Waves and Partial Differential Equations via CNN. IEEE Transactions on Circuits and Systems 42(10) (October 1995); Part I: Basic Techniques
16. Kozek, T., Chua, L.O., Roska, T., Wolf, D., Tetzlaff, R., Pufferand, F., Lotz, K.: Simulating Nonlinear Waves and Partial Differential Equations via CNN. IEEE Transactions on Circuits and Systems, Part 11 42(10) (October 1995)
17. Arena, P., Basile, A., Bucolo, M., Fortuna, L.: Image processing for medical diagnosis using CNN. Nuclear Instruments and Methods A 497(1), 174–178 (2003)
18. Szabo, T., Barsi, P., Szolgay, P.: Application of Analogic CNN algorithms in Telemedical Neuroradiology. Journal of Neuroscience Methods 170(7), 2063–2090 (2005)
19. Kek, L., Karacs, K., Roska, T.: Cellular Wave Computing Library (Templates, Algorithms and Programs ver.2.1), Cellular Sensory Wave Computers Laboratory, Hungarian Academy of Science (2007)

20. Trapp, B.D., Ransohoff, R., Rudich, R.: Axonal pathology in multiple sclerosis: relationship to neurologic disability. Current Opinion in Neurology 12(3), 295–302 (1999)
21. Keegan, B.M., Noseworthy, J.H.: Multiple sclerosis. Annual Review of Medicine 53, 285–302 (2002)
22. Lassmann, H.: Cellular damage and repair in multiple sclerosis. In: Lazzarini, R.A. (ed.) Myelin Biology and Disorders, pp. 753–762. Elsevier, Amsterdam (2004)
23. Ozturk, A., Smith, S., Gordon-Lipkin, E., Harrison, D., Shiee, N., Pham, D., Caffo, B., Calabresi, P., Reich, D.: Mri of the corpus callosum in multiple sclerosis: association with disability. Multiple Sclerosis 16(2), 166–177 (2010)
24. Kurtzke, J.F.: Rating neurologic impairment in multiple sclerosis: an expanded disability status scale (edss). Neurology 33(11), 1444–1452 (1983)
25. Gronwall, D.M.: Paced auditory serial-addition task: a measure of recovery from concussion. Perceptual and Motor Skills 44, 367–373 (1977)
26. Young, K., Schuff, N.: Measuring Structural Complexity in Brain Images. NeuroImage 39(4), 1721–1730 (2008)
27. Giorgio, A., Palace, J., Johansen-Berg, H., Smith, S.M., Ropele, S., Fuchs, S., Wallner-Blazek, M., Enzinger, C., Fazekas, F.: Relationships of brain white matter microstructure with clinical and MR measures in relapsing-remitting multiple sclerosis. Journal of Magnetic Resonance Imaging 31(2), 309–316 (2008)
28. Wonderlick, J.S., Ziegler, D.A., Hosseini-Varnamkhasti, P., Locascio, J.J., Bakkour, A., Van Der Kouwe, A., Triantafyllou, C., Corkin, S., Dickerson, B.C.: Reliability of mri-derived cortical and subcortical morphometric measures: effects of pulse sequence, voxel geometry, and parallel imaging. NeuroImage 44(4), 1324–1333 (2009)
29. Wu, Y., Warfield, S.K., Tan, I., Wells, W.M., Meier, D.S., Van Schijndel, R., Barkhof, F., Guttmann, C.R.: Automated segmentation of multiple sclerosis lesion subtypes with multichannel mri. NeuroImage 32(3), 1205–1215 (2006)
30. Akselrod-Ballin, A., Galun, M., Gomori, J.M., Filippi, M., Valsasina, P., Basri, R., Brandt, A.: Automatic segmentation and classification of multiple sclerosis in multichannel mri. IEEE Transactions on Biomedical Engineering 56(10) (October 2009)
31. Zharkova, V., Jain, L.: Introduction to pattern recognition and classification in medical and astrophysical images. In: Artificial Intelligence in Recognition and Classification of Astrophysical and Medical Images. SCI, vol. 46, pp. 1–18. Springer, Heidelberg (2007)
32. Brzakovic, D., Luo, X.M., Brzakovic, P.: An Approach to Automated Detection of Tumors in Mammograms. IEEE Transactions on Medical Imaging 9(3) (September 1990)
33. Ertas, G., Gulcur, H.O., Osman, O., Ucan, O.N., Tunaci, M., Dursun, M.: Breast MR segmentation and lesion detection with cellular neural networks and 3D template matching. Computers in Biology and Medicine 38, 116–126 (2008)
34. Brem, M.H., Lang, P.K., Neumann, G., Schlechtweg, P.M., Schneider, E., Jackson, R., Yu, J., Eaton, C.B., Hennig, F.F., Yoshioka, H., Pappas, G., Duryea, J.: Magnetic resonance image segmentation using semi-automated software for quantification of knee articular cartilage-initial evaluation of a technique for paired scans. Radiology 38, 505–511 (2009)

35. Filippi, M., Yousry, T., Baratti, C., Horsfield, M.A., Mammi, S., Becker, C., Voltz, R., Spuler, S., Campi, A., Reiser, M.F., Comi, G.: Quantitative assessment of MRI lesion load in multiple sclerosis. Brain 119, 1349–1355 (1996)
36. Souplet, J.C., Lebrun, C., Chanalet, S., Ayache, N., Malandain, G.: Approaches to segment multiple-sclerosis lesions on conventional brain MRI. Revue Neurologique 165(1), 7–14 (2009)
37. Shiee, N., Bazin, P.L., Ozturk, A., Reich, D.S., Calabresi, P.A., Pham, D.L.: A topology-preserving approach to the segmentation of brain images with multiple sclerosis lesions. NeuroImage 49(2), 1524–1535 (2010)
38. Larochelle, H., Bengio, Y., Louradour, J., Lamblin, P.: Exploring Strategies for Training Deep Neural Networks. Journal of Machine Learning Research 10, 1–40 (2009)
39. Bishop, C.M.: Pattern Recognition and Machine Learning. Springer, Heidelberg (2006)
40. Bilotta, E., Pantano, P.: Cellular Non-Linear Networks as a New Paradigm for Evolutionary Robotics. In: Frontiers in Evolutionary Robotics, Hitoshi Iba, Vienna, Austria, pp. 87–108 (2008)
41. Bilotta, E., Pantano, P., Stranges, F.: A Gallery of Chua Attractors. International Journal of Bifurcation and Chaos 1, 1–60 (2007)
42. Bilotta, E., Pantano, P., Stranges, F.: A Gallery of Chua Attractors. International Journal of Bifurcation and Chaos 2, 293–380 (2007)
43. Bilotta, E., Pantano, P., Stranges, F.: A Gallery of Chua Attractors. International Journal of Bifurcation and Chaos 17(3), 657–734 (2007)
44. Bilotta, E., Pantano, P., Stranges, F.: A Gallery of Chua Attractors. International Journal of Bifurcation and Chaos 17(4), 1017–1078 (2007)
45. Bilotta, E., Pantano, P., Stranges, F.: A Gallery of Chua Attractors. International Journal of Bifurcation and Chaos 17(5), 1383–1511 (2007)
46. Bilotta, E., Pantano, P., Stranges, F.: A Gallery of Chua Attractors. International Journal of Bifurcation and Chaos 17(6), 1801–1910 (2007)
47. MacDonald, A.E., Lee, J.L., Sun, S.: QNH: Design and test of a quasi-nonhydrostatic model for mesoscale weather prediction. Monthly Weather Review 128, 1016–1036 (2000)
48. Anand, A.J., Shattuck, D.W., Pantazis, D., Li, Q., Damasio, H., Leahy, R.M.: Optimization of landmark selection for cortical surface registration. In: CVPR 2009, pp. 699–706 (2009)
49. Joshi, A., Leahy, R., Toga, A.W., Shattuck, D.: A Framework for Brain Registration via Simultaneous Surface and Volume Flow. In: Prince, J.L., Pham, D.L., Myers, K.J. (eds.) IPMI 2009. LNCS, vol. 5636, pp. 576–588. Springer, Heidelberg (2009)
50. Schneider, P., Andermann, M., Wengenroth, M., Goebel, R., Flor, H., Rupp, A., Diesch, E.: Reduced volume of Heschl's gyrus in tinnitus. NeuroImage 45(3), 927–939 (2009)
51. Ylipaavalniemi, J., Vigrio, R.: Analyzing consistency of independent components: an fMRI illustration. NeuroImage 39(1), 169–180 (2008)
52. Wachs, J.P., Stern, H.I., Edan, Y., Gillan, M., Handler, J., Feied, C., Smith, M.: A gesture-based tool for sterile browsing of radiology images. Journal of the American Medical Informatics Association 15(3), 321–324 (2008)
53. Thompson, P.M., Vidal, C., Giedd, J.N., Gochman, P., Blumenthal, J., Nicolson, R., Toga, A.W., Rapoport, J.L.: Mapping adolescent brain change reveals dynamic wave of accelerated gray matter loss in very early-onset schizophrenia. The Journal of Neuroscience 21(22), 8819–8829 (2001)

54. Wang, Y., Zhang, J., Gutman, B., Chan, T.F., Becker, J.T., Aizenstein, H.J., Lopez, O.L., Tamburo, R.J., Toga, A.W., Thompson, P.M.: Multivariate tensor-based morphometry on surfaces: application to mapping ventricular abnormalities in HIV/AIDS. NeuroImage 49(3), 2141–2157 (2010)
55. Tosun, D., Prince, J.L.: A geometry-driven optical flow warping for spatial normalization of cortical surfaces. IEEE Transactions of Medical Imaging 27(12), 1739–1753 (2008)
56. Vernon, A.C., Johansson, S.M., Modo, M.M.: Non-invasive evaluation of nigrostriatal neuropathology in a proteasome inhibitor rodent model of Parkinson's disease. BMC Neurosci. 11(1) (2010)
57. Labate, A., Gambardella, A., Aguglia, U., Condino, F., Ventura, P., Lanza, P.: Temporal lobe abnormalities on brain MRI in healthy volunteers: A prospective case-control study. A. Neurology 74(7), 553–557 (2010)
58. Leemput, K.V., Maes, F., Vandermeulen, D., Colchester, A., Suetens, P.: Automated segmentation of multiple sclerosis lesions by model outlier detection. IEEE Transactions on Medical Imaging 20(8), 677–688 (2001)
59. Freifeld, O., Greenspan, H., Goldberger, J. (eds.): Lesion detection in noisy MR brain images using constrained GMM and active contours (ISBI 2007), 4th IEEE International Symposium on Biomedical Imaging (2007)
60. Aït-Ali, L.S., Prima, S., Hellier, P., Carsin, B., Edan, G., Barillot, C.: STREM: A Robust Multidimensional Parametric Method to Segment MS Lesions in MRI. In: Duncan, J.S., Gerig, G. (eds.) MICCAI 2005. LNCS, vol. 3749, pp. 409–416. Springer, Heidelberg (2005)
61. Bricq, S., Collet, C., Armspach, J.P. (eds.): 5th IEEE International Symposium on Biomedical Imaging Lesion detection in 3D brain MRI using trimmed likelihood estimator and probabilistic atlas (ISBI 2008), (2008)
62. Garcia-Lorenzo, D., Prima, S., Collins, D., Arnold, D., Morrissey, S., Barillot, C. (eds.): Combining robust expectation maximization and mean shift algorithms for multiple sclerosis brain segmentation (MIAMS 2008), MCCAI Workshop on Medical Image Analysis on Multiple Sclerosis (2008)
63. Harmouche, R., Collins, L., Arnold, D., Francis, S., Arbel, T. (eds.): 18th International Conference on Pattern Recognition Bayesian MS lesion classification modeling regional and local spatial information (ICPR 2006) (2006)
64. Ramasamy, D.P., Benedict, R., Cox, J.L., Fritz, D., Abdelrahman, N., Hussein, S., Minagar, A., Dwyer, M.G., Zivadinov, R.: Extent of cerebellum, subcortical and cortical atrophy in patients with ms: a case-control study. Journal of the Neurological Sciences 282(1-2), 47–54 (2001)
65. Beltrame, F., Koslow, S.H.: Neuroinformatics as a megascience issue. IEEE Transactions on Information Technology in Biomedicine 3, 339–340 (1999)
66. Bota, M., Arbib, M.A.: The NeuroHomology Database. In: Arbib, M.A., Grethe, J. (eds.) Computing the brain: A guide to neuroinformatics, pp. 337–351. Academic Press, New York (2001)
67. Burns, G.A.P.C., Stephan, K.E., Ludäscher, B., Gupta, A., Kötter, R.: Towards a federated neuroscientific knowledge management system using brain atlases. Neurocomputing 38(40), 1633–1641 (2001)
68. Shattuck, D.W., Leahy, R.M.: Graph Based Analysis and Correction of Cortical Volume Topology. IEEE Transactions on Medical Imaging 20(11), 1167–1177 (2001)
69. Megalooikonomou, V., Ford, J., Shen, L., Makedon, F., Saykin, F.: Data mining in brain imaging. Statistical Methods in Medical Research 9, 359–394 (2000)

70. Mazziotta, J.C., Toga, A.W., Evans, A.C., Fox, P., Lancaster, J.: A probabilistic atlas of the human brain: theory and rationale for its development. NeuroImage 2(2), 89–101 (1995)
71. Caponetto, R., Fortuna, L., Frasca, M.: Advanced Topics on Cellular Self-Organizing Nets and Chaotic Nonlinear Dynamics to Model and Control Complex Systems. World Scientific Series on Nonlinear Science, vol. 63 (2008)
72. Holland, J.: Adaptation in natural and artificial systems. University of Michigan Press, Ann Arbor (1975)
73. Warfield, S.K., Rexilius, J., Huppi, P.S., Inder, T.E., Miller, E.G., Wells III, W.M., Zientara, G.P., Jolesz, F.A., Kikinis, R.: A binary entropy measure to assess nonrigid registration algorithms. In: Niessen, W.J., Viergever, M.A. (eds.) MICCAI 2001. LNCS, vol. 2208, pp. 266–274. Springer, Heidelberg (2001)
74. Dice, L.R.: Measures of the amount of ecologic association between species. Ecology 26(3), 297–302 (1945)
75. Barkhof, F., Van Waesberghe, J.H., Filippi, M.: T(1) hypointense lesions in secondary progressive multiple sclerosis: effect of interferon beta-1b treatment. Brain 124, 1396–1402 (2001)
76. Paty, D.W., Li, D.K.: Interferon beta-lb is effective in relapsing remitting multiple sclerosis: II. MRI analysis results of a multicenter, randomized, double-blind, placebo-controlled trial. Neurology 57 (1993)

An Evolutionary Algorithm for Skyline Query Optimization

Marlene Goncalves, Ivette Martínez, Gabi Escuela,
Fabiola Di Bartolo, and Francelice Sardá

Abstract. Skyline queries have emerged as an answer to the need of solving queries that involve user preferences. Although the evaluation of the Skyline operator is costly, its efficient incorporation into an execution plan may decrease the execution time of SQL queries. This process is known as Skyline Query Optimization. Several solutions for Skyline Query Optimization have already been presented. These solutions are most often based on Dynamic Programming (DP), which means that all alternative plans are exhaustively enumerated. This approach loses effectiveness as the search space size increases. On the other hand, stochastic search algorithms have shown to be successful in solving optimization problems arising from standard queries. Our previous studies have shown that Evolutionary Algorithms (EAs), implemented in the form of *eaSky*, may outperform DP approaches for Skyline Query Optimization. Such a comparison between EAs and DP is necessary, because DP is used in most database management systems despite its aforementioned scaling problem. In this chapter, we present experimental results of *eaSky* and show that *eaSky* can achieve better performance than DP for large queries. An extended version of *eaSky* has been developed, and experimental results show further improvements in its performance as compared to the original *eaSky*.

Marlene Goncalves · Ivette Martínez · Gabi Escuela
Universidad Simón Bolívar, Departamento de Computación y T.I..
Caracas, Venezuela
e-mail: `{mgoncalves,martinez,gescuela}@usb.ve`

Fabiola Di Bartolo · Francelice Sardá
Universidad Simón Bolívar, Grupo de Inteligencia Artificial,
Caracas, Venezuela
e-mail: `{fabiola.francelice}@gia.usb.ve`

1 Introduction

Structured Query Language (SQL) [1] is a standard language for defining questions or queries about data in a relational database. An SQL query corresponds to an expression which consists of a set of Relational Algebra operators such as *Selection*, *Projection*, *Join*, and *Cross-product* [2]. During an SQL query execution, a Database Management System (DBMS) enumerates several alternative execution plans to evaluate an expression. An execution plan is a binary tree generated by a DBMS defining an order among the operators, where each tree node represents an operator and its implementation. The execution plans may differ in terms of the operators' implementation and evaluation order. Both differences may affect a query evaluation cost. Consequently, a DBMS must estimate the costs of each enumerated plan and choose the plan with the least estimated cost. This process is known as Query Optimization. Finding an optimal evaluation plan is a combinatorial problem [3]. In practice, computing an optimal evaluation plan might take longer than evaluating a suboptimal one, therefore DBMSs typically use heuristics to find low-cost plans, or, at least, to avoid poor plans.

Recently, efforts have been made by different researchers to express user preferences in queries. Skyline is an SQL extension for user-preference based queries which is used in many applications, including multi-criteria decision making. A Skyline query finds a set of interesting results while satisfying a set of possibly conflicting conditions.

The complexity for Skyline query evaluation is high. It is in $O(n^2)$ in the worst case, where n is the input dataset size [4]. Nevertheless, several solutions have been proposed for improving the runtime problem in Skyline evaluation [5, 6]. A Skyline query may be translated into an SQL query, for instance. However, this evaluation strategy is expensive [5]. A better strategy may be to use and implement Skyline as an integrated operator into a DBMS [5, 6].

Due to its semantics, Skyline should be evaluated after other operators because it is an operator that is placed at the end of an evaluation plan. However, using some algebraic rules, such a so-called canonical plan may be transformed by inserting Skylines within other operators [2, 7]. Such plans could have a lesser cost than a canonical plan because they may considerably reduce the cardinality of intermediate results, thus decreasing the Skyline query execution time [8]. Also, considering such transformed plans for execution, i.e., a wider subset of the search space, allows a DBMS to select a better evaluation plan.

Existing solutions for Skyline Query Optimization are based on Dynamic Programming (DP) [7, 9, 10]. These solutions enumerate all alternative plans. Hence, they become less effective as the search space size increases [3]. On the other hand, stochastic search algorithms, particularly Evolutionary Algorithms (EAs), have revealed to be successful in solving optimization problems for standard queries [3, 11, 12].

In this chapter, we integrate the EA for Skyline Query Optimization proposed in [13] into a real DBMS known as PostgreSQL [14]. We call this algorithm *eaSky*. Our experiments have shown that *eaSky* outperforms a DP algorithm (*dpSky*) on large Skylines over real data. We refer to Skylines as large if they involve a significant number of base relations. It was confirmed that optimization methods which are able to change the position of Skyline operators could find better plans than those which cannot move them.

We also propose an extension to *eaSky*, the extended *eaSky*, which incorporates a crossover operator and three new mutation operators. Our initial experiments over synthetic data show that the extended *eaSky* can produce better results than the original one.

The rest of this chapter is organized as follows. In Section 2, a motivating example for the Skyline Query Optimization problem is presented, and the Skyline operator and its algebraic rules, as well as cost formulas for Join and Skyline operators' estimations are defined. Section 3 describes the state-of-the-art approaches for optimizing classical and Skyline queries. In Section 4, the proposed Skyline optimization algorithm *eaSky*, based on EAs, is presented. In Section 5, the performance of the proposed technique is empirically evaluated against state-of-the-art solutions. Finally, Sections 6 presents conclusions and future work in this area.

2 Skyline Query Optimization

SQL is a standard language to query relational databases. It is a declarative language where a user expresses a query and the database management system (DBMS) determines a procedure or plan to retrieve an answer for the user's query [1]. This plan may be selected based on estimated cost during a query optimization process.

A plan may be represented using relational algebra expressions [2] which comprise relational operators such as Selection (σ), Projection (π), Cartesian Product ($\times$), Union ($\cup$), Intersection ($\cap$), Difference ($-$) and Join ($\bowtie$) [2]. Any relational algebra expression may be represented by an operator tree and each relational operator may have several implementations or physical operators in a DBMS.

The estimated cost of a plan is the sum of all estimated costs of its physical operators. The Query Optimization goal is to find the plan with the best estimated cost. Unfortunately, searching for the best plan may be highly expensive. The DBMS must identify a good plan whose execution time is acceptable for the user. A good plan produces less data in intermediate query phases which may reduce the execution time.

Query optimization has been widely studied on SQL queries [2] and this process may also be applied on Skyline queries. Similarly, a Skyline query may be expressed in relational algebra including Skyline as a new operator, denoted as Sk.

2.1 Motivating Example

Let us consider a website to purchase a personal desktop computer. Any user might be interested in a cheap computer characterized by high RAM memory capacity, high hard disk capacity, and a good processor. Although these goals are desirable for users, computers tend to be more expensive as they have more memory, storage space, and faster processors. As a website engine may not determine nor even decide which the best computer is, it must identify all interesting ones. A computer is interesting if and only if there is no other cheaper computer with better RAM memory, hard disk, and processor. The set of interesting computers is known as Skyline. Thus, a Skyline query finds a set of interesting results, satisfying a set of possibly conflicting conditions. A query example in SQL may be expressed as follows:

```
select * from RAM r, Processor p, Disk d
[where ... ]
skyline of price min, r.size max, p.speed max, d.size max
```

To illustrate Skyline Query Optimization, let us suppose that computer components have a price and are stored in three different relational tables: RAM, Processor, and Disk. A possible relational algebra expression that represents a query to identify interesting computers may be Sk_{P} (RAM $\times$ (Processor $\times$ Disk)) where P corresponds to user criteria, i.e., the best price, RAM memory, hard disk and processor. Figure 1(a) shows a tree structure associated with this expression where a Skyline Sk_P was placed at the end of the tree in order to be executed on the whole dataset. In this case, the DBMS builds all possible computer configurations (RAM $\times$ (Processor $\times$ Disk)) and then takes the best ones in terms of price, RAM memory, hard disk, and processor applying the Sk_{P} operator placed at the end.

In [2], the authors have defined rules that allow constructing equivalent expressions. Thus, the expression represented in Figure 1(a) may be transformed into the expression shown in the tree structure of Figure 1(b) using algebraic rules in order to push down the Skyline [9]. For this expression, the engine selects the best options for each brand memory, hard disk and processor to build cheaper computer configurations following these specifications. Intuitively, in Figure 1(b), preferences have been separated (Sk_{P1}, Sk_{P2} and Sk_{P3}) and computed by desirable goals for users; and specifications are joined to retrieve a final answer. In this way, each Skyline operator reduces the cardinality of the intermediate results which, in turn, decreases the Skyline query execution time. Let us, for example, suppose that each relational table has 10 components and five brands exist for each component.

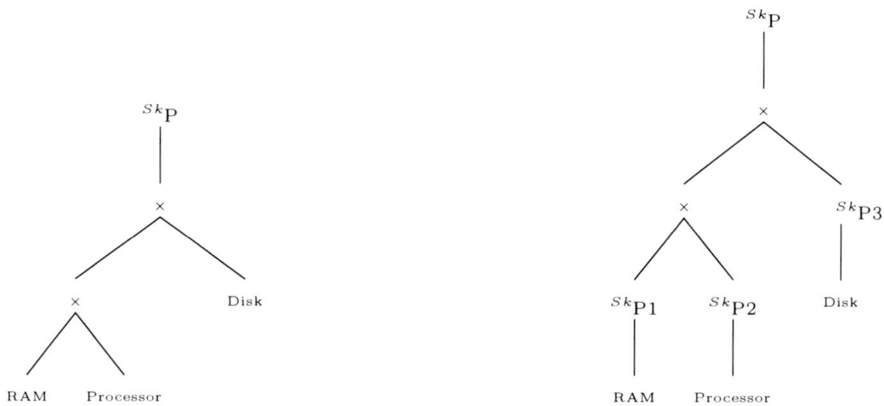

(a) Initial Plan (b) Transformed Plan

Fig. 1. Tree structures

This means that there are 1000 configurations (10 RAMs × 10 Processors × 10 Disks) and two best components for each table. The first plan, $Sk_{\rm P}$ (RAM × (Processor × Disk)), must calculate 1000 configurations on the Skyline. In the second plan, the Skyline is to be evaluated on 10 components producing only two best ones, and the final Skyline must compute 8 components (2 RAMs × 2 Processors × 2 Disks).

Finally, a complete search space exploration for building good plans may be worse than executing the worst plan. Consequently, several techniques [11, 15, 16, 17] are used to test subsets of the whole search space in order to reduce query execution time. These techniques are based on DP, genetic algorithms, EAs, tabu search, etc. Techniques based on DP have been used in many relational DBMSs and were popularized by System-R [15]. Subsequently, techniques based on pseudo-random algorithms began to be established because their search time does not increase as much as the runtime of dynamic programming algorithms when the search space enlarges. Taking into account these findings, we are proposing the development of EAs for the Skyline Query Optimization problem.

2.2 Skyline Operator

SQL was extended in [5] to express desirable conditions, but not precisely, over data. In Skyline queries, conditions may be conflictive. Therefore, several optimal answers may be retrieved. Syntactically, a Skyline query may be specified as follows:

```
select *
from table_1, table_2, ... table_n
[where ... ]
skyline of att1 [max|min|diff] [, att2 [max|min|diff], ...] ...
```

A SKYLINE OF clause is composed by a list of criteria or a multi-criteria function definition. Each criterion contains an attribute or dimension used to rank a dataset. Each dimension may be an integer, float, or a date and may be annotated with directives: min, max, and diff. The directives min and max indicate minimum or maximum values, whereas diff is used to retain the best choice with respect to every distinct value for an attribute.

A Skyline query retrieves a set of interesting or non-dominated tuples. A tuple dominates another one if it is better or equal than the other one in every dimension and better in at least one. Given two tuples t and p from a table R, and a multi-criteria function defined by m_1 **min,** m_2 **min, ...,** m_p **min,** M_{p+1} **max,** M_{p+2} **max, ...,** M_q **max,** d_{q+1} **diff,** d_{q+2} **diff, ...,** d_n **diff**, t dominates p if:

- $t_i \preceq p_i$ with $i = 1, 2, \ldots, p$
- $t_i \succeq p_i$ with $i = p + 1, \ldots, q$
- $t_i = p_i$ with $i = q + 1, \ldots, n$

2.3 Estimating Cost

DBMSs use formulas that estimate the execution cost of physical operators. Given two relations, R and S; the number of pages of R, M; the number of tuples per page of R, p_R; the number of pages of S, N; the number of tuples per page of S, p_S; the number of index levels, X_B; the size of input dataset, n; and the number of Skyline criteria, d, Table 1 presents the formulas used in this work to estimate the cost of Join and Skyline, where the Skyline cardinality $\hat{S}_{n,d}$ is: $\hat{S}_{n,d} = \frac{1}{n}\hat{S}_{n,d-1} + \hat{S}_{n-1,d}$.

Table 1. Cost formulas for Join and Skyline operators

Operator	Cost formula
Nested Loops Join [2]	$M + p_R * M * N$
Block Nested Loops Join [2]	$M + N$
Index Nested Loop Join [2]	$M + (p_R * (X_B + 1))$
Sort Merge Join [2]	$M + N +$ cost of sorting R and S
Hash Join [2]	$3(M + N)$
Block Nested Loop (BNL) Skyline [9]	$\approx \sum_{j=2}^{n} \frac{\hat{S}_{j-1,d}}{j-1} \hat{S}_{j-1,d+1}$
Sort Filter Skyline (SFS) [9]	$\approx \sum_{j=2}^{n} \frac{\hat{S}_{j-1,d-1}}{j-1} \hat{S}_{j-1,d}$

2.4 Algebraic Rules

Given two relations R and S; P_r and P_s be the preference subsets defined on R and S, respectively; and c, the join condition $R.x = S.x$, $Sk[p](R)$ is denoted as the Skyline on R with a preference set p. Some of the algebraic rules used in this work are shown in Table 2. Condition C_0 is true if each R tuple joins with at least an S tuple.

Table 2. Algebraic Rules between Skyline and Join operators

$$\text{Sk}_{[P_r]}(\text{R} \bowtie_c \text{S}) \equiv \text{Sk}_{[P_r]}(\text{R}) \bowtie_c \text{S} \qquad \textbf{if } C_0$$
$$\text{Sk}_{[P_r]}(\text{R} \bowtie_c \text{S}) \equiv \text{Sk}_{[P_r]}(\text{Sk}_{[P_r \,\cup\, \{\text{R.x diff}\}]}(\text{R}) \bowtie_c \text{S})$$
$$\text{Sk}_{[P_r \,\cup\, P_s]}(\text{R} \bowtie_c \text{S}) \equiv \text{Sk}_{[P_r \,\cup\, P_s]}(\text{Sk}_{[P_r \,\cup\, \{\text{R.x diff}\}]}(\text{R}) \bowtie_c \text{S})$$

3 Related Work

Skyline computing corresponds to the classic problem of maximal vector computing [4]. Skyline establishes a partial order between input dataset elements. This ordering is induced by a multi-criteria function that maximizes and/or minimizes dimensions simultaneously.

In the database area, Skyline emerges as an extension to the SQL standard language for allowing users to specify their preferences [5]. Lately, much effort has been made to define physical Skyline operators in the area of relational database. The BNL (Block Nested Loop) [5], the SFS (Sort Filter Skyline) [4], and LESS [4] are three scan-based physical operators that allow to identify Skylines in relational database systems. The algorithm [6] by Tan et al. and BBS (Branch-and-Bound Skyline) [18] are index-based operators that progressively return each Skyline tuple without necessarily scanning all tuples.

In [4], the authors have been studying the complexity of several Skyline physical operators. In the worst case, the Skyline operator evaluation is in $O(n^2)$, where n is input dataset size. For that reason, the Skyline operator evaluation may be very costly as each input dataset element must be compared against each other.

On the other hand, Skyline query optimization in a database management system may significantly reduce the Skyline query execution cost. Since the Skyline set may be smaller than the input dataset [9] and Skyline operator evaluation costs may be very high [4], there is a need for a process that determines when a Skyline must be executed in order to reduce the query execution time. This process is known as query optimization. Previous works have shown the advantages of Skyline Query Optimization. In [9], the Microsoft SQL server was extended for optimizing Skyline queries characterized by a relation or a join. Chomicki [19] and Hafenrichter et al. [7] proposed a set of algebraic rules for equivalence between winnow operator queries [19] (winnow is a more general operator than Skyline). In [9], Chadhuri et al. defined a set of formulas to estimate Skyline evaluation cost. BNL [5] and SFS [4] have been integrated into a database management system [9, 10, 20]. The Skyline operator cardinality has also been estimated in [9, 21, 22]. An EA for Skyline-Join optimization was proposed by Di Bartolo et. al in [13]. This article showed that an EA with three mutation operators that alter the positions of Skylines in queries may find better Skyline plans and within less iterations than an EA that does not alter the positions of Skylines. It is also shown that the results of the Skyline EA are of better-quality than the ones achieved with the DP approach for Skyline Query Optimization.

4 Our Approach

In this section, a mathematical description of the Skyline optimization problem is presented. The basic elements of an EA for solving this problem are described. The proposed algorithm is called *Extended eaSky*. The algorithm modules that we will define here are the individual representation (or chromosome encoding), the creation of initial population, the fitness function, the selection and replacement operators, and the variation operators (mutation and crossover). Three mutation operators from *eaSky* (JoinOrderMutation, JoinImplementationMutation, and InsertSkylineMutation), three new operators for the Extended *eaSky* (DeleteSkylineMutation, CombineSkylineMutation, and DivideSkylineMutation), and the added crossover operation are described in detail.

4.1 Mathematical Problem Description

In this section, preliminary definitions needed to describe the Skyline Query Optimization problem are provided.

Definition 1. *A physical operator is an implementation of a logical (algebra relational or Skyline) operator into a DBMS.*

Definition 2. *A plan p of a query Q is a sequence of m physical operators $o_1, \ldots, o_m$. Each operator o_j has a cost c_j.*

Thus, given a Skyline query Q; a set $P = \{p_1, \ldots, p_n\}$ of equivalent plans for query Q where p_i is a sequence of physical operators $o_1, \ldots, o_m$. Then, the

Skyline Query Optimization problem SKQ is to select a plan p_k in P with a minimum $\sum_j C_{o_j}$

4.2 Evolutionary Algorithm Design

An EA called Extended *eaSky* is being introduced as an approach to deal with the SKQ problem. *eaSky* supports all database Join and Skyline operators listed in Table 1. Algorithm 1 shows a general structure for the *eaSky* approach. For its first version [13], the authors used three mutation operators: two affected Join operators and one which was implemented to insert a Skyline operator. This first version of *eaSky* was then integrated into PostgreSQL. In order to improve its performance, a crossover and three mutation operators were added to the original *eaSky* and a set of experiments were performed to determine the impact in terms of fitness and execution time for these new operators in the Extended *eaSky*. A detailed description of *eaSky* is presented in forthcoming sections.

Algorithm 1. The eaSky main cycle

```
popSize ⟵ determineSize();
population ⟵ initializePopulation();
repeat
    parent1,parent2 ⟵ selection(population) ;
    child1, child2 ⟵ crossover(parent1,parent2);
    mutation(child1,child2);
    calculateFitness(child1,child2);
    addToPopulation(child1,child2);
until terminationCondition;
return theBest
```

4.2.1 Chromosome Encoding

As encoding, the chromosome extension proposed by Bennett et al. [3] is used, where each individual is represented by a variable length gene array. Each gene represents an operation that involves a Join or Skyline operator with its corresponding tables and attributes. Each individual refers to a left-deep strategy, represented as a left linear tree. A left linear tree is an expression tree, where the right (inner) child of every Join node represents a single table in the query. These trees are translated in an array using a pre-order traversal. Figure 2 shows a tree that represents a plan and its corresponding chromosome representation. $J_{A,B}^{BNLJ}$ is a Join operator implemented as $BNLJ$ between tables A and B. In order to simplify the notation, it is assumed that all Join conditions are over attribute x. $S_{\mathrm{A.a\ max,\ B.b\ max}}$ is a Skyline which maximizes attributes a and b from tables A and B respectively.

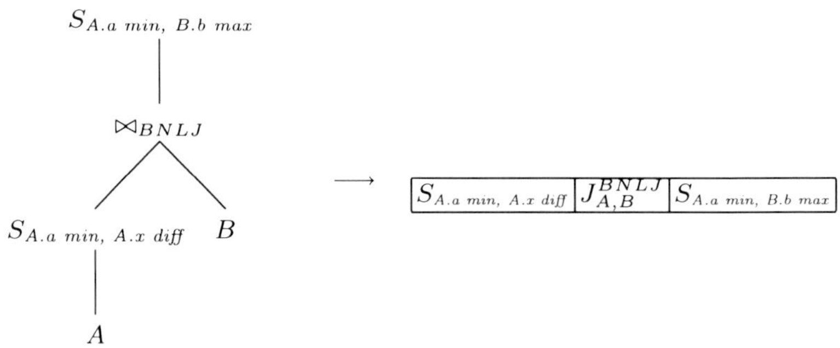

Fig. 2. Encoded individual

4.2.2 Population Initialization

The initial population is filled in with canonical individuals. In order to create a canonical individual, g Joins are created for the initial query and put into the first g genes of the chromosome. The defined Skyline used for the initial query is then located at gene $g+1$. Therefore, canonical plans are those whose Skylines are at the evaluation tree root. The population sizes are set depending on the number of tables involved in the query. We choose to set the population size to ten times the number of tables. Only valid plans (individuals) are considered, i.e., those which are semantically equivalent to the original query. This means the search space consists of all valid combinations of Joins and Skylines.

4.2.3 Selection and Replacement

Individuals are selected proportionally to their fitness. New, mutated individuals replace all individuals to produce the next generation. However, the best individuals are also preserved and copied to the new generation.

4.2.4 Fitness Function

The fitness of an individual is the estimated cost of the plan represented by it. This cost is calculated using the following equation:

$$Fitness = Cost_{plan} = \sum_{i=1}^{n} Cost(gene_i) \tag{1}$$

where n is the chromosome length and $Cost(gene_i)$ is the operation cost of $gene_i$. The estimated cost for an individual is calculated using the cost models presented in Section 2.3. It is important to notice that the cost of an operator depends on the results of the previously applied operators.

4.2.5 Mutation

Genetic optimization approaches, which only apply to standard queries based on Join operators, do not require checking the individuals after mutation and crossover [3]. However, the Skyline operator is neither commutative nor associative, and the genetic operators must be carefully designed in order to avoid invalid individuals.

During the mutation process, a set of individuals is randomly selected from the population. For each individual, a mutation operator is chosen. The first version of *eaSky* implemented three mutation operators: *JoinOrderMutation*, *JoinImplementationMutation*, and *InsertSkylineMutation*. In this version, only Nested Loops Join, Index Nested Loop Join, and Sort Merge Join were considered. Each one of implemented operators modifies individuals in a different way, by changing the query operator order or by modifying or adding new genes to a chromosome.

JoinOrderMutation: For a given individual (plan) P, this operator randomly selects two Join genes. If there is not a Skyline between them and the result of the left Join's left child contains the first relation of the second Join, then both Joins positions are swapped to produce a new plan P^m. Figure 3 shows a *JoinOrderMutation* between Joins at positions 1 and 2. The restrictions over swapping are checked in order to generate valid individuals.

P:	$J^{BNLJ}_{A,B}$	$J^{SMJ}_{B,C}$	$J^{BNLJ}_{B,D}$	$S_{A.a\ max,\ C.c\ max}$
	0	1	2	3

P_m:	$J^{BNLJ}_{A,B}$	$J^{BNLJ}_{B,D}$	$J^{SMJ}_{B,C}$	$S_{A.a\ max,\ C.c\ max}$
	0	1	2	3

Fig. 3. *JoinOrderMutation* example

JoinImplementationMutation: This operator randomly selects a Join gene and changes its implementation for a different one. The Join implementation could be: Nested Loops Join (NLJ), Block Nested Loop Join (BNLJ), Index Nested Loop Join (INLJ), Sort Merge Join (SMJ), and Hash Join (HJ). Figure 4 shows an example of the *JoinImplementationMutation* operator. From plan P, the Join in position 0 is selected to be mutated, then its implementation BNLJ is replaced by SMJ to produce P_m.

P :	$J^{BNLJ}_{A,B}$	$J^{SMJ}_{B,C}$	$J^{BNLJ}_{B,D}$	$S_{A.a\ max,\ C.c\ max}$
	0	1	2	3
P_m :	$J^{SMJ}_{A,B}$	$J^{SMJ}_{B,C}$	$J^{BNLJ}_{B,D}$	$S_{A.a\ max,\ C.c\ max}$
	0	1	2	3

Fig. 4. *JoinImplementationMutation* example

InsertSkylineMutation: This operator randomly selects a Skyline gene that can be mutated. The selected gene is modified using one of the algebraic rules for Skylines and Joins presented in Section 2.4. This operator may transform a Skyline operator by creating a new Skyline gene with a condition that is a sub-condition of the original Skyline. Then, the new gene is pushed into a chromosome. Pushing is done by adding the new Skyline gene into the position where all involved tables have appeared. Figure 5 shows an example of the *InsertSkylineMutation* operator. In this case, the Skyline in position 3 is selected from P, then from a random selection of algebraic rules, the third rule of table 2 was applied. To form P_m, a new Skyline is inserted at the first chromosome position.

P :	$J^{BNLJ}_{A,B}$	$J^{SMJ}_{B,C}$	$J^{BNLJ}_{B,D}$	$S_{A.a\ max,\ C.c\ max}$
	0	1	2	3

P_m :	$J^{BNLJ}_{A,B}$	$S_{\text{A.a max, A.x diff}}$	$J^{SMJ}_{B,C}$	$J^{BNLJ}_{B,D}$	$S_{A.a\ max,\ C.c\ max}$
	0	1	2	3	4

Fig. 5. *InsertSkylineMutation* example

For the extended version of the *eaSky* algorithm, three mutation operators were added to Skyline operators, they are as follows:

DeleteSkylineMutation: This operator randomly selects a Skyline gene that can be deleted. The selected gene is deleted if none of the algebraic rules for the Skylines and Joins presented in Section 2.4 are violated. Notice that the Skyline located at the tree root may not be deleted according to these rules. Figure 6 shows an example of this operator, where the Skyline in position 1 is deleted from chromosome.

CombineSkylineMutation: This operator randomly selects a Join gene characterized by two child Skyline operators, and then combines them into a Skyline. The new Skyline operator is placed over the selected Join operator

	0	1	2	3	4
P :	$J^{BNLJ}_{A,B}$	$S_{\text{A.a max, A.x diff}}$	$J^{SMJ}_{B,C}$	$J^{BNLJ}_{B,D}$	$S_{A.a\ max,\ C.c\ max}$

	0	1	2	3
P_m :	$J^{BNLJ}_{A,B}$	$J^{SMJ}_{B,C}$	$J^{BNLJ}_{B,D}$	$S_{A.a\ max,\ C.c\ max}$

Fig. 6. *DeleteSkylineMutation* example

with a condition that corresponds to the union of two child Skyline conditions. Figure 7 shows how this operator changes chromosome inserting the Skyline operator in position 3 which combines the Skylines at positions 0 and 2.

	0	1	2	3	4
P :	$S_{A.a\ max,\ A.id\ diff}$	$J^{SMJ}_{A,C}$	$S_{C.c\ max,\ C.id\ diff}$	$J^{BNLJ}_{C,D}$	$S_{A.a\ max,\ C.c\ max}$

	0	1	2	3	4	5
P_m :	$S_{A.a\ max,\ A.id\ diff}$	$J^{SMJ}_{A,C}$	$S_{C.c\ max,\ C.id\ diff}$	$S_{A.a\ max,\ C.c\ max,\ C.id\ diff}$	$J^{BNLJ}_{C,D}$	$S_{A.a\ max,\ C.c\ max}$

Fig. 7. *CombineSkylineMutation* example

DivideSkylineMutation: This operator divides a Skyline operator that affects a Join operator into two Skyline operators. It is the only mutation that can be applied to a Skyline operator located at the tree root. In order to perform this mutation, a Skyline operator over a Join is randomly selected and the tables for the Skyline operator are obtained. Then, two new child Skyline operators are inserted into a chromosome at the corresponding table positions. The Skyline operator at the tree root keeps its position. Figure 8 shows an example of the application of this operator to P, where the Skyline operator in position 1, affecting a Join operator over tables C and A, is divided into two Skylines, resulting in P_m.

	0	1	2	3	4
P :	$J^{SMJ}_{C,A}$	$S_{\text{A.a max, C.c min A.x diff}}$	$J^{SMJ}_{A,D}$	$J^{BNLJ}_{B,D}$	$S_{A.a\ max,\ C.c\ min}$

	0	1	2	3	4	5
P_m :	$S_{\text{C.c min C.x diff}}$	$S_{\text{A.a max, A.x diff}}$	$J^{SMJ}_{C,A}$	$J^{SMJ}_{A,D}$	$J^{BNLJ}_{B,D}$	$S_{A.a\ max,\ C.c\ min}$

Fig. 8. *DivideSkylineMutation* example

4.2.6 Crossover

As individuals have shown to be variable in length, operators based on homologous crossover [23] have been implemented, that is, operators that tend to preserve the positions of the genetic material. In order to do so, the operator gets hold of a common region from two selected parents, defined as the region where two plans have the same table order. If two parents have more than one common region, one is randomly selected. In the common region, a single cut off point is randomly chosen and genetic material is exchanged, creating two children. In a post-processing phase, each child is repaired in order to avoid inconsistencies. This repairing process always builds feasible individuals. Figure 9 shows an example for the crossover operator. In this case, the common region between $Parent_1$ and $Parent_2$ are tables B, A, D. If the cutoff point is A, then operators affecting B and A are interchanged to obtain $Child_1$ and $Child_2$.

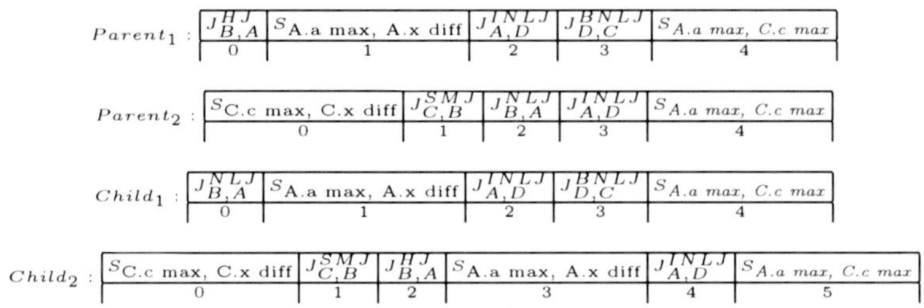

Fig. 9. Crossover example

5 Experimental Results

We conducted an experimental study with the purpose of analyzing the effectiveness of the proposed technique. First, the scalability of DP, EAs in the Skyline Query Optimization problem, and the pushing Skyline operator's impact are reported. Second, the impact of new mutation and crossover operators for improving the performance of EA is shown.

5.1 Scalability and Impact of Pushing Skyline

The first technique we studied optimizes Skyline queries by means of DP and is called dpSky; the second uses an EA and it is named *eaSky*. The execution time rate as well as the optimization and execution times resulting from these two techniques were computed.

5.1.1 Experimental Design

Dataset and Query Benchmark: The study was conducted on randomly generated datasets following a uniform distribution. Sixty tables were created and loaded. The table cardinality varied from 1000 to 20000 tuples.

The Join selectivity was 0.0006. Sixty chained queries were generated and executed in order to compare the two techniques. Chained queries were randomly generated and characterized by the following properties: a) 4, 6, and 8 tables in the FROM clause; b) 3, 5, and 8 joins; c) 4, 6, and 7 criteria. With the evolutionary techniques, 20 runs were executed over each query.

Evaluation Metrics: The Optimization time, total execution time (TET), and execution time rate ($ETR = \frac{TET_{t1}}{TET_{t2}}$) between two techniques $t1$ and $t2$ are reported. Experiments were evaluated on an Intel(R) Pentium(R) D CPU 3.20GHz, 1024Kb cache, and 1Gb memory.

Implementation: The experimentation environment is PostgreSQL 8.1.4. The parser of this DBMS was modified to be able to analyze Skyline queries. The physical operators BNL and SFS were incorporated into the system. Our techniques were developed in the C programming language and plugged into the optimizer. All data were stored in relational tables in PostgreSQL 8.1.4 databases. The PostgreSQL 8.1.4 Analyze Tool was executed in order to gather statistics on the tables.

5.1.2 dpSky and eaSky Scalability

The optimization and evaluation times for these two techniques are illustrated in Figure 10. For each number of tables involved in the FROM clause (4, 6 and 8 tables), the time average over 20 queries was measured. Since the difference between time magnitudes for each number of tables was too large, the y-axis (time) is logarithmically scaled.

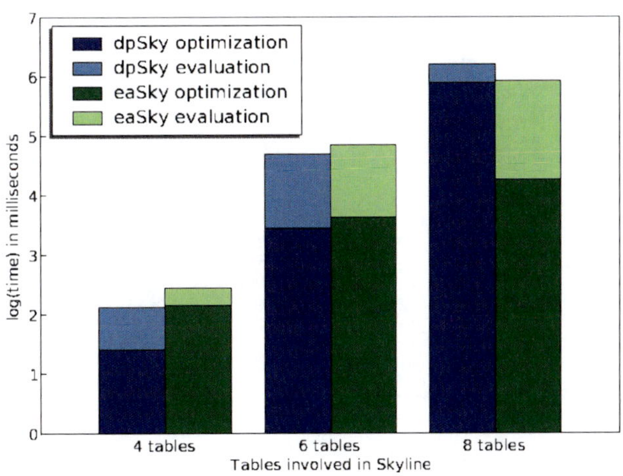

Fig. 10. dpSky vs eaSky performance

We observed that for four-table queries, the execution time of *dpSky* was lower than *eaSky*'s runtime. The optimization time of *eaSky* was also almost one magnitude order higher than the optimization time required by *dpSky*. We noted that the evaluation time of *eaSky* was lesser than *dpSky*'s evaluation time. For six-table queries, the times needed for optimization and evaluation were similar for both techniques. For eight-table queries, the optimization time of *dpSky* was almost two orders of magnitude higher than *eaSky*'s optimization time. Even when *dpSky*'s evaluation time was lower than *eaSky*'s evaluation time, the execution time of *eaSky* was lower than the execution time of *dpSky*.

Hence, we confirmed that *dpSky* did not scale up well to queries that involve many tables while *eaSky* can better deal with a higher number of tables in a query. *dpSky* failed when the number of tables in a query increased because the search space was greater and much time was spent to find out a good plan during the query optimization. Thus, *dpSky* did not scale and its performance declined as the number of tables increased in a query. Additionally, *eaSky* reduced the total execution time when more tables were added to queries because stochastic algorithms are less sensitive to search space size. Therefore, the optimization time of *eaSky* did not significantly rise and thus, it scales up better than *dpSky*.

5.1.3 Skyline Pushing Performance Impact

Figure 11 shows Skyline pushing impact over *eaSky* performance.

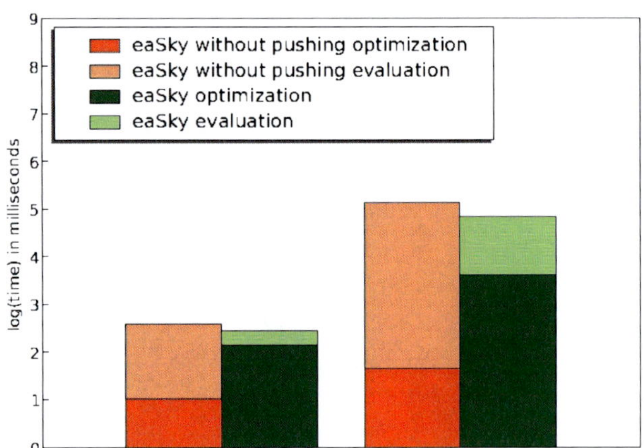

Fig. 11. *eaSky* Performance, with and without pushing

In this experiment, Skyline pushing improved the Skyline query evaluation time even though the time needed for optimization increased. The evaluation time of Skyline pushing was smaller than the ones obtained without pushing because better plans were found. As shown in our motivating example, this experiment has confirmed that Skyline-pushing based solutions achieve better performances than those without it.

In Table 3, the total time rate depending on the number of tables in the queries is presented. It was observed that all rates were higher than 1. *eaSky* has improved 30% (for 6 tables) and 106% (for 4 tables) to *dpSky* which indicates that *eaSky* has better performance than *dpSky*.

Table 3. Rate $TET = T_{eaSky\ without\ pushing}/T_{eaSky\ with\ pushing}$

Number of Tables	Rate
4	2.060584608
6	1.309546621

5.2 *New Operators Performance Impact*

An algorithm without being integrated into PostgreSQL was executed in order to determine performance impact of the *eaSky* crossover and three new mutation operators. The objective was to verify if *eaSky*'s new conditions allowed the algorithm to find better solutions while keeping good execution times.

Query Benchmark: The study was conducted on 40 generated queries with different characteristics. The number of tables involved in the query we 4 or 6, the number of table attributes was 10, the table cardinality ranged from 10 to 70, the tuple size was 10, the number of different table attributes spanned from 30 to 225, and a table could either have an index or not. Like in the previous experiments, 20 runs were executed over each query.
Evaluation Metrics: The averaged best fitness and execution time between different eaSky configurations were reported. The experiments were carried out on an Intel(R) Pentium(R) 2 Duo CPU 2.20GHz, 2Gb memory.
EA parameters: Table 4 shows settings used for these experiments.

In order to select the main genetic parameters for the extended *eaSky*, we varied the crossover and mutation rates. The results of the experiment are presented in Figure 12. Figure 12 shows the average fitness and time (20 runs for each query). Two mutation rate values (0.2 and 0.5) and three crossover rate values (0.0, 0.5 and 0.8) were considered on the four-tables queries. As shown, a high mutation rate (0.5) benefited the algorithm to quickly reach a

Table 4. Parameter values settings

Parameter	*eaSky*	**Extended** *eaSky*
EA Parameters:		
Number of Individuals	10× query relations	
Replacement Model	Generational and Elitism	
Selection approach	Fitness Proportional	
Mutation	0.2	0.5
JoinOrderMutation	0.5	0.3
JoinImplementationMutation	0.1	0.1
InsertSkylineMutation	0.4	0.3
DeleteSkylineMutation	0.0	0.1
CombineSkylineMutation	0.0	0.1
DivideSkylineMutation	0.0	0.1
Crossover	0.0	0.5
Simulation Parameters:		
Termination condition	1000 generations or no change in 60 generations	no change in 60 generations
Run numbers	20 per experiment	

good solution. A crossover operator seemed to be useful at a medium value (0.5) in combination with a high mutation rate (with respect to frequently used values in other EAs). This is especially true when the query instances processed were particularly difficult, as in queries 16, 17, and 18.

Figures 13 and 14 show the performance of the extended *eaSky* compared to the original *eaSky* behavior. Considering the obtained results in previous experiments, crossover and mutation rates of 0.5 were used in the extended *eaSky*. In the figures, we can observe that for all queries (no matter the number of tables involved), the extended *eaSky* has the lowest estimated plan cost (fitness) and optimization time.

Table 6 shows statistics for the *eaSky* compared to the extended *eaSky* over six-tables queries. Even when both averages (fitness and time) were better for the extended *eaSky*, their differences did not seem to be statistically significant. These numbers show a 6% improvement on fitness and 15% on time for the extended *eaSky* over the original one.

Table 5 shows statistics for the *eaSky* compared to the extended *eaSky* over the four-tables queries. We display the average, standard deviation, and Wilcoxon test results[1]. These numbers showed a 9% improvement on fitness and 15% on time for the extended *eaSky* over the original one. As this analysis has shown, even when the extended *eaSky* fitness average was better than the *eaSky*, this difference was not statistically significant. However, the

[1] A non-parametric statistical hypothesis test, similar to the t-test without the normal data assumption.

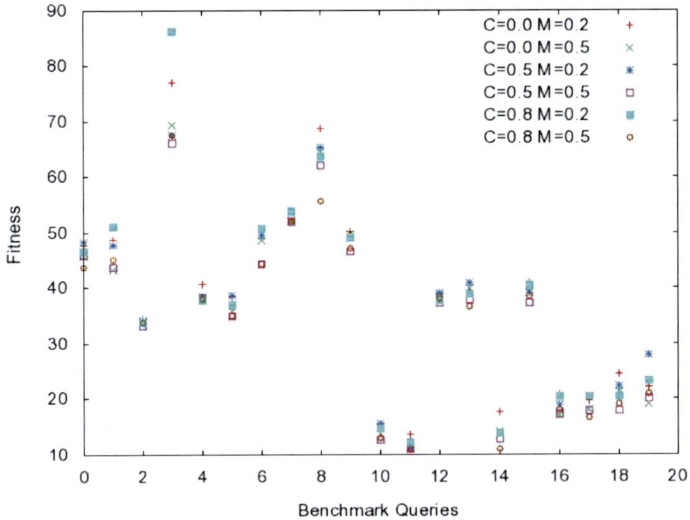

(a) fitness comparison

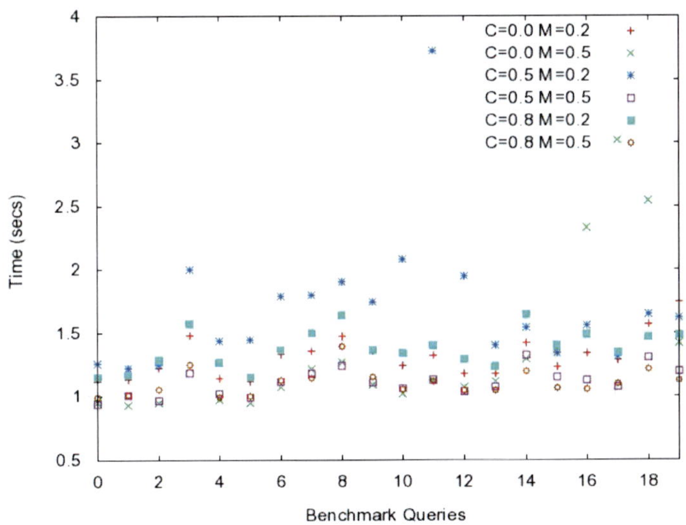

(b) time comparison

Fig. 12. eaSky performance using different crossover and mutation rates

extended *eaSky*'s estimated evaluation time was better than *eaSky*'s time and this difference was highly significant (at least 99.99% level).

Over all queries (four and six involved tables) the varied mutation operator types seemed to help the algorithm in the searching process jointly with the

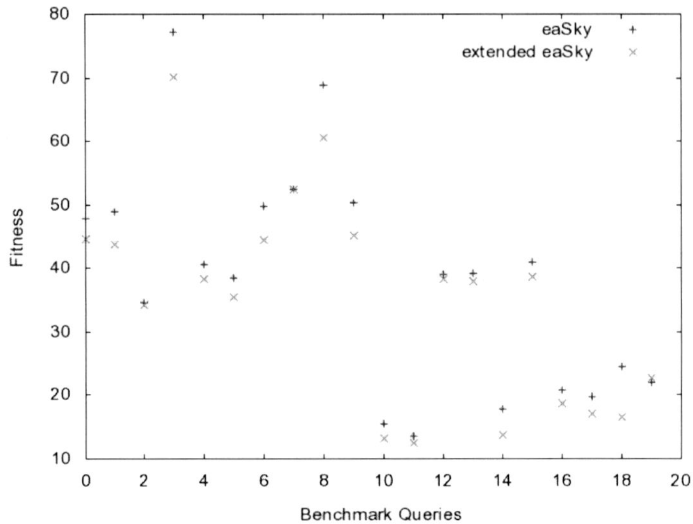

(a) fitness comparison

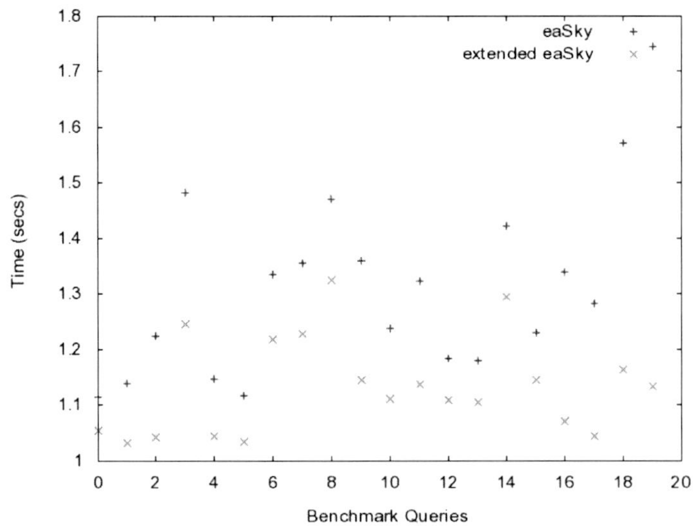

(b) time comparison

Fig. 13. Extended eaSky vs eaSky for query instances with 4 tables

crossover operator, under the described conditions. In spite of the extended eaSky version seemed to outperform *eaSky* by obtaining better solutions in less time, their differences in our experiments were not always statistically significant. Therefore, we need to perform more experiments on the subject.

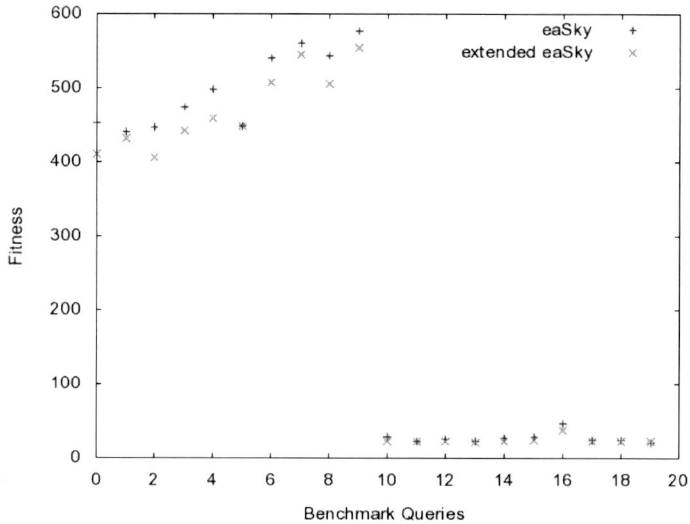

(a) fitness comparison

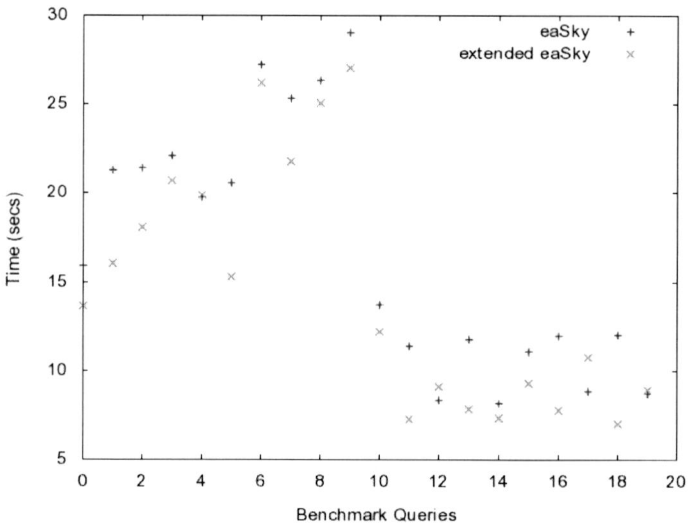

(b) time comparison

Fig. 14. Extended eaSky vs eaSky for query instances with 6 tables

Table 5. Statistics for comparison of eaSky vs. extended eaSky over 4 tables

	Fitness		Time	
	eaSky	**extended**	**eaSky**	**extended**
Average	38.08183	34.91329	1.31302	1.13429
Stdev	5802.64834	5115.67430	0.51455	0.14937
Wilcoxon-test	p-value=0.4094		p-value=0.00011	

Table 6. Statistics for comparison of eaSky vs. extended eaSky over 6 tables

	Fitness		Time	
	eaSky	**extended**	**eaSky**	**extended**
Average	262.67975	247.78816	16.74726	14.55973
Stdev	1136372.27	1026309.47	935.99806	901.75517
Wilcoxon-test	p-value=0.3507		p-value=0.2503	

6 Conclusions and Future Work

An EA, *eaSky*, for the Skyline Query Optimization problem has been presented and tested. The first version of *eaSky* included three mutation operators and was integrated into a real DBMS (PostgreSQL). The DBMS PostgreSQL with a Skyline Query Optimization algorithm based on DP, called *dpSky*, was also extended. Thus, the Skyline operator was integrated into an optimizer of a real DBMS, giving significant evidences of performance improvement in Skyline query execution. Results comparing *eaSky* against *dpSky* have shown that *eaSky* has better performance as the number of tables in a query increases. In addition, Skyline pushing introduced a performance improvement in terms of evaluation time. *eaSky* has also been extended by adding new genetic operators such as a single-point crossover and three additional mutation operators for Skyline operators. Using this version and increasing the mutation rate, better solutions can be obtained in less time. Setting a high mutation rate was found to be a key factor for obtaining good solutions. Furthermore, the contribution of the crossover operator seemed to be especially important for large queries.

In future work, we aim to incorporate further Skylines operators. We also will integrate an extended *eaSky* version into PostgreSQL. Additionally, we would like to address the Skyline Query Optimization problem using classical metaheuristics such as simulated annealing or tabu search.

References

1. Date, C.: A Guide to the SQL Standard. Addison-Wesley Longman Publishing Co., Inc., USA (1989)
2. Ramakrishnan, R., Gehrke, J.: Database Management Systems. McGraw-Hill Higher Education, New York (2000)
3. Bennett, K., Ferris, M., Ioannidis, Y.: A genetic algorithm for database query optimization. In: Proceedings of the Fourth International Conference on Genetic Algorithms, California, USA, pp. 400–407 (1991)
4. Godfrey, P., Shipley, R., Gryz, J.: Maximal vector computation in large data sets. In: Proceedings of the VLDB, Trondheim, Norway, pp. 229–240 (2005)
5. Borzsonyi, S., Kossmann, D., Stocker, K.: The skyline operator. In: IEEE Conf. on Data Engineering, Heidelberg, Germany, pp. 421–430 (2001)
6. Tan, K., Eng, P., Ooi, B.: Efficient progressive skyline computation. In: VLDB 2001: Proceedings of the 27th International Conference on Very Large Data Bases, pp. 301–310. Morgan Kaufmann Publishers Inc., San Francisco (2001)
7. Hafenrichter, B., Kießling, W.: Optimization of relational preference queries. In: Proceedings of the 16th Australasian Database Conference, Newcastle, Australia, pp. 175–184 (2005)
8. Chomicki, J.: Semantic optimization techniques for preference queries. Inf. Syst. 32(5), 670–684 (2007)
9. Chaudhuri, S., Dalvi, N., Kaushik, R.: Robust cardinality and cost estimation for skyline operator. In: Proceedings of the ICDE, Atlanta, USA, pp. 64–74 (2006)
10. Eder, H.: On extending postgresql with the skyline operator. Master's thesis. Vienna University of Technology (2009)
11. Hong, J., Kao, C., Liu, B.: A genetic algorithm for database query optimization. In: Proceedings of the First IEEE World Congress on Computational Intelligence, Orlando, USA, pp. 350–355 (1994)
12. Ioannidis, Y.: Query optimization. ACM Computing Surveys 28(1), 121–123 (1996)
13. Di Bartolo, F., Goncalves, M., Martínez, I., Sardá, F.: An evolutionary algorithm for skyline-join query optimization. In: Proceedings of Portuguese Conference on Artificial Inteligence, Guimaraes, Portugal, pp. 276–287 (2007)
14. Stonebraker, M., Rowe, L.: The design of postgres. In: SIGMOD 1986: Proceedings of the 1986 ACM SIGMOD International Conference on Management of Data, pp. 340–355. ACM Press, New York (1996)
15. Astrahan, M., Blasgen, M., Chamberlin, D., Eswaran, K., Gray, J., Griffiths, P., King, W., Lorie, R., McJones, P., Mehl, J., Putzolu, G., Traiger, I., Wade, B., Watson, V.: System r: Relational approach to database management. ACM Trans. Database Syst. 1(2), 97–137 (1976)
16. Bayir, M., Toroslu, I., Cosar, A.: Genetic algorithm for the multiple-query optimization problem. IEEE Transactions on Systems, Man and Cybernetics, Part C 37, 147–153 (2007)
17. Michalewicz, Z.: Genetic Algorithms + Data Structures = Evolution Programs. Springer-Verlag GmbH, Germany (1996)
18. Papadias, D., Tao, Y., Fu, G., Seeger, B.: An optimal and progressive algorithm for skyline queries. In: SIGMOD 2003: Proceedings of the 2003 ACM SIGMOD International Conference on Management of Data, pp. 467–478. ACM Press, New York (2003)

19. Chomicki, J.: Preference formulas in relational queries. ACM Trans. Database Syst. 28(4), 427–466 (2003)
20. Brando, C., Goncalves, M., González, V.: Peaqock: A postgresql extension with evaluation algorithms for skyline and top-k skyline queries. In: Proceedings of the CLEI, San José, Costa Rica, pp. 1–11 (2007)
21. Bentley, J., Kung, H.T., Schkolnick, M., Thompson, C.D.: On the average number of maxima in a set of vectors and applications. Journal of the ACM 25(4), 536–543 (1978)
22. Godfrey, P.: Cardinality estimation of skyline queries. Technical Report CS-2002-03, York University (2002)
23. Poli, R., Langdon, W., McPhee, N.: A Field Guide to Genetic Programming. Lulu Press (2008)

A Bio-inspired Approach to Self-organization of Mobile Nodes in Real-Time Mobile Ad Hoc Network Applications

Cem Şafak Şahin, Elkin Urrea, M. Ümit Uyar, and Stephen Gundry

Abstract. In this chapter, we study the applicability and effectiveness of an evolutionary computation approach to a topology control problem in the domain of mobile ad hoc networks (MANETs). We present formal and practical aspects of convergence properties of our force-based genetic algorithm, called FGA, which is run by each mobile node to achieve a uniform spread. Our FGA is suitable for MANET environments since mobile nodes, while running the FGA, only use local neighborhood information. An inhomogeneous Markov chain is used to analyze the convergence speed of our bio-inspired algorithm. To demonstrate our topology control algorithm's applicability to real-life problems and to evaluate its effectiveness, we have implemented a simulation software system and two testbed platforms. The simulation and testbed experiment results indicate that, for important performance metrics such as the normalized area coverage and convergence rate, the FGA can be an effective mechanism to deploy mobile nodes with restrained communication capabilities in MANETs operating in unknown areas. Since the FGA adapts to the local environment rapidly and does not require global network knowledge, it can be used as a real-time topology controller for realistic military and civilian applications.

Cem Şafak Şahin · Elkin Urrea
Department of Electrical Engineering,
The Graduate Center of The City University of New York, NY USA
e-mail: {csahin,eurrea}@gc.cuny.edu

M. Ümit Uyar · Stephen Gundry
Department of Electrical Engineering,
The City College of The City University of New York, NY, USA
e-mail: {uyar,sgundry00}@ccny.cuny.edu

1 Introduction

Evolutionary algorithms (EAs) [1] are a family of optimization techniques inspired by the natural selection concept called *survival of the fittest* in biological systems. They promise outstanding adaptation, reliability, and robustness in dynamic environments such as mobile ad-hoc networks (MANETs). EAs differ from the traditional optimization algorithms (e.g., hill-climbing [2], tabu search [3]) in that they typically use a *population* of potential solutions instead of a single solution. EAs, intuitively, are proper choices for multi-optimum search problems due to their intrinsic parallel search mechanisms.

For problems with no known deterministic solution, there has been an increasing interest in applying evolutionary and hereditary principles based on special selection methodologies to generate solutions. Genetic algorithms (GAs) [1, 4] are a set of EAs that use techniques motivated by evolutionary biology such as *selection*, *crossover*, and *mutation* operators. GAs are suitable for solving a wide range of complex problems requiring automated, adaptive, and self-learning computational techniques across military, commercial, and scientific applications. For example, decentralized intelligence is used where a large number of simple robots in multi-robot systems coordinate to obtain a collaborative behavior for completing a difficult task [5, 6, 7]. GAs are one of the most suitable approaches in finding solutions to these types of collaborative and complex task-oriented problems, including the NP-hard ones.

A MANET is a wireless network paradigm which encompasses autonomous systems of mobile nodes connected using wireless links with limited ranges. These mobile entities dynamically establish a network without any need of pre-existing structure or an administrator. Each mobile node, in most circumstances, moves and operates in a distributed peer-to-peer mode, generating independent data and acting as a router to provide multi-hop communications (i.e., communication between far away mobile nodes can be established via multi-hop routing). The mobility of MANET nodes can lead to frequent and unpredictable topology changes due to mobile nodes leaving and/or joining the network without notice. MANET applications span a wide spectrum ranging from commercial to military applications such as clearing mine-fields, spreading military assets in unknown areas (e.g., robots, mini-submarines, etc.), controlling unmanned vehicles and transportation systems, emergency and rescue operations (e.g., hurricane, earthquake, tsunami, etc.), and environmental cleaning operations. These applications typically require a uniform distribution of autonomous mobile nodes controlled by active running software agents over an unknown geographical area [8, 9, 10]. Dynamically changing topology, lack of centralized authority, nodes' selfishness, and unknown deployment terrain present difficulties in self-deployment algorithms for mobile entities in MANETs [11, 12, 13].

In this chapter, the main objective for our EA is to uniformly distribute mobile nodes over a two dimensional area while maintaining their communication connectivity. One of the best ways to maintain and improve network connectivity is to provide mobile nodes with the ability to adjust their speed and movement direction without a centralized control unit. However, these skills represent a challenging problem since: *(i)* the conditions of geographical terrain may change dramatically in a short time span (although there are many factors that affect the performance and success of mobile nodes distribution in an unknown terrain, the environmental configuration has one of the largest impacts on the speed and success of mobile nodes separation), *(ii)* the number of mobile nodes may change dynamically due to malfunctions and/or loss of communications, *(iii)* mobile nodes may not have access to navigation maps or GPS devices, but have only limited information collected from local neighbors, *(iv)* mobile nodes may be deployed into a terrain from a single entry point (more realistic but more difficult than random or other types of initial distributions often seen in the existing research), and *(v)* the real-world MANET applications may not have persistent and reliable network connections.

We have introduced different GA-based approaches for topology control problems in MANETs [14, 15, 16]. In this framework, a force-based GA (FGA) [17] is constructed as a decentralized topology control mechanism among active running software agents to achieve a uniform deployment of autonomous nodes over an unknown geographical area. Using only local neighborhood information, a GA guides each mobile node to select a *fitter* speed and movement direction among an exponentially large number of choices, converging towards a uniform node distribution. We have developed a simulation software system and implemented several testbed platforms to evaluate the effectiveness of our GA-based approach using network performance metrics of node density and convergence speed towards uniform distribution.

The rest of this chapter is organized as follows. In Section 2, we review prior research on the use of GAs on mobile autonomous node deployment, target localization in MANETs, swarm robotics, and mathematical methods for EAs used in deployment. In Section 3, we briefly explain the mobile model and outline our FGA approach. Section 4 introduces a formal model for convergence properties of the FGA. Simulation and testbed experiment results are presented in Section 5.

2 Literature Review

Our FGA is inspired by the force-based distribution in physics where each molecule attempts to remain in balanced position and to spend minimum energy to protect its own position. The FGA is run by each mobile node as a standalone topology control application to uniformly distribute mobile nodes in an unknown terrain [14, 16, 17]. Compared to other techniques, our GA-based approach presents encouraging results by converging towards a uniform node distribution as shown in [15]. In our research, we used discrete-time walk model adapted from [18]. In [19, 20], we analyzed the convergence rate of our GA-based approach by using homogeneous Markov chains (called hMC_{FGA}). In this book chapter, we study the convergence rate of our FGA with inhomogeneous Markov chain (called iMC_{FGA}). The transition matrix of iMC_{FGA} is different at each time unit. Therefore, the convergence rate can be exercised more accurately than with hMC_{FGA}. Furthermore, iMC_{FGA} also provides the time-based behavior of our topology control approach with respect to different environmental conditions (e.g., the number of neighbors and the positions of obstacles) rather than the average behavior presented in [19, 20, 21]. These are valuable observations since we can improve our algorithm based on different scenarios.

GA-based approaches are popular for topology control and deployment of MANET nodes. For example, [22] proposes a GA for path planning of a mobile robot, which incorporates the domain knowledge into its specialized operators and is capable of finding an optimal or near-optimal robot path in both complex static and dynamic environments. Compared to single node in [22], the system of multiple nodes in [23] has various advantages for environment exploration. [23] describes how a parallel computing GA can be applied for allocating target points to multiple mobile nodes, such as robots, appropriately so that the overall area exploration time is minimized. [24] discusses GAs to explore the space of path finding algorithms in training and three dimensional naval real-time strategy games rather than two dimensional area given in [22, 23]. Various topology control approaches have been studied [25, 26]. For example, autonomous dispersion of mobile nodes using a random diffusion method are considered in [25]. A potential-field approach is used to deploy mobile sensors in [26]. The fields are constructed such that each sensor is repelled by both obstacles and by other sensors, thereby forcing the network to spread itself throughout the environment.

Several promising results in swarm robotics are recently reported. Multi-agent collaboration for swarm robots for distributed missions to fetch and retrieve objects is presented in [27]. In [28], cooperative exploration strategy for mobile robots are presented based on the sensor-based random trees. [29] provides a basis for a distributed robotic system capable of constructing any given planar structure. [30] discuss various applications and a prototype for different scenarios including self-powered and self-governing swarm robotic platforms. Abstract models for rescue operations using swarm robots are studied in [31]. In [32], swarm optimization is used for route planning for unmanned aerial vehicles using a fitness function based on both flight time and safety. Similarly, a particle swarm optimization algorithm for path optimization of soccer playing robots is proposed in [33].

Markov chains are widely used to provide a formal structure for analyzing stochastic algorithms including GAs. For example, fundamental properties of a finite Markov chain, including graph theoretic considerations for transient and non-transient, recurrent and non-recurrent cases are discussed in [34]. [35, 36, 37] provide a formal structure for analyzing the convergence of a simple

GA using Markov chains. GA convergence models using Markov chains show that GAs applied to large-scale problems should avoid convergence towards an unwanted solution or a local optimum in [35]. In another study [36], finite discrete-time Markov chains are used to model and understand the complex dynamics of a simple GA. In [37], a modified elitist strategy is used to generate a current population from the reserved individual with the highest fitness value and the rest from the previous generation. [38] proposes an algorithm for the control of autonomous swarms using the Gibbs sampler simulated annealing processing.

There are fundamental differences between our approach and the existing research cited above. In our bio-inspired algorithm, there is no difference in mobile node's privileges (i.e., no leader or follower). The FGA only utilizes information from neighboring nodes and local terrain to make movement and speed decisions to converge towards a uniform distribution of mobile nodes; there is no central controller unit or global knowledge of the entire network. In fact, as will be shown in Section 5, the FGA adapts to its immediate environment rapidly and does not require global network knowledge, and hence it can be used as a real-time topology controller for realistic military and civilian applications. Furthermore, the decentralized characteristic of FGA makes it resilient to mobile node losses. Another significant difference is that no prior knowledge of the geographical area is needed for the FGA. It is also important to note that the self-deployment is more challenging in MANETs compared to sensor networks since, unlike sensor networks, there are no stationary nodes are present in MANETs.

3 GA-Based Approaches for Topology Control

GAs are a subset of EAs and mimic nature such that the hereditary transfer of biological trait information is used as a role model for stepwise improvement and development of a population of candidate solutions. The desired individuals are selected according to a specified fitness function among all candidate solutions. GAs are typically applied to problems where deterministic methods are not present or cannot provide satisfactory results. Furthermore, there are GA-based applications to deterministic problems when the runtime of an exact, deterministic method is too high to be feasible. Candidate solutions (represented by chromosomes) with better fitness values have higher probability to be selected for the breeding process of GAs. A set of individual chromosomes form a *population* in a GA. To create a new, and eventually better, population from an old one, GAs use biologically inspired operators, such as crossover (a new generation of individuals are created from different parents), and mutation (random changes to children to provide diversity in a population) [1, 4].

GAs have been used to solve a broad variety of problems in a diverse array of fields including aircraft design, circuit design, price prediction in financial markets, antenna design, protein sequence prediction, computer games, and others. GAs are chosen to solve complex problems since: *(i)* GAs are intrinsically parallel, *(ii)* large problem spaces can easily be scanned due to parallelism, *(iii)* GAs do not get trapped at local optimum points, and *(iv)* GAs can easily handle multi-objective optimization problems with proper fitness functions. However, the probability of success of a GA application largely depends on defining its fitness function and its parameters (i.e., the chromosome structure) properly.

3.1 Mobility Model

In MANETs, mobile nodes can move in many different ways. A mobility model attempts to mimic the movements of real mobile nodes. There are different mobility models for ad hoc networks such as the random walk mobility model, the gauss-markov mobility model, the column mobility model, and others [39]. In our approach, the mobility model is adapted from [18]. In self-organizing autonomous mobile nodes without global knowledge, the mobility model has an important effect on the performance analysis since each mobile node decides its own movement direction and speed. Therefore, when we design our mobility model, it not only aims at describing an individual node's motion behavior, but also considers the collective motion of all the mobile nodes over time. Our mobility model reflects the behavior of an individual mobile node with respect to its neighboring nodes and surroundings (e.g., obstacles, area borders, etc.). To reflect the autonomous nature of individual nodes, in our model, there is no notion of collective movement by all of the mobile nodes with reference to any particular point.

Fig. 1 shows an example with six mobile nodes. The area is partitioned into logical hexagonal cells. Each mobile node can move into six different directions (i.e., d_0 through d_5), if the mobile node is not on the boundary, from a neighboring cell within one time unit.

A wireless link between two mobile nodes is represented by a vector whose dimensions are in terms of layers. One layer is equal to the center-to-center distance between two neighboring cells. In general, for a mobile node in location $< 0, 0 >$ and another mobile node in location $< x, y >$, the link state between these mobile nodes is $< x - 0, y - 0 >=< x, y >$. For example, in Fig. 1, for a mobile node N_1 in location $< 0, 0 >$ and another node N_6 in location $< 1, 2 >$, the vector representing wireless link between these nodes is $< 1 - 0, 2 - 0 >=< 1, 2 >$.

Let d be the center-to-center distance between two neighboring cells, while x_a and y_a are non-negative integers that are equivalent to the state of $< x, y >$ given in previous paragraph. Two nodes are communicating with each other if and only if $d \leq R_{com}$, where $d = x_a + y_a$ and R_{com} is a positive integer representing the communication range of a mobile node. For simplicity, we accept all mobile nodes have the same communication range in this chapter (however, it is easy to generalize all solutions to varying communication ranges). For example, in Fig. 1, N_1 communicates with N_2, N_3, N_5, and N_6 if R_{com} is set to 3. The number of available wireless links for a node is called the node's *degree*. The degree of N_1 is 4. N_1 cannot communicate with N_4 since the distance between them is greater than the communication range. The hexagonal cells $< 3, -4 >$ and $< 4, -4 >$ are not located at any mobile node's communication range.

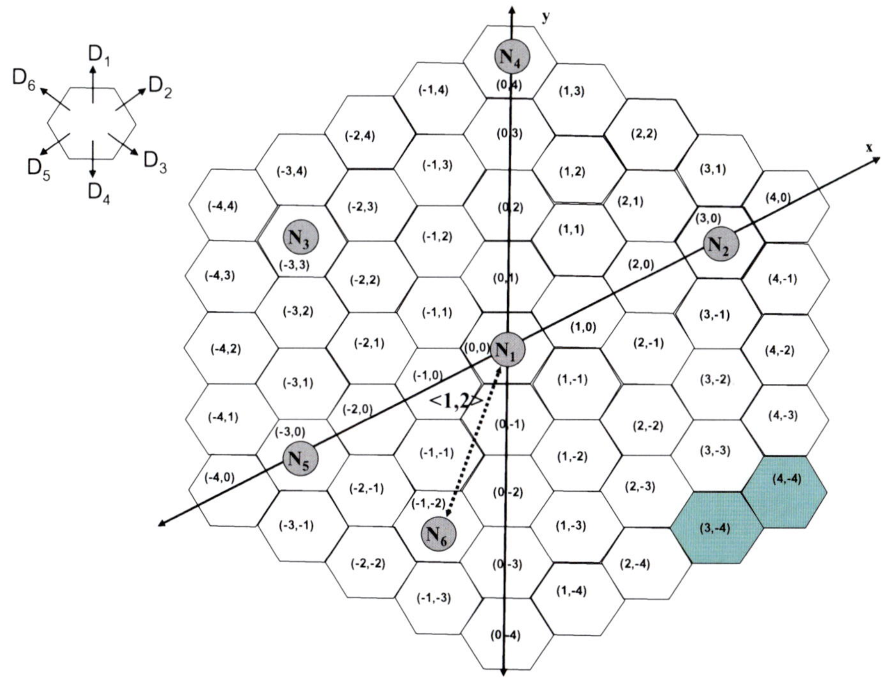

Fig. 1. Six nodes distributed within a 8×8 hexagonal area partitioned into logical cells. $R_{com} = 3$

As mentioned earlier, the direction and speed of movement for all the nodes are determined by the FGA running on each node. In our mobility model, without loss of generality, a mobile node is not allowed to move beyond the area boundaries. For example in Fig. 1, N_4 can only move in the directions of d_3, d_4, or d_5.

3.2 The Force-Based Genetic Algorithm (FGA)

In our earlier work, we introduced an FGA [14, 16, 17] inspired by repulsive forces in physics [12]. In our FGA, each mobile node is applied a force by its near neighbors (i.e, the nodes located within its communication range, R_{com}), which should sum up to zero at the equilibrium. If this force is not zero, the FGA uses this non-zero force value in its fitness calculation to find the mobile node's next speed and movement direction.

In the FGA framework, each node maintains a *neighborhood table* to keep records for its neighboring mobile nodes. Every ΔT time units, a mobile node N_i runs its GA-based software agent to find a better location to move (if such a location exists) based on the information from its neighborhood table; if it cannot find a location to improve its fitness, the node stops moving momentarily (the details of FGA fitness function are presented in Section 3.2.1).

Algorithm 1. Pseudo-code of our FGA

```
begin
    while !(stopCrit) do
        for (all neighbors) do
            if Neighbor is in the neighborhood table then
                Update the neighbor's information
            else
                Add the neighbor's information into table
        g ← 0 // generation counter
        Initialize population P(0)
        Evaluate population P(0)
        while !(evolveDone) do
            g ← g + 1
            Select P(g) from P(g-1)
            Crossover P(g)
            Mutate P(g)
            Evaluate P(g)
            if localPositionFound then
                evolveDone :=true
        if betterLocationFound then
            Move to new location
        else
            Wait (ΔT)
            Update neighborhood table
        Update stopCrit
end
```

Algorithm 1 presents the pseudo code of our FGA. First the neighborhood table is updated by the information received from the nodes in its communication range. Then a population of N individuals, called the initial population, is randomly generated where each individual represents a speed and movement direction for the node. Each individual (i.e., chromosome) is then evaluated using a fitness score and sorted based on its fitness value. Since our FGA is posed as a minimization problem, individuals are sorted in a decreasing order of fitness scores, representing virtual forces applied to them. By the selection and crossover operators in Algorithm 1, individuals are paired for breeding purposes. The mating probability is proportional to their fitness scores. At generation g, the offspring are added to a pool as candidate solutions for a new population $P(g)$ based on the old one $P(g-1)$. After the offspring in the pool are evaluated, only the better performing individuals are accepted into the newly created population of $P(g)$. Mutation occurs on randomly selected individuals of a new population to protect the populations against local optimum points. The population evolves using this process for many generations until a termination criterion (i.e., the variable betterLocationFound in Algorithm 1) is satisfied (e.g., convergence tolerance of the best individuals reaches a certain limit, the fitness value drops below a predefined value, or the generations limit is exceeded). If the FGA evolves a better speed and movement direction to minimize the total virtual force on the corresponding node, the mobile node adapts this new speed and direction. Otherwise, it stops. A node repeats running the FGA in this manner until the condition called *stopCrit* is satisfied when the node obtains an acceptable level of uniformity in its vicinity.

3.2.1 Genetic Operators and Fitness Function in the FGA

Our GA-based approach uses one-point crossover with the probability of $\mu_c = 0.9$, roulette wheel selection, and single-bit mutation with the probability of $\mu_m = 0.01$ as genetic operators. A mobile node gathers information about its neighboring environment (e.g., mobile nodes and obstacles located within its R_{com}) direction, and location, and then, using the fitness function (given in Eq. 2), proceeds to run our FGA to generate new chromosomes representing candidate solutions for the next generation. These candidates are ordered according to their fitness values from the lowest to the highest. The lowest fitness corresponds to the solution representing the least amount of force applied to a mobile node, and hence, the best one among the candidate solutions for that generation. The FGA chooses the new speed and movement direction such that the total force on the corresponding mobile node will be lowered.

In the FGA framework, each mobile node maintains a neighborhood table to keep records for its neighboring mobile nodes. Every ΔT time units, a mobile node N_i runs its GA-based software agent to find a better location to move (if such a location exists) based on the information from its neighborhood table; if it cannot find a location to improve its fitness, the node stops moving momentarily.

A fitness function is used to measure the quality of a chromosome within a solution space. The FGA's fitness function is based on the virtual forces applied to a node by its neighboring nodes [12]. The force between two nodes depends on the distance between them and the number of other nodes within their vicinities (i.e., communication range). In general, the force from a closer node is greater than the force from a farther one. The force exerted on node N_i by its neighboring node N_j is calculated as:

$$F_{ij} = \begin{cases} F_{max} & \text{for } d_{ij} = 0 \\ \sigma_i \left(d_{th} - d_{ij}\right) & \text{for } 0 < d_{ij} < d_{th} \\ 0 & \text{for } d_{th} \leq d_{ij} \leq R_{com} \end{cases} \quad (1)$$

where, d_{ij} is the Euclidean distance between nodes N_i and N_j, d_{th} is the threshold to define the local neighborhood, and σ_i is the *mean node degree* for node N_i. The mean node degree is the expected number of node degree to maximize the coverage. σ_i, which depends on the geographical area size, the number of nodes, and the communication range, is analytically derived in [15].

The fitness function f_i is given as the sum of all the partial forces that node N_i exerts on its k neighboring nodes:

$$f_i = \sum_{j=1}^{k} F_{ij} = \sum_{j=1}^{k} \sigma_i \left(d_{th} - d_{ij}\right) \text{ for } 0 < d_{ij} \leq d_{th} \quad (2)$$

Eq. 2 is used by our GA-based algorithm on each mobile node as the fitness function to maximize the area coverage (see Section 3.3), to keep network connectivity, and to provide a near optimum number of neighbors on a MANET.

Notice that the goal of FGA is to find a set of parameter values (i.e., chromosomes) such as a speed and direction that minimizes the fitness function f_i in Eq. 2. The best fitness value (i.e., the lowest value of force) between node i and j is obtained when the two nodes are d_{th} units apart from each other. Similarly, the worst fitness value corresponds to two nodes that are next to each other (i.e., $d_{ij} \approx 0$).

3.2.2 Chromosome in the FGA

In a GA, the chromosome refers to a candidate solution. In the FGA, it is encoded as a set of binary strings representing the parameters of a mobile node (i.e., a mobile node's speed and direction). Let Ω denote the search space, and n the cardinality of Ω. Then, a fixed-length binary string is represented as $\Omega = \{0,1\}^{\ell}$, where ℓ is the string length. We identify the elements of Ω with the integers in the range of $0, \cdots, n-1$.

In our mobility model, a mobile node can move in one of six different directions with (assumedly) four different speeds. For example in the case of $\ell = 5$, three and two bits can be used to represent, respectively, the direction and speed. A node's movement directions are encoded as **0** = (000) representing North, **1** = (001) for Northeast, **2** = (010) for Southeast, **3** = (011) for South, **4** = (100) for Southwest, and **5** = (101) for Northwest. Binary strings **6** = (110) and **7** = (111) are considered invalid and are not used as part of candidate solutions. Also, suppose a node's possible speeds are encoded as **0** = (00) (or immobile), **1** = (01), **2** = (10), and **3** = (11). The speed implies the number of hexagonal cells that a node will move at one time unit in a given direction. For example, the chromosome **10** = (10100) means that the mobile node should move two positions (or hexagonal cells) heading Southwest.

3.3 Normalized Area Coverage (NAC)

Uniform distribution of mobile nodes in a geographical terrain relates to the issue of how well each point in a node's communication range is covered. One of the main objectives is to deploy mobile nodes in strategic ways (e.g., uniformly) such that a maximum area coverage is achieved according to the needs of the underlying applications. Therefore, normalized area coverage (NAC) is an important performance metric for the FGA as it measures the success of node distribution over a geographical terrain at specific time instants and during time intervals. NAC is defined as the ratio between the union of areas covered by each node and the total terrain. The areas with duplicated coverage by multiple nodes are not included in the NAC calculation. Each mobile node covers the area within its communication range of R_{com}. The total area covered by all mobile nodes is given as:

$$\text{NAC} = \frac{\bigcup_{i=1}^{n} A_i}{A} \tag{3}$$

where A_i is the area covered by node N_i, and A is the total geographical area. NAC is a positive real number with an upper bound of one (one means that the area is fully covered by the mobile nodes).

If a node is located well inside the terrain, the full area of a circle around the node with a radius of R_{com} is counted as the covered region (i.e., πR_{com}^2) (assuming that there are no other nodes overlapping with this node's coverage). If, however, a node is located near the boundaries of the geographical area, then only the partial area of the terrain covered by that node is included in the NAC computation.

For example, Figs. 2 (a) and (b) show two nodes with a separation distance of $d_1 = R_{com}/2$ and $d_2 = R_{com}$, respectively. Let $A_T(2)$ denote the total area covered by the two nodes shown in Fig. 2 where the shaded region is counted only once. $A_T(2)$ gets larger and the overlapping region becomes smaller as the separation distance d between nodes N_1 and N_2 increases. The largest value for $A_T(2)$ is reached when $d = 2R_{com}$. However, this separation distance is infeasible because the two nodes are too far apart and cannot communicate with each other. Therefore, the allowable separation distances among mobile nodes are bounded as $d \leq R_{com}$.

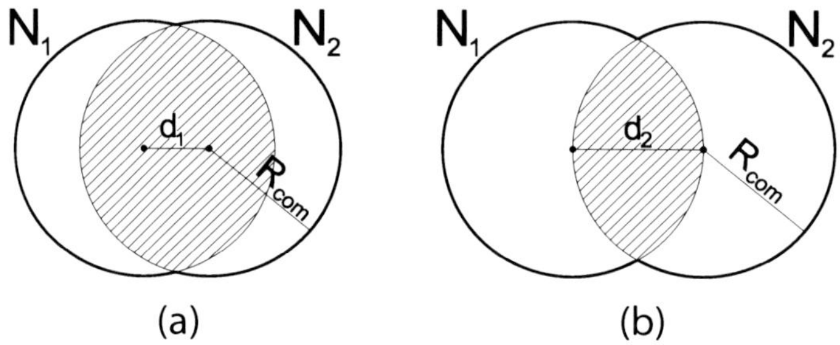

Fig. 2. Intersections of two nodes (a) $d_1 = R_{com}/2$, (b) $d = R_{com}$.

Let A_1 and A_2 be the covered area of two nodes as shown in Fig. 2, then A_T is given as:

$$A_T(2) = \bigcup_{i=1}^{n} A_i = A_1 + A_2 - (A_1 \cap A_2). \tag{4}$$

For the general case, the area covered by n nodes is

$$\begin{aligned} A_T(n) &= \bigcup_{i=1}^{n} A_i = \sum_{r=1}^{n} (-1)^{r-1} S_r \\ &= S_1 - S_2 + \cdots + (-1)^{n-1} S_n \end{aligned} \tag{5}$$

where

$$\begin{aligned} S_1 &= A_1 + \cdots + A_n \\ S_2 &= \sum_{i=1}^{n-1} \sum_{j=i+1}^{n} A_i \cap A_j, \text{ and} \\ \cdots \\ S_n &= A_1 \cap \cdots \cap A_n \end{aligned}$$

As explained in the previous section, we designed the fitness function for our FGA such that if the distance between two nodes gets closer to R_{com}, NAC improves.

4 Convergence of the Inhomogeneous Markov Chain Representation of FGA

A Markov chain is a suitable probability model for certain systems where the observation at a given time maps to the category into which an individual falls. This mapping is done by using a *stochastic matrix* (i.e., *transition matrix*) which contains the transition probabilities of a Markov chain over a finite state space X. If x_{ij} shows the probability value of moving from state i to state j (where $i, j \in X$) in one time unit, the transition matrix P is given by using x_{ij} as an element at i^{th} row and j^{th} column. P must satisfy the property of $\sum_j x_{ij} = 1$. For example:

$$P = \begin{pmatrix} x_{11} & x_{12} & \cdots & x_{1n} \\ x_{21} & x_{22} & \cdots & x_{2n} \\ \cdots & \cdots & \cdots & \cdots \\ x_{n1} & x_{n2} & \cdots & x_{nn} \end{pmatrix}$$

The FGA, like all GA-based approaches, uses different sets of chromosomes in every population (see Section 3.2) and, therefore, can be modeled by Markov chains since the state of population at time $t + 1$ depends only the state of population at time t. Since the FGA is run by each mobile node in a MANET as a decision mechanism for next speed and movement direction, the movement result assigned by our FGA at each time unit represents a state in iMC_{FGA} given in Fig. 3.

We designed the mobility model of our FGA as a collection independent finite-state Markov chains capturing the operational behavior of each mobile node. In our approach, the values of iMC_{FGA} are determined through estimation instead of exact representation as theoretically represented in [40]. Using the direct approach for iMC_{FGA} quickly becomes unwieldy and computationally unfeasible. Even if we assume that the fitness function of FGA generates a fitness with a resolution of 10^4 levels, the number of states in a Markov chain would be 10^6 using 10 different speeds and six directions. Therefore, in this chapter, we present a simplified inhomogeneous Markov chain to illustrate the convergence properties of FGA. In this simplified model a mobile node has six directions on the hexagonal lattice (up, up-right, down-right, down, down-left, or up-left), two values for fitness (good or bad), and two different speeds (mobile or immobile) as seen in Fig. 3. For simplicity, we assumed that mobile nodes are capable of changing their directions arbitrarily without stopping. Different values of the node fitness are merged into two distinct values as either good or bad (shown as 1 or 0 in Fig. 3, respectively). Let $d(i)$ be the number of neighbors for a mobile node i and $\overline{\text{N}}$ the ideal number of neighbors [15] to maximize the area coverage. The state where $(\overline{\text{N}} - 1) < d(i) < (\overline{\text{N}} + 1)$ is marked as *ideal* in iMC_{FGA}; otherwise, it is marked as *non-ideal* (shown as *Non* in Fig. 3).

In this book chapter, the estimation of a mobile node's state is found by using the empirical probability of traversing from one state to another obtained experimentally (i.e., the relative frequency of moving from one state to the other is recorded while running the FGA at each time instant). As the FGA determines the mobile node's speed and movement direction using local neighborhood information including neighbors and obstacles, we assume that a mobile node traverses from one state to another in our Markov chain model shown in Fig. 3. By conducting a large number of experiments, statistical anomalies can be smoothed out resulting in a model that closely approximates the behavior of a mobile node in a MANET.

As can be seen in Fig. 3, iMC$_{\text{FGA}}$ has 15 states. If a mobile node is moving in one of the six directions, its state must be one of the 12 states based on its number of neighbors: six directions with the ideal number of neighbors, and six directions with non-ideal number of neighbors. Speed and fitness are inherently covered by including direction into the state information. The remaining three states are: (stop, non, 0) state where the node is immobile due to the non-ideal number of neighbors and zero fitness, (stop, ideal, 0) state where the mobile node is immobile, the fitness is 0 in spite of the ideal number of neighbors, and the (stop, ideal, 1) state where the mobile node does not move because of having an ideal number of neighbors with a fitness of 1 (the desired final state in our problem). If a mobile node reaches the final state, the mobile node has the desired number of neighbors at the correct locations using Eq. 2 and stops moving (perhaps until another node comes and disrupts its equilibrium).

Extending our earlier hMC$_{\text{FGA}}$ [19, 20], we present here an inhomogeneous Markov chain with a Markov kernel. In this model, the Markov kernel (i.e., transition matrix) is different for every time step (i.e., $P_t = P_1, P_2, \cdots$, where $t = 1, 2, \cdots$) for a given finite state space X, with any initial distribution of ν. The distribution of states $x \in X$ at times $t \geq 0$ is given by $P^{(t)}(x_0, \cdots, x_t) = \nu(x_0)P_1(x_0, x_1) \cdots P_t(x_{t-1}, x_t)$. This model has the benefit over the hMC$_{\text{FGA}}$ by preserving the time-based precision of experimental data shown below and Section 5. To prove the convergence of iMC$_{\text{FGA}}$, the Dobrushin contraction coefficient method [41] is used to derive limit theorems for the Markov chains.

Dobrushin states that an inhomogeneous Markov chain on a finite set X will have a limiting distribution as long as the sum of the total variation between its one-step output distributions is finite and that the contraction coefficients for the transition matrices go to zero for any starting point i ($i = 1, 2, \cdots$) [41]. Based on this explanation we assert that our iMC$_{\text{FGA}}$ will converge to a stationary behavior:

Theorem 1. *The set of inhomogeneous transition matrices for i*MC$_{\text{FGA}}$ *fulfills both conditions: (i)* $\sum_t ||\mu_{t+1} - \mu_t|| < \infty$ *and (ii)* $\lim_{t\to\infty} c(P_i \cdots P_t) = 0 \ \forall\ i \geq 1$*, therefore, it will converge to a stationary distribution.*

Proof. (*sketch*) It is shown in [21] that the set of inhomogeneous Markov kernels for our FGA has a finite sum in the one step total variation of output

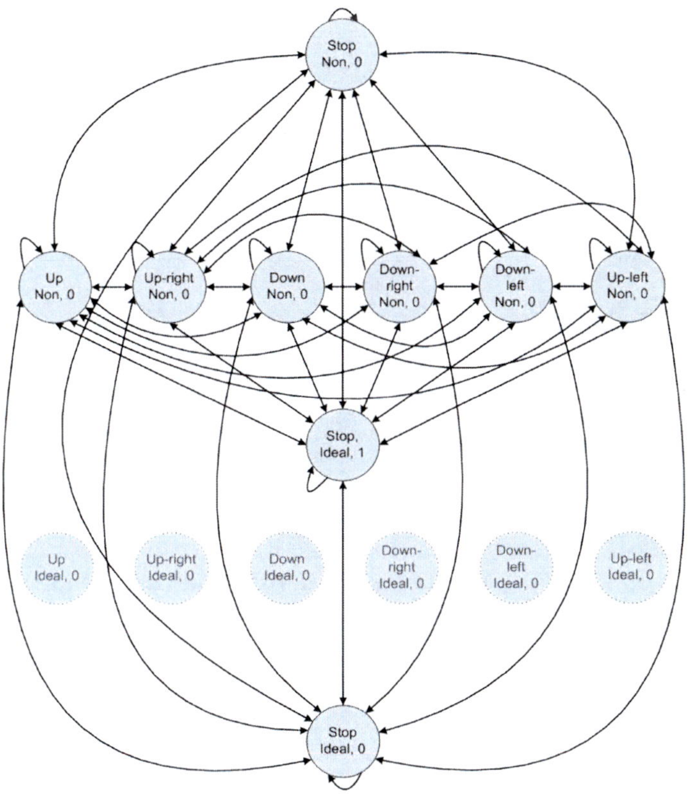

Fig. 3. Markov chain model for our FGA(each state is connected to each of the states in dotted lines, which are not shown for simplicity)

distributions and that the contraction coefficient of the transition matrix will converge to zero from any starting point. Therefore, using Theorem 4.6.2 in [41], it will converge to a stationary distribution. □

5 Simulation Software and Testbed Implementations for Bio-inspired Self Organizing Algorithms

In this section we present experiment results from our simulation software and two different testbeds to analyze the effectiveness and convergence of our FGA. In Section 5.1, we use simulation software to study convergence speed of our FGA by using iMC$_{\text{FGA}}$ (see Section 4). Two different testbed implementations are presented in Section 5.2 to provide the convergence and effectiveness of our GA-based approach for a uniform distribution of mobile nodes in an unknown terrain (see Section 3.3). Section 5.2 also validates our simulation software with real-life scenarios. For simplicity, we assume that all mobile nodes have the same capability including communication range (R_{com}) and movement capabilities (i.e., speed and directions) in the simulation experiments.

5.1 Simulation Software for Genetic Algorithms

5.1.1 Implementation

In order to study the convergence and effectiveness of our GA-based framework for a uniform distribution of knowledge sharing mobile nodes, we implemented simulation software in Java, using Eclipse SDK version 3.2.0 as development environment, and Mason, a fast discrete-event multi-agent simulation library core developed by the George Mason University ECJLab, as the visualization tool (i.e., GUI) and multi-agent library. Although this is small according to industry standards, our simulation software currently has more than 4,500 lines of algorithmic Java code for the evolutionary algorithms and the mobility model. This system is designed such that a programmer can easily add new features (e.g., different types of crossover operators, or different rules for mutation operators, etc.) and new evolutionary approaches. Our simulation software runs as a multi-agent application which imitates a real-time topology control mechanism. Therefore, the results from our simulation software match closely to those from our real testbed experiments [42]. User-defined input parameters for our software include:

- Total number of mobile nodes (N),
- Communication range of a mobile node (R_{com}),
- Type of evolutionary algorithm,
- Maximum number of iterations (T_{max}),
- Initial deployment type: currently there are three different initial deployment strategies for the mobile nodes: (*i*) start from the northwest corner, (*ii*) place the nodes randomly in a given area, and (*iii*) start from a given coordinate (e.g., the center of the area) in the terrain,
- Size of the geographical terrain (d_{max}),
- Obstacle inclusion (on user defined locations),
- Random node failures,
- Silent mode (i.e., no communication among mobile nodes for given time periods).

The initial deployment of autonomous mobile nodes in this chapter starts from the northwest sector of a given terrain. Note that a corner initial deployment option represents a more realistic approach of the topology control problem for the knowledge sharing mobile nodes than other deployment options over an unknown terrain. For example, in an earthquake rescue, a mine clearing mission, a military mission in a hostile area, or a surveillance operation, all mobile nodes may be forced to enter the operation area from the same vicinity rather than random or central node deployment.

Our simulation software also has the ability to run experiments using a previously used initial mobile node distribution and initial conditions from previous runs (i.e., the initial data for each mobile node includes a starting coordinate, speed, and direction). This ability is important since each experiment is repeated many times to eliminate the noise in the collected data and provide an accurate stochastic behavior of GA-based algorithms.

5.1.2 Convergence Experiments for Our FGA

For each experiment, the area of deployment is set to be 100×100 square units with all nodes initially placed in the northwest of the terrain each with random speed and direction. We ran experiments for networks with $N = 125$ and $R_{com} = 10$. To reduce the noise in the outcomes, each simulation experiment is repeated 50 times with the same initial values for node speed, R_{com}, direction, and with the same initial node deployment.

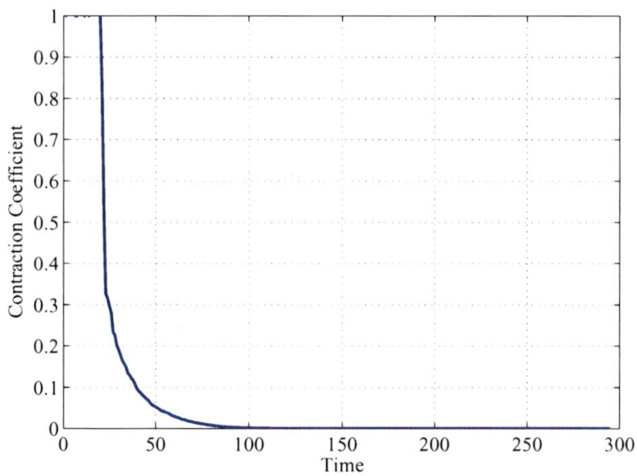

Fig. 4. Contraction coefficients when $t \to \infty$

Figs. 4 and 5 present convergence characteristics of the Markov kernel's states for the experiments where mobile nodes perform our FGA. Fig. 4 shows Dobrushin's contraction coefficients (see Section 4), which provide a rough measure of the orthogonality between distributions in the Markov kernel, for $R_{com} = 10$ when time goes to infinity. The Markov kernel for our FGA reaches its final state for $R_{com} = 10$ after approximately 80 time units. Another important observation from Fig. 4 is that the system evolves to a stationary distribution as time goes to infinity. It must be noted that any initial distribution will converge to the same stationary distribution based on Theorem 1. The only difference using varying initial distribution will be the number of steps that the system takes to reach the stationary distribution. In fact, this makes practical sense when considering the manner that the mobile agents are initially deployed. If the initial node distribution is dispersed such that the nodes are close-to-uniformly spatially distributed over the geographical area, then they will take very few steps to achieve a uniform distribution. In our experiments, mobile agents are placed using a worst case scenario in terms of mobile node deployment where all of the mobile nodes are clustered in a single corner. In this case, many mobile entities will initially be trapped between other mobile nodes and the boundaries of the geographical area. This will increase the time required to reach spatial uniformity. The importance of the relationship between initial distributions of the Markov chain and the initial dispersion of mobile nodes is that iMC$_{\text{FGA}}$ accurately represents experimental behavior.

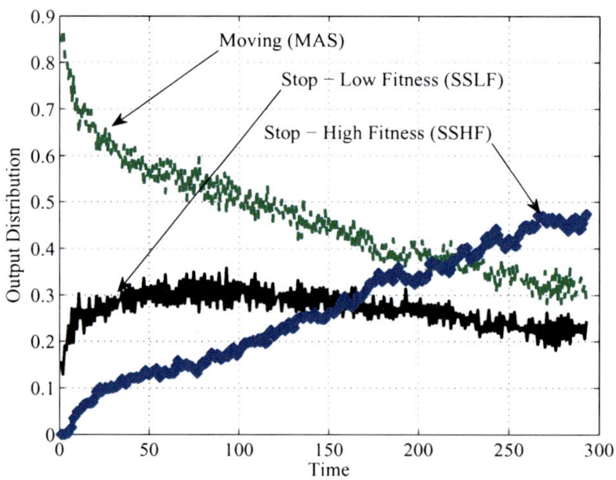

Fig. 5. Output distribution of Markov chain when $t \rightarrow \infty$

Fig. 5 represents the outcome distribution of each state in our Markov model: the stop state with high fitness (SSHF), the stop state with low fitness (SSLF), and the aggregation of all moving states (MAS). As seen by the SSHF plot in Fig. 5, the probability of stop with high fitness increases when time goes to infinity. More mobile nodes find the desired number of neighbors in the correct position in which the aggregate force on a corresponding mobile node is approximately zero and stay immobile. When time reaches 300, nearly half of the mobile nodes have good fitness and are immobile. On the other hand, the probability of mobile nodes reaching and/or staying at the moving states (with any direction and any speed) drops when time passes. A little more than one third of the mobile nodes are in any moving state when $t = 300$. The stopped state with low fitness increases until $t = 100$ and decreases afterward as seen by the SSLF plot in Fig. 5. Initially, the mobile nodes are located at the northwest corner of the geographical terrain and most of them cannot move because of the overcrowded vicinity. After some time passes, there are enough available hexagonal cells to move as assigned by the FGA. The final stationary distribution at $t = 300$ verifies the experimental behavior of our FGA where mobile agents achieve a distribution that is close to the uniform distribution. Some nodes continue to move slightly; these nodes exert small external forces on neighbors who in turn readjust themselves to return to ideal fitness.

5.2 Testbed Implementations for the FGA

Most of the research in wireless ad-hoc networks is based on software tools simulating network environments under strictly controlled conditions rather than implementing realistic testbeds due to their extreme cost of design, operation and difficulty of adapting real-time topology changes. To study the effectiveness of our GA-based algorithms and to prove the results of our simulation software, we implemented two different testbeds using virtual machines, laptops and PDAs.

5.2.1 Testbed Implementation with Virtual Machines

Using VMware virtualization, we implemented a testbed to create configurable multiplicity emulation that overcomes the scarce availability of computing resources/platforms and to scale down the deployment cost for large scale experimentation. We use VMware technology for virtualization. VMware is a product of the VMware Corporation and leverages virtualization technology to emulate a wide variety of x86-based operating systems on x86-based physical hardware. The operation of the system on physical hardware is accurately replicated by VMware. This testbed is programmed in C++ and runs in both Windows and Linux operating systems. For simplicity, all mobile nodes in this testbed are configured with identical capabilities. This testbed emulates realistic node mobility and wireless features of MANETs including, but not limited to, autonomous mobility, wireless communication characteristics, and periodic heartbeat messages (periodically broadcasted to the maximum allowed communication distance by a mobile node in order to inform neighboring nodes about its genetic material).

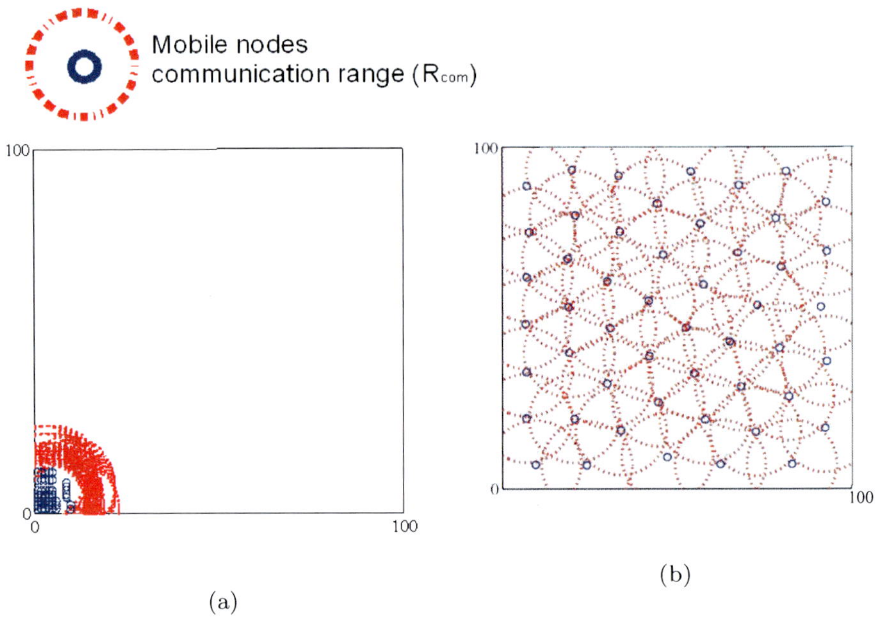

Fig. 6. (a) A screen shot of the initial mobile node distribution (N=50 and R_{com} = 20) in 100x100 area, (b) a screen shot from a final mobile node distribution (N=50 and R_{com} = 20) in 100x100 area after 300 time units

In this testbed architecture, from the users' point of view, each mobile node uses its own neighborhood table to initiate the GA-based operators. Genetic materials are exchanged upon successful acknowledgments between the neighboring nodes. Time duration of the each experiment, time to start GA-based application, maximum communication range (R_{com}), and speed are the configurable parameters in our testbed. There is natural latency in wireless network communication in real-time applications. Our installation of VMware takes into consideration different types of latencies including late-arriving acknowledgments and delayed responses. If a corresponding node does not receive a heartbeat message from a neighboring node within a predefined time, then the neighbor is purged from the neighborhood table. Node mobility is emulated as follows: When a node receives a heartbeat message from another one (containing node location), it compares the distance between itself and the sender node. If this distance is greater than R_{com}, the message is dropped by the receiver. Our testbed also contains random heartbeat message losses to characterize a wireless communication medium.

Our emulation experimentation resides on a single computer with a configurable number of virtual machines interconnected by a virtual switch, simplifying and allowing our mobile nodes running FGA experimentations on the single computer as though they run on a real network. For experiments requiring more mobile nodes than we can handle by a single computer (approximately seven nodes for FGA applications), our testbed can be also run on multiple computers. VMware is used in the scope of our study solely to help provide deployment configurations by multiplexing a single physical host into multiple independent virtual machines, hence presenting a different usage for virtualization in the context of distributed computing.

Our testbed implementation is not aware of platform differences or whether it is actually running on a physical or a virtual machine. This helps to facilitate a flexible deployment paradigm. All virtual machines are connected to the network through a virtual switch. Typically, all nodes on this network use the TCP/IP protocol suite, although other communication protocols can be used. A host virtual adapter connects the host computer to the private network used for network address translation (NAT). Each virtual machine and the host have assigned addresses on the private network. This is done through the DHCP server included with the VMware Workstation. NAT provides a way for virtual machines to use most client applications over almost any type of network connection available to the host. The only requirement is that the network connection must support TCP/IP. NAT is useful when there is a limited supply of IP addresses or connected to the network through a non-Ethernet network adapter. When a virtual machine sends a request to access a network resource, it appears to the network resource as if the request is coming from the host machine. The host computer has a host virtual adapter on the NAT network (identical to the host virtual adapter on the host-only network). This adapter allows the host and the virtual machines to communicate with each other for such purposes as file sharing. The NAT device never forwards traffic from the host virtual adapter.

Experiment results using our testbed for the FGA are shown in Figs. 6 (a)-(b) and 7 (a)-(b). In these experiments, we consider the following scenario. A team of mobile nodes equipped with cameras enter an unknown terrain to collect information on a disaster area for rescue recovery and survey purposes. As Fig. 6 (a) shows, at time $t = 0$, all mobile nodes are located at the southwest of the terrain.

Fig. 6 (b) illustrates the mobile nodes deployment after 300 time units. We can observe that, in spite of the lack of global knowledge and a global controller, the mobile nodes using our FGA obtain an almost uniform coverage of the area in a relatively short time period.

Figs. 7 (a)-(b) display the convergence of our FGA with respect to the NAC value through time for different network densities. As seen from Fig. 7 (a), a total of 40 mobile nodes successfully deploy themselves in an unknown geographical area and achieve 38%, 82%, and 99% area coverage when $T \approx 250$, 200, and 150 time units for $R_{com} = 10$, 15, and 20, respectively.

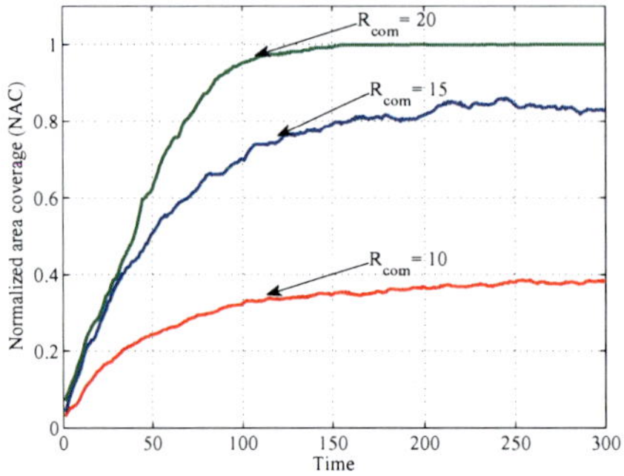

(a)

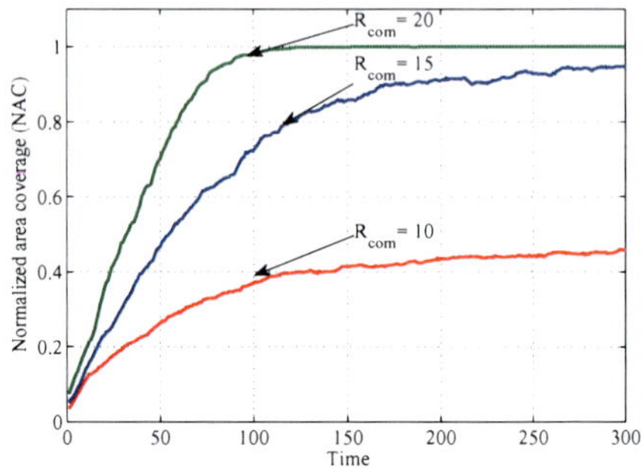

(b)

Fig. 7. (a) Normalized area coverage of N=40 with R_{com} = 10, 15, and 20 in 100x100 area, and (b) Normalized area coverage from a final mobile node distribution of N=50 with R_{com} = 10, 15, and 20 in 100x100 area

Fig. 7 displays the NAC values of 50 mobile nodes for the same communication ranges. The mobile nodes approximately cover 45%, 95%, and 99% for R_{com} = 10, 15, and 20, respectively. To achieve and stay at maximum area coverage, it takes 125 time units for the communication range of 20, and more than 200 time units for the communication ranges 10 and 15. Clearly, mobile nodes perform better separation in an unknown terrain when their communication range is larger since a wider communication range represents a denser network. Another reason for obtaining a better area coverage with larger communication range is that a mobile node with a wider communication range can collect more neighborhood information compared to a node with smaller range.

5.2.2 Testbed Implementation with Laptops and PDAs

In this testbed implementation, each laptop and personal digital assistant (PDA) runs an identical copy of our GA-based topology control application to obtain a uniform node separation in an unknown terrain. We implemented a translator for converting the FGA outputs for movement directions and speed to vocal and visual commands for the users. This way, each user moves in a given number of steps (emulating different speeds) in a given direction. Fig. 8 (a) shows the initial deployment for this experiment where all students are placed together at the bottom-right part of the area. The convergence towards a uniform distribution is displayed in Figs. 8 (b)-(d) over 30 time units.

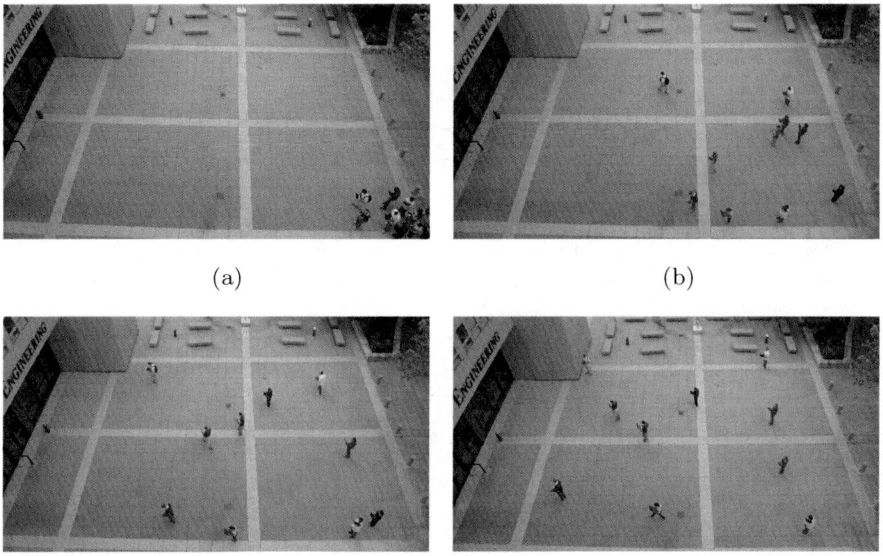

(a) (b) (c) (d)

Fig. 8. Node spreading experiments using laptops and PDAs (a total of 30 time units elapsed)

6 Conclusions

In this chapter, we have outlined a GA-based topology control approach for efficient, reliable, and effective self-deployment of mobile nodes in MANETs. Our FGA controls the mobile node's speed and direction using local neighborhood information without a global controller. We presented inhomogeneous Markov chains to formally analyze the convergence of our GA-based approach. Experimental results from our simulation software and two different testbed implementations showed that the FGA delivers promising results for uniform node distribution of knowledge sharing mobile nodes over unknown geographical areas in MANETs. Future work will include the application of FGA to transportation systems and mini-submarines.

Acknowledgements. This work has been supported by U.S. Army Communications-Electronics RD&E Center. The contents of this document represent the views of the authors and are not necessarily the official views of, or are endorsed by, the U.S. Government, Department of Defense, Department of the Army, or the U.S. Army Communications-Electronics RD&E Center. This work has been supported by the National Science Foundation grants ECS-0421159 and CNS-0619577.

References

1. Mitchell, M.: An Introduction to Genetic Algorithms. MIT Press, Boston (1998)
2. Yuret, D., de la Maza, M.: Dynamic hill climbing: Overcoming the limitations of optimization techniques. In: The Second Turkish Symposium on Artificial Intelligence and Neural Networks, pp. 208–212 (1993)
3. Glover, F., Laguna, M.: Tabu Search. Kluwer Academic Publishers, Dordrecht (1997)
4. Holland, J.H.: Evolutionary swarm robotics: Evolving Self-organizing Behaviors in groups of Autonomous Groups. University of Michigan Press, Ann Arbor (1975)
5. Bekey, G., Agah, A.: A genetic algorithm-based controller for decentralized multi-agent robotic systems. In: Proc. of the IEEE International Conference of Evolutionary Computing, pp. 431–436 (1996)
6. Miryazdi, H.R., Khaloozadeh, H.: Application of genetic algorithm to decentralized control of robot manipulators. In: ICAIS 2002: Proceedings of the 2002 IEEE International Conference on Artificial Intelligence Systems (ICAIS 2002), p. 334. IEEE Computer Society, Washington, DC, USA (2002)
7. Ping-An, G., Zi-Xing, C., Ling-Li, Y.: Evolutionary computation approach to decentralized multi-robot task allocation. In: International Conference on Natural Computation, vol. 5, pp. 415–419 (2009)
8. Song, P., Li, J., Li, K., Sui, L.: Researching on optimal distribution of mobile nodes in wireless sensor networks being deployed randomly. In: International Conference on Computer Science and Information Technology, pp. 322–326 (2008)
9. Heo, N.: An intelligent deployment and clustering algorithm for a distributed mobile sensor network. In: Proceedings of the IEEE International Conference on Systems Man And Cybernetics, pp. 4576–4581 (2003)
10. Chen, Y.M., Chang, S.-H.: Purposeful deployment via self-organizing flocking coalition in sensor networks. International Journal of Computer Science & Applications 4(2), 84–94 (2007)
11. Cayirci, E., Coplu, T.: Sendrom: Sensor networks for disaster relief operations management. Journal Wireless Networks 13, 409–423 (2007)
12. Heo, N., Varshney, P.K.: A distributed self spreading algorithm for mobile wireless sensor networks. IEEE Wireless Communications and Networking (WCNC) 3(1), 1597–1602 (2003)
13. Wang, H., Crilly, B., Zhao, W., Autry, C., Swank, S.: Implementing mobile ad hoc networking (manet) over legacy tactical radio links. In: Military Communications Conference, MILCOM 2007, pp. 1–7. IEEE, Los Alamitos (2007)
14. Sahin, C.S., Urrea, E., Umit Uyar, M., Conner, M., Ibrahim, G.B., Pizzo, C.: Genetic algorithms for self-spreading nodes in manets. In: GECCO 2008: Proceedings of the 10th Annual Conference on Genetic and Evolutionary Computation, pp. 1141–1142. ACM, New York (2008)
15. Urrea, E., Sahin, C.S., Umit Uyar, M., Conner, M., Hokelek, I., Bertoli, G., Pizzo, C.: Bioinspired topology control for knowledge sharing mobile agents. Ad Hoc Netw. 7(4), 677–689 (2009)
16. Sahin, C.S., Urrea, E., Umit Uyar, M., Conner, M., Bertoli, G., Pizzo, C.: Design of genetic algorithms for topology control of unmanned vehicles. International Journal of Applied Decision Sciences 3(3), 221–238 (2010)

17. Sahin, C.S., Urrea, E., Umit Uyar, M., Conner, M., Hokelek, I., Bertoli, G., Pizzo, C.: Uniform distribution of mobile agents using genetic algorithms for military applications in manets. In: IEEE International Conference on Military Communications Conference (IEEE/MILCOM), pp. 1–7 (2008)
18. Hokelek, I., Umit Uyar, M., Fecko, M.A.: A novel analytic model for virtual backbone stability in mobile ad hoc networks. Wireless Networks 14, 87–102 (2008)
19. Sahin, C.S., Gundry, S., Urrea, E., Umit Uyar, M., Conner, M., Bertoli, G., Pizzo, C.: Markov chain models for genetic algorithm based topology control in mANETs. In: Di Chio, C., Brabazon, A., Di Caro, G.A., Ebner, M., Farooq, M., Fink, A., Grahl, J., Greenfield, G., Machado, P., O'Neill, M., Tarantino, E., Urquhart, N. (eds.) EvoApplications 2010. LNCS, vol. 6025, pp. 41–50. Springer, Heidelberg (2010)
20. Sahin, C.S., Gundry, S., Urrea, E., Umit Uyar, M., Conner, M., Bertoli, G., Pizzo, C.: Convergence analysis of genetic algorithms for topology control in manets. In: 2010 IEEE Sarnoff Symposium, pp. 1–5, 12–14 (2010)
21. Sahin, C.S.: Design and Performance Analysis of Genetic Algorithms for Topology Control Problems. PhD thesis, The Graduate Center of the City Univeristy of New York (2010)
22. Hu, Y., Yang, S.X.: A knowledge based genetic algorithm for path planning of a mobile robot. In: Proc. of the 2004 IEEE Int. Conference on Robotics & Automation (2004)
23. Ma, X., Zhang, Q., Lip, Y.: Genetic algorithm-based multi-robot cooperative exploration. In: Proceedings of the IEEE International Conference on Control and Automation, pp. 1018–1023. IEEE Computer Society Press, Guangzhou (2007)
24. Leigh, R., Louis, S.J., Miles, C.: Using a genetic algorithm to explore a*-like path finding algorithms. In: IEEE Symposium on Computational Intelligence and Games, pp. 72–79. IEEE, Honolulu (2007)
25. Winfield, A.F.: Distributed sensing and data collection via broken ad hoc wireless connected networks of mobile robots. Distributed Autonomous Robotic Systems 4, 273–282 (2000)
26. Howard, A., Mataric, M.J., Sukhatme, G.S.: Mobile sensor network deployment using potential fields: A distributed, scalable solution to the area coverage problem. In: Proceedings of the International Conference on Distributed Autonomous Robotic Systems, pp. 299–308 (2002)
27. Tang, F., Parker, L.: Asymtre: Automated synthesis of multi-robot task solutions through software reconfiguration. In: Proceedings of the IEEE International Conference on Robotics and Automation, pp. 1501–1508 (2005)
28. Franchi, A., Freda, L., Oriolo, G., Vendittelli, M.: A randomized strategy for cooperative robot exploration. In: Proceedings of the IEEE International Conference on Robotics and Automation, Roma, Italy, pp. 768–774 (2007)
29. Stewart, R.L., Russell, A.: A distributed feedback mechanism to regulate wall construction by a robotic swarm. Adaptive Behavior 14, 21–51 (2006)
30. Joordens, M.A., Shaneyfelt, T., Nagothu, K., Eega, S., Jaimes, A., Jamshidi, M.: Applications and prototype for system of systems swarm robotics. In: Proceedings of the IEEE International Conference on Systems, Man and Cybernetics, Singapore, pp. 2049–2055 (2008)

31. Xue, S., Zeng, J.: Sense limitedly, interact locally: the control strategy for swarm robots search. In: Proceedings of the IEEE International Conference on Networking, Sensing and Control, pp. 402–407 (2008)
32. Werfel, J.: Robot search in 3d swarm construction. In: Proceedings of the First International Conference on Self-Adaptive and Self-Organizing Systems, pp. 363–366. IEEE Computer Society, Washington, DC, USA (2007)
33. Saska, M., Macas, M., Preucil, L., Lhotska, L.: Robot path planning using particle swarm optimization of ferguson splines. In: Proceedings of the IEEE Conference on Emerging Technologies and Factory Automation, Prague, pp. 833–839 (2006)
34. Jarvis, J.P., Shier, D.R.: Graph-theoretic analysis of finite Markov chains. CRC Press, Cambridge (2000)
35. De Jong, K.A., Spears, W.M., Gordon, D.F.: Using markov chains to analyze gafos. In: Foundations of Genetic Algorithms, vol. 3, pp. 115–137. Morgan Kaufmann, San Francisco (1995)
36. Horn, J.: Finite markov chain analysis of genetic algorithms with niching. In: Proceedings of the 5th International Conference on Genetic Algorithms, pp. 110–117. Morgan Kaufmann Publishers Inc., San Francisco (1993)
37. Suzuki, J.: A markov chain analysis on a genetic algorithm. In: Proceedings of the 5th International Conference on Genetic Algorithms, pp. 146–154. Morgan Kaufmann Publishers Inc., San Francisco (1993)
38. Baras, J.S., Tan, X.: Control of autonomous swarms using gibbs sampling. In: CDC – 43rd IEEE Conference on Decision and Control, vol. 5, pp. 4752–4757. IEEE, Los Alamitos (2004)
39. Camp, T., Boleng, J., Davies, V.: A survey of mobility models for ad hoc network research. Wireless Communications and Mobile Computing (WCMC): Special issue on Mobile Ad Hoc Networking: Research, Trends and Applications, 483–502 (2002)
40. Rudolph, G.: Convergence analysis of canonical genetic algorithms. IEEE Transactions on Neural Networks 5 (1994)
41. Winkler, G.: Image Analysis, Random Fields and Markov Chains Monte Carlo Methods. Springer, Heidelberg (2006)
42. Dogan, C., Sahin, C.S., Umit Uyar, M., Urrea, E.: Testbed for node communication in manets to uniformly cover unknown geographical terrain using genetic algorithms. In: Proc. of the NASA/ESA Conference on Adaptive Hardware and Systems (AHS 2009), pp. 273–280 (2009)

Author Index

Alavi, Amir Hossein 343
Arantes, Marcio S. 59

Banceanu, Alex 167
Becerra, David 167
Bilotta, Eleonora 377
Blum, Christian 1

Cerasa, Antonio 377
Chiong, Raymond 1
Clerc, Maurice 1
Coello Coello, Carlos A. 245
Cui, Hao 281

Dasgupta, Dipankar 167
Davarynejad, Mohsen 245
De Jong, Kenneth 1
Di Bartolo, Fabiola 413

Escuela, Gabi 413

França, Paulo M. 59

Gandomi, Amir Hossein 343
Garrett, Deon 167
Goncalves, Marlene 413
Gundry, Stephen 437

Hochmuth, Christian A. 95

Ibrahimov, Maksud 31, 143

Lässig, Jörg 95

Ma, Guohua 205
Martínez, Ivette 413
Michalewicz, Zbigniew 1, 31, 143
Mohais, Arvind 31, 143
Mollahasani, Ali 343
Morabito, Reinaldo 59

Neri, Ferrante 1
Nino, Fernando 167

Pantano, Pietro 377
Pasemann, Frank 305

Quattrone, Aldo 377

Rempis, Christian 305

Şahin, Cem Şafak 437
Sardá, Francelice 413
Schellenberg, Sven 31, 143
Staino, Andrea 377
Stramandinoli, Francesca 377

Thiem, Stefanie 95
Toledo, Claudio F.M. 59
Turan, Osman 281

Urrea, Elkin 437
Uyar, M. Ümit 437

van den Berg, Jan 245
Vrancken, Jos 245

Wagner, Neal 31, 143
Weise, Thomas 1

Zhang, Fu 205